Now Media

Now in its fourth edition, this book is one of the leading texts on the evolution of electronic mass communication in the last century, giving students a clear understanding of how the media of yesterday shaped the media world of today.

Now Media, Fourth Edition (formerly *Electronic Media: Then, Now, Later*) provides a comprehensive view of the beginnings of electronic media in broadcasting and the subsequent advancements into 'now' digital media. Each chapter is organized chronologically, starting with the electronic media of the past, then moving to the media of today, and finally, exploring the possibilities for the media of the future. Topics include the rise of social media, uses of personal communication devices, the film industry, and digital advertising, focusing along the way on innovations that laid the groundwork for 'now' television and radio and the Internet and social media. New to the fourth edition is a chapter on the amazing world of virtual reality technology, which has spawned a 'now' way of communicating with the world and becoming a part of video content, as well as a discussion of the impacts of the COVID-19 pandemic on media consumption habits.

This book remains a key text and trusted resource for students and scholars of digital mass communication and communication history alike.

The new 'now' edition also features updated online instructor materials, including PowerPoint slides and test banks. Please visit www.routledge.com/cw/medoff to access these support materials.

Norman J. Medoff, Ph.D., is a professor in the School of Communication at Northern Arizona University. He has taught and served as an administrator at three different universities, produced numerous television and corporate video projects, and overseen the productions of many students. Dr. Medoff has authored articles in scholarly journals as well as trade and consumer magazines. He has also written textbooks on the Internet and mass media, television production, and electronic media. He is a past president of the Broadcast Education Association. Currently he is co-director of the Advanced Media Laboratory where students create virtual reality and augmented reality projects and compete on the NAU Esports team.

Barbara K. Kaye (Ph.D., Florida State University) is a professor in the School of Journalism & Electronic Media at the University of Tennessee-Knoxville. Her research focuses on the influence of the Internet, blogs, social media, and other digital technologies on political viewpoints and how they have changed media use behavior and attitudes. She also studies the uses and effects of profanity on broadcast and cable television programs. She has co-authored five textbooks, published almost 80 journal articles and book chapters, and presented her research at 100 academic conferences. But closest to her heart is teaching abroad in Italy, Austria, and Taiwan.

Now Media: The Evolution of Electronic Communication

Formerly 'Electronic Media: Then, Now, Later'

FOURTH EDITION

Norman J. Medoff and Barbara K. Kaye

NEW YORK AND LONDON

Fourth edition published 2021
by Routledge
605 Third Avenue, New York, NY 10158

and by Routledge
2 Park Square, Milton Park, Abingdon, Oxon OX14 4RN

Routledge is an imprint of the Taylor & Francis Group, an informa business

© 2021 Taylor & Francis

The right of Norman J. Medoff and Barbara K. Kaye to be identified as authors of this work has been asserted by them in accordance with sections 77 and 78 of the Copyright, Designs and Patents Act 1988.

All rights reserved. No part of this book may be reprinted or reproduced or utilized in any form or by any electronic, mechanical, or other means, now known or hereafter invented, including photocopying and recording, or in any information storage or retrieval system, without permission in writing from the publishers.

Trademark notice: Product or corporate names may be trademarks or registered trademarks, and are used only for identification and explanation without intent to infringe.

First edition published by Focal Press 2004
Third edition published by Routledge 2017

Library of Congress Cataloging-in-Publication Data
Names: Medoff, Norman J, author. | Kaye, Barbara K, author.
Title: Now media : the evolution of electronic communication /
Norman J. Medoff and Barbara K. Kaye.
Other titles: Electronic media
Description: Fourth edition. | London; New York : Routledge, 2021. |
Includes bibliographical references and index.
Identifiers: LCCN 2020050570 | ISBN 9780367897215 (paperback) |
ISBN 9780367896751 (hardback) | ISBN 9781003020721 (ebook)
Subjects: LCSH: Broadcasting–History. | Mass media–History. |
Digital media–History. | Telecommunication–History.
Classification: LCC HE8689.4 .M44 2021 | DDC 384–dc23
LC record available at https://lccn.loc.gov/2020050570

ISBN: 9780367896751 (hbk)
ISBN: 9780367897215 (pbk)
ISBN: 9781003020721 (ebk)

Typeset in TimesTen
by Newgen Publishing UK

Access the companion website: www.routledge.com/cw/medoff

As always thanks to my wife, Lynn Medoff, for her patience with me while I paced around trying to find the best way of writing phrases without repeating myself.
Norman J. Medoff

Hugs and kisses to my husband, Jim McOmber, for keeping me supplied with gallons of Starbucks' iced tea whenever I needed a caffeine jolt to keep writing, and hugs to my funny cat, Jacky Paper, who kept me laughing and wrote parts of the book by simply walking across the keyboard. To Jane, with love, the best mom ever. RIP.
Barb K. Kaye

Contents

Preface ix
Acknowledgments xii

1 Opting Into 'Now' Media 1

2 From Marconi to Mobile Listening 23

3 Television: From Analog to Digital to 8K 51

4 Radio and Television Programs and Programming 83

5 Interconnected by the Internet 123

6 Social Media: Private Conversations in Public Places 157
   Jason Stamm

7 Digital Devices: Up Close, Personal, and Customizable 183
   Mary Beadle

8 XR: Inside of Media, Inside the Mind 211
   Charles P. ('Chip') Linscott

9 Advertising: From Clay Tablets to Digital Tablets 237

10 Audience Measurement: Who's Listening, Who's Watching, Who's Surfing? 279

11 The Business of Entertainment and Media Ownership 307

12 Working Behind the Scenes in Media 333

13 Feature Films: 'The Movies' 367
   Ross Helford and Paul Helford

14 The Personal and Social Influence of Media 391

Index 425

Preface

The coronavirus-induced lockdown orders in spring 2020 taught the world that despite the use, or overuse, of 'now media' (a term coined by this book) to communicate with others, humans crave in-person, one-to-one contact. Before the stay at home orders, people were content to communicate with others via social media, texting, or email. Make a telephone call? "What! Are you kidding?" Work on a project with a fellow student at the library? "What! What a pain, it's easier to use Google Drive." Go to class? "What! Why can't my professor just put everything online?" Stop to talk to someone while walking across the campus green? "What! I'll look at my phone, pretend I don't see her." Pre-pandemic, people used their 'now media' devices so they wouldn't have to socialize in person.

By mid-March 2020, most students and professors had been ordered off campus. Students packed up their dorm rooms to move back home with their parents, and professors sat in their in-home offices transitioning face-to-face classes to online delivery. At first, staying at home was a welcome respite from our hectic, over-scheduled lives, but as time wore on we began to miss our classmates, colleagues, friends, and family members. The isolation was especially hard on those who were sheltering by themselves. To mitigate our loneliness and fulfill our need to be around others, we turned to 'now media.' Suddenly, meetings, classes, parties, happy hours, Easter brunches, Passover Seders, weddings, and even funerals took place on Zoom. We were 'zooming' all over the place. Zoom usage soared by over 300%, from about 10 million users in 2019 to about 300 million users in the first five months of 2020. Zoom was not the only 'now media' application that 'zoomed' in usage. Facebook, Twitter, HouseParty, WhatsApp, TikTok, Skype, and other social apps soared in number of users and time used, and even voice calls were up over 40% since the start of the lockdowns.

'Now media' use not only increased for social reasons but also for informational purposes. The world anxiously connected online to follow the number of coronavirus outbreaks and deaths and where they were occurring. Whether through our smartphones, tablets, or computers, online traffic to media sites jumped. Online readership skyrocketed on newspaper sites like *The New York Times*, *San Francisco Chronicle*, *The Washington Post*, and *The Seattle Times*, and for television sites like CNBC.com and Foxnews.com. Even traditionally delivered television evening newscasts gained an audience. Moreover, sites hosted by the Center for Disease Control, Johns Hopkins University, and others that posted daily coronavirus updates were a strong draw for the online public.

'Now media' also kept us entertained while sheltering-in-place. Netflix, Hulu, Disney+, Amazon Prime, and other streaming services collectively grew by over 30% just during the first three weeks of shelter-in-place orders. Even traditional broadcast and cable television viewership increased.

It is hard to imagine what sheltering-in-place would have been like without 'now media.' The Spanish flu of 1918 swept across the world killing 675,000 Americans and 30–50 million globally. And just like with the coronavirus, people in the U.S. were ordered to wear masks, and schools, stores, manufacturing plants, and other businesses were shuttered, and people were told to stay home. People only had daily newspapers to find out what was happening. They did not have 'now media' for instant updates from a broad range of perspectives, or for keeping in touch with friends and loved ones. People in quarantine were truly isolated and alone.

Our modern-day experience with the coronavirus underscores the importance of media, especially 'now media' in our lives. We relied heavily on traditionally delivered radio and television and online sites for information about the virus. Streaming media kept us entertained and social media kept us connected to friends and loved ones.

Maybe the coronavirus has taught us to value human relationships and appreciate the time we spend together. Although Zoom parties are fun, people grew weary of them and hankered for a hug. Perhaps, post-pandemic we will socialize less online, and trend to more face-to-face interactions. Maybe the virus has also taught us a new-found appreciation for our smart devices but also a realization that overuse and addiction, ironically, leads to social isolation, the very state of mind that we relied on our smart devices to keep us from during the pandemic.

The changes to higher education stemming from the coronavirus-forced closing of college campuses and the subsequent move to online delivery, have yet to be fully realized. Although we tend to think of online course delivery as a new method of teaching it has been around since the late 1990s. But in the last decade or so, new course delivery systems have made online courses more efficient and attractive. You might even be using this textbook for an online course.

Now Media: The Evolution of Electronic Communication, is the fourth edition and new title for *Electronic Media: Then, Now, and Later*. The book is rooted in the notion that studying the past not only facilitates understanding the present, but also helps predict the future. Just as we can show how broadcast television spawned the cable industry, we can trace how the cable industry led to the satellite industry and how both have led to a digital world – one in which convergence has blurred the lines separating media functions and in which old-style broadcasters have expanded, consolidated, and adapted to the multiplatform system of contemporary 'now' media.

The study of 'now' media addresses more than just the delivery systems used to reach mass audiences. Personal electronic devices that deliver information and entertainment selected by individual consumers are covered as well. Devices such as smartphones and tablets, which allow users to surf the Internet, record and send video images, play music, and interpersonally communicate with voice or text, have changed the modern lifestyle to the point that they must be included in any discussion of the digital electronic media revolution. Online connections open the world to on the go, anytime entertainment.

This book links the traditional world of broadcasting to the contemporary universe of digital electronic 'now' media, which offer increasingly greater control over listening, viewing, and electronic interaction. The book informs students of these changes and increases their understanding of technology's enormous cultural impact and how newer digital devices and functions will affect the future of the industry.

ORGANIZATION OF THE TEXT

With the knowledge that what comes next is based on what came before, this book follows the structure of journalist Edward R. Murrow's programs, *Hear It Now* (1950–1951) and *See It Now* (1951–1958). Each chapter of the book is organized chronologically into these sections:

- *See It Then* focuses on the rise of a new medium, idea, or technology and traces its development up to about the new millennium.
- *See It Now* discusses activities and developments from about the year 2000 to the present.
- *See It Later* takes the present into the future and predicts what will happen in the digital world of tomorrow.

Chapter one summarizes the history of electronic media, introduces industry terms, and discusses current trends in media, such as digitization, convergence, and consolidation. Chapter two covers the history of radio from electrical telegraphy to satellite and Internet delivery. The chapter includes biographies of early radio pioneers and delves into laws and regulations regarding the transmission of radio content. Chapter three focuses on the development of television from the early days of broadcasting to cable, satellite, and online delivery systems. Chapter four concentrates on radio and television programs and programming. This chapter explains radio station formats and the various types of television program genres, such as dramas and situation comedies, and includes program origination and how shows make it to the 'airwaves.' It also covers the production, distribution, and delivery, including streaming of radio and television programming. Chapter five moves away from radio and television and focuses on the Internet. The chapter provides a history of the Internet – how it developed, how it works, and how it is used. The chapter specifically looks at the Internet from a media perspective and how it is used for information delivery, whether in text, graphic, audio, or visual forms. Chapter six, Social Media, starts with a history of social media, then focuses on the most popular social media sites, and then concludes with discussion of the personal and social benefits and consequences of social media. Social media are in the forefront of news dissemination and personal communication, and thus deserve a chapter of their own. Chapter seven describes newer mobile and customizable technologies and platforms, such as cell phones, tablets, Netflix, Hulu, Sling TV, satellite radio, Pandora, and DVRs, through which users design their own communication and viewing experience. Chapter eight about virtual reality is a new chapter for this fourth edition. The chapter takes you on a journey to fantasy worlds. The chapter is all about XR, the fascinating world of immersive media: How it developed, the impact that it has on us, and the applications of it now and in the future.

Although we love the content side of media, there is no escaping the business side and so Chapter nine is about advertising – the primary way media earn revenue. Chapter ten is an extension of the advertising chapter in the sense that by selling advertising time and online space the media are actually selling their viewers, listeners, and users to advertisers. Chapters eleven and twelve are both about the business of media. The chapters cover business models, ownership structures, and operations of the various types of media and delivery systems. Chapter eleven is more broadly focused, whereas Chapter twelve concentrates on working in the media – radio and television stations, cable and satellite companies, and content distribution. Chapter thirteen is about the film industry, which has strong ties to other forms of electronic media and to the Internet. The chapter is all about 'Hollywood' – studio system, film creation, origination, distribution, and movie stars. Finally, Chapter fourteen examines the social, cultural, and personal effects of media. The chapter discusses various effects theories, and describes how mediated violence and positive and negative depictions affect audience behavior, emotions, and thought processes.

Acknowledgments

The authors send many thanks and much appreciation to Charles P. ('Chip') Linscott, Ph.D., at Ohio University for adding the new, and very 'now' chapter, 'XR: Inside of Media, Inside the Mind,' and to Ross Helford for writing Chapter thirteen, 'Feature Films: "The Movies."'

A special thank you to Jason Stamm, Ph.D., at The University of Tennessee for authoring Chapter six, 'Social Media: Private Conversations in Public Spaces,' and to Mary Beadle, Ph.D., at John Carroll University, for penning Chapter seven 'Digital Devices: Up Close, Personal, and Customizable,' and for their contributions to Chapter fourteen.

The authors are also indebted to Jeffery S. Wilkinson, Ph.D., at Florida A&M, and Glenn T. Hubbard, Ph.D., at East Carolina University, for their marvelous editing and updating of Chapters one, two, three, eleven, and twelve, and for keeping the authors up on the latest in 'now' media.

We also thank Rebecca Lind, Ph.D., University of Illinois at Chicago, for creating the Instructor's Manual and Test Bank.

We would like to thank the following individuals with Routledge Taylor & Francis for initiating this edition and for moving it from inception to publication: Ross Wagenhofer (editor); Margaret Farrelly (editor); Priscille Biehlmann (editorial assistant); Gareth Toye (cover designer). Additional thanks to Rachel Carter (copy-editor); Kelly Winter (project manager); Geraldine Lyons (proofreader); Emma Brown (permissions).

The authors are very appreciative of the following individuals for their 'Career Tracks' stories: Jennifer McClure-Metz, Trey Fabacher, John Montuori, Glenn Reynolds, Jay Renfroe, Maria Hechanova, Rajah Maples, Dave Zorn, Erin Bigelow, Ryan Kloberdanz, Doug Drew, and Norm Pattiz.

Special thanks for their thoughtful comments and help with various parts of the book also go to: Gerald S. Adler, Alan Albarran, Patrick Parsons, Larry Patrick, Greg Stene, Mickey Gardner, Esq., Cassidy Zimarik, and Paul Helford.

Our special thanks to these people who contributed content to previous editions: Greg Pitts, Grant Guillory, Dale Hoskins, and Greg Newton.

We also gratefully acknowledge the assistance of Carolyn Stewart at the MZTV Museum of Television in Toronto. Thanks for the wonderful images.

Chapter one

Opting Into 'Now' Media

The authors thank Jeffrey Wilkinson, Ph.D. (Florida A&M University)
for his contributions to this chapter.

NOW MEDIA, NEW MEDIA, OLD MEDIA

Old, or traditional media, are still fully functional in today's world. What was once considered new media, however, are now getting old and stepping aside for '**now media**' (a term coined by this book), which are current digital variations of existing media. Although the lines between old and new media often blur, there are important distinctions. The term 'old media' refers to the delivery, programs, and formats that came directly from early broadcast radio and television. In radio, for example, old practices that persist include top-of-the-hour station IDs and station formats (e.g. rock, country). In television, programs are still usually 30 or 60 minutes in length including a set number of minutes for commercials, introductory and closing credits, and a theme song.

The Internet is often thought of as a 'new' medium, but it has been around in one form or another since the early 1960s – about 60 years. The World Wide Web was 'born' in 1993, making it 27 years old (as of 2020), and the first social media platform, *Six Degrees*, was created in 1997. Like everything, what was once 'new' becomes old and must evolve to stay relevant.

The convergence of old, new, and 'now media' is made possible by modern technology and contemporary distribution models. 'Now media' encompasses various types of contemporary digital devices, content, and delivery. An example of 'now media' is the combination of streaming television shows through a digital device (iPad) or through a television set connected to a streaming device such as a Roku. Viewers can watch a network television show like *Young Sheldon* when it is broadcast on television, or they can stream it whenever they want on CBS All Access, Netflix, Amazon Video, over any Internet-connected device. Now media.

'Now media' extends to viewing programs on social media or video sites, such as YouTube. In a sense, the terms 'old' and 'new' and are more relevant to the delivery system than the content, whereas 'now media' is about connecting content to a viewing device, and connecting audiences to new types of digital programming and information. Radio was once delivered only over-the-air, then it was added to the 'new' Web. But 'now media' music is not only streamed to any number of devices, but algorithms select users' favorite songs and suggest new songs they might like based on past selections. Podcasting is a 'now medium.' Creative tech-savvy people take advantage of digital delivery to come up with new, and very now, types of audio programming.

We love to sit back and be entertained. Special broadcasts like *Super Bowl*, *Olympics*, and *Grammy Awards* give us cultural currency to talk with friends and even strangers. Shows like *Empire* or *America's Got Talent* help us relax at the end of the day. In between we catch our favorite music countdown show, watch our favorite YouTuber, or latest streaming program on Netflix or Disney+. We have several choices whether to be entertained, enlightened, or even be alone with our thoughts. Digital media provide structure and reassurance about the world and our place in it. Our media world is almost unlimited in terms of what we can get at any time, any place, in any form.

Media are ubiquitous and constantly reach out to us. In a 'now' world of 'now' media, we are inundated with messages sent by broadcast, broadband, cellular, and satellite. We use multiple platforms like smartphone, tablet, laptop, or big-screen HDTV/UHDTV in whatever screen size works best at that moment. And perhaps the 'nowest of now' is virtual reality that envelops our senses and sweeps us into another world – yet we have not gone anywhere.

'Now media' are an integral part of everyday life. Through media we learn about the world. Our phones or *Alexa* tell the weather forecast so we know what clothes to wear. We listen to radio or television while we have breakfast, or surf the Web for the latest news and information.

We use our phones or tablets to read about products, send texts, listen or watch audio and video programs, surf countless Web sites, go on social media, and maybe play a video game, all in pursuit of information and news. And yesterday's information is not as satisfying as 'now' information. We want to know what is happening as it is happening or as it just happened. We want to know – NOW.

This book 'opts in' and examines the many aspects of electronic media and how it has evolved into 'now media.' It looks at the history, structure, delivery systems, economics, content, operations, regulation, and ethics of electronic communication from the perspectives of what happened in the past (See It Then), what is happening now (See It Now), and what might happen in the future (See It Later).

SEE IT THEN

Origins of Electronic Communication

The desire to communicate is part of being human. We first learned to express ourselves using combinations of verbal and non-verbal behaviors. Facial expressions and gestures conveyed meaning, as did grunts, sighs, and yells. Drawing on cave walls and flat surfaces evolved from art to storytelling and became written language. The earliest records of written communication are estimated to be 5,000 to 6,000 years old. With written language, we no longer had to rely solely on memory.

As early as 4000 BCE, people were writing on clay tablets using pictograms, a simple picture to symbolize an object or word, to communicate some basic information about everyday events. These tablets were portable and durable records. One thousand years later, about 3000 BCE, the Egyptians used the fibrous plant papyrus as a type of primitive paper. At the time, a form of picture writing called *hieroglyphics* evolved. About 2000 BCE, the

Figure 1.1 A 1,500-year-old cave painting from South Africa.
Source: Photo courtesy of iStockphoto. © Skilpad, image #10277331

Figure 1.2 Hieroglyphics from inside a temple in Egypt.
Source: Photo courtesy of iStockphoto. © Tjanze, image #10353358

Egyptians developed an alphabet of 24 characters. In the western United States, early Native Americans carved pictographs in rocks to show others what they saw and how they lived their lives.

Through the ages, people have developed innovative ways to communicate faster, over longer distances. Smoke signals, horns, flags, and pigeons were once commonly used to send messages great distances. Each system worked when conditions allowed, but were otherwise limited. Smoke signals and flags needed good weather and daylight to be seen; horns had to be heard; and pigeons could carry messages but were vulnerable to natural predators and severe weather. Even human messengers had drawbacks because they could be intercepted, captured, and killed.

In many early civilizations, people called scribes were trained to read and write in order to create permanent written records. Scribes served rulers and facilitated government and business by recording events using written language. Scribes attended and recorded what happened in public and private transactions such as the exchange of goods and services for money. They were an important part of business and government until the advent of printing.

Wood blocks and ink were first used in China around 200 AD to make copies of documents, but the process was slow and cumbersome. It took human effort to carve the block and make copies one at a time.

The age of **mass media** arrived in the mid-1400s when the printing press was invented. Johannes Gutenberg, a metalworker in Europe, developed movable type using a modified wine press. Gutenberg could print pages by coating hand-carved three-dimensional alphabetic letters with ink, which were then pressed onto paper. The result was a printed page that could quickly be duplicated as many times as needed. For the first time, *one* individual with a printing press could reach *many* people with printed material instead of handwritten copies of a book or manuscript.

Thanks to the printing press, books became common and reading became important to people. In Europe, handbills and flyers became newspapers in the 1500s as citizens increasingly sought to know about significant current events. As books and

Figure 1.3 These petroglyphs from native American cultures date back to 1500 AD.
Source: Photo courtesy of Getty image #172711400

Figure 1.4 The first printing press was built in the 15th century.

newspapers became popular, the practice of communicating to many people at once became common. This **one-to-many model** of communicating was not a balanced two-way model, however. The sender of a message could reach an audience (the *many*) through the printed word using various distribution methods. This one-to-many model became known as *mass communication*. But since it was not easy or fast to communicate back to the sender, the element we call *feedback*, was limited. The mass media became known as the senders of messages (individual, company, or even government) that use a mechanical device (e.g. a printing press) or electronic device (e.g. broadcast transmitter) to deliver those messages to a mass audience.

The electronic age was brought about when the telegraph was invented in 1844 by Samuel F. B. Morse. It used electricity to send messages almost instantaneously over long distances through wires. Telegraph messages used a system of dots and dashes – short on/offs and long on/offs – to spell out words one letter at a time. Using Morse code, the telegraph helped to coordinate the nation's railroads, newspapers, and financial hubs because it provided timely person-to-person information. Messages were called telegrams and averaged around a dozen words in length. Telegrams were relatively expensive for that time and remained popular until the early 1900s. Around the world, telegraph offices were established in key cities, where an office housed equipment and trained operators who could translate, send, and receive the messages.

One-to-one distance communication changed radically with the invention of the telephone. In 1876 Alexander Graham Bell unveiled a device enabling conversation between people in two separate locations. Both the telegraph and the telephone facilitated *person-to-person* (or *one-to-one*) communication over distances.

Gutenberg's printing press made modern mass communication possible in the 1400s and rudimentary forms of newspapers appeared in Europe in the 1500s. In North America, the first general-information pamphlet was published in 1690. The publication of *Publick Occurrences, Both Foreign and Domestick*, is considered the precursor to the modern-day newspaper. In its first and only issue, the publisher stated, "the country shall be furnished once a month with an account of such considerable things as have arrived unto our notice" (National Humanities Center, 2004). Ironically, the government

FYI: COMMUNICATION MODELS

MODELS EXPLAINING COMMUNICATION; SHANNON AND WEAVER MATHEMATICAL MODEL

Coinciding with the rise of electronic media and the advent of television, scholars proposed models of mass communication to help us understand how it worked. One of the earliest models of mass communication was proposed in 1948 by political scientist Harold Lasswell. Lasswell deconstructed media propaganda by asking a series of questions in the form of Who? Said What? In Which Channel? To Whom? With What Effect?

Around the same time another significant model was developed by Shannon and Weaver (1949), based on their work at the telephone company, AT&T. A transmission model of communication helped explain the process by which information is sent and received. That model, also known as a *linear model*, still works well to explain telephone communication.

In both models, key elements are the *information source*, the *transmitter*, the *channel*, and the *destination*. The information source (a person) uses a transmitter (a telephone) to send a signal through a channel (telephone wires) that is delivered to a receiver (another telephone) and then heard at the destination (a person). In mass communication, the information source (say, a weathercaster at a television station) uses a broadcast television transmitter to send a signal using broadcast waves through the air (channel) that is delivered to a television receiver and then seen and heard by the viewer (destination). Additional concepts, such as noise that can interfere with the process, were added to improve the model.

SCHRAMM–OSGOOD COMMUNICATION MODEL

Schramm and Osgood (Schramm, 1954) used a simplified model to explain communication. Using only three basic elements – a *message*, an *encoder*, and a *decoder* – their model elegantly demonstrates the reciprocal nature of interpersonal communication, such as a conversation or a debate. It shows how communication is a two-way process in which participants are both senders and receivers of information.

SCHRAMM MASS COMMUNICATION MODEL

Schramm (1954) also proposed a model for mass communication, built around source, message, and audience. In this model, the source represents an *organization* that sends out *many identical messages* to the *audience* composed of many individual receivers. The receivers are connected to groups of others who can pass along information about the messages from the initial receiver. The dotted lines in the model represent *feedback* from the receivers, which in that era was delayed and not explicit. Back then, the organization had to infer the meaning of the feedback (such as ratings for a program) and act accordingly.

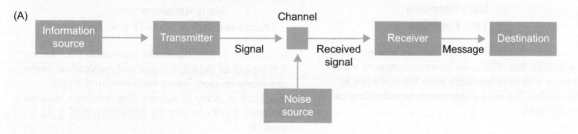

Diagram 1.1a Channel flowchart.
Source: Based on Shannon and Weaver, 1949

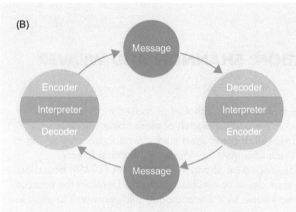

Diagram 1.1b Message flowchart – circles.

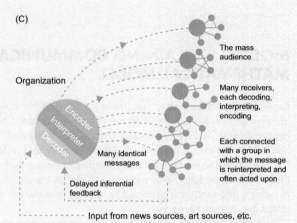

Diagram 1.1c Encoder flowchart – circles.
Source: Based on Schramm, 1954

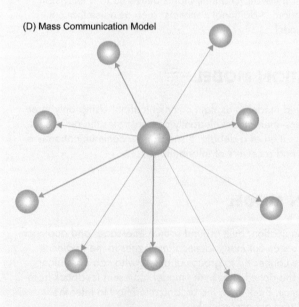

Elite Sender
Many Receivers
Little Feedback

Diagram 1.1d This model has an elite sender in the middle that distributes communication to many at once with little feedback from the audience to the sender. The model represents broadcasting or newspapers.

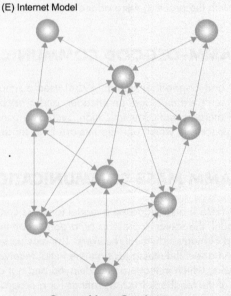

One or Many Senders
Many Receivers
Much Feedback in Many Directions

Diagram 1.1e This model represents the Internet, it is typical of digital/Internet communication: there are many senders, many receivers, and much feedback in many directions. The feedback may be delayed, but it can also be immediate and quite explicit.

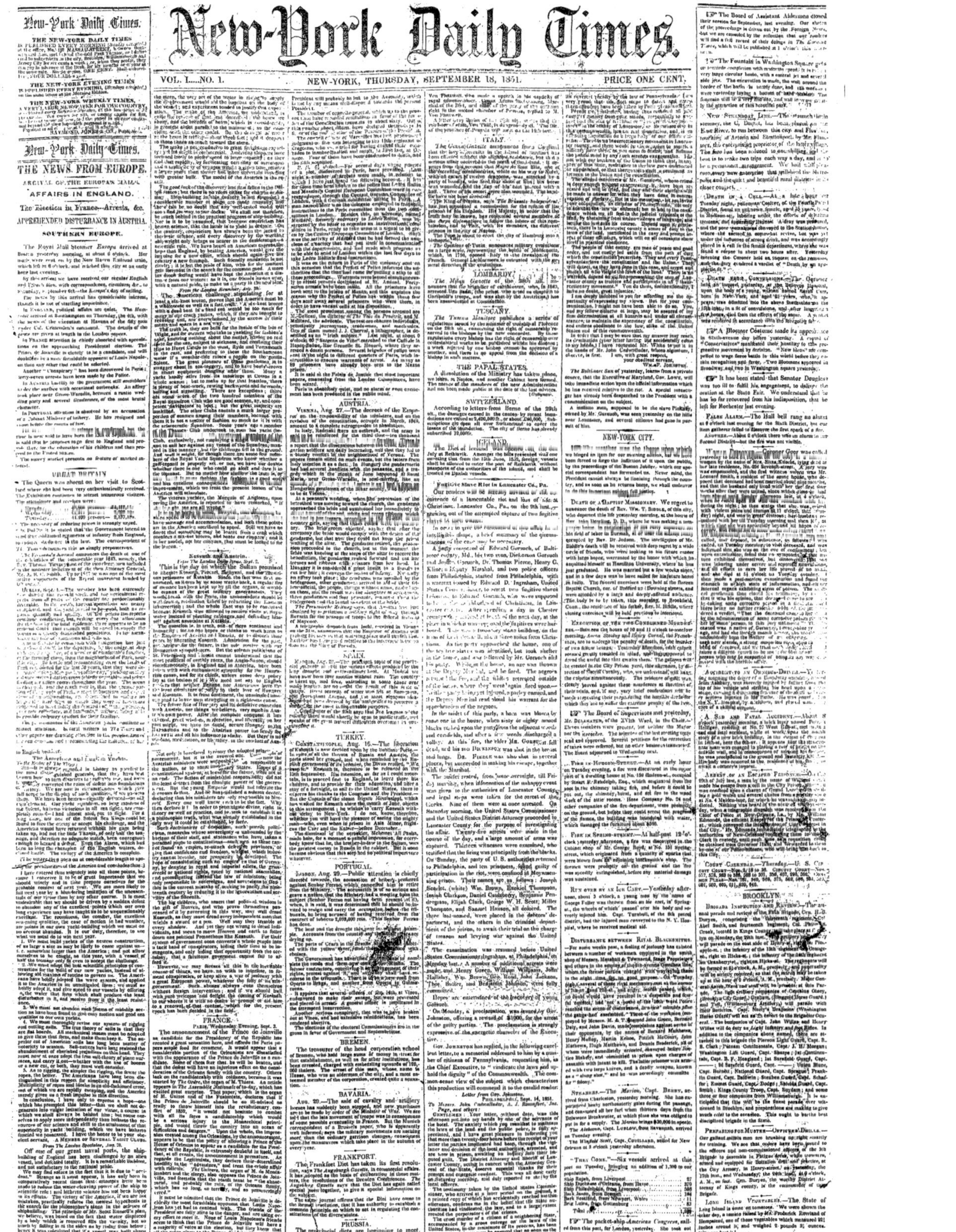
New-York Daily Times.

VOL. I....NO. 1. NEW-YORK, THURSDAY, SEPTEMBER 18, 1851. PRICE ONE CENT.

Figure 1.5 The front page of the first edition of *The New York Daily Times*, which sold for a penny in 1851. *The New York Daily Times* eventually became *The New York Times*.

did not allow a second issue for two reasons. First, it had been published without legal authority (permission), and, second, it contained a story about incest in the French royal family (salacious content). Various pamphlets/newspapers came and went until 1833, when the *New York Sun* established itself as the first American daily newspaper created for the mass audience. Publisher Benjamin Day set the price of his paper, the *New York Sun*, at one penny, launching the term 'penny press.'

While print was the new mass medium of the early 1800s, there were also several experiments to develop electronic wireless machines. By the early 20th century, Guglielmo Marconi successfully demonstrated *radio telegraphy*, sending a signal from a radio transmitter to a radio receiver, a system known as point-to-point delivery. Radio telegraphy worked in a similar fashion to Morse's telegraph, but without the wires. Soon after the advent of radio telegraphy, other inventors produced a system for transmitting the human voice, and, by popular demand, music.

After World War I, radio stations and receivers grew rapidly in popularity, signaling the dawn of broadcasting as a commercial electronic medium. All over the country, local newspapers, magazines, civic clubs, and schools promoted local radio stations and their programs. Excitement was fueled by radio's warm and personal appeal. Listeners felt connected to the personalities and were captivated by its spontaneity. By the late 1920s, radio's popularity was established. The live medium delivered news, original and dramatic programs, live music, and other forms of entertainment directly into the home.

For more than 30 years AM radio was the electronic hearth that connected the coasts and the heartland across the nation. At that time, radio was the only instantaneous electronic medium and it capitalized on the 'live' aspect to successfully develop programming formats like situation comedies, police dramas, and soap operas which later moved to television. Radio enjoyed tremendous financial success, and was a mainstay in American culture.

Radio's stature changed after World War II, when television broadcasting got off to a roaring start. Many of the popular shows on network radio shifted over to television, providing the new medium with an audience already familiar with those programs.

Television emerged as the dominant national electronic medium after World War II. By the mid-1950s it had surpassed radio's influence on American culture, and by 1960 it was a standard piece of furniture in most homes. We still refer to many forms of serial visual programs like sitcoms or dramas as simply 'TV.' Radio and television have always had a close relationship with advertising, and several persuasion techniques were first developed through electronic media ad campaigns. Besides subscription and pay per view, television programs continue to make money through paid advertising. Television programs are attractive to advertisers who place commercials inside episodes in order to reach the viewers.

From the late 1940s until the 1980s, the three major networks – ABC, CBS, and NBC – claimed almost 90% of the national prime-time viewing audience. Since then, the viewing audience has fragmented, first migrating to cable and satellite and now to the Internet-based streaming services. In early 2020, Disney's streaming channel, Disney+, surpassed ten million subscribers in its first day of operation.

The companies historically known as broadcast networks remain a dominant force in the media world even as they continue migrating to Internet-based streaming platforms. These companies still collectively show huge audience shares for their diversified media properties. These media properties include legacy media (newspapers, radio, magazines, television stations, movie studios) as well as new media (cable systems, streaming services, and other digital properties). Consider that, for almost a century, the networks have produced high-quality program content, employing many of the best writers, actors, and production people. The most popular programs continue to generate significant revenue either on broadcast networks, cable channels, satellite systems, the Web, or some combination.

Today, the television industry (broadcast, cable, satellite, streaming) is diverse and fragmented, despite numerous cross-ownership arrangements and both vertical and horizontal integration. Pay services like HBO in the 1970s proved they could produce high-quality shows that rivalled the networks. Netflix and Disney+ compete with other online video services to grab viewers' attention. The Internet has proven to be a *paradigm breaker* – a medium that does not neatly fit into previous models of electronic media. It has grown more rapidly than any other medium in history and recombines different aspects of all our previous models.

The Beginnings of Media Regulation and Free Speech

The development of mass media in this country has been shaped by many factors, but especially law and regulation. Law and regulation have set

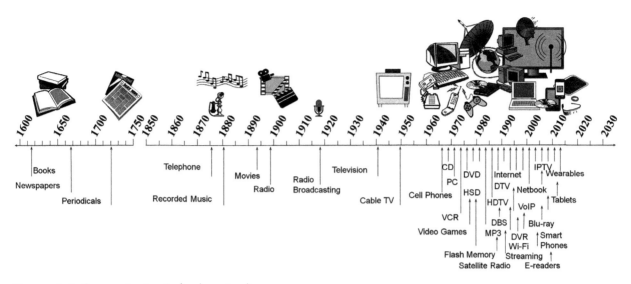

Figure 1.6 Communication technology timeline.
Source: Grant and Meadows, 2020

the parameters for how the media industries were created, how they grew, and how they could operate and even innovate. The first and perhaps most important regulatory guideline for media in the U.S. comes from the United States Constitution.

The First Amendment to the Constitution was adopted in 1791 as one of ten amendments that make up the Bill of Rights. Although originally offered third, the First Amendment states:

> *Congress shall make no law respecting an establishment of religion, or prohibiting the free exercise thereof; or abridging the freedom of speech, or of the press; or the right of the people peaceably to assemble, and to petition the Government for a redress of grievances.*

This provision for free speech undergirds our belief that access to accurate information leads to the best decisions. This anti-censorship view was first articulated by English writer and poet John Milton in his 1644 book *Areopagitica*. According to Milton, people should be exposed to the widest variety of opinions and ideas so they can apply reason and select the very best ideas. Even bad ideas and inaccurate information should be available because they present opinions and materials worth considering. Milton stated, "Let [Truth] and Falsehood grapple; who ever knew Truth put to the worse in a free and open encounter?" This concept, the **self-righting principle**, states that people will ultimately choose the best path forward. In other words, 'truth' will win out over 'untruth' in a free and responsible society.

Closely related to Milton's self-righting principle is the concept of a **marketplace of ideas**. This term refers to a social marketplace, where ideas can be sampled just like goods presented in a marketplace. In the marketplace of ideas, all ideas are deemed worthy of public exposure, but the public then decides which ideas are good and which to forget about. It is assumed that, since people are both rational and inherently good, they make the right decisions. These concepts underlie the First Amendment to the U.S. Constitution.

The founders of this country could not have envisioned the amazing innovations in mass media that have occurred since then. For example, in the 1920s unregulated radio stations were broadcasting on whatever frequency they wished, often conflicting with each other. The federal government was forced to intervene and assign stations to specific frequencies on the electromagnetic spectrum. By the mid-1930s, when radio networks could reach the entire country at once with a single program, the national security implications became obvious. In 1945, the federal government forced NBC to divest itself of roughly half its radio stations (creating ABC), arguing that monopolistic power does not serve the public interest.

The right to speak freely is expected, and often taken for granted, by people who live in the United States. The First Amendment guarantees the right to free speech even though that right has never been absolute. Time, place, and manner restrictions make it illegal to threaten a public official or yell 'fire' in a movie theater. Electronic media regulation is still guided by the *scarcity principle*, because both radio and television use parts of the electromagnetic spectrum, and that space is limited. Only a certain number of radio and television stations can broadcast inside that range of frequencies without interfering with each other. Since the 1920s, the U.S. Congress has held that it has the right to protect that spectrum space on behalf of the American people. Since the 'airwaves' belong to the people, it is in everyone's best interests to protect children from inappropriate materials and messages.

The right to regulate broadcasting because of scarcity was upheld by the U.S. Supreme Court in the 1969 case, *Red Lion* v. *FCC*. The Court also upheld *the fairness doctrine*, which required stations to provide on-air rebuttal time for opposing political views. Stations complied but complained about the government requirement. The fairness doctrine was abandoned in the late 1980s when the government embraced deregulation across several sectors of business and industry, which included electronic media. The argument was that there were enough voices in the marketplace (in other words, a 'lack of scarcity'), so there was no need for balancing views at the program or station level.

Some of those new voices included the first 24-hour cable news network, CNN, which was launched in 1981. CNN brought a wide variety of news, information, and new perspectives into homes. After a decade of declining listenership, AM radio experienced new popularity in the late 1980s through a cultural phenomenon sparked by conservative talk show host Rush Limbaugh. As new voices were added (Fox News and MSNBC both in 1996), America entered an age of partisan electronic media which has since grown through the Internet.

The FCC still treats subscription-based electronic media differently than broadcast media. Broadcast signals are freely heard or watched by anyone at any time. Therefore, the government has an interest in protecting citizens (and their children) from inappropriate material. The infamous 'Janet Jackson Wardrobe Malfunction' which occurred during the 2004 Superbowl halftime show would not have been as controversial on cable. Multichannel services, direct broadcast satellite (DBS), and cable and Internet-based subscription services enjoy stronger First Amendment protection because we pay – give them permission – to bring their content into our home.

The Internet has the full protection of the First Amendment, and both Congress and the FCC

ZOOM IN 1.1

The electromagnetic spectrum is the range of electromagnetic energy that radiates in waves from a source. At different frequencies the waves have different properties and can be used for a variety of purposes.

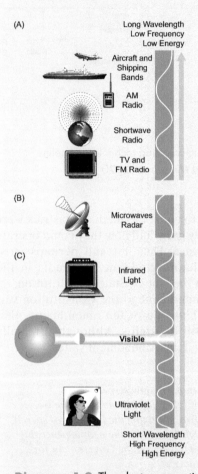

Diagram 1.2 The electromagnetic spectrum.

Learn more about the spectrum at: www.thinglink.com/scene/726160490195058688

essentially allow Internet content to be unregulated. Still, there is potential legal liability because all published or posted material is subject to various laws related to speech. There have been efforts by Congress to regulate some kinds of inappropriate content. For instance, the Telecommunications Act of 1996 originally included the Communications Decency Act (CDA), which tried to protect children from indecent material on the Internet. But the CDA was removed from the Act because it was ruled overly broad in restricting free expression and, therefore, unconstitutional.

Listening and Viewing Behavior

Electronic media can influence us in profound ways. Three common categories of media influence are Cognitive effects (what we think), Emotional effects (what we feel), and Behavioral effects (what we do).

- *Cognitive effects:* Electronic media expose us to a flood of information. News reports and stories teach us about our world and open our minds to new ideas. The media are our eyes to places we may actually never visit, but can learn about. Through media, we can vicariously experience the thrill of skydiving, the drama of legislative debate, enjoy the view from the top of Mt. Everest, and reflect on the serenity of an isolated white-sand beach on the shores of the Caribbean Sea.
- *Emotional effects:* Electronic media touch our feelings. Close-up images of hungry children, lonely elderly, and even crippled animals accompany calls for us to give to charities. Music and mood are intertwined and we use certain songs to motivate ourselves, or pause and reflect. Music is 'the soundtrack of our lives,' evoking powerful memories of people, places, and personal events.
- *Behavioral effects:* The electronic media can influence where we go, what we do, and what we buy. We meet with friends to watch election returns. A new product is advertised and we buy it. After a natural disaster, we act, giving money to a GoFundMe page or a media-sponsored charitable organization.

When television became popular, radio was forced to narrow its focus to attract specialized audiences. Today, modern media managers understand that the *niche audience* is the most effective and efficient way to reach people with targeted messages. General radio formats have become more narrowly targeted, and also more eclectic. Instead of broad

Figure 1.7 Small transistor radios became ubiquitous in the late 1950s and brought radio to parks, beaches, and almost everywhere people went to listen to their favorite stations.
Source: Courtesy M7TV

labels like 'Rock,' we can choose classic rock, alternative rock, and other subtypes. Audiences today are studied across a host of characteristics including age, income, ethnicity, gender, and even religion.

During the 1950s, radio listening moved outside the home to make way for television. Thanks to small lightweight portable transistor radios and relatively inexpensive cars equipped with AM radios, a vibrant teen culture emerged that facilitated the rise of popular music like that of Elvis Presley and the Beatles.

Innovations like the transistor helped redesign television sets from large wooden living room consoles to small plastic bedroom sets, and the shift from black and white to color enriched the viewing experience. As television sets became smaller and prices dropped, multi-television households became the norm, and viewing shifted from a family event to an individual activity.

SEE IT NOW

Characteristics of Mass Media

The Internet has transcended any particular model of media communication and, in fact, may be the foundation that ties together all of our current models. All our communication devices use the Internet for distribution, storage, and reaching audiences. Globally, nations continue to upgrade network capacity and capability in order to migrate increasing goods and services onto the Internet. But the upgrade comes at a cost, and concerns continue to emerge about safety, security, and over-reliance on the Internet.

We still find that each mass medium has particular strengths and is best suited for specific types of communication. For example, television broadcasts effectively reach large, geographically and demographically diverse audiences; radio is an engaging personal medium that offers immediate local information. But now we add social media and devices such as tablets, smartphones, and computers because they are used in ways that complement radio and television programs. The primary strength of audio and video music, news, and entertainment programming is reaching targeted audiences. To better understand how people use and rely on media messages, we need to consider six important characteristics: audience, time, display, distribution, distance, and storage.

Audience

A good reminder is that audiences are made up of individual persons. Your mother, your teacher, the cashier in the convenience store are all people with feelings, needs, and dreams. Each person uses media in different ways, but *they use each medium the way they prefer*. Traditionally media played to large, undifferentiated, and anonymous audiences, believing that 'bigger is better.' Radio and television were single-source media that could reach large audiences simultaneously. We still witness this aspect during events of global interest such as election returns, sports championships, and awards ceremonies.

The Internet, however, is a unique medium enabling individual as well as mass communication. Since online activity can be tracked, *media analytics* help us better understand and value smaller, more homogenous viewers or listeners. Defining and reaching these target audiences enable media to remain valuable and relevant.

The final point about audiences is that the smartphone has given individual users the power of choice. Smartphones afford us the opportunity to decide what to see or hear. The smartphone enables the user to be a creator-sender of content in any capacity, be it person-to-person, person-to-group (apps like groupme), and even mass media (via Twitter or Instagram). Each of us can now decide what audience to be a part of, and engage with those who make up our audience, those who follow us.

Time

Another characteristic of media is that it can be experienced in the moment (*synchronous*) or consumed later at a more convenient time (*asynchronous*). Recorded or downloaded media such as podcasts, DVDs of films, or using a DVR are all asynchronous media. Listening to 'live' content on radio, watching a live sports event, and conversing on the telephone are examples of *synchronous* communication. The concept is useful but gets complicated when several media are used at the same time. We listen to music on the radio (synchronous) played from a digital file (asynchronous), as we send a text message (synchronous) to a friend who has to look something up before responding (asynchronous). Or consider that we schedule a tweet to be sent later (asynchronous), which is read immediately by our followers (synchronous).

The early days of radio and television were mostly – if not entirely – synchronous. Radio stations employed live bands, and television shows were broadcast live before studio audiences. After World War II, inventions like the audio tape recorder and the videocassette recorder made it possible for programmers and the audience to *time shift*. The broadcaster could air recorded programs later, and also the audience could record the show and watch it at a more convenient time. Today, radio and television programs are commonly downloaded onto a digital device or simply accessed through online services such as Netflix or Amazon Prime Video.

Display

Media differ in how they display information. Display refers to the technological means (e.g. video, audio, text) used to present messages to audiences or individual receivers. Print is often produced to be seen from a certain distance from the eyes. Audio, though not seen, is displayed via speakers. Consider the displays for home speakers from Google, Amazon, and Apple, or traditional speakers placed either on the floor, hung on walls, or hanging from the ceiling.

Most of our displays involve the presentation of visual images. We have entered the age of 'any screen, any size,' which means delivering content to the various devices available and factoring in the actual distance from our eyes. Consider the average screen sizes of each of these devices: Smartphone (5-inch screen), tablet (10-inch screen), laptop (14–16 inches), computer (22–24 inches), and television (23–75+ inches). At the 2020 Consumer Electronics Show, Samsung unveiled a 219-inch video screen and a 150-inch 8K video screen. These larger screens are targeted for commercial use or exclusive home theater setups. Many companies are also experimenting with VR headset optical devices such as the Oculus

Figure 1.8 The Oculus Rift was one of the first VR headsets produced for the mass audience.

Rift. These display devices tend toward gaming, education, and entertainment rather than general media use.

Distribution

Distribution refers to the method used to carry information to receivers. Television's audio and visual images have historically relied upon broadcasting, cable, microwave, or direct broadcast satellite for distribution. Radio is still transmitted over the air, making it convenient to listen in the car, at home, or when out with friends. But in today's highly competitive media environment, radio stations look to alternative forms of distribution.

For example, many radio stations have aggressively moved to accessing home speaker digital assistants such as Alexa, Google, and Siri. These and other digital audio and video display devices are delivered over the Internet and streamed live on-demand. Information and programs are available from a server (*host*) and distributed to a home speaker, computer, or smartphone (*client*). Individuals connect to the Internet via *cable modem* or *digital subscriber line* (*DSL*), a 4G or 5G cell phone connection, or public or private *Wi-Fi* (wireless local area network). Mobile devices like tablets and smartphones play music from radio stations (generally through apps like Tune In or iHeart Radio) and video from program suppliers such as Netflix simply by using the app to access a library of movies or television programs.

Interactivity

Interactivity is not only the way but also the degree to which we use our devices to communicate. Today, we can be as interactive as we want to be.

Our smartphone is an integral part of who we are. We take it everywhere (even to bed) and use it throughout the day to interact with friends, family, and do business. While waiting for an elevator we go online, listen to music, or watch a TikTok video. We interact with our smartphones throughout the day.

The Internet has varying degrees of interactivity. Some Web sites are strictly informational and do not try to engage the visitor. But the best, most savvy Web sites are filled with interactive games, puzzles, and useful information so users will keep coming back.

Social media are, by definition, highly interactive. By posting and making comments, users invite others to post and make comments. Social media reflect the Schramm model of communication where the encoder posts something that is commented on by another (decoder), which leads to another posting, and so on.

Sometimes we do not want to interact. Sometimes we simply want to sit and watch, or listen to whatever is playing. Radio and television broadcasts fulfill this function very well. Interactivity is still an option; listeners can call radio request lines if they wish and television viewers can call or text a vote (e.g. *The Voice*). But these probably better reflect *feedback* rather than true interactivity.

Sending texts and emails to a radio program can start a conversation leading to an immediate response. Pandora listeners can fast forward on a song they do not like or give it a thumbs down and delete it instantaneously from the playlist. The key point here is that users have choice whether to interact and traditional electronic media can help us to disengage.

Distance

The Internet has changed the definitions of local, regional, or national media, enabling us to send and receive messages from across the room or across the globe. Our smartphones allow us to transcend distance. You can send a selfie to a friend in Tokyo while having a video chat with your parents in London, then text message your BFF to wait for you at the apartment. Messages travel at the speed of light with almost no signal delay.

Some media are better suited for long-distance delivery and others for short or local transmission. Print media need to be physically delivered to their destinations, a process that can be cumbersome and expensive over long distances. Electronic media deliver messages quickly and inexpensively through both wired and wireless transmission. But print has greater permanence; electronic messages, once sent, are gone (unless they are recorded and stored).

Storage

It is a digital world and digital storage capacity continues to grow exponentially. Our phones and flash drives today hold tens of gigabytes of data. Our computers hold a terabyte or more. Soon we will talk regularly in terms of petabytes.

In terms of scale, text is relatively small and can be saved as files measured in bytes and kilobytes. High-quality photos and audio are often stored in megabytes. Video data is huge and stored in gigabytes. High-quality recorded media like film are stored on DVDs, and Ultra HD Blu-ray DVDs (aka 4K Ultra HD) hold upwards of 100 gigabytes of data. We can also store large numbers of our favorite programs on DVRs from vendors like TiVo, Dish, Amazon, and others.

To manage global data storage, an entire industry has emerged offering to archive our collective data away from our homes and our work. With so-called '*cloud computing*,' data are not really held in a cloud but in buildings filled with endless rows of storage drives and servers in a secure, climate-controlled setting. Uploading our data to the cloud frees us from taking up physical space in production studios, businesses, and private homes. Cloud facilities are spread out in multiple locations and cloud users can access their information from anywhere, using any computer. Cloud storage can be bought or leased from a *hosting* company such as Google or Amazon (AWS) that provide cloud services.

Why Digital?

Digitization of mass mediated content is probably the most revolutionary innovation since the printing press. Digitization allows information to be compressed and endlessly reproduced, and converted and stored in very small but exceptionally high-quality forms. And digitization makes media searchable.

These aspects have transformed media and the ways consumers use the media. Users no longer have to search through torn pages or garbled video- and audiotapes to find specific information. Digitized material can be saved in a variety of formats and stored on laptops, smartphones, or flash drives. With compressed digitization, a smartphone can store a small library of books, hundreds of songs, and favorite films or television shows.

Digitization has also improved our music listening. Satellite-direct digital radio service began in 2002 with XM Satellite Radio and Sirius Satellite Radio. The two merged in 2008 to become Sirius XM. Satellite radio takes advantage of geosynchronous orbit to send signals unobstructed to radio receivers anywhere in the coverage area. Sirius XM subscribers can drive cross-country and listen to the same station the entire trip. Satellite radio is subscription based, providing more than 200 channels of music, talk, sports, and news.

Digital terrestrial radio, also known as **HD radio (high-definition radio)**, is another digital audio option. Digital broadcasting (technically called *in-band on-channel, IBOC*) lets a radio station use the same frequency to broadcast its analog and digital signals, which translates to clearer, static-free radio, and it is compatible with both analog and digital sets. Better yet, stations can offer more than one 'channel' in their allotted frequency. In other words, an HD FM radio station that is broadcasting at 103.1 MHz can program several subchannels within its frequency allocation with a variety of formats.

Trends and Terminology

The last major regulatory overhaul was the 1996 Telecommunications Act, which paved the way for but did not foresee the rise of tech giants like Facebook, Twitter, Google, and Amazon. The rules and regulations regarding media technology ownership and practices always lags behind current practices as the media industry struggles to keep up with new devices, the ever-changing uses of existing devices, and the legal and social challenges brought on by digitization.

Audio podcasts have tremendous variation in style, format, and content, and though a podcast sounds like an over-the-air 'radio' program, it is digitized audio. Digital video is creative, it can be any length, and does not have to adhere to strict format conventions involving storylines, cast, content, and language. The freedom to experiment with a variety of storytelling techniques is possible via the Internet and streaming, which are not subject to broadcast regulations, defy traditional production techniques, and reach niche audiences.

Program Selection: Linear, Recommendations, Random Access

Beginning in 1920 when radio became a source of free entertainment for the audience, daily newspapers listed the shows and channels for that night, or the Sunday paper would have program listings and descriptions for the week. These guides informed the radio audience about the programs and when to gather around the radio to listen. With a linear-designed program order listeners always knew when to tune in so they would not miss an episode and possibly lose interest. For example, scheduling two or three situation comedies in a row would appeal to the audience members who had an inclination for that type of entertainment that night. Keeping an audience tuned in for more than one show was both a science and an art. Programmers had to decide which programs attracted a large enough audience that would stay tuned in for the subsequent programs.

Today, television programs can be watched on various venues, such as a broadcast or cable network, the network's Web site, a streaming service such as Hulu or Netflix, or video on demand (VOD) through a cable company. Keeping viewers watching has become more complex than simply adhering to a regular airing schedule – programs now need promotion on social media and throughout the Internet. Algorithms keep a program up front in a viewer's mind by 'recommending' a similar show to the one that was just streamed. When viewers scroll through the Netflix menu they are alerted to a list of similar shows in content or style, perhaps even country of origin (e.g. British comedies), to the program they watched earlier.

The important point is that what a user watches leads to recommendations to try other programs. Past behavior is a helpful predictor for future program selections. This simple form of machine learning is used to assist viewers as well as programmers and advertisers.

Integration: Vertical and Horizontal

Vertical Integration Vertical integration is when a company expands its business into areas that are at different points on the same production/distribution path, such as when a company like Comcast owns a broadcast network (NBC) as well as film studios and theme parks (Universal) to create programs for its network, cable channels, and Web site. Vertical integration helps companies reduce costs and improve efficiency, for example, by decreasing distribution expenses and simplifying negotiations. All the major broadcast networks are now part of large media companies, that own other companies (or divisions of the main company) that produce and distribute programming, as well as other businesses that have nothing at all to do with media.

Horizontal Integration Horizontal integration is the means by which a company increases its audience share and revenue by acquiring or merging with competing outlets, or by expanding internally. For example, CBS is part owner of The CW Television Network and owner of the Showtime cable channels. Disney owns Disney+ but also sells merchandise through Disney Apparel, an example of internal horizontal integration.

Convergence

Convergence refers to the blurring of boundaries between all types of electronic communication media as they merge together on devices such as a smartphone, tablet, or smart TV. Traditionally we once read news printed on paper, watched video on television, and heard music on the radio. We now refer to television, telephone, and computers as simply *platforms*, each being regulated with separate laws.

Increasingly, content is being streamed, uploaded, and distributed through the Internet and the Web to whatever screen a user is on at that moment. Depending on the service plan and subscriptions to various content services, listeners can hear their favorite iTunes music or online radio station, watch their favorite YouTube videos, post to Snapchat or Instagram, or check on weather or road conditions using the right app. There are apps for almost everything. For example, Whatsapp or WeChat users can place a call or even video conferencing with another person anywhere in the world. During the call, users can also send text messages, upload pictures, and surf the Web. Smartphones are increasingly used to manage household features like thermostats, lighting, and home security.

Digitization has also changed the way we read books, even though the printed book market remains far larger than that for e-books. In fact, 2019 printed book sales are higher than in 2004. In the U.S., less than 30% of book sales involve an iPad, Kindle Fire, or other type of tablet or e-reader device, which allows users to download digital copies of books and other printed materials. Digitization and convergence have blurred the lines helping us distinguish one medium from another. Clearly, the traditional definitions of these media need to be evaluated.

Deconvergence

The trend toward media convergence was energized by digitization. Large media companies combined their print and media operations to facilitate the sharing of stories, information, personnel, and infrastructure. However, there are also circumstances when companies *deconverge*, where a large media company may 'spin off' their print business from their electronic media business. In 2014, Gannett Company separated its broadcast operation from its newspaper business, yielding two separate companies, Gannett Company (print) and Tegna Inc. (broadcast and digital). Then, in 2019, Gannett was purchased by GateHouse Media to create the nation's largest newspaper publisher. In 2020, Gray Television announced it was interested in purchasing Tegna, then withdrew the bid a month later.

While technological convergence via the Web has been a reality in mass media, convergence between print culture and electronic media culture has often been difficult. The style of journalism between print and electronic versions varies greatly, and the personalities of those who create content and the goals of the platforms often differ as well. In addition, business practices and investment strategies have encouraged media companies to allow both print and electronic media to operate in their own style and with their own goals.

Consolidation

Deconvergence has not stopped companies from buying and trading media properties. In 2019, CBS and Viacom merged to become ViacomCBS Inc. for $11.7 billion. The deal brought CBS and Viacom program catalogs together, totaling around 140,000 episodes of television series and 3,600 film titles. ViacomCBS announced their combined platforms accounted for 22% of all television viewing in the U.S., ahead of Comcast (18%) and Disney (14%) (Littleton, 2019).

Consolidation concentrates media ownership, giving a smaller number of owners more media outlets and a larger share of the audience. Media companies are spinning off parts of the conglomerates to concentrate their business efforts in specific media platforms.

When one media company merges with or acquires another media company, the overall ownership of media outlets is consolidated. This type of business *consolidation* was facilitated by the Telecommunications Act of 1996, which relaxed many of the limits on media ownership. Consolidation has occurred in the radio business, with the top ten radio station groups owning thousands of stations and reaching two-thirds of the national radio audience.

In 2017, the FCC helped legacy media stay competitive by eliminating the decades-old restrictions on *cross-media ownership*. The two rules had limited newspaper–broadcast station cross ownership as well as limited radio–television station cross ownership. These rules were seen as unnecessary in today's media marketplace where there are numerous sources for entertainment, news, and information.

Localism Broadcast stations are local because the signals are limited to the communities where they are licensed. These areas can be relatively small, like the city limits for a college radio station, or cover much of a state like with a VHF television station. Broadcast networks, on the other hand, have no such requirement and can be distributed to the entire country. Cable channels such as ESPN or HBO also have no such requirement and can be distributed nationally. In television, the concept of localism is a bit more complex because television station signals cover a larger area than most radio stations, and also are often bundled to distant markets via cable television systems. For example, television stations in Phoenix, AZ reach many smaller communities around Arizona, including some that are 150 miles from Phoenix. Television stations in Tallahassee, Florida, are an important local voice for north Florida as well as nearby south Georgia. Localism is still an important concept in radio, because a radio station best serves its community with local-interest entertainment programming, news, weather, sports, and community affairs and information. Small businesses can afford to advertise on radio to reach potential customers in that community.

It Is Called Show Business, Not Show Art

Most media consumers focus on the messages, the news or entertainment programs that people can see or hear, but, conversely, those in the media are focused on the business bottom line. Streaming companies like Hulu, cable operators, broadcast stations, satellite channels, and Web sites like *The New York Times*, *Huffington Post*, and the *Drudge Report* cost money to operate and often have investors who demand a return on their investment. For decades there was the illusion that media simply provided content that catered to the tastes of viewers or listeners. No longer. Today the environment is hyper-competitive. For any program provider there are several competitors vying for the same eyeballs or ears. Ownership convergence means there are fewer companies owning more stations. Many of these ownership groups are publicly held and seek to please stockholders first.

Buying a successful station is expensive, and the purchasers typically incur a huge debt that takes many years to pay off. Station owners entice the largest possible audience to their programming so they can sell advertising for the highest possible price to bring in enough profit to support the enterprise and pay back the investors and debt holders. In a weakened economy, such as the aftermath of the coronavirus outbreak in 2020, advertisers cut back, often leaving stations in dire straits and ripe for takeover or purchase by a media conglomerate.

Monetization

The primary goal of media companies is to monetize content or generate revenue from the newer delivery systems (e.g. Web sites, podcasts). Media companies are faced with the challenge of reaching target audiences who may be distracted by other entertainment outlets. When the audience moved to the Internet, media created their own Web sites. As the audience moved to smartphones and social media, companies followed with programs and promotional content. As the audience moved to social network sites, media companies started their own social network groups. Radio and television stations find themselves chasing their listeners and viewers from one media landscape to another, because if they lose them, the advertising dollars also go and with them the ability to finance new digital efforts to lure the audience back.

Big Data

Big data refers to sets of data that are so large that traditional data-processing procedures are inadequate. Big datasets are generated by numerous sources and are too large to be stored on a single computer. For example, server logs from a popular Web site can show information such as who visited, when, and for how long that person stayed on the Web site. Big data would also include if the person clicked on a story or banner ad, and what Web sites the person went to afterward. Big data would look at these behaviors over time and include purchases, searches, and other online activity. In other words, as compared with typical audience measurement as supplied by a company like Nielsen Global Media, digital media provide much more information about the viewers' behavior. This type of information is a result of being connected to the Internet and can involve millions of people and require much work to capture, analyze, share, visualize, and interpret. Big data yields reams of information and requires talented and educated analysts and complex computer programs to interpret the data. As the world of interconnected devices becomes more common (referred to as the Internet of Things or IoT), so will the influence and applications of big data.

Audience Demographic Changes

The overall U.S. population is projected to grow by 100 million people to over 430 million by the year 2050. As multi-ethnic families continue to populate the U.S., minorities will make up around 50% of the population by 2050. Latin and Asian populations are expected to triple within the next 30 years. These broad changes will affect media and media programming, presenting opportunities to develop entertainment and information programs to reach the increasingly diverse and multicultural U.S. population.

SEE IT LATER

This book provides an overview of 'now' media and the evolution of electronic communication and electronic media, which have traditionally been defined as mass media that are not in printed form. The book also presents an expanded view of now media including newer personal digital devices. When attempting to predict the future of media, it is important to consider the concept of *disruptive technologies*. In his book *The Innovator's Dilemma*, author Clayton Christensen (1997) coined the term 'disruptive technologies' to describe a technology that displaces an existing technology or a product that creates a new industry. Personal computers changed the way people 'type,' severely tamping the market for typewriters and paving the way for changes in how people write. Email changed the way we communicate with others, supplanting letter writing and disrupting the flow of mail via the postal service. Related industries that produced personal stationery and greeting cards have been changed forever.

Innovations in personal media have prompted noticeable changes in the way people listen to music and watch television. Specifically, listeners are moving away from broadcast radio and toward Internet radio, satellite radio, and personal music services, like Spotify. This movement has caused a shakeup in the radio and music industry and the transition in media consumption is difficult to measure. Radio listening has declined some in the past 20 years, and U.S. weekly listenership across all age groups is now somewhere between 90–92%, according to Nielsen Global Media.

Paying to stream television shows and music is displacing buying DVDs and CDs. According to the Recording Industry Association of America (RIAA), streaming music accounted for 80% of music industry revenues in 2019. Total 2019 revenues of $11.1 billion were up 13% versus $9.8 billion the prior year (Glazier, 2020), and paid subscriptions to on-demand streaming services produced revenues of $6.8 billion, up 25% from 2018. Vast libraries of media are now available from a variety of providers, which has created a need for 'now media.' People no longer want to wait for a program to air, nor do they desire to travel to a store to purchase a CD or DVD.

New ways of accessing television programming have dramatically changed the television industry. Because of streaming services, there are hundreds of shows available at any given time. This appetite for content is a golden age for pitching ideas for shows. To meet demand, program ideas go to not only the big four television networks and cable networks such as USA, TNT, HBO, and Showtime, but also streamers like Netflix or Amazon Prime Video. It is becoming less important whether programs like *Stranger Things* and *13 Reasons Why* are shown only through streaming, especially now that the traditional television networks have launched their own streaming platforms.

Individuals are now generating their own content and displaying it on the Internet. In 2020, nearly two billion people each month watched up to five billion YouTube videos each day (Rutnik, 2019; Spangler, 2019). Since 2005, people from all over the world continue to upload content, and today it is estimated that more than 500 hours of short video clips are uploaded each minute (Rutnik, 2019). YouTube is a favorite worldwide, with content of any type, any length, and any viewpoint. Public domain films and television shows can be found as well as the latest clips from network late night comedy shows, squeezed between animal cameos and how-to videos for any conceivable task. Since being purchased by Google in 2006, YouTube has made a concerted effort to monetize the site with advertising.

Although most of the videos can be described as amateur, YouTube's popularity has led to a cottage industry of influential posters called YouTubers. YouTubers have their own YouTube channel and regularly post content. They solicit viewers to

Figure 1.9 The GoPro camera is a small but high-quality video camera designed for use in almost any location.

subscribe, helping to bring the site more influence and in turn more advertising. Today, programs are being created and produced using high-quality but inexpensive video cameras and laptop computers. For example, the 2018 Steven Soderbergh thriller, *Unsane*, was shot entirely with iPhones. These examples demonstrate how today's computers and production software enable people to shoot and edit video and audio on their personal laptops and tablets and thus produce television or radio programming in their own homes. GoPro cameras placed on a surfboard, mountain bike handlebars, or helmet provide amazing live-action video in HD quality. These high-quality, miniature cameras that cost under $400 are changing adventure videography forever.

The ease of producing and displaying homemade video is causing the networks and cable channels to rethink their scheduling and programming strategies. Networks and cable channels are looking for ways to reach and keep viewers. Producers are seeking to find some balance between subscribing, advertising, and product placement to satisfy all parties involved.

The ability to produce programs with personal computers and inexpensive high-quality video cameras and audio equipment has increased the need for *media literacy*. Media literacy is the ability to access, critically analyze, evaluate, and create media in a variety of forms. Media literacy is understanding the meaning of the content of the media as well as the power, the intent, and influence of the media. The ubiquity of media content in all forms has resulted in scares and conspiracy theories. The use of mis- and disinformation has helped polarize communities. The case of Cambridge Analytica and the 2016 election demonstrated how foreign powers can use social media to distract, confuse, and mislead Americans. During the coronavirus pandemic of 2020, some well-known names tried to sell fake cures, demonstrating the importance of knowing the difference between what is true and what is false.

The importance of media literacy is underscored by our increasing reliance on *the Power of Search*. Professional media content in the past could only be produced by large companies on expensive equipment. Today, distributing content is not a technological problem. Distribution via the Internet is becoming the primary means of delivering media into the home or on mobile devices. Therefore, the challenge is not how the program is distributed, but rather, *being found*. Collectively the decades of past programming available as edited clips or entire shows, coupled with new shows and program forms, means hundreds if not thousands of options to choose from. It is becoming more difficult to find an old favorite show or select a new favorite. Search engine optimization for music and video is rapidly evolving but to date there is no one solution. And posting user-generated content on Facebook, YouTube, or Vimeo adds to this dilemma of 'too many choices.' How individual consumers spend their limited time choosing from almost unlimited program options is the next challenge facing audiences, producers, broadcasters, cable channels, and movie studios.

ACCELERATING CHANGE

A final consideration in looking forward to the future of media is the concept of *accelerating change*. Accelerating change is the increase in the rate of technological progress.

Technological change is occurring at an ever-increasing pace, as are the resulting social and cultural changes. Two forces are driving this acceleration. One is explained through Moore's law, an observation made in 1965 by Gordon Moore, co-founder of the Intel Corporation, which first posited that the number of transistors in a dense integrated circuit doubled approximately every two years. Moore's law held up well until around 2018, yet still captures the idea of dramatic improvements in relatively short periods of time. Another obvious application of the law is seen in the capacity of digital memory devices. As computing speeds and storage capabilities improve, the ability of media outlets to send programming out to many people, either synchronously or asynchronously, is greatly enhanced. A practical example of the benefit of enhanced storage capability is the ability of companies such as Netflix or Disney+ to make their enormous libraries of television programs and feature films available to subscribers anywhere and at any time.

A second driving force behind accelerating change is the power of the billions of people and billions of 'things' that are now 'connected' to the Internet. Ideas and information are being shared constantly among people who can change and improve almost anything. In his book *Digital Wisdom*, futurist author Shelly Palmer suggests that inventors and innovators can be assisted by billions of people who share their abilities and experiences. Apps are created daily to help us perform almost any task. The saying "There is an app for that" epitomizes the reality that technological advances alter our lives by changing the way we behave, entertain ourselves, and communicate with others. Our media world is changing rapidly, and according to futurists like Shelly Palmer, the rate of change continues unabated.

QUESTIONS YOU SHOULD ASK YOURSELF ABOUT STUDYING 'NOW' MEDIA

The questions posed here are good questions to ask yourself about your attitudes and beliefs about 'now' media:

1. How much time do you spend each week with various forms of media, both old and new? As noted at the beginning, the average time spent with media is over 11 hours per day. That is longer than your time sleeping, eating, talking with friends, doing homework, or exercising.
2. How has your use of media changed now that you are in college? How do you split your time between your phone, your computer, and a television? What types of things do you watch alone compared to what you watch with others? How has that affected your ability to establish and maintain relationships?
3. Compare the time you spend listening to music you own versus on the Web versus on the radio. What do you listen to in your car? While working out? At work? Hanging out with friends? How much time each day do you listen to music with

ear buds compared to through speakers (so anyone can also hear).

4. How familiar are you with the icons of pop culture, and where do you get most of the information about your favorite celebrities? How well do you conform to 'your generation' and the label affixed to your age group? Compare those you consider pop culture icons with those of your parents or even those now in high school.

5. What social benefit do you get from watching and listening to electronic media? Media users often talk about what they see on television and hear on the radio: *American Idol*, *Black-ish*, *Hunters*, *DCs Legends of Tomorrow*, *Dancing With the Stars*, *Wheel of Fortune*, and the weather forecast. They talk about television programs and movies they have downloaded, streamed, and seen on television or their phone. Music and musical performers are common discussion topics at home, work, and school.

6. How do you think 'now' media represent our nation and our culture? American electronic media content is pervasive in many parts of the world. That means that the perceptions that people in other countries have formed about us are often based on what they have seen in the movies and on television.

7. How has watching/listening to electronic media shaped your views about celebrity and the importance of *getting attention*? We know so much about media personalities because celebrities get quite a bit of news coverage. Some shows are dedicated to news about the media and movie personalities. People are fascinated with the lives of prominent and famous people. Shows like *Entertainment Tonight*, *TMZ*, and *Access Hollywood* get more viewers when famous people commit foolish or scandalous acts that generate entertainment news stories. And now we have the rise of influencers referred to by the *Oxford English Dictionary* as *YouTubers*.

8. How do 'now media' actually influence you?
   - Speech – We learn new phrases and meanings for words and slang.
   - Customs and traditions – The portrayal of holiday festivities, like the dropping of the 'apple' on New Year's Eve, shapes how we observe these holidays.
   - Styles of clothing, cars, and technology – We see and hear about these products through electronic media, and we are tempted to try them out.
   - Sense of ethics and justice – We view many stories of good and evil and even experience real courtroom dramas by viewing the many courtroom shows on television.
   - Perceptions of others in our society and distant countries – *National Geographic* programs show us how people in South America or remote Alaska live.
   - Lifestyles – We learn about other people's lives and our own by watching talk shows, self-help shows, and advice shows.

9. How are you different from previous generations of college students because of electronic media? By the time you finish high school, you have seen and heard thousands of hours of electronic media. What effect does that have on you? Are you different from your parents or grandparents because you have used so much electronic media? Do electronic media have a quick and direct effect on you or a slow, subtle, cumulative effect? People who study media, including psychologists and sociologists, believe that contemporary digital media equipment sets young people apart from older people (e.g. those over 30 years old). There has never been a time when you were not connected and available to your peers and to the world. Entertainment has always been customized and individualized. You are comfortable at multitasking and can seek out and even create your own entertainment content. You expect change and innovation at a much faster pace than older generations.

10. How does studying 'now media' help you sort fact from fiction, truth and falsehood? Media literacy helps you make good media choices. By knowing more about the history, structure, economics, and regulation of electronic media, you can better understand and even predict the future of media. Understanding the storytellers that create media content helps us realize the effects that the stories have upon audiences. It also helps you understand how the constant connectivity of today will influence us in the future.

11. Will studying 'now media' help you in your career? Few industries have undergone and continue to undergo the dramatic technological and business changes that we have seen in electronic media in the past ten years. The electronic media are always changing, and as they change, so do you. For college students interested in a career in electronic media, knowing about these changes will present appropriate strategies for job seeking.

Figure 1.10 These children were able to watch the launch of the historic SpaceX mission by using 'now media' streamed to their laptop computer.
Source: Photo courtesy of Sarah Clover

SUMMARY

At first, the number of people we could communicate with was limited to those we could see face to face. Then came writing, and then, since the mid-19th century, electricity enabled us to communicate to one person or to many people over long distances with one message. Through the use of electronic media – radio, television, cell phones, and now broadband wireless Internet – we now communicate with almost anyone, anywhere, at any time. The number of people we can reach instantaneously is almost unlimited.

Traditional mass media share characteristics such as audience, time, display, distribution, distance, and storage. Electronic media are not constrained by time and distance. Electronic media can have cognitive, emotional, and behavioral effects on the audience that can influence us and can change our lives.

The Internet has emerged as an all-encompassing mass medium which combines or converges with various mass media. The Internet enables communication with a large audience for low cost and short turnaround time. The process of digitization has simplified the format through which information is transmitted. Numerous trends are changing the media industry and how we relate to and use electronic media. Convergence is the combining of media and thus the blurring of the distinctiveness among them. Deconvergence is the process of splitting media companies to better focus on individual media platforms and conform with regulation. Consolidation involves fewer companies owning more electronic media stations and businesses. Some trends have resulted directly from changes in technology. For example, desktop production has been fueled by digital technology and faster computers, which allow individuals to create content for electronic media on a single computer.

Technological change is closely related to changes in the media. As computers get faster and memory storage becomes less expensive and easier, we have access to more media. We can rent or use media without having to buy it. We can get media at any time, practically from anywhere. These changes, both technological and cultural, alter our ways of creating and using media.

The study of electronic media is important not only as a field of intellectual pursuit but also as a means of preparing for a successful career in a media-related field. Because electronic media are so pervasive, we must critically analyze the content which consumes so much of our time. Electronic media provide a window for the world to view many different cultures. Finally, electronic media are dynamic forces in society that are constantly changing. We need to study the changes and understand that they affect us deeply.

BIBLIOGRAPHY

Christenson, C. (1997). *The innovator's dilemma*. Boston: Harvard Business Review Press.

Dizard, W. (2000). *Old media, new media*. New York: Longman.

Dominick, J. (2013). *The dynamics of mass communication: Media in translation* (12th ed.). New York: McGraw Hill.

Eggerton, J. (2010, January 20). Kaiser: First drop for real-time TV viewing among youth. *Broadcasting and Cable*.

Federal Communications Commission. (2017, January 17). FCC Broadcast Ownership Rules. *FCC.gov*. Retrieved from: www.fcc.gov/consumers/guides/fccs-review-broadcast-ownership-rules

Ga, B. (2004, September 19). How people spend their time. *Google*. Retrieved from: http://answers.google.com/answers/threadview?id=403376

Glazier, M. (2020, February 25). RIAA Releases 2019 Year-End Music Industry Revenue Report. *RIAA.com*. Retrieved from: www.riaa.com/reports/riaa-releases-2019-year-end-music-industry-revenue-report/

Grant, A., & Meadows, J. (2020). *Communication technology update and fundamentals* (17th ed.). New York and London: Technology Futures, Inc., and Focal Press Taylor & Francis.

Harris, B. (2020). Publick Occurences Both Foreign and Domestick. *Encyclopaedia Britannica*. Retrieved from: www.britannica.com/topic/Publick-Occurrences-Both-Foreign-and-Domestick

How many online? (2001). *NUA surveys*. Retrieved from: www.nua.ie/surveys [January 14, 2002]

Jayson, S. (2010, February 23). The net set. *The Arizona Republic*.

Katzmaier, D. (2020, January 6). Samsung's 'The Wall' TV might be the biggest screen we've ever seen. *CNET.com*. Retrieved from: www.cnet.com/news/samsung-tv-the-wall-biggest-screen-weve-ever-seen/

Kaye, B. K., & Johnson, T. J. (2003). From here to obscurity: The Internet and media substitution theory. *Journal of the American Society for Information Science and Technology*, 54(3), 260–273.

Kaye, B., & Medoff, N. (2001). *The world wide web: A mass communication perspective*. Mountain View, CA: Mayfield.

Kelly, B. (2019, June). *Audio Today 2019: How America Listens*. The Nielsen Company. *Nielsen.com*. Retrieved from: www.nielsen.com/us/en/insights/report/2019/audio-today-2019

Larsen, L. (2020, March 13). The most-subscribed-to YouTube channels. *Digitaltrends.com*. Retrieved from: www.digitaltrends.com/web/biggest-youtube-channels/

Littleton, C. (2019, August 13). CBS, Viacom Reach Long-Awaited Merger Agreement to Reunite. *Variety.com*. Retrieved from: https://variety.com/2019/biz/news/cbs-viacom-merger-shari-redstone-1203262058/

Longley, R. On an 'average' American day. *About*. Retrieved from: http://usgovinfo.about.com/od/censusandstatistics/a/averageday.htm

Markus, M. (1987). Toward a 'critical mass' theory of interactive media: Universal access, interdependence and diffusion. *Communication Research*, 14(5), 491–511.

Medoff, N., Tanquary, T., & Helford, P. (1994). *Creating TV projects*. White Plains, NY: Knowledge Industry.

Palmer, S. (2012). *Digital wisdom: Thought leadership for a connected world*. Stamford, CT: York House Press.

Parker, O. (2018, March 16). Print books vs. e-books – an update. *Graphic Arts*. Retrieved from: https://graphicartsmag.com/articles/2018/03/print-books-vs-e-books-update/

Rutnik, M. (2019, August 11). YouTube in numbers: Monthly views, most popular video, and more fun stats! *Androidauthority.com*. Retrieved from: www.androidauthority.com/youtube-stats-1016070/

Schramm, W. (1954). *The process and effects of mass communication*. Urbana: University of Illinois Press.

Schramm, W. (1988). *The story of human communication: Cavepainting to microchip*. New York: Harper Collins.

Shane, E. (1999). *Selling electronic media*. Boston, MA: Focal Press.

Shannon, C., & Weaver, W. (1949). *The mathematical theory of mass communication*. Urbana: University of Illinois Press.

Spangler, T. (2019, May 3). YouTube now has 2 billion monthly users, who watch 250 million hours on TV screens daily. *Variety.com*. Retrieved from: https://variety.com/2019/digital/news/youtube-2-billion-users-tv-screen-watch-time-hours-1203204267/

Sterling, C., & Kittross, J. (2002). *Stay tuned: A history of American broadcasting*. Mahwah, NJ: Erlbaum.

Straubhaar, J., & LaRose, R. (2004). *Media now* (4th ed.). Belmont, CA: Wadsworth.

Turow, J. (2014). *Communication in a converging world* (5th ed.). New York and London: Taylor & Francis.

Chapter two

From Marconi to Mobile Listening

The authors thank Glenn T. Hubbard, Ph.D.
(East Carolina University) for his
contributions to this chapter.

SEE IT THEN

Early Inventors and Inventions

The invention of the Gutenberg printing press in the mid-1400s brought enormous change to human communication, leading to increases in literacy in the western world and ultimately to political, religious, and social change previously unimaginable in human history. Even so, it took another 400 years for additional technological developments to have such an impact on communication. The 19th century brought the Industrial Revolution, a time of tremendous technological growth around the world. England led the way to the Industrial Revolution, which took off in the United States just after the Civil War, and continued into the early 1900s. It was during this period that 'mass' communication emerged and began to resemble more closely the connected world we know today.

However, change remained slow at first. Printed books, newspapers, and magazines were still the main ways of communicating to the public at large in the 1800s, but much of the population was semi-literate, and printed publications were expensive to purchase, so they were a communication luxury for the elite. Inventors and progressive thinkers were experimenting with new ways of transmitting messages. They were not sure of where it would all lead but were hoping to somehow connect people from across the globe.

This chapter focuses on the development of radio, the first electronic medium, from the discovery of electromagnetic radiation – the broadcast 'airwaves' – to how it became a mass medium. It also covers technological developments that have reinvented the medium of radio more than once – developments that have consistently left their mark on society and continue to evolve in today's world of 'now' media.

Electrical Telegraphy

Samuel F. B. Morse, a well-known American painter and inventor, was the first to send an electromagnetic signal through electrical wires. He invented what he named the '**electric telegraph**' in 1835. The original instrument used pulses of current to deflect an electromagnet, which moved a marker to produce a written code on a strip of paper. Within a year the system was changed so that it instead embossed paper with a system of dots and dashes, which later became known as **Morse code**.

Morse code works by assigning each letter of the alphabet to a series of dots and dashes. For instance, the letter H is represented by a series of four dots, an E is one dot, an L is a dot, a dash, and two dots, and an O is three dashes. To transmit the dots and dashes, an operator would use one finger to push down a lever connected to the telegraph. To send the message "hello," the operator would tap out the representative dots and dashes for each letter, with the dots requiring a short tap and the dashes a long tap. The code was universally accepted and simple enough that telegraph operators could quickly tap out and decode messages. Telegraph messages were sent across wires strung from pole-to-pole. The system was not very secure, and messages did not always get to the receiver. Thunderstorms would disrupt the electrical current, and high winds would blow over the poles or break the wires. Also, outlaws would sometimes deliberately cut the wires to block transmission.

Figure 2.1A The first telegraph, circa 1840.
Source: Photo courtesy of iStockphoto © Jonnysek, image #3785353

Figure 2.1B Portrait of American inventor and painter Samuel Morse with his telegraph.
Source: Photo courtesy of Shutterstock 244389145

Samuel Morse patented the telegraph in 1840, and the U.S. government provided funds for a public demonstration. An electrical line was strung between Washington DC and Baltimore. After dealing with some technological problems (some of which were remedied by Ezra Cornell, for whom Cornell University is named), Morse sent the first official message, "What hath God wrought," on May 24, 1844. The message was taken from the Bible to suggest that the telegraph would change not only communication but also the world forever – and maybe not for the best.

Realizing the telegraph's potential in high-speed communication, the U.S. government permitted private businesses to set up their own telegraph services from which to develop an electrical communication industry – a policy that would be repeated in later years with other communication technologies. By 1861, a transcontinental line was in place, and telegraph messages buzzed across the United States.

Despite its shortcomings, the telegraph was the fastest way to communicate across distance and was quickly adopted by the public. Its impact was especially significant in the field of journalism, because it allowed reporters to file stories from long distances and newspapers anywhere in the country to share stories of national interest. As a result, the concept of 'national news' developed substantially, along with a 'wire service' called the Associated Press, which used the telegraph to disseminate stories around the country almost instantaneously. An event that happened in New York could be in the next day's morning paper in Chicago and even San Francisco. The impact of rapid sharing of information over long distances about social norms and politics is impossible to overstate. Among the effects of the telegraph and the Associated Press was the development of the inverted pyramid news story format, placing the most essential information first, in case the telegraph signal failed before the whole story could be transmitted. But there were much bigger impacts, not least of which was the U.S. involvement in the Spanish American War in 1898, a result of newspapers carrying dispatches from a distant location, shifting public opinion in favor of American intervention. Suffice it to say, the impact of electronic media on world history would only increase as time went on.

Even though the electrical telegraph conquered the problems of distance and speed, it still presented some challenges:

1   It required building a costly system of wires.
2   It worked only as long as the wires were in place.

3   It was a restricted system that required trained telegraph operators who knew how to send and receive in Morse code. Western Union was the dominant company in the business, and it controlled all messages.
4   Once a message was decoded, delivering it to the appropriate receiver sometimes proved problematic. The local address of the receiver had to be found, and a courier had to physically travel from the telegraph office to the receiver's home or place of business. Ironically, physically delivering a message to a nearby receiver took a great deal more time than sending the message across the country.
5   The cost to send a telegraph message was determined by the length of the message; therefore, messages were often short and somewhat cryptic. Because of the need for brevity, messages often lacked specific meaning or emotion.

Electrical Telephony

Considering the times, the telegraph was quite a remarkable communications invention, but it lacked voice transmission. Stemming from a lifelong interest in hearing and speech, **Alexander Graham Bell**, whose mother and wife were deaf, sought a better method of two-way communication that could be used by individuals without special training.

Bell combined two scientific principles to achieve electrical conduction of sound wave patterns that were converted to electrical patterns and sent through a wire, thus inventing the electrical telephone. Instead of tapping out a signal that needed decoding, a message could be spoken and transmitted by sound waves. On March 10, 1876, Bell made the first 'telephone call' to his assistant – "Mr. Watson, come here; I want to see you." Word about Bell's invention spread quickly and within one year the first telephone line was strung between Boston and Somerville, Massachusetts. The telegraph and telephone used different types of conduction wires, so though both types of wires were strung pole-to-pole, two different systems were created.

Electrical telephony was an astounding new invention. For the first time ever, people could speak to one another across distances without a special coding/decoding system. The costs and logistics of building a telephone system inhibited widespread use at first. As in the case of the telegraph, telephone calls could only be made between places that were wired to receive and transmit them. Even areas that had been wired sometimes had problems, as when bad weather caused telephone lines to break and created lapses in service. Despite these drawbacks, however, the growth of distance communication was assured, and the telephone soon became an integral part of life.

Point-to-Point Electrical Communication

Both electrical telegraphy and electrical telephony were designed and used as systems to facilitate point-to-point communication. Using either of these systems, a person could send a message to another person at a distant location. The speed at which the signal traveled through the wires was the same as the speed of light (186,000 miles per second), which meant the message reached its destination almost instantaneously.

Electrical point-to-point communication proved its value in many situations, such as announcing the arrival of incoming stagecoaches, trains, and ships into port. By the late 1800s, telegraph messages and telephone calls were commonly used to alert manufacturers and merchants about when materials, supplies, and products would be arriving, and telegraphed and telephoned information about incoming storms and weather forecasts was invaluable to farmers.

The U.S. economy depended somewhat on maritime shipments of goods and products, but there was no way to contact ships when they were at sea. A way of sending electrical messages without wires was needed. Given the importance of the shipping business and the potential for profit, solutions were sought from scientists, innovators, and inventors to develop a wireless system of ship-to-shore communication.

Wireless Transmission

James Clerk Maxwell   Many scientists from around the world contributed to the development of radio. As early as 1864, James Clerk Maxwell, a Scottish physicist, predicted that signals could be carried through space without the use of wires. In 1873, he published a paper that described invisible radiant waves. These waves, which were later called radio waves, were part of *electromagnetic theory* that suggested that wireless signals could travel over distances and carry information. The theory used mathematical equations to demonstrate that electricity and light are very similar and both radiate at a constant speed across space.

Heinrich Hertz   Heinrich Hertz, a physicist from Germany, expanded Maxwell's electromagnetic theory to build a crude detector of radiated waves in 1886. Hertz set up a device that generated high-voltage sparks between two metal balls. A short distance away, he placed two smaller electrodes. When the large electric spark jumped across the gap between the two large balls, Hertz could see that a smaller spark appeared at the second set of metal balls. It was proof that electromagnetic energy had traveled through the air, causing the second spark. Hertz never pursued the idea of using the waves to transmit information, but his work is considered crucial to the use of electromagnetic waves for communicating. In fact, the basic unit of electromagnetic frequency, the **hertz**, was named after him.

In the late 1890s, English physicist Sir Oliver Lodge devised a way to tune both a transmitter and a receiver to the same frequency to vastly improve signal strength and reception. Russian scientist, Alexander Popoff, further contributed to the efforts by improving the wave detector and antenna in the 1890s. Interestingly, Popoff's work was dedicated to finding a better way to detect and predict thunderstorms.

Guglielmo Marconi Italian inventor Guglielmo Marconi is generally credited with the first practical demonstration of the wireless transmission of signals. After reading about Hertz's experiments, Marconi used electromagnetic waves to transmit Morse code signals. But he also noted that having an antenna above ground improved signal transmission. Marconi also improved Hertz's wave detector and discovered that an antenna could strengthen signal transmission. Because of the success of his inventions, Marconi is often incorrectly identified as the 'inventor of radio.' He is most correctly understood to be the inventor of the **wireless telegraph**, but his contributions to the development of radio place him at or near the top of the list of key inventors of the technology.

When 20-year-old Marconi approached the Italian government for a patent and financial support, they expressed no interest. Fortunately for Marconi, his mother came from an Irish family with connections in Great Britain, which was a strong maritime power and thus very interested in developing a system to contact ships at sea. Marconi contacted the head of the telegraph office of the British Post Office, William Preece, who had also done some wireless experimentation. Preece was very receptive to the idea of sending wireless signals and Britain eventually provided Marconi with a patent and the financial support he needed to further develop his wireless system.

Toward the end of the 1890s, Marconi carried out a series of demonstrations showing that his wireless radio system could work over short ranges. In 1899, he successfully sent a signal across the English Channel, the longest distance yet. Just two years later, he sent a signal (the letter *S* in Morse code) across the Atlantic Ocean from Great Britain to North America, convincing many that wireless communication across great distances was possible.

Marconi's system could carry only dots and dashes, or Morse code, and not the human voice. As the telephone had been invented more than 25 years earlier, the public believed that radio would not be very useful unless it, too, transmitted sound.

American inventor Nathan Stubblefield is often credited as the first person to successfully transmit the human voice using radio waves, although he used ground conduction rather than transmitting through the air. Regardless, Stubblefield supposedly communicated the words "Hello Rainey" to his assistant during an experiment near Murray, Kentucky in 1892.

Reginald Fessenden Reginald Fessenden, a Canadian electrical engineer, worked with Ernst Alexanderson, an engineer with General Electric, to construct a high-speed alternator (a device that generates radio energy) to carry voice signals. On Christmas Eve, 1906, Reginald Fessenden transmitted the first radio broadcast using modulated continuous electromagnetic waves carrying sound wave patterns instead of Morse code.

From his home at Brant Rock, Massachusetts, Fessenden sent out sounds of violin music, vocal Bible readings, and general season's greetings that were picked up by ships at sea along the East Coast of the United States.

Lee de Forest In 1899, American inventor Lee de Forest earned a Ph.D. from Yale University. His dissertation investigated wireless transmissions. One year later, he invented the **triode audion tube**, which could amplify sound transmission, thus eliminating the need to wear headphones as required by Fessenden's system. De Forest's audion was a step up from both Fessenden's system and the *Fleming valve* or *diode tube*, which was a radio wave detector within a sealed glass tube. Invented by English engineer John Fleming, the diode tube, which looked like a household light bulb, detected voice waves but could not amplify sound. By adding a third element to Flemings' diode tube, de Forest's audion could amplify sound transmitted over radio waves.

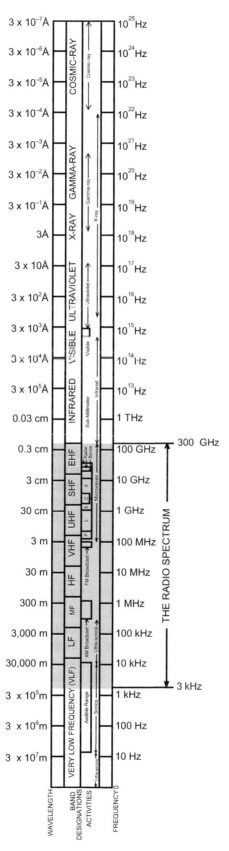

Figure 2.2 The radio bands represent a portion of the electromagnetic spectrum.

Figure 2.3 Guglielmo Marconi at work with his wireless radio.
Source: Photo courtesy of MZTV Museum

Figure 2.4 Lee de Forest with his audion tube.
Source: Photo courtesy of MZTV Museum

De Forest filed for a patent for the audion in 1906, but he knew that to make money from the audion he first had to sell the notion of amplified voice transmission to the public. De Forest generated publicity with events like playing a phonograph record from the top of the Eiffel Tower in Paris and transmitting the sound across radio waves for 500 miles. Despite his attention-getting stunts, de Forest was not a skillful businessman. He spent much of his time trying to create a place for himself in the burgeoning radio industry, but he suffered many financial setbacks and legal problems with patents.

His biggest legal battle regarded the 1914 invention of the regenerative or feedback circuit, which achieves higher amplification than the audion. Edwin H. Armstrong, another radio inventor, was credited for inventing the circuit while he was an undergraduate at Columbia University, but de Forest claimed that he was the real inventor. De Forest prevailed legally even though while testifying in court, he could not explain exactly how the system worked. Regardless, historians often see it the other way and give Armstrong, who later invented FM radio transmission, credit for the regenerative circuit.

The Basis for Regulatory Power Over Broadcasting

Congress, as the law-making body of the U.S. government, was deemed to have legal authority to regulate broadcasting based on the argument that the airwaves belong to the people. Unlike newspaper companies that printed and distributed newspapers wherever and whenever they liked to as many people as possible, broadcasters used the electromagnetic spectrum to transmit their information. In other words, newspapers created their own medium, whereas broadcasters used an existing natural resource and limited medium – the electromagnetic spectrum – to reach their audience. The important point to remember about the electromagnetic spectrum is that different portions of it are used to transmit different types of signals. Only a small portion of the spectrum is usable for over-the-air broadcasting. The government considered the spectrum a scarce resource, like water in some areas of the country. Use of the electromagnetic spectrum was therefore subject to regulation by the federal government on behalf of the people. In particular the government wanted to prevent monopolistic use and squandering of the electromagnetic spectrum. Hence, in the early days of radio, the government stepped in and began regulating how broadcasting could use the spectrum.

A Tragic Lesson

In the early 1900s, radio was somewhat of an oddity. For wireless radio to work, signals are sent from a transmitter to a receiver. Although many hobbyists were experimenting with radio transmitters and wireless was demonstrated to the public at fairs and department stores, only two companies were making parts for radio receivers. Beginning as early as 1903, government representatives of the leading industrial countries began holding annual conferences (called Radio Conferences) to discuss humanitarian and international uses of wireless radio. By 1910, many of these nations had established regulations to guide the use of radio, particularly in terms of maritime uses. The U.S. Congress passed the **Wireless Ship Act of 1910**, which required a ship with more than 50 passengers to carry a radio that could send signals from its transmitter to another ship's radio receiver within 100 miles. A ship also needed to have at least one person aboard who was capable of operating the transmitter and the receiver.

Unfortunately, the requirements set forth by the Wireless Ship Act were not enough to save the victims of *Titanic*. In mid-April 1912, the 'unsinkable' luxury liner set out on its maiden voyage from England, bound for the United States. Its passengers, mostly wealthy and well-known people, set out on a luxurious trip across the Atlantic Ocean on the newest and most sophisticated ocean liner ever built. *Titanic* was equipped with the most modern technology available at the time, including a wireless radio and trained operators.

Late at night on April 15, *Titanic* collided with a huge iceberg in the North Atlantic Ocean, ripping open the hull and causing the ship to rapidly take on water. Supposedly, the ship's radio operator had received warnings about icebergs dangerously close to the ship's path, but he did not heed them. Instead, the operator requested that other ship radio operators clear the airwaves to allow *Titanic* to send personal messages from the ship's famous passengers to friends and family in Europe and the United States.

After *Titanic* collided with the iceberg and began to sink, the radio operator on board frantically sent out SOS signals. Unfortunately, the collision occurred late at night, and most of the wireless operators on other ships in the area had already gone off duty. One ship, the *U.S.S. California*, was less than ten miles away and could have rescued survivors from the icy waters, if only its wireless operator had been on duty ten minutes longer. Because the Wireless Ship Act of 1910 failed to require 24-hour staffing of wireless systems (as agreed to in the 1906 international agreements), *U.S.S. California* had no one on duty to receive *Titanic*'s SOS signal. Only one operator, on the ship *Carpathia*, picked up the SOS signal and sped to help *Titanic*. Although *Carpathia* was able to rescue more than 700 passengers, over 1,300 perished when *Titanic* went down.

At the time *Titanic* was sinking, a young wireless operator named David Sarnoff was stationed inside Wanamaker's Department Store in New York City. He claims to have picked up *Titanic*'s distress signals and the responses from *Carpathia*. Sarnoff says he stayed at his wireless station for the next 72 hours, receiving information about survivors. Sarnoff relayed the events of the sinking to other wireless operators and newspapers. As news of *Titanic* spread, the government ordered that the airwaves be cleared of other wireless operators, leaving Sarnoff as the only one to send messages to coordinate rescue traffic and to pass along exclusive information to the *New York American*, an influential daily newspaper of the time.

Some historians, however, dispute Sarnoff's claims that he was the sole operator who picked up the *Titanic*'s distress signals, and claim that Sarnoff greatly exaggerated his role in informing the public about the ship's fate. Whatever the case, Sarnoff went on to become president of Radio Corporation of

Figure 2.5 A young David Sarnoff at his wireless station in Wanamaker's Department Store.
Source: Photo courtesy of MZTV Museum

Figure 2.6 The news reported about the sinking of the *Titanic* in April 1912 came from wireless radio transmission. Source: Photo courtesy of Getty Images #80680699

America (RCA), and the story marks an interesting and important milestone in the development of the wireless. The sinking led to government scrutiny of the role of wireless radio and to the provisions of the Wireless Ship Act of 1910. Governments worldwide sought to increase wireless conformity and compatibility. Congress amended the Wireless Ship Act of 1910 to require two trained radio operators, an auxiliary power supply, and the ability of the radio operator to communicate with the bridge commanders of the ship. Also, the Act was extended to include not only ships at sea but also those on the Great Lakes.

With radio emerging as an important communication tool for the government, business, and the public, radio operators, who were largely anonymous, were transmitting signals as far and wide as they could without oversight. As a way to limit radio transmission, Congress passed the **Radio Act of 1912**, which required the licensing of radio operations used for the purpose of interstate commerce. It also required licensed operators to be citizens of the United States and that licenses must be obtained from the U.S. Secretary of Commerce and Labor, who had jurisdiction over commercial radio use in this country. The Act also required that ships had to be equipped to send out a distress signal at any time and that all ships must monitor distress frequencies continuously. The Radio Act of 1912 stood until it was amended by the Radio Act of 1927.

It is worthwhile to point out that government regulation of any private industry can be politically controversial, not to mention legally questionable. Legislative action in response to the *Titanic* disaster established that owning and operating radio technology carried with it a responsibility to serve the public interest. This public safety function, along with the spectrum scarcity and public-airwaves arguments became the legal basis for broadcast regulation that exists to the present day.

Radio Becomes a Mass Medium

While Reginald Fessenden and Lee de Forest were working mainly on the mechanics of radio, Charles D. 'Doc' Herrold was thinking about what kinds of content could be transmitted over the airwaves. Three years after Fessenden's Christmas Eve broadcast, Herrold came up with the idea of regularly scheduling transmissions. Students at his Herrold College of Wireless and Engineering in San Jose, California, began transmitting music, speeches, and other types of broadcasts at scheduled times. In a sense, Herrold created the first radio station, specifically the first college station. Herrold's station was one of the first to operate regularly after the Radio Act of 1912, and it continued broadcasting until World War I. The station went back on the air in early 1922 with the call letters KQW. The station was later sold and then moved to San Francisco, where it became KCBS, a station that still broadcasts today.

Station **KDKA** in Pittsburg, PA, however, is officially considered the first commercial broadcast station because it was the first one formally licensed by the Secretary of Commerce on November 2, 1920. Dr. Frank Conrad, an engineer at Westinghouse, formed KDKA in 1916. Conrad sent both voice and music programs from his home in Pittsburgh to the Westinghouse plant located about five miles away. Conrad devised a system for broadcasting music by placing a microphone next to a phonograph record as it played. Conrad's broadcasts were so popular that he regularly scheduled them on Wednesday and Sunday evenings so listeners would know when to tune in.

Wanting its place in history, station 8MK-WWJ in Detroit, claims it was the first licensed station on the air, presumably because it broadcast news updates as early as August of 1920, whereas KDKA famously carried election returns in November of that year. An amateur station, WWJ began broadcasting with the call letters 8MK, and first went on the air from a 'radio phone room' in the *Detroit News* building. The license for this station was eventually issued by the Department of Commerce to the *Detroit News* on March 3, 1922.

Radio Broadcasting

Early radio transmissions reached a small segment of the general population. Few people had a radio receiving set, and there were not enough programs on the air to induce people to buy one. Moreover, there was not a universally accepted word for what to call these transmissions. Eventually, the word 'broadcasting' was used to describe the mass transmission of radio programs to the public. Broadcasting is actually an agricultural term that describes a way to plant seeds by casting them in all directions using a circular hand and arm motion rather than planting them in rows (Lewis, 1991).

When radio was still in its infancy, no one was really sure of the direction it should take. David Sarnoff, the person who supposedly received and transmitted news about the *Titanic*, had by 1916 become commercial manager of the American Marconi (a subsidiary of British Marconi). Sarnoff thought that money could be made from radio, but he just had to figure out how to do it. He came up with an idea that he penned in a memo addressed to the manager of American Marconi. In it, Sarnoff outlined the essence of what he thought radio broadcasting should become.

*A plan of development, which would make radio a 'household utility' in the same sense as the piano or phonograph. The idea is to bring music into the house by wireless.... The problem of transmitting music has already been solved in principle and therefore all the receivers attuned to the transmitting wavelength should be capable of receiving such music. The receiver can be designed in the form of a simple '**Radio Music Box**' and arranged for several different wavelengths, which should be changeable with the throwing of a single switch or pressing of a single button.... The box can be placed on a table in the parlor or living room, the switch set accordingly and the transmitted music received. There should be no difficulty in receiving music perfectly when transmitted within a radius of 25 to 50 miles.... The same principle can be extended to numerous other fields as, for example, receiving of lectures at home, which can be made perfectly audible; also, events of national importance can be simultaneously announced and received.*

Benjamin, 1993

In addition, Sarnoff's memo suggested that large profits could be gained from the sale of radio receivers to the general public. In retrospect, it seems that the idea of a 'radio music box' would have been adopted immediately, but American Marconi and other companies ignored it. The idea of Sarnoff's 'radio music box' simply did not catch on for many reasons, including the need for listeners to wear headphones to hear mostly static-filled broadcasts. Additionally, the radio receiving sets were large, heavy, and expensive, and the public was generally uninterested. Also, engineers and executives who held the power in the radio industry were tied to more serious uses for radio than entertaining the masses.

World War I (1917–1918)

World War I started in Europe in 1914, but the United States did not formally get involved until April 6, 1917. For security reasons, the U.S. Navy took over all commercial radio enterprises and began recruiting amateur radio operators and experimenters to provide the military with knowledgeable and experienced radio personnel.

Radio was a strategically essential part of military communication for the government, the armed forces and its allies. The federal government took over the operation of all high-power radio stations in the country, including the point-to-point sending and receiving stations owned by American Marconi (the American subsidiary of Marconi's Wireless Telegraph and Signal Co.) and it prevented foreign radio operators from transmitting in the United States. Moreover, in 1917 all amateur and experimental stations were ordered off the air, which stopped the progress of radio as an entertainment medium.

The government even took control of all patents related to wireless communication and placed them in a 'patent pool' for all scientists and engineers to access. This pooling of patents helped the war effort by stimulating the technological development of radio for military purposes. In turn, these developments helped stimulate the growth of the radio industry after the end of the war.

When the war ended in November of 1918 and recruits became civilians again, the Navy realized that it lacked the experience and expertise to maintain its control over the radio industry, so it ceded control to other federal government entities.

The U.S. government considered continuing its control of the radio industry, but it lacked the skilled operators needed to do so. Considering the vicious competition between telephone and telegraph companies, the monopolistic leanings of radio companies, and examples in Europe of government control, governmental control seemed like the best move to guide and regulate radio in the U.S. But opposition from American Telephone and Telegraph (AT&T), the Marconi Company, General Electric, and other companies that contributed patents to the government patent pool during the war, along with amateur radio operators, was strong enough to convince the government to back off, and allow the radio industry to be guided by private enterprise. Experimenters, hobbyists, and commercial radio companies like AT&T and Marconi pressured Congress and President Woodrow Wilson to return radio to citizens and private industry. On July 11, 1919, the president acquiesced, and the military takeover of radio ended eight months later on March 1, 1920.

Radio Corporation of America (RCA)

After World War I, the British-owned Marconi Company sought to strengthen its position as the leader in long-distance radio communication. It tried to buy a large number of the powerful Alexanderson alternators, which were produced by General Electric (GE) for its American subsidiary and were an early way of amplifying AM radio signals before tube technology replaced them. Such an arrangement essentially would have given Marconi a near monopoly in transatlantic radio. Because the U.S. government did not want foreign control of radio in this country, it was opposed to selling equipment to British-owned Marconi.

GE bought a controlling interest in American Marconi, but soon thereafter GE decided that its strength and expertise were in manufacturing, so it formed RCA in 1919 as the radio communication side of the business. RCA took over the radio stations that were formerly owned by American Marconi. In the next few years, much legal wrangling occurred among AT&T, RCA, Westinghouse, and GE over which company controlled broadcast equipment patents. From 1919 through 1921 these companies signed agreements to pool their patents, leading to a consortium of companies that would move the technology and business of broadcasting to a national level. By 1922, GE, Westinghouse, AT&T, and United Fruit Company (a small company that held desirable patents for radio equipment) had become the corporate owners of RCA. Approximately 2,000 patents for radio equipment were pooled, and an effective manufacturing and marketing plan was enacted in which radio receivers would be manufactured by GE and Westinghouse but sold exclusively by RCA; in turn, AT&T would be allowed to charge for the use of its phone lines in radio broadcasting, such as to deliver national network programming content to local stations.

The 1920s

Broadcasting as it is known today began during the 1920s. Some of the groundwork for commercial broadcast radio was laid as early as 1909 with Doc Herrold's and Frank Conrad's stations. By mid-1920, Conrad had convinced his superiors at Westinghouse that the company could make money by selling premanufactured radio sets that could pick up programming from a radio station operated by Westinghouse. The inaugural broadcast for station KDKA was the presidential election of 1920. KDKA's programs consisted mostly of music, much of which came from live bands that performed in a tent on the roof of the building that housed the station. After high winds destroyed the tent, a purpose-built studio was constructed so bands could play indoors with much better sound reproduction.

In 1920, Westinghouse was manufacturing and selling radio sets and was looking for ways to increase radio set sales. The company promoted and broadcast a program each evening in the hope that people who were listening at their neighbor's house or in public places would buy their own radios and get into the habit of listening nightly. The real goal for Westinghouse was to promote itself and its programming so it could sell its receivers to the public. As part of its aggressive marketing strategy, Westinghouse made deals with appliance and department stores to advertise and sell its radios. For example, a store in Pittsburgh ran an ad in *The Pittsburgh Press* to convince people that it was worth spending $10 (a lot of money in 1920) on a wireless set, when they did not know what was being broadcast and had not yet conceptualized in-home entertainment

FYI: 1920 PRESIDENTIAL ELECTION

Until 1920, the sounds of presidential campaigns had been heard only by phonograph record. That changed on election night, November 2, 1920, when returns from the presidential election were broadcast live. Moreover, through a process created by Westinghouse Electric & Manufacturing Company and its subsidiary the International Radio Telegraph Company, the election returns within a radius of 300 miles of Pittsburgh were received and transmitted by wireless telephone. The returns were received directly from an authoritative source and sent by a wireless telephone stationed in East Pittsburgh. Anyone with a receiving set, even one as basic as a crystal detector, a tuning coil, a pair of telephone receivers, and a small antenna, could find out who was winning the election. And if a two-stage amplifier was attached to a phonograph speaker, anyone within a few feet of the receiving set could also hear the returns without needing to wear headphones. With broadcasting the election returns, radio became a force in the political process, bringing the live events and real sounds of political campaigns directly to the voters.

At first the number of stations that went on the air after KDKA grew slowly. In fact, only 30 stations had been granted licenses to broadcast by January 1, 1922. But that changed, and by May 1, the number of stations increased to 218 and by March 1923, there were 556 licensed stations on the air. During 1923, many thousands of radio receivers were sold, which increased the demand for programming and thus stations.

Many department stores and large hotels that were housed in tall buildings took advantage of the height by placing radio antennas on their roof. Not only were they able to receive radio signals but they also set up their own radio stations to transmit their own original content.

Live music was the most popular form of radio entertainment, and some sporting events, like heavyweight prize fights and World Series baseball games, drew audiences as large as half a million listeners. Political programming also abounded on the airwaves, like President Warren G. Harding's 1921 Armistice Day speech from Arlington Cemetery near Washington, DC.

In the early 1920s, there were few programs on the air, and people were hesitant to buy a receiving set until more programs were broadcast. Factory-built receiving sets were expensive, a sophisticated receiver could cost $60 and a simple one $10. As the daily pay for the average worker at that time was about $1, the less-expensive models were the best sellers, but without many programs, it was difficult to convince the public to purchase a set. Westinghouse came up with a plan to boost sales of its own receiving sets – it broadcast programs in the towns where it had manufacturing plants, and nearby retail outlets sold the sets.

Radio sales increased after the 1922 invention of the superheterodyne receiver. Edwin Armstrong's superheterodyne was a sophisticated system that filtered radio signals more effectively than standard technology of the time, and it pushed signals out over longer distances than ever before. Later that same year, thanks to the superheterodyne, a broadcast originating from London was received at station WOR in New York City.

By 1923, more than 600 licensed stations were broadcasting, although many of them went off the air after only a short time. Most of the owners of one or more stations were radio receiver manufacturers and dealers and businesses involved with electrical device repair. Many of the radio stations were put on the air as a sideline to the main business of the company that held the broadcast license. At about this same time, many colleges and universities put stations on the air in the hope that doing so would help supplement the education of their students.

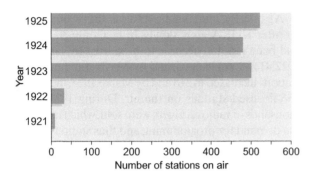

Diagram 2.1 The increase in radio stations on the air began in 1921 and exploded in 1922–1923, when the number went from 30 to more than 550. A slight decrease occurred after that boom, followed by another increase in 1924–1925.
Source: Sterling and Kittross, 2002, p. 827

The year 1923 is also known as the year of the call letters. To differentiate one radio broadcaster from another, the federal government adopted the four-letter call sign rule, such that stations west of the Mississippi River were assigned 'K' as the first letter and stations east of the Mississippi were assigned 'W' as the first letter, with very few exceptions. KDKA in Pittsburgh, for example, is an exception because it was licensed shortly before this rule went into effect.

FYI: A RADIO IN EVERY HOME, THE INTERNET IN EVERY HUT

The rapid technological advancement of early radio led H. P. Davis, a Westinghouse vice president (in 1922), to state, "A receiving set in every home, in every hotel room, in every schoolroom, in every hospital room.... It is not so much a question of possibility, it is rather a question of how soon" (Hilliard & Keith, 2001, p. 33).

Interestingly, President Bill Clinton made a similar statement regarding access to the so-called Information Superhighway. In October 1996, Clinton stated, "Let us reach a goal in the 21st century of every home connected to the Internet and let us be brought closer together as a community through that connection" (Clinton Unveils Plan, 1996). He also later stated, "Our big goal should be to make connection to the Internet as common as the connection to telephone today" (Internet in Every Hut, 2000). In 2009, President Barack Obama stated that a major component of putting the American Dream within reach of the American people is by expanding broadband lines across America to give everyone the chance to get online.

The Rise of Commercial Radio

In the early 1920s, radio stations were often started for the purpose of promoting and selling a product or service offered by the station owner, often the sale of home appliances such as radios. But there was a limited demand for more than one radio in a home, and at that, most people did not see any reason to buy even one radio. But as the number of radio programs increased, the number of radio receivers sold in the United States grew dramatically. In just one year, from 1923 to 1924, sales of radio sets jumped from 0.5 million to more than 1.25 million. This phenomenal growth was due in part to manufacturers making inexpensive sets, which were affordable to most people. At this point, radio had truly become a mass medium, and soon the demand for radio receivers exceeded the supply. But it was quickly learned that a large listening audience did not guarantee a station's success. The radio industry had yet to come up with a way to make radio pay for itself, let alone make a profit.

It was time-consuming and expensive for a station to create its own programming. Lectures, news, political information, weather announcements, and religious sermons were the most common. Sports broadcasting came along later when transmitting equipment became portable. Politicians were among the first to seize the opportunity to reach the public. Politicians preferred that voters hear their messages directly, not as interpreted (or edited) by newspaper writers. Going live on the radio was the best way for local, state, and federal politicians to speak directly to their constituents with an immediacy never before achieved. While politicians, musicians, lecturers, and others gained prominence from being on the air, the stations themselves needed a way to profit monetarily.

WEAF

In 1922, WEAF – the AT&T-owned station in New York – came up with a novel way to generate revenue. Using part of the telephone economic model, WEAF acted as a *common carrier*, a company that transports goods or services for the public. The station sold time to advertisers, a method of generating revenue it called *toll broadcasting*, which was similar to that used when a long-distance call was made and charged to the caller. Like the caller, the advertiser would pay a toll for the time used on the air. The concept of toll broadcasting was a critical part of the new economic model for supporting the radio industry. Toll broadcasting brought in money from advertisers to pay station expenses and it kept broadcast programming free to the audience; moreover it helped advertisers reach an audience to sell their products or services.

In retrospect, it would seem that toll broadcasting should have been an instant sensation among broadcasters, but it did not catch on immediately, partly because of cross-licensing agreements among the companies that had shared in the patent pooling of World War I, which gave AT&T the sole right to 'charge' for messages. At a radio conference in 1922, U.S. Secretary of State Herbert Hoover disparaged the idea of toll broadcasting, stating that a service with as much promise as radio should not "be drowned in advertising chatter" (Hilliard & Keith, 2001, p. 30).

Station Interconnection or Chain Broadcasting

WEAF also pioneered the interconnection of stations. Just after the beginning of 1923, WEAF sent a musical performance over the telephone lines (owned by AT&T, its parent company) to a station in Boston, and both stations broadcast the program simultaneously. This interconnection was called *chain broadcasting*, and though this term is not commonly heard today, it still appears in legal documents. The more common term used now is 'network.' Both terms refer to stations sharing programs simultaneously. In some cases 'networked' stations are under the same ownership, but not necessarily.

The Copyright Act of 1907

For years, artists, writers, and composers had envied the legal protection given to inventors for their patents. Similar protections were extended to inventors, engineers, and scientists for their 'intellectual product' when Congress passed the Copyright Act of 1907. The American Society of Composers, Authors, and Publishers (ASCAP) was established in 1914 to collect 'royalty fees,' payment for the right to use a product, such as a song, on behalf of the composers and authors of songs and other owners of copyrighted material.

As radio programming included more and more recorded (phonograph) music, musicians, composers, and lyricists began to complain that radio stations were broadcasting their work without permission (and, more importantly, without the recording artists receiving any compensation). The stations felt, however, that broadcasting copyrighted phonograph music actually benefitted the artists by promoting their work. This logic is still used today by stations and audio services when negotiating with copyright holders (usually songwriters and publishers).

The Network System

By 1926, individual stations were having a difficult time filling airtime. There were few phonograph recordings to play, so performers had to come to the station, which was burdensome. It became apparent that a system of shared programming broader than AT&T's existing chain broadcasting was needed. RCA, GE, and Westinghouse together formed the National Broadcasting Company (NBC), which was established as a programming network. NBC's main purpose was to provide programs to stations. NBC then bought AT&T-owned WEAF, which essentially took AT&T out of the ownership and chain broadcasting business and gave NBC a programming monopoly. Instead of just supplying stations with programs they could air at any time, NBC came up with the idea of establishing station affiliation. Stations would affiliate with NBC and agree to broadcast programs simultaneously by means of telephone lines connecting local stations with the network headquarters in New York. NBC initially affiliated with 19 stations. NBC's first live network program featured live orchestras and singers. The broadcast was carried by the affiliated stations and reached millions of listeners. By the end of 1926, NBC was successfully operating two major networks. The original NBC network was renamed NBC Blue, and the newly acquired AT&T WEAF chain broadcasting system was named NBC Red. NBC Blue and NBC Red – along with their later-formed lone network competitor, CBS – dominated broadcasting for the next 15 years.

The radio industry had now established a system in which a radio station (an entity that broadcasts over the airwaves), could affiliate with a network (a company that sends programs to local stations), to receive network-generated programming which, due to higher budgets and access to national talent, could be of a higher quality than most local stations were capable of producing. In our modern era, it is easy to confuse terminologies such as 'station' and 'network' – or 'channel' – but this early history of network broadcasting should clarify these terminologies for generations that think of all broadcasting entities as similar options to be found on a television remote, a smart phone, or desktop computer.

The financial success of NBC led to other program networks. In 1927 the United Independent Broadcasters (UIB) started a network but it had limited success, primarily because it was not well funded. In fact, AT&T would not lease interconnecting lines to UIB because of the fear of non-payment due to the company's high costs and financial struggles. Columbia Phonograph Company rescued UIB. The Columbia Phonograph Company and the Victor Phonograph Company were in direct competition. Victor was about to merge with RCA (the parent company of NBC), a move that worried Columbia because of RCA's name recognition and business power and the danger that Victor might gain a competitive edge by having its records played on NBC stations. To hold off Victor, Columbia merged with UIB to form the Columbia Phonograph Broadcasting System (CPBS) so that it could play its records on its own network.

At first CPBS was financially successful mostly because it attracted big dollar advertisers, but the network quickly encountered financial difficulties. Realizing that CPBS could become a network powerhouse, William S. Paley, a wealthy cigar company magnate, bought a controlling share of the network. Paley became president of the network and changed its name to Columbia Broadcast System (CBS). Paley controlled CBS until 1983, becoming one of the most well-known electronic media moguls in the United States.

The Radio Act of 1927

When radio went beyond point-to-point communication to point-to-multipoint communication, or *broadcasting*, the 1912 legislation was no longer adequate to regulate commercial radio. The government was unprepared to deal with the proliferation of radio stations in the early 1920s. Instead of assigning each station to its own frequency on the dial, all stations were put on the same wavelength: 360 meters, about 830 kHz on the AM dial. With all of the stations in a given reception area operating on the same frequency, listeners could not hear any one station clearly; instead they heard a jumble of several stations whose signals were interfering with each other. What was needed was a technological way for each station to broadcast on its own frequency or channel.

To contain the interference on the airwaves, Congress passed the Radio Act of 1927, which formed the Federal Radio Commission (FRC). The FRC developed regulations for stations and station networks. According to the 1927 Act, the U.S. Secretary of Commerce was authorized to inspect radio stations, license operators of stations, and assign call letters.

The FRC's specific responsibilities included issuing licenses, redesigning the use of the electromagnetic spectrum, and providing 107 channels for radio stations. Each station was then assigned its own frequency and power level, which significantly reduced interference among stations.

It was mandated that within 60 days of the Act's passage, all existing radio licenses were to became null and void, thus all stations were forced to reapply for licenses. On relicensing, the FRC assigned to each station a particular broadcast frequency with the intent of minimizing interference and bringing some order to the chaos of the radio band. The process of designating frequencies was not always fair, as stations with a powerful signal were given desirable frequencies, while the less powerful stations were given less desirable frequencies. Other stations, such as college stations, which had a weak signal or power in a business or political sense, were simply forced off the air or bought out by commercial stations.

The Act addressed the equality of transmission facilities and issues of reception and service. It also restated the concept that the public owns the airwaves, but individuals and corporate entities could be licensed to own and operate stations. Criteria for ownership also were established because the number of applicants competing for frequencies exceeded the number of frequencies available on the electromagnetic spectrum. The criteria – of operating in the 'public interest, convenience, and necessity' – set the tone for case law in subsequent legal disputes over who controlled radio and why. In addition, it was made clear that government censorship of radio programs was not allowed.

The Communications Act of 1934

By 1933, several government agencies – including the Interstate Commerce Commission, the FRC, and the Department of Commerce – were regulating various aspects of electronic communication, but they were not working closely together. Realizing that radio needed better government oversight, President Franklin D. Roosevelt established a committee to study the nine different government agencies involved in public, private, and government radio. At the end of 1933, the committee recommended that one agency should regulate all radio and related services. The result of that recommendation and subsequent bills sponsored in Congress was the **Communications Act of 1934**.

The Communications Act of 1934 incorporated most of the Radio Act of 1927 but made

Figure 2.7 Advertisement for Atwater Kent Radio, 1927.

several updates. A new governing body, the **Federal Communications Commission (FCC)**, replaced the FRC. The Act established that the president of the United States would appoint seven FCC commissioners, no more than four from any one political party, and each would serve a seven-year term.

Substandard and Political Programming The FCC first set about tackling substandard radio programming, which ranged from fortune telling, huckstering medicinal cure-alls, and excessive advertising, to airing obscenity and programs that promoted religious intolerance. Although it could not censor content, the FCC had the right to evaluate stations' policies and programs to make sure they were operating in the public interest. Between 1934 and 1941, the FCC examined many stations, but only two licenses were revoked, and only eight others were not renewed. Perhaps more important, the FCC made radio stations aware that it was watching them, although the Commission took relatively little action at first. A letter of inquiry from the FCC, often referred to as a 'raised eyebrow' letter, would usually prompt the stations to initiate action to correct the problems.

By 1934 there were three radio networks operating in the U.S.: NBC Red, NBC Blue, and CBS. Radio receivers were highly desirable and in about 15 million American homes. Moreover, automobile manufacturers were putting radio receivers in vehicle dashboards. In-vehicle radio began a love affair between drivers and stations that boosted radio listenership. The public had become enamored with radio, so anyone on the air was highly regarded. Politicians especially took advantage of the ability to reach potentially thousands of viewers. But some stations only allowed favored candidates on the air, thus limiting listeners' exposure to other candidates.

One provision of the Communications Act of 1934 regulated political programming on the radio. *Section 315* of the Act stated that any radio station that allowed a candidate for an elected office to use the station's time for political purposes had to allow all bona fide candidates for the same office an equal opportunity for airtime, such that all candidates must have the opportunity to buy an equal amount of time on all shows. News and public-affairs programs were excluded from this provision.

Sixty days before an election, candidates could purchase time at the lowest unit rate (cost per spot) available to any station advertiser. For example, if one candidate bought 25 60-second announcements in a time of heavy listenership, then all the other candidates also must be given the opportunity to buy 25 60-second announcements in a similar time slot at the same price. *Section 312(a)(7)* stated that stations had to make a reasonable amount of time available to candidates for federal office. Candidates now had equal opportunity to have their voices heard on the airwaves.

The Communications Act of 1934 clarified government regulation of radio in the United States, and later underwent many revisions at the emergence of television, cable, satellite, microwave transmissions, and the Internet. The Act was the single most important piece of legislation in terms of how it shaped the development of electronic media until the 1996 Telecommunications Act rewrite.

A New Network

By 1934 the public had warmed to radio and the demand for new and innovative programming increased, and hence the need for a new network. The Mutual Broadcasting System (MBS) was formed as a cooperative programming network, which, unlike the other networks, did not own any stations. The programming was created by affiliated stations and sent out over the network. *The Lone Ranger*, *The Adventures of Superman*, and *The Shadow* were Mutual's best-known radio shows. *The Lone Ranger* originated at member station WXYZ in Detroit, MI and was broadcast by all stations affiliated with the Mutual network. Mutual did not have the financial means to hire big name radio stars who could attract advertisers, and so it was considered a 'second-class' network. Mutual however, went on to broadcast games for Major League Baseball, the National Football League, and it carried Larry King's show until its demise in 1999.

Press/Radio War

As radio gained in popularity, newspaper readership and the number of daily newspapers in the United States began to decline. Although many factors likely brought about newspapers' hardship, news over the radio was a major contributor. Radio newscasts began in the 1920s, and NBC started a regular network nightly newscast with Lowell Thomas, a well-known newscaster, on its Blue network in 1930. NBC's initiation of nightly news signaled the beginning of a serious effort to expand radio's influence and use.

Newspapers, already wary of radio stealing their readers, tried to force radio stations and networks to limit their newscasts. Both newspapers and radio stations employed staff reporters, but they were also highly reliant on the wire services for the bulk of their news. The wire services had more money than the stations or networks and thus could send many more reporters out into the field. The newspaper industry, however, did not like that the wire services were sending stories to radio stations and thus in 1933 began pressuring the news wire services to release news reports only to the papers and to cease delivering stories to the stations. The newspapers started the press/radio war by persuading the wire services to align with them and against the radio stations. The newspapers also refused to print radio program schedules without charge. The radio industry retaliated by hiring more freelance reporters to help its small news staff gather and report the news.

ZOOM IN 2.1

Radio and newspapers settled the press/radio war in 1933 by signing the Biltmore Agreement (named after the hotel in New York where the agreement was signed), which included the following restrictions for broadcasting radio news:

1. Stations could air only two newscasts per day: one before 9:30 a.m. and one after 9:00 p.m. (to protect the morning and evening editions of newspapers).
2. Radio reports were limited to commentary and soft news (features and stories that are not necessarily time sensitive), rather than hard (e.g. crime and time-sensitive) news reporting.
3. Stations could get news only from the Press-Radio Bureau, which would supply the networks and stations with news through subscriptions.
4. Stations could not have their own news-gathering operations.
5. Stations could not sell advertising sponsorships of news shows.
6. At the end of each radio newscast the announcer had to say, "You can read more about it in your local newspaper."

The Biltmore Agreement did not last long. Radio stations felt constrained by the limitations put forth by the agreement, but more so they felt they should continue doing their civic duty of gathering and reporting the kind of hard news that its audience demanded. Stations found ways to work around the agreement. For example, newscasters now became *commentators*, and technically not journalists, so they could still 'report' the news but as a commentator they skirted around the new regulations. Even the two major wire associations (International News Service and United Press) soon found work arounds and eventually broke the agreement by tailoring their news feeds for broadcast. Even though the Biltmore Agreement was short-lived, it is a classic example of efforts by an existing medium to slow the growth of, and competition from, a newer medium.

FYI: DISNEY V. SONY

In 1976, a different type of press/radio war was waged, but this time it pitted movies against home videocassette recorders (VCR). Disney Studios sued Sony, the first maker of VCRs, for copying, distributing, and selling recorded Disney movies to the public, who then watched them on their home VCRs. Disney claimed Sony was infringing copyright laws and demanded royalties from Sony. In a 5–4 decision, the U.S. Supreme Court ruled that the sale of VCRs was not considered 'contributory infringement' as alleged, a ruling that cleared the way for the home video industry to flourish in the 1980s.

As in the case of the Biltmore Agreement, attempts to stop new media technologies from developing have been mostly unsuccessful.

After the press/radio war, the radio industry became free to develop its news operations and cultivate broadcast journalists. Edward R. Murrow, perhaps one of the most articulate and popular radio journalists of all time, started as a young executive with CBS who did not initially intend to work on the air. As Europe rapidly descended toward war in the late 1930s, Murrow established a news bureau in London to cover the developing events. He lined up 'talks' by experts on foreign affairs, but as war began it became increasingly difficult and dangerous to get people to a studio for live broadcasts. Despite intense nervousness on the air from which he never fully recovered – even after becoming a national celebrity – Murrow began writing his own stories and reporting on the events in Europe himself. In addition, with his role as a CBS executive, he hired a group of reporters later nicknamed 'Murrow's Boys' to cover the war. Several of these reporters continued to influence broadcast journalism for the remainder of the 20th century and beyond. Murrow himself became one of the best reporters of his time. After the war broke out, he took to the rooftops and reported on the live bombings of London. He also pioneered a newscast format in which multiple reporters at remote locations contributed stories introduced by a host in a studio (later referred to as an 'anchor'). The technological feat of delivering live reports from war zones in the late 1930s and early 1940s was significant, as was the style, delivery, and accuracy of the information presented on the CBS news programs. Despite having a deep, pleasant radio voice himself, Murrow was more interested in the quality of journalism than the sound or presentation styles of the reporters he hired. He brought on seasoned print journalists, openly defying the complaints of CBS upper management about their squeaky voices and regional accents. Shortly after the war, Murrow hosted the radio show *Hear It Now*, which later moved to television and was renamed *See It Now*. Murrow set the tone for radio news reporting, which was emulated by other successful newscasters such as Walter Cronkite and Dan Rather, who started in radio but became famous on television.

World War II

After the Japanese attack on Pearl Harbor on December 7, 1941, the radio networks interrupted their regular programming to tell the world about the atrocity. President Franklin D. Roosevelt went on the air the next day calling the bombing "a date which will live in infamy." His emotional broadcast in which he asked Congress for a declaration of war against Japan was listened to by 62 million people and is one of the most replayed speeches from that era.

Upon entering World War II, the U.S. government immediately took steps to support the overall war effort. A small number of new stations got on the air between 1942 and 1945, but the government curtailed most of the growth of the industry.

ZOOM IN 2.2

Hear a clip from the audio of President Roosevelt's speech on December 7, 1941, by going to www.archives.gov/education/lessons/day-of-infamy

Despite the hardships faced by Americans during the war years, the radio industry thrived because it provided both escape and information.

AM Radio Evolves

Until the 1960s, most radio stations aired on the AM band from 540 kHz to 1610 kHz, which uses amplitude modulation (AM) to carry voice communication. AM combines audio with the basic carrier wave that is sent from the broadcast antenna to a receiving antenna. AM combines the audio with the carrier wave by varying the size or height of the wave (its amplitude) in a manner consistent with how air moves when sound is made. For instance, a 440 hertz tone (the musical note A) would cause the amplitude of a beam of electromagnetic radiation (the carrier frequency) to vary at a rate of 440 times per second.

AM radio's popularity diminished as television's popularity grew. Although the networks continued to supply programming to radio stations into the early 1950s, most radio comedy and dramas were being refashioned for television broadcasting. If a show was good enough to hear, it was probably good enough to be seen as well. Audiences eventually preferred the television versions, which made the radio versions unprofitable and left radio to come up with a way to fill the many hours of daily airtime.

Radio of the 1950s was no longer the daytime companion and nighttime focus of attention, as television became the dominant medium. By 1955, AM radio had reinvented itself by becoming more music oriented, with an in-station announcer spinning records. Radio stations' in-station announcers began taking air shifts in time blocks – for example, from 6:00 to 10:00 a.m. each morning – playing music, announcing song titles and artists, and reading weather or brief news reports. These announcers became known as disk (or disc) jockeys, or DJs, because most of the time they were playing phonograph records, or *discs*, on the air. Many stations, trying to differentiate themselves in a competitive market, selected a specific style of music and played it most of the time. The result was that stations specialized in musical genres such as country-and-western music, African American-influenced music (known as rhythm and blues), classical music, popular music (Top 40), and so on. Other stations, like KFAX in San Francisco, adopted an all-news format, and KABC, an ABC-owned station in Los Angeles, adopted an all-talk format.

In 1993, the AM band was expanded from the spectrum range of 550 to 1600 to include 1610 to

1700 kHz. This addition allowed local low-power stations (e.g. college, religious, non-English, and government-owned stations in locations like airports and national parks) to operate on the AM band without interfering with existing stations.

FM Radio Captures the Audience

AM is susceptible to the static caused by thunderstorms and electrical equipment, which creates noise distortion on the receiving end. The fidelity (or sound quality) is limited such that AM cannot reproduce very-high-frequency sounds (often called 'treble,') which humans hear as clarity or shrillness, or very-low-frequency sounds (such as the low notes from a bass drum), which we perceive as fullness, warmth, or even larger size.

Edwin H. Armstrong, the inventor of the superheterodyne radio receiver, sought to eliminate the static and improve the fidelity of the radio signal. After many years of experimentation, Armstrong's patents were finally granted in 1933. In 1935, Armstrong gave a public demonstration of FM (frequency modulation) radio. He explained that FM's audio quality was superior to AM radio because the *frequency* of each wave was modulated by sound rather than the *amplitude* of each wave.

One important discovery that Armstrong made was that frequency modulation required more bandwidth. Instead of the 10 kHz channel used by AM broadcasting, FM required 20 times more space for each channel, or 200 kHz. The government set aside the 42 to 50 MHz band for FM radio beginning January 1, 1941. By the end of 1941, there were about 40 FM stations on the air, but many of the stations were not fully powered, and some were experimental. Further, the FM audience was limited because by 1941 only about 400,000 receivers had been sold that could pick up FM signals.

During World War II, interest in FM waned, and the FM band was later reassigned to the 88 to 108 MHz band in 1945, and FM radio broadcasting in the 42 to 50 MHz band ceased in 1948. As a result, many listeners owned what were supposed to be FM radios but that no longer received FM signals. FM stations did not operate profitably for some time and total national FM revenues did not pass $1 million until 1948.

In 1961 the FCC authorized FM *stereo broadcasting*. When listeners started to notice the superior sound quality of FM, it signaled the slow decline of AM as a music broadcaster and the rise of FM music. AM radio was better suited for talk than for music and FM radio for music rather than

FYI: THE ARMSTRONG/ SARNOFF CONFLICT

Edwin H. Armstrong's FM radio invention seemed like a natural for the radio networks: less static, better sound, and a receiver that picked the strongest signal on the frequency without interference. Despite those technological advancements, David Sarnoff, the head of RCA and a friend of Armstrong's, decided against supporting FM. Rather, he wanted to spend more time and energy on the development of television and to avoid having to pay Armstrong for his invention. Later, Sarnoff testified in court that "RCA and NBC have done more to develop FM than anybody in this country, including Armstrong" (Lewis, 1991, p. 317). Armstrong fought Sarnoff and his company for patent infringement, vowing to continue "until I'm dead or broke" (Lewis, 1991, p. 327).

The Armstrong/Sarnoff conflict started when Armstrong sued RCA and NBC in 1948 and continued through 1953. By then, Armstrong had run out of money to pay his lawyers, and the prospect of receiving damages from RCA in the near future (lawyers estimated it would take until 1961) seemed remote. On January 31, 1954, despondent over a dispute with his wife and the continuing battle with Sarnoff and RCA, Armstrong jumped to his death from his tenth-story bedroom window, "the last defiant act of the lone inventor and a lonely man" (Lewis, 1991, p. 327).

talk. The counterculture of the 1960s, the rise of rock 'n' roll, and high-fidelity stereo systems boosted FM listenership.

It took until the late 1970s and early 1980s before FM radio gained an equal footing with AM radio. In 1978, the FM audience surpassed the AM audience for the first time. By the late 1980s, the FM audience was much larger than the AM audience, commanding almost 75% of radio listeners. Almost all car and portable radios then had an easy-to-tune FM receiver, which made FM as easy to find and listen to as AM.

SEE IT NOW

Now Audio

Radio transmission has changed greatly in the last several decades. Digitization has brought about new ways to delivery audio content, and with it new

How AM and FM differ

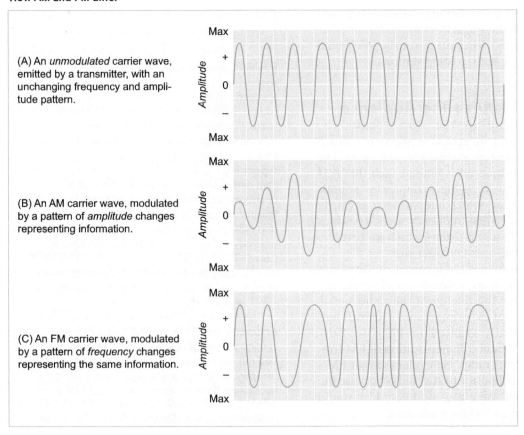

(A) An *unmodulated* carrier wave, emitted by a transmitter, with an unchanging frequency and amplitude pattern.

(B) An AM carrier wave, modulated by a pattern of *amplitude* changes representing information.

(C) An FM carrier wave, modulated by a pattern of *frequency* changes representing the same information.

Diagram 2.2 These three graphs show the differences in AM and FM waves.

concerns about technological boundaries and copyright. This chapter next covers satellite radio, HD radio, Internet audio, and copyright issues, especially as they pertain to digital audio services, such as Pandora.

Satellite Radio

The biggest disadvantage to over-the-air radio (also referred to as terrestrial radio) is the limited signal range. Some stations are more powerful than others, and AM typically covers a larger area than FM, but when driving long distances, signals are often lost between cities. Long-distance drivers have to frequently retune to local stations. Satellite-delivered radio, however, solves the range problem because satellite signals cover huge distances.

XM Satellite Radio started operating in 2001and **Sirius Satellite Radio** began operations in 2002 as competitors for the new satellite radio market. Both companies beamed radio signal up to a satellite equipped with a special satellite receiver, which is then downlinked to satellite receivers, either stationary, as in a home, or mobile, as in a vehicle. In addition to satellites, small terrestrial transmitters called *repeaters* augment the signal to ensure that listeners in big cities can pick up a satellite signal even if it is blocked by tall buildings.

XM was formed with the help of investors from automobile manufacturing, broadcast radio, and satellite broadcasting, namely, General Motors, American Honda, Clear Channel (the nation's largest radio group at the time), and DirecTV. Both services provided hundreds of channels of audio service, including commercial-free music and news, sports, talk, and children's programming. Slow subscriber rates led to the merger of Sirius and XM in 2008. The combined company is now known as Sirius XM Radio.

Radio Goes Digital

Radio broadcasting has traditionally been an analog medium and, compared to analog television broadcasting, uses far less of the electromagnetic spectrum for each channel. AM radio channels are

10 kHz wide and FM radio channels are 200 kHz wide. Compared to the size of a television channel (6 MHz, or 600 times the width of an AM channel), radio is a more efficient use of the spectrum.

Digital broadcasting allows several separate signals to use the same space as one analog signal. In addition, digital broadcasting provides better signal reproduction and potentially better sound, depending on the extent to which the digital data stream is 'compressed' to reduce file size.

Although the FCC has not mandated that radio stations change to digital, many stations have initiated the change on their own. Both the FCC and the radio industry wanted a switch to digital because its efficient use of bandwidth increased both sound quality and the amount of programming content a station could put out. However, it was also deemed important to keep all existing radios from becoming obsolete (which is what happened when the FCC mandated the FM band to change frequencies in 1945). A system called **IBOC (in-band, on-channel)**, enables broadcasters to use their existing frequencies to broadcast in digital and analog at the same time. The digital signal is generally of higher quality than the existing analog service. Audience members who are happy with analog can stay with analog. iBiquity Digital, which licenses the software to radio stations (and consumer electronics manufacturers) under the brand name HD Radio, owns IBOC technology.

HD radio has been slow to make inroads, as the lack of a mandated change has presented broadcasters and listeners with a 'chicken or egg' quandary. In 2009, less than 15% of all the radio stations in the United States were broadcasting a digital signal. By 2018, more than 3,500 stations were broadcasting in HD. Although a significant portion of all radio listening is on stations broadcasting with HD radio, many listeners are unaware of the difference between analog radio and HD radio because all of the main channels for radio can be received on any radio receiver, HD-capable or not.

Digital or HD radio provides a station with space in their allotted channel to program additional signals or stations. More stations give broadcasters more airtime to sell for advertising, thus providing additional revenue streams for the station. The size of the bandwidth available to licensed broadcasters and the ability of digital broadcasting to make very efficient use of the band through compression, additional signals, known as digital subchannels, are available (although the more subchannels that are created, the smaller the available bandwidth, which can affect sound quality). For example, one classic rock station can offer a narrowly targeted classic 1970s channel, a 'deep cuts' album track channel, and a 1990s 'classic alternative' station, all on the same frequency that previously allowed only one analog signal.

Consumers in many markets have very little incentive to purchase digital radio receivers because their old radios receive the main signal of all radio stations. The relative lack of consumer interest slowed the rollout of both home and car radios capable of receiving digital signals, and auto manufacturers were at first slow to make the radios available in new cars. By 2016, all major car manufacturers supported HD radio technology with optional HD radios. Since most of the HD radio stations are located in larger markets, many consumers in medium, small, or rural markets do not gain any radio signals by spending the extra money for an HD-capable radio.

Internet Radio

Internet radio began in the late 1990s, when streaming via the Internet became a technologically feasible method of transmitting music. Broadcast stations added an online component to their over-the-air signal to provide their listeners additional ways of connecting to their stations. Internet radio can be heard via computers, tablets, and smartphones.

In addition to licensed broadcast stations streaming their signals over the Internet, many non-licensed 'radio' stations have been streaming music, talk, news, and entertainment. Notable among these 'netcasters' are music services like Pandora and Spotify. These services have a strategy of enabling the listeners to interact with the service to create individual playlists and personal 'radio stations.' SiriusXM has an Internet and smartphone app component for its subscribers to provide its music, talk, sports, and entertainment programming anywhere.

A study done in 2016, found that two-thirds of all smartphone owners streamed music daily (Washenko, 2016b), and if anything, the popularity of streaming has only increased since. In addition,

ZOOM IN 2.3

Go to hdradio.com for more information about terrestrial broadcast HD radio. For examples of other radio/audio services that are Web based, go to www.pandora.com or www.accuradio.com

podcasting has gained significantly in popularity. A large portion of the listening audience is getting audio service through their smartphones, creating a challenge for traditional radio stations.

ZOOM IN 2.4

Not all 'radio' stations are broadcasters. Many stations are Internet only.
Finding radio stations on the Internet is easy. Try this site for an up-to-date list of stations: www.internet-radio.com

New Radio Technology and Copyright Issues

Until the late 1990s, radio was an over-the-air service that was programmed by radio station programmers and listened to by a passive audience. Radio began to change when digital technology spawned online radio stations, audio services, peer-to-peer music file sharing, and podcasting.

Beginning in the late 1990s, online file sharing of copyrighted music became a serious problem for the music industry. Internet users could go to various *peer-to-peer sharing* sites, such as Napster, and download music by copying files from other users who had connected to the site. The music industry claimed that it lost substantial revenue because so many people were getting free music online instead of buying CDs. Despite the threat of legal action, individual users have continued to download music and even feature-length movies without paying for them. The music and movie industries continue to pursue copyright violators. Streaming services such as Spotify and Pandora do bring some revenue to the music industry, approximately $8.8 billion according to recent estimates. Audiences have warmed to the idea of paying a small subscription fee to listen to music, but the music industry earns only a small fraction of what it did when its products were exclusively physical (records, tapes, CDs) and the radio industry effectively promoted new releases.

Copyright law originated in Article I, Section 8(8) of the U.S. Constitution, which allowed authors rights to use their 'writings and discoveries' for their own benefit, protecting those rights from infringement by others. Copyright law has since been revisited and revised, and it has been upheld in the courts, generally protecting the creators of original works from unauthorized use, such as illegal downloading.

Specifically, an artist of an original work receives copyright protection for his or her lifetime plus 70 years for work created after January 1978 and 95 years for work created before that time.

Sometimes authors and others allow a licensing agency to collect usage fees for them. In the case of musicians who create original music, certain organizations negotiate and collect fees from others who wish to use their music. In the United States, there are two large music-licensing agencies that perform this function – ASCAP (started in 1914) and Broadcast Music, Inc. (BMI) (started in 1940). BMI was formed during the height of network radio popularity to compete with ASCAP and be friendlier to the radio stations and networks, because it was formed by broadcasters themselves and charged less for the use of musical compositions.

Both ASCAP and BMI negotiate *blanket fees* with users such as broadcasters and production companies. A blanket fee is determined by using a formula to calculate the yearly amount that a radio station will be charged to use all of the music licensed by ASCAP and BMI). That amount is based on factors such as the percentage of ASCAP and BMI music the station plays per week, the size of the station's market, and the station's overall revenue. Large stations in large markets pay more than small stations in small markets.

If a copyright expires, then anyone can use the materials without asking permission or paying a fee. Once a copyright has expired or if a work was never copyrighted, the material is considered to be in the *public domain*. Advertisers, performers, and writers like to use material that is in the public domain for their projects, because no permission or payment is necessary. Public-domain material is particularly attractive to those producing low-budget projects.

Educators and others can use copyrighted material without getting permission or paying a fee if their use of the material is non-commercial and limited. This allowance falls under Section 107 of the 1976 Copyright Act and is referred to as **fair use**, as determined by the following four criteria: (1) the purpose or use of the material, (2) the characteristics of the original work, (3) the amount of the original work used, and (4) the possible impact that the use might have on the market for the original work. It is important to point out, however, that fair use is a legal defense that is argued in court after someone has been sued for copyright infringement. The safest choice of all in terms of copyright is to create one's own work and only use the work of others with permission.

The **Recording Industry Association of America (RIAA)** is the trade organization that represents the people and companies that produce 90% of the recorded music in this country. It aggressively attempts to identify people who illegally share files containing copyrighted music, and when it does, it often prosecutes them for copyright violation. In late 2003, the RIAA filed hundreds of lawsuits against individual music file sharers.

The **Digital Millennium Copyright Act (DMCA)**, passed in 1998, was designed to protect creative works in this digital era. It prohibits the manufacture and distribution of devices or procedures that are designed to violate copyright law in the digital environment. In addition, this law requires Internet service providers (ISPs) to identify their customers who violate copyright law by using file-sharing services so RIAA can take legal action.

SoundExchange is an organization that represents record labels similarly to how ASCAP represents composers, authors, and publishers for music licensing. SoundExchange was originally a division of the RIAA that was formed to collect royalties resulting from the DMCA. SoundExchange was spun off and became an independent non-profit organization in late 2003.

An 'intellectual property' right closely connected to copyright is performance right. Performance rights provide the performer with the legal protection of their product, such as the performance of a musical piece or an announcer's voice in a commercial. Traditionally, performance rights were negotiated in a performer's contract with a show's producer. In some instances, such as commercials, actors as performers were granted residuals (similar to royalties) if the airing of the commercial exceeded a set number of airings. However, in the early part of this century, commercial announcers demanded extra payment for commercials produced for broadcast that aired on the Internet. With the widening distribution of performers' works on the Internet, through audio streaming services as well as video sites such as YouTube, performers are now demanding intellectual property right protection of their work, similar to copyright and patents. This extension of intellectual property rights protection into the area of performance rights, especially on the Internet, appears to be the next 'copyright' issue.

SEE IT LATER

Internet Radio

Competing services online, as well as satellite radio, now supply much of the content that radio has offered for the last 90 years. Music is heard over the Internet 24 hours a day, seven days a week, from thousands of 'now media' sources all over the world. Internet and satellite providers mimic licensed radio stations free from the content restrictions imposed on broadcast stations by the FCC.

Until several years ago, online radio could not compete with broadcast radio because it was not portable. But now with wireless Internet signals (Wi-Fi) and mobile devices (smartphones, tablets, iPods), Internet radio is a now media that can be listened to almost anywhere. Broadcast radio stations need a strategy for maximizing their potential via the Internet. Even the automobile, once the exclusive domain of broadcast radio, now offers online radio stations and audio services.

Broadcast radio is going through the same challenge that it did in the 1950s, when television took over its audience. Radio once again must reinvent itself to ensure its viability. Digital technologies are important parts of that process, but the key is developing new content. Radio stations need to offer audiences programming and other services that they cannot get anywhere else. A return to more local content is one option, as is a general emphasis on talent rather than narrow and formulaic music offerings. The broadcasting industry still argues that over-the-air radio is still needed, especially during emergency situations, such as during and after hurricanes, when Internet service usually goes down. During times of crisis, radio stations continue to broadcast. The public safety function of broadcasting is as important as ever, but radio has stiff competition from digital platforms that entertain the audience.

Other Technological Considerations

Although broadcast radio has always been portable, it has been virtually blocked from reaching mobile users on their cell phones. Most smartphones have the technological capability to receive radio broadcast signals, but very few phone users have an activated chip in their smartphones necessary for FM reception.

Smartphone apps, however, may be the answer broadcasters are looking for to reach the mobile audience. One such app, TuneIn Radio, lets users access local stations or distant stations with a few taps on their phones. Smartphone owners are avid music listeners – 68% stream music daily and spend about 45 minutes each day listening on their phone.

ZOOM IN 2.5

CHANGING AUDIO BUSINESS

The business of audio has been changing, due at least in part to technology, but also in part to audience behavioral changes. Industry leader Norm Pattiz began Westwood One in the 1970s to take advantage of distributing syndicated audio programs to radio stations via satellite rather than tape or LP. At first Westwood One distributed its programming to its Mutual Broadcast System affiliates. The business grew dramatically and eventually served more than 5,000 radio stations, providing more than 150 news, sports, music, talk, and entertainment programs, features, and live events. Westwood One also provided local news, sports, and weather to over 2,300 stations.

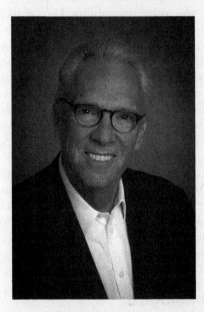

Figure 2.8 Norm Pattiz.

Pattiz's more recent venture is called Podcast One. Taking advantage of an audience that is comfortable with talk entertainment about a large variety of topics, Podcast One distributes talk shows that program to very specific audiences as well as general audiences. This business model avoids using radio stations as the middleman, essentially going 'over the top' in audio by bringing podcast programs directly to the listening audience. Podcasts about sports, comedy, politics, news, society and culture, television and film, and health topics are all available to audiences through one Web site: podcastone.com

And app use is big too. Out of three hours a day on a smartphone, 2 hours and 42 minutes are spent using apps.

In-car radio listening has long been the domain of terrestrial radio. Although radio listening in the car is still strong despite competition over the years from 8-track tapes, cassette tapes, satellite radio, CDs, and MP3s, the connected car (with a mobile Internet connection) threatens that dominance. If listening to a digital Internet-based system becomes as easy as pushing a button on the car radio, broadcast radio audience size may suffer additional setbacks.

Radio will probably experience a continued decline in the amount of time listeners spend on broadcast delivery because of competition from the many competing audio services and other entertainment and information options. However, mobile delivery through smartphones, tablets, and other devices could actually boost radio listening.

The Government's Role in Radio

The FCC has long been interested in preserving radio's localism. Over the years, the agency has encouraged radio stations to serve their communities. Local broadcast television and radio stations enjoy a competitive advantage over satellite services, because they provide the programming, news, and talk important to people in their community.

The FCC has also granted licenses to several hundred low-power FM (LPFM) stations across the United States. These stations offer non-profit organizations, such as religious groups, academic institutions and schools, and community organizations the opportunity to reach local audiences with just 100 watts of power, enough power to reach listeners within a few miles of the station.

So far, SiriusXM has not been required to provide local stations to subscribers, although it does provide some local traffic news and weather forecasts in larger cities. The FCC could decide to require SiriusXM to provide local programming, commercials, and even news. If the FCC does make such a decision, it could change the revenue stream for all electronic media and have serious implications for local broadcasters. Local broadcasters would be forced to compete for local advertising dollars with SiriusXM, and some siphoning off of local dollars would certainly occur.

Thus, local broadcasters would have to fight even harder to keep their revenues from shrinking.

Traditional radio stations do enjoy one advantage over newer digital services. Because of a quirk in U.S. copyright law, traditional radio stations only pay the songwriter and music publisher. On the other hand, satellite radio, Internet music services, and digital cable music providers all pay two separate performance royalties for the music they play, the first to the songwriter and music publisher and the second to the record labels and recording artists. As music sales dropped in the early years of this century, the music industry scrambled for new revenue streams and, in an echo of the 1920s copyright battles, the RIAA began actively lobbying Congress to revise the law and introduce an additional royalty for broadcast radio that would be paid to the labels and artists. The rule for broadcasters requiring payment only to the songwriter and music publisher was maintained because Congress believed that playing a song on the air encouraged people to buy the music; the promotional value was thus considered to be adequate compensation to the performers because the record labels and artists made money from the sales of their albums.

Financial Outlook

Online 'now media' services like Pandora and Spotify continue to gain strength in subscriber revenues and advertising revenues. Spotify first reported profits in early 2019. Satellite radio continues to aggressively seek program content to attract subscribers and advertisers. As audience targeting becomes more sophisticated, advertisers seek listeners regardless of the audio content or the audio delivery system. Radio station owners continue to pursue listeners to try to convince sponsors that they can deliver the desired target audiences.

SUMMARY

Electronic media communication has changed over the years in response to the human desire to go beyond face-to-face contact. Since the beginning of the 20th century, humankind has developed the technology to reach people over long distances in a matter of seconds. First using the wired telegraph and telephone and then using radio telegraphy and telephony, people have been able to communicate both one-to-one and one-to-many. The ability to communicate one-to-many using radio signaled the beginning of electronic mass media. The excitement generated by broadcasting lured many people to experiment with radio, both transmitting and receiving. Many hobbyists built their own radio receivers, and a number of them also dabbled in radio transmitting.

Entrepreneurs and inventors such as Guglielmo Marconi, Lee de Forest, Edwin H. Armstrong, Frank Conrad, and David Sarnoff propelled radio from an experimental system to an industry and storehouse of American culture. From its modest audience size in 1920 to its peak in 1950, radio was the dominant mass medium.

The U.S. government has played a role in the development of the radio industry by ensuring that control stayed in the hands of American companies, as evidenced by its seizure of all powerful radio transmitters during World War I. Rather than keep control after the war, the radio industry became a commercial enterprise, guided by market factors more so than government intervention. At first, radio stations experienced numerous problems with technology, some of which stemmed from all stations broadcasting on the same frequency. The government corrected that problem by establishing separate frequencies for stations in the same market and region with the Radio Act of 1927 and the establishment of the FRC. The government also established the philosophy that the airwaves belong to the people and that broadcast stations must operate in the 'public interest, convenience, and necessity.' The print media embraced radio to a certain extent. For example, many newspapers added a section for radio programming schedules, discussions of programs, and even technical tips for better reception. There was a definite ambivalence displayed by newspapers when radio began broadcasting live news reports, a practice that newspapers tried unsuccessfully to impede during the press/radio wars in the early 1930s.

Radio exposed the American audience to the concept of free entertainment and information programming (once the initial price for the receiver was paid). Although newspapers were very inexpensive, radio programming was free and could be enjoyed in unlimited amounts by the audience. Moreover, it did not require literacy. Radio also encouraged people to stay home and listen to free programs rather than go to vaudeville shows at their local theaters. 'Talking' motion pictures, a product of the late 1920s, drew large audiences but did not seem to slow down radio's growth. The phonograph record industry was forced to cope with the fact that their customers could now receive free music on the radio, rather than pay for phonograph records that were expensive and had lower-quality sound.

Radio also exposed listeners to the voices of politicians, celebrities, sports heroes, and even common people. Audiences heard different regional dialects and accents. Politicians embraced radio as a means of reaching their constituency with messages that were tailored to their audience.

Networks NBC Blue and NBC Red and CBS provided radio programming from the late 1920s through the 1940s. These networks' programming innovations set the stage for many years of audience loyalty and appreciation. In fact, many of the program types developed during these years made the transition to television and continue to the present. Radio strongly influenced American society by providing free entertainment and information and exposing listeners to voices of celebrities and government officials. Radio also siphoned some of the interest away from newspapers.

AM radio lost its network entertainment programming and prominence in the minds of the audience when television was introduced after World War II. But AM radio reinvented itself by developing music formats hosted by disc jockeys. FM radio, which rebounded from a serious setback when the FCC changed its band location, gained dominance in musical programming after the introduction of stereo broadcasting in the 1960s. By the 1980s, the FM audience was larger than the AM audience. Once again, AM had to reinvent itself, which it did by concentrating programming more on talk, news, and religion instead of music.

The Telecommunications Act of 1996 triggered a dramatic increase in broadcast station owner consolidation, because it relaxed ownership rules, and this consolidation has led some industry watchers to criticize radio for losing its localism. Alternative delivery systems that can deliver 'now media' have further fragmented the radio audience and, along with various economic and regulatory issues, created challenges for the radio industry. However, many of these services cannot compete with traditional radio when it comes to providing locally oriented news and entertainment using a technology that is both portable and without direct cost to the audience. Changes in audience behavior signal a change in how the industry will continue to operate. Podcasting may change the audience for talk radio. A recent study found that more than two-thirds of smartphone owners spend time streaming music daily. Most people acquire radios when they buy a car, but few radios are bought for homes or dorm rooms. The future points to a world where mobile listening is done on 'now media' devices that deliver audio on a smartphone via streaming or in an automobile, where the dashboard has become Internet connected and therefore allows easy access to non-broadcast audio services.

BIBLIOGRAPHY

Abbot, W. (1941). *Handbook of broadcasting* (2nd ed.). New York: McGraw-Hill.

Banning, W. (1946). *Commercial broadcasting pioneer: WEAF experiment, 1922–1926*. Cambridge, MA: Harvard University Press.

Barnouw, E. (1966). *A tower in Babel: A history of broadcasting in the United States Vol. 1: to 1933*. New York: Oxford University Press.

Beauchamp, K. (2001). *History of telegraphy*. London: Institute of Electrical Engineers.

Benjamin, L. (1993, summer). In search of the 'radio music box' memo. *Journal of Broadcasting and Electronic Media*, 325–335.

Broadcasters start digital radio service. (2003, June 17). *USA Today*. Retrieved from: www.usatoday.com/tech/news/2003–06–17-digital-radio_x.htm [July 1, 2003]

Clinton: 'Internet in every hut.' (2000). *Reuters Wired News*. Retrieved from: www.wired.com/news/print/0,1294,34065,00.html [July 15, 2002]

Clinton unveils plan for 'next generation of Internet.' (1996). *CNN*. Retrieved from: www.cnn.com/US/9610/10/clinton.internet [July 15, 2002]

Daly, C. B. (2012). *Covering America: A narrative history of a nation's journalism*. Amherst: University of Massachusetts Press.

Faruk, I. (2014, April 23). 2 headwinds for Sirius XM radio you shouldn't ignore. *The Motley Fool*. Retrieved from: www.fool.com/investing/general/2014/04/23/2-headwinds-for-sirius-xm-radio-you-shouldnt-ignor.aspx

Ferguson, D., & Greer, C. (2011). *Local radio and microblogging: How radio stations in the U.S. are using Twitter*. UK and Europe: Broadcast Education Association, Taylor & Francis.

Floherty, J. (1937). *On the air: The story of radio*. New York: Doubleday, Doran & Co.

Godfrey, D., & Brinson, S. (2015). *Routledge reader on electronic media history*. New York: Routledge.

Goldman, D. (2010, February 2). Music's lost decade: Sales cut in half. *CNN.money.com*. Retrieved from: www.money.cnn.com/2010/02/02/news/companies/napster_music_industry/

Goodman, M., & Gring, M. (2003). The radio act of 1927: Progressive ideology, epistemology, and praxis. In M.

Hilmes (Ed.), *Connections: A broadcast history reader* (pp. 19–39). Belmont, CA: Wadsworth.

Gordon McLendon and KLIF. (2002). *Encyclopædia Britannica*. Retrieved from: www.britannica.com/%20/%20facts/5/71004/KLIF-as-discussed-in-Gordon-McLendon-and-KLIF

Gross, L. (2003). *Telecommunications: Radio, television, and movies in the digital age*. New York: McGraw-Hill.

Hayward, A. (2019, August 16). techradar. iTunes shutting down: when and why it's happening. Retrieved from: www.techradar.com/news/itunes-shutting-down-when-and-why-its-happening

Heine, P. (2009, July 19). Stream it like you mean it. *Media Week.com*. Retrieved from: www.adweek.com/as/content_display/special-reports/other-reports/e3i4d0b1b4303c8399766735f5b52963ebe

Hendricks, J., & Mims, B. (2015). Keith's *radio station: Broadcast, internet, and satellite* (9th ed.). Burlington, MA: Focal Press.

Hewitt, L., Krause, A., & North, A. (2014). *Music selection behaviors in everyday listening*. USA and North America. Broadcast Education Association, Taylor & Francis.

Hilliard, R., & Keith, M. (2001). *The broadcast century and beyond* (3rd ed.). Boston, MA: Focal Press.

Lessing, L. (1956). *Man of high fidelity: Edwin Howard Armstrong*. Philadelphia, PN: Lippincott.

Lewis, T. (1991). *Empire of the air: The men who made radio*. New York: HarperCollins.

Pandora, You Tube, AM/FM, Spotify compete for ears as American listenership evolves: Infinite Dial. (2016). *Rain News*. Retrieved from: http://rainnews.com/pandora-youtube-amfm-spotify-compete-for-ears-as-Americanlistenership-evolves

Pitchfork.com. (2019). Streaming made up 80 percent of music revenue in 2019. Retrieved from: https://pitchfork.com/news/streaming-made-up-80-of-music-industry-revenue-in-2019-riaa-says [May 24, 2020]

Public interest, convenience, and necessity. (1929, November). *Radio News*. Retrieved from: www.antiqueradios.com/features/frc.shtml

Radio is officially America's number one mass research medium. (2015, June 24). *New York: Radio Advertising Bureau*. Retrieved from: www.rab.com

Settel, I. (1960). *A pictorial history of radio*. New York: Citadel Press.

Siepmann, C. (1946). *Radio's second chance*. New York: Little, Brown and Company.

Sterling, C., & Kittross, J. (2002). *Stay tuned: A history of American broadcasting*. Mahwah, NJ: Erlbaum.

Washenko, A. (2016a, March 4). Radio's digital revenue breaks $1 billion in RAB 2015 survey. *Rain News*. Retrieved from: rainnews.com/radios-digital-revenue-breaks-1-billion-in-rab-2015-survey

Washenko, A. (2016b, March 11). Two-thirds of smartphone owners stream music daily (study). *Rain News: Chicago*. Retrieved from: rainnews.com/two-thirds-of-smartphone-owners-stream-music-daily-study/

Webster, G. (1998). *The Roman imperial armies of the first and second centuries* (3rd ed.). Norman: University of Oklahoma Press, p. 255.

Whitmore, S. (2004, January 23). Satellite radio static. *Forbes.com*. Retrieved from: www.forbes.com/2004/01/23/0123Whitmore.html

Wurmser, Y. (2019, May 30). Emarketer. U.S. Time Spent with Mobile 2019. Retrieved from: www.emarketer.com/content/us-time-spent-with-mobile-2019

Chapter three

Television: From Analog to Digital to 8K

The authors thank Glenn T. Hubbard, Ph.D. (East Carolina University) for his contributions to this chapter.

SEE IT THEN

The Experimental Years

Television as we know it today came about after years of extensive research and testing. Early inventors had conceived of the idea of some sort of way to bring distance ('tele') into our sight ('vision'). Thus, the notion of 'television' was born, but it took years to bring it to fruition. Inventors toiled day and night trying to come up with a system that could broadcast video signals through the airwaves. At first it seemed that a mechanical system was the way to go, but further experimentation uncovered various flaws and provided an inroad for a more sophisticated electronic system. Even after the first programs astonished the viewing audience, it took another 50 or so years to sharpen the picture from blurry black and white to sharp color, capable of showing every hair of a man's beard, the sweat on a baseball pitcher's brow, and a woman's every eyelash. Thanks to the innovators' tenacity and knowledge, we now enjoy 'now' television picture clarity beyond anything imagined in the late 1800s when experimentation first began.

Early Innovations

While some experimenters and inventors worked with radio waves to send audio across distances, others were more interested in transmitting live pictures. Experiments in television began in the 1880s. Early thinking about how to send pictures using electricity was divided between two methods: Mechanical scanning and electronic scanning.

In Germany in 1884 Paul Nipkow developed the first mechanical scanning system. Mechanical scanning utilized a spinning metal disc system that used one disc to record a visual image and another disc to view the image. Mechanical scanning was a primitive system that produced rather rough and undefined images. It was not until 1926 that a workable system for sending live images was perfected by Scottish engineer John Logie Baird. The British Broadcasting Corporation (BBC) adopted Baird's mechanical system and began broadcasting television programming in 1936.

By today's standards, the John Logie Baird system was primitive, using only 30 horizontal lines of resolution, each line consisting of minuscule black-and-white dots that form the picture. The biggest problem with the mechanical system was that at only 30 lines of resolution the picture was a bit blurry and there did not seem to be a way to improve it. But in 1922, Philo T. Farnsworth, an Idaho high school student, sketched out a system for electronic television that did not need to use the spinning discs of mechanical television. While the UK was standing behind mechanical scanning, attention in the U.S. was looking to electronic scanning, which held promise of delivering a clearer picture. Eventually, the UK dropped Baird's mechanical system in favor of electronic scanning, which yielded more than 30 lines of resolution. The more lines, the clearer the image.

Farnsworth joined the Navy after high school but left for the University of Utah after he learned that if

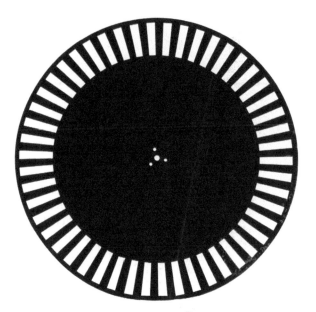

ZOOM IN 3.1

For more information about Philo T. Farnsworth and the beginnings of television go to: www.thehistoryoftv.com/mztv-ptf-tour

Figure 3.1 The Baird disc was part of the Baird mechanical television system.
Source: Photo courtesy of MZTV Museum

Figure 3.2 An actor performs in an experimental television studio in 1928.
Source: Photo courtesy of MZTV Museum

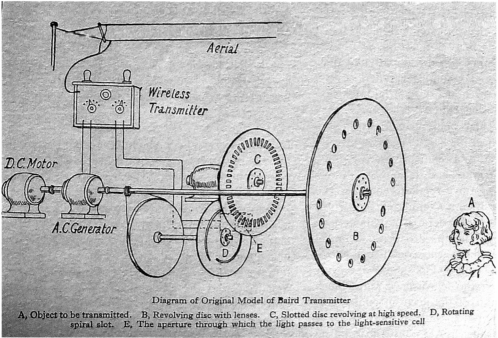

Diagram of Original Model of Baird Transmitter
A, Object to be transmitted. B, Revolving disc with lenses. C, Slotted disc revolving at high speed. D, Rotating spiral slot. E, The aperture through which the light passes to the light-sensitive cell

Figure 3.3 British family watching a 1930 Baird Televisor, a mechanical scanning television.

he pursued electronic television while in the military, any patents he developed would become government property. Farnsworth made many connections and eventually lined up backers to fund his research into electronic television. He set up a lab in Los Angeles, where he continued his work developing an electronic scanning system.

With more experience than Farnsworth, Westinghouse researcher Vladimir K. Zworykin beat out Farnsworth by coming up with a working electronic television scanning system in 1923. Zworkin's system produced a clearer picture by using an *iconoscope*, a special photosensitive camera tube that converted light into electrical energy.

By the early 1930s, Zworykin had been granted a number of patents that improved the electronic system, but he could not get his superiors at Westinghouse to pay much attention to his work.

RCA managers, however, were interested and recruited Zworykin to work for them. RCA was also eyeing Farnsworth's work and tried to recruit him as well, but Farnsworth instead joined the Philco company and moved to Philadelphia. Farnsworth stayed at Philco for only a few years, and he eventually returned to the laboratory in Los Angeles. Zworykin and Farnsworth were competing head to head to develop the best system and ended up in several patent battles. Although he gained some smaller victories, Zworykin lost out to Farnsworth when RCA agreed to pay to license Farnsworth's patents for a workable electronic scanning system. But Zworykin was still researching television. He developed the iconoscope, a cathode ray that substantially improved picture brightness.

In 1930, the leaders in radio technology – RCA, GE, and Westinghouse – joined forces to develop electronic television. Zworykin worked with engineers from RCA and GE, and by 1936, an experimental television station – W2XF in New York – began transmitting television pictures.

Development of electronic television continued throughout the 1930s. In 1939, a 441-line resolution electronic picture had been developed, and television made its debut when President Franklin D. Roosevelt gave the first presidential television address from the World's Fair.

Although RCA had been broadcasting at a different standard since 1939, by 1941, the television picture had improved to a relatively sharp-looking 525-horizontal-line picture, and stations were transmitting programming on a regular schedule. The National Television System Committee (known as the NTSC) advised the Federal Communications Commission (FCC) about the broadcast technical standards for operation, allowing commercial television broadcasting to begin by FCC approval on July 1, 1941. Compared to radio, television required much more space (bandwidth) on the electromagnetic spectrum. Television required 30 times as much space as FM and 600 times as much as AM radio. The technical standard for broadcast television required 6 MHz of spectrum compared to AM radio that requires only 10 kHz and FM which requires just 200 kHz.

Broadcast Television and World War II

Commercial television broadcasting was ready to begin business in 1941, but U.S. involvement in World War II essentially halted its development. In early 1942, the federal government noticed that the manufacturing of television stations and receivers

Figure 3.4 Vladimir K. Zworykin with the iconoscope, the cathode ray television tube he invented.
Source: Photo courtesy of MZTV Museum

used materials and equipment that could be used for the war effort, especially in the production of radar equipment, and so it stopped television manufacturing and thus television broadcasting almost entirely.

As World War II drew to a close, resources became plentiful, and restrictions on television set manufacturing were gradually removed. Television, which had been talked about by many but seen by few, was about to get a real test in the marketplace. Yet even after the war ended, it took almost two years to resume television station construction, set manufacturing, and broadcasting.

Off to a Slow Start

In 1945, there were only six television stations on the air, and three years later, on January 1, 1948, there were only 16. There were many reasons for the initial slow growth of television, but perhaps the most important was that building a television station was difficult. Television added pictures to the sounds, making the construction of a broadcast facility much more complicated. Television required more space, more equipment, and more personnel than radio. Investors were concerned that not many people

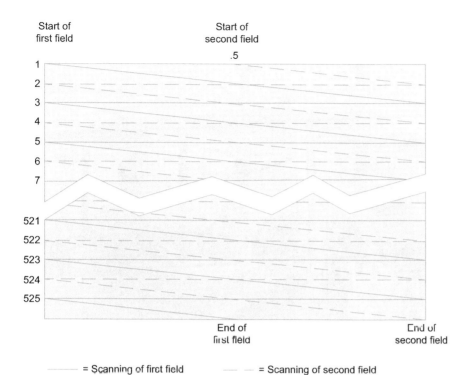

Diagram 3.1 The diagram shows the scanning lines developed in electronic scanning for NTSC television. Each frame of the video was composed of two fields, or half frames with 262.5 lines. The first field began in the upper left and ended in the middle of the screen at the bottom. The second field began in the middle of the screen at the top and continued to the bottom of the screen at the lower right. The frames were shown at 30 per second, fast enough to not be noticeable to the viewer.

Figure 3.5 Philo T. Farnsworth with an early television pickup tube.
Source: Photo courtesy of MZTV Museum

owned television sets and viewers were not going to buy a set unless there were good programs to watch. It was not until about 1948, when the U.S. economy began to boom after World War II and prices of television sets became more affordable, that the number of new programs increased substantially, which set the stage for massive growth in the television industry.

By late 1948, there were still only 34 stations on the air, but numerous applications were being submitted to the FCC for new licenses. Many of these applications came from AM radio broadcasters who wanted to start television stations. Some newspaper companies were also interested in adding a television station to their holdings. Several newspaper companies built powerful stations that have stayed on the air for many years. For instance, WGN (whose call letters are also an acronym for the **W**orld's **G**reatest **N**ewspaper) in Chicago was founded by the Tribune company, then owner of the *Chicago Tribune* and other papers. Station WTMJ was established by the owner of the *Milwaukee Journal*, and WBAP in Ft. Worth, Texas, was founded by the Ft. Worth *Star Telegram*.

Similar to what happened in the early days of the feature film industry, the new television industry offered opportunities for many people. Veterans returning from the war who had radar experience often became television engineers. Others moved from camera operation to director to producer in a matter of months. The race was on to provide as

Figure 3.6 David Sarnoff, president of NBC, makes an introductory speech for live television at the 1939 World's Fair.
Source: Photo courtesy of MZTV Museum

Figure 3.7 A 1939 RCA television.
Source: Photo courtesy of MZTV Museum

many television programs as possible, but the talent pool of people qualified to work in television was still quite small.

The broadcast system for television was similar to that of radio: A signal was sent out from a single antenna to many receivers in a given geographic area; the one-to-many model of mass communication. Television stations were interconnected or networked from a single source, with the television network headquarters typically in New York. The network sent out the programming signal to the affiliated stations via telephone wire. Individual stations received the signal and broadcast it via a transmitter and antenna. Within a given time zone, the signal was sent out simultaneously to many stations, and the program was aired at the same time on all stations. Television stations in the western part of the U.S. received a later version of a live show.

Figure 3.8 This image of Felix the Cat was the result of early electronic scanning experiments.
Source: Photo courtesy of MZTV Museum

From 1946 until the late 1950s, a show that was broadcast live in the eastern half of the U.S. had to be reperformed for broadcast in western states. For example, a live variety show on the air at 8:00 p.m. EST was reshot for live broadcast at 8:00 p.m. PST. This practice changed when videotape was used to record a program and hold it for broadcast later in the Mountain and Pacific time zones.

The Big Freeze (1948–1952)

The post-World War II audience demand for television sets and programming was the catalyst for the increased number of new applications for television licenses, especially in the large markets, where available channels were scarce because there were only 12 channels located in the VHF band for television. The number of new station applicants overwhelmed the FCC, and though it had a set procedure for allocating radio stations to markets, it simply was not prepared to deal with licensing television stations. And so, in September 1948, the FCC essentially threw up its hands and yelled, "Freeze!"

The FCC put a six-month freeze on the application process to reconsider the licensing process and other issues associated with the industry. Owners whose applications had been approved before the freeze, however, were allowed to continue constructing their stations and to begin broadcasting. The FCC expected that it would take six months to resolve the issues, but it actually took about 3.5 years. The *Sixth Report and Order* was signed on April 15, 1952, officially ending the freeze. Three major issues were settled as follows:

1 **Additional UHF channels**. Before the freeze, all stations were licensed to the very-high-frequency band, or VHF channels, which were channels 2 through 13. The *Sixth Report and Order* made the ultra-high-frequency band (UHF), consisting of channels 14 through 83, available to television stations across the United States. The high end of the UHF band, channels 70 through 83, is reserved for *translators*, which receive television signals from stations and retransmit them on different frequencies, and *repeaters* that receive a television signal then retransmit it to improve reception in cities distant from the city of license.

Broadcasting television signals on the UHF band seemed like a great idea, except that none of the television sets manufactured up to that time could receive those channels without using a UHF converter. VHF tuners clicked into place on channels with each turn of the dial, but UHF converter dials were often not easy to operate, and they required time-consuming and sometimes frustrating channel adjusting. Stations that were given licenses to broadcast on the UHF band were disadvantaged in comparison to their VHF rivals simply because most television sets could only receive VHF programs, and viewers were reluctant to buy clumsy-to-use converters. Moreover, UHF signals do not travel as far as VHF signals. Even though UHF stations pumped out more power than VHF stations just to reach the same geographical area, in many cases viewers also had to buy a special UHF antenna to pick up the signal. Exasperated viewers found themselves up on their roofs adjusting their VHF antenna in one direction and their UHF antenna in another, and even combo VHF and UHF antennas required multi-directional adjusting.

Congress passed the All-Channel Receiver Act in 1962. The Act authorized the FCC to mandate that all television sets manufactured in 1964 and beyond have the capacity to receive both VHF and UHF stations without a UHF converter. Despite this legislation, UHF stations were still thought of as inferior to VHF. In fact, a common saying was that getting a VHF license was like getting a "license to print money," while UHF stations often lost money.

2 **Educational channels.** As part of *Sixth Report and Order*, the FCC reserved 242 channel spaces for educational television, which made up about

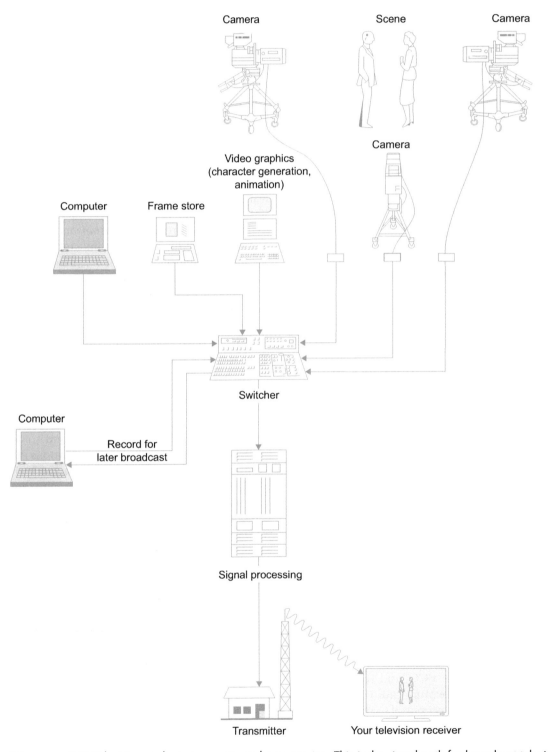

Diagram 3.2 Television studio to transmitter to home service. This is the signal path for broadcast television.

12% of the 2,053 channels allocated for television station use across the United States. Unlike the allocation set aside for FM non-commercial radio (i.e. 88.1 to 91.9 MHz), these channels were not located within one part of the television band but rather were spread throughout the UHF and VHF bands. Although commercial entities opposed this generous allotment of educational channels on the grounds that it made commercial channels less available in many cities, the ruling was upheld.

3 **Color television**. In the 1950s, while the public was just beginning to get used to the idea of black-and-white television, the networks were experimenting with full-color broadcasting.

Two systems for color television emerged and both fit into the existing 6 MHz of channel space that was allotted for every television station. The CBS system used a mechanical color wheel that transmitted a color signal. There were several drawbacks to the CBS color system. Any program that was broadcast in color using the CBS system could not be seen on a black-and-white set, even in black and white. Despite these disadvantages, in 1950 the FCC announced support of the CBS color system. The public, however, was not quite ready to buy color sets, nor had CBS or its manufacturing partners even produced them in the months following the FCC's decision. Very few programs had been produced for color broadcasting, and very few viewers could afford color television sets, which were very expensive.

The FCC Rethinks Color Television

Even the FCC's support of the CBS system did not last long. A little over two years later, the FCC reversed its decision in favor of RCA's electronic color system. No doubt David Sarnoff, head of RCA, also lobbied the FCC for acceptance of his company's system. Because almost all television cameras were capable of producing only black-and-white images, very few programs were made for broadcasting in color. Also, neither CBS nor ABC were strongly motivated to support the RCA system because RCA owned their rival network, NBC. The strong point of the RCA system was that programs broadcast in color could be picked up by black-and-white sets, though the picture was still in black and white. Because of this compatibility, viewers did not have to run out and buy color sets, and most did not. The viewing public was generally satisfied with

Figure 3.10 The first color television was made by RCA in 1954 and sold for around $1,000 – or about $6,000 in today's money.
Source: Photo courtesy of MZTV Museum

Figure 3.9 In some cities, multiple antennas were needed to receive the broadcast stations. In this picture, there is a VHF antenna to receive channels 2–13 and several directional UHF antennas to receive channels 14–83.
Source: Photo courtesy of shutterstock.com/XL1200

receiving color-produced programs in black and white on their black-and-white sets and early color sets required some fine tuning on each channel. It took more than 20 years for the quality of color to improve and the price of color sets to come down for color television to become standard in most homes. The NTSC (RCA) system remained in use until the U.S. replaced it with digital television in 2009.

Domination of the Networks

Beginning in the early days of radio and continuing into the era of television, the networks exerted quite a bit of control over their affiliated stations. An **affiliated station** provides its airtime (known as *clearance*) and its audiences to the network, and the network provides a dependable schedule of high-quality shows to the local affiliate for most of the broadcast day. In addition, the affiliate is paid for its airtime. This practice, known as *station compensation*, is based on the size and the demographic make-up of the audience delivered by the station to the network, its advertisers, and the competition for the station's affiliation. But with contemporary affiliations, sometimes the stations pay the networks for the privilege of airing network programs.

Network programming provided big-name stars from Hollywood and New York. Without the big-name stars and high-quality programs, a non-network local station was limited to showing station-produced programs that drew small audiences. Left to their own programming, local stations were nothing special – often a so-called mom-and-pop operation owned by a group of small businesspeople who could not meet the viewers' needs. As had been the case in the 1930s and 1940s when the networks dominated radio programming, they also controlled television programming from its inception. The fortune the networks made from radio was used to bankroll television, which ironically cannibalized radio's audience and its advertising dollars and revenue.

Network Relationships With Affiliates

During the four years of the freeze, existing stations scrambled to affiliate with the two powerful networks, CBS and NBC. Thus, two years after the freeze ended, CBS and NBC had more than three-quarters of all stations that were affiliates. ABC, which was formed after NBC divested its Blue radio network in 1943, was always a distant third in number of stations and audience size. ABC was so financially strapped that in 1951 it merged with United Paramount Theaters to receive a cash infusion and stay in business. A fourth network, the DuMont network (owned by a television set manufacturing company), affiliated with stations mostly in medium- and small-sized markets, but without large audiences it could not sustain itself and ceased operation in 1955.

Independent stations were forced to either produce shows on their own or pay for programs from independent producers or syndicators. In the early 1950s, quality programming came almost entirely from the television networks. It was easier and less expensive to get programs from a network than to produce them at the station or to get them from other sources. More important, network programs were usually of higher quality than locally produced or independently produced programs. Some local television stations produced news, public affairs, children's, and sports programming, but drama, situation comedies, and even variety shows were too expensive for most local stations and thus were produced by the networks.

Network affiliation, thus, was highly valued. The networks had their choice of stations in a given market and were in a very strong bargaining position with their affiliates. Most television markets only had three stations. If there was an independent station, it was probably a newcomer to a market or broadcast on a less desirable UHF channel. The networks also had quite a bit of freedom from government regulation. Although local stations were regulated directly by the FCC, the networks themselves were not under the purview of the FCC as such, because only stations used publicly owned airwaves. The affiliations between local stations and networks were renewable every year. However, from the post-World War II years until recently, an affiliation with a network usually lasted for many years. Often, the relationship between a network and an affiliate began in the very early days of the station's existence and remained unchanged.

In the early days of television, NBC and CBS were both financially strong, had highly rated VHF affiliates, and an inventory of programs that were set to make the transition from radio to television. The shows that moved from radio often took their sponsors and listeners with them to television and created an instant audience of loyal television viewers.

If there was a problem with network programming, it was that television was really radio with

Figure 3.11 Advertisements for early television sets.

Figure 3.11 Cont.

pictures. Many of the same programs, with the same stars, switched from radio to television. Although this transition was comfortable for the audience, it did not encourage much experimentation or the development of new program types and styles. Nevertheless, the years after World War II were considered the Golden Age of television, in terms of programs, when audiences and advertisers flocked to the tube. In the 1950s, television was severely restricted by technological factors. Cameras were large and heavy, and strong lighting (which generated quite a bit of heat) was required to get a good video image. Portable video cameras did not exist. Although about one-quarter of the prime-time programs were recorded on film, or by the poorly reproduced Kinescope, most shows were produced live.

FYI: KINESCOPES

In the early days of television, Kinescope recording was the most common way to preserve live television programs for airing across time zones. Developed by Du Mont, NBC, and Kodak, Kinescopes stored visual images by aiming a film camera at a television monitor as the show was airing live. The recorded broadcast could then be re-shown in a different time zone. Kinescope recordings were of poor quality and did not hold up as well as programs recorded on film.

After videotape became available in 1956, Kinescopes ceased to be a viable medium for storing video programs. Kinescopes are now a collectors' item.

See Kinescope images on this short video: www.youtube.com/watch?v=N_TavkpMaXg

Blacklisting and Broadcasting

After World War II, the U.S. government and the public in general became very aware of the growing power and nuclear capabilities of the communist-controlled Soviet Union. In addition, Americans feared that communism was spreading in many parts of the world. The general attitude toward communism was not just that it was different from the American system but that it was a political ideology that would be used to take over the world. Politicians seized on these fears and used them for political gain. In the early 1950s, a small group of former Federal Bureau of Investigation (FBI) agents published a newsletter called *Counterattack*. Its purpose was to encourage Americans to identify and even shun people who demonstrated sympathy or ideological agreement with communism. Another publication, *Red Channels: The Report of Communist Influence in Radio and Television*, described the communist influence in broadcasting and named 151 people in the industry who supposedly had communist ties. The names of these people were put on a blacklist, and they were essentially no longer permitted to work in broadcasting or related industries. *Red Channels* was published just as North Korea, a communist country, invaded South Korea. The United States got involved in the conflict as it sent American troops to Korea, and hence the nation's role in stopping communism began.

The Korean War, which lasted from 1950 to 1953, reinforced many Americans' beliefs that communism had to be stopped, and the idea that communism had infiltrated broadcasting triggered efforts to expose and eliminate communist influence. Although no celebrity or individual employed by the television industry could be linked to the Communist Party, the mere listing of a person's name in *Red Channels* prevented him or her from continuing a career in broadcasting. The networks and advertising agencies even employed individuals to check the backgrounds of people working in the industry. If a person's name showed up in any list of communist sympathizers, he or she would not be hired to work for a network, an advertising agency, or any project or program. New employees were expected to take a loyalty oath to the U.S. before working. People who were suspected of communist activities were often forced to confess (even if they were not a communist) and name their communist associates.

The most prominent of the politicians who used Americans' fear of communism to strengthen his own political power was Joseph McCarthy, a U.S. senator from Wisconsin. He was a savvy manipulator of news media, timing his revelations of supposed communist infiltration in the U.S. government to coincide with newspaper deadlines around the country, maximizing the national news value of his nebulous claims. He also used televised congressional hearings about communism in the U.S. Army to further his notoriety. Eventually, McCarthy's tactics caught up with him. Edward R. Murrow devoted an entire episode of his news magazine show *See It Now* to exposing McCarthy's unfair tactics, which often had the effect of damaging the reputations of private citizens. In Murrow's March 9, 1954 program, excerpts of speeches given by McCarthy, as well as his relentless questioning of apparently innocent witnesses, were replayed to uncover his inconsistencies and his attacks. In the end, McCarthy was shown to be a bully who ignored fact and used innuendo to level accusations at his adversaries and innocent people who worked in the media. McCarthy himself was censured by his fellow senators and ended up a pariah.

ZOOM IN 3.2

Watch this short video about blacklisting. See the top Hollywood actors of the day who were accused of communist activities fight against McCarthyism.

www.youtube.com/watch?v=Slga_nUrpYg

The *Red Scare* created a chilling atmosphere for broadcasting and its employees. The cloud of blacklisting continued until 1962, when radio comedian John Henry Faulk, who had been blacklisted in 1956, won a multimillion-dollar lawsuit against AWARE, a group that had named him as a communist sympathizer. The effect of the lawsuit was that the blacklisting practices went from up front and public to secretive and private. The *blacklist* became a *graylist*, which was used in broadcasting to identify people who might have subversive ideas, especially those who embraced communism. The practice of graylisting lasted into the 1960s.

Cable Television: Television by Wire

The development of cable is best understood by going back to the late 1940s, when the television industry was just beginning. Television found an eager audience for its programs, many of which were taken directly from radio and modified for the screen. Viewers wanted to *see* their favorite radio stars and *watch* the new shows, such as the *Texaco Star Theater*, starring Milton Berle, and *Your Show of Shows*, with Sid Caesar, Imogene Coca, and Carl Reiner.

The public heard all about television and viewed some programs in their neighbors' homes, appliance stores, or bars. They were eager to buy sets and begin watching in their own homes. The problem was that television stations were located primarily in large cities such as New York, Philadelphia, and Chicago.

Broadcast television signals were (and still are) sent from a station's antenna to receiving antennas connected to television sets. Television antennas located in rural areas and other smaller markets outside major metropolitan areas sometimes received a signal from one or more major market stations with the help of small transmitters located nearby. These retransmitters, called translators, received the television signal from a station and then retransmitted it to antennas in the rural area. Translators would send the signal out on a special frequency to avoid interfering with the signal of the originating station.

Translators were not used in the very early days of television, however, and many of the people who lived in areas far from big cities or where the television signal was blocked by hills or mountains could not receive television signals. The desire for television led to the birth of an alternative delivery system: Community antenna television, commonly known as cable television.

Community Antenna Television

As the saying goes, 'Necessity is the mother of invention,' and in this case, it was the lack of a television signal that led to invention. The idea of sharing or distributing television signals first started in New York City when apartment dwellers found that other buildings blocked the signals and they could only get television by placing an antenna on the roof of the building. Sharing an antenna system solved these problems. The *master antenna system* was used to receive the television signal and to distribute it to the apartments through a system of wires. This type of strong antenna was the forerunner of cable television delivery (CATV).

There are competing stories about who first came up with a specific solution to the problem of

Figure 3.12 Advertisement for the 1958 Philco Predicta.

reception of distant television signals. Some contend that John Walson, an appliance storeowner in rural Pennsylvania, was the first to bring *community antenna television* (CATV) to his community via cable, yet others claim that L. E. Parsons of Astoria, Oregon, was the first to do so. Essentially, both of these men came up with the idea of placing a television antenna on top of a hill or mountain to bring in television signals from distant stations. In 1948 Parsons set up a television antenna on top of a hotel in Astoria to receive the signal of a Seattle station. He then connected a long wire from the antenna to his apartment. Local interest grew considerably when word got out that Parsons had the only television set within 100 miles that could receive a signal. He also ran a wire from the antenna to the television set in the lobby of the hotel. Consumer demand led to Parsons wiring many homes, which would have been most easily accomplished by stringing wires on utility poles, though he did not initially have permission to use them. Eventually, Parsons obtained permission to run wire through the underground conduits to businesses in downtown Astoria. The result of Parsons' work was a somewhat primitive delivery system that nonetheless became a viable business – the Radio and Electronics Company of Astoria. Parsons consulted with other entrepreneurs who wanted to set up similar systems elsewhere in the country. But bringing in television was just part of the plan; making money was the other incentive.

Appliance store owners in smaller cities and towns had a hard time selling television sets if there were no signals to pick up. Bob Tarlton, another early community antenna entrepreneur, connected his appliance store in Lansford, Pennsylvania, to an antenna on top of a nearby hill, presumably to demonstrate television so he could sell more sets. He was the first to conceptualize CATV as a local business.

Microwave Transmission

CATV systems eager to expand their reach came up with yet another way of doing so – microwave transmission. By using microwave signals, a CATV system imported television signals from distant large markets rather than just from the closest market. Microwave signals were especially important in the western part of the United States where cities are far apart. For example, a CATV system in Flagstaff, AZ could pick up signals from a television station in Phoenix (144 miles away). Bringing in distant signals gave the audience more programming choices and thus a reason to buy a set and subscribe and pay for CATV.

Community Antenna Becomes Cable Television

Community antenna television was an idea that grew out of strong demand for television from an audience hungry for entertainment. In some ways, broadcast television was simply 'hacked.' Instead of a simple community antenna system that picked up the broadcast signals from a nearby town, community antenna systems became complex cable television systems that added signals from distant markets and eventually added non-broadcast television channels to their offerings. Community antenna television was the forerunner to the more sophisticated cable television of today.

Reining in Cable

Whether or not a local station was carried by a local cable system was a decision made on the whim of the cable system's owner. As more and more viewers subscribed to cable, local stations not carried by the system were losing their audience and, thus, advertising revenue. Local stations lobbied the FCC to compel cable systems to transmit local station programming. The FCC agreed, and in 1965 it issued the **must-carry rule**, which stipulated that cable systems must carry signals from all 'significantly viewed' stations in their market. In other words, cable systems had to carry local broadcast stations on their systems.

The cable industry argued that FCC restrictions hampered its development by forcing it to use its limited number of channels to carry certain stations even though they were not in demand by viewers. The FCC capitulated to cable and eased its regulations, dropping most *carriage rules* for small cable systems (fewer than 3,500 subscribers). Dropping the carriage rules allowed unrestricted importation of foreign-language and religious programs. In 1972, the FCC adopted a new policy that was a result of the commission's continuing desire to preserve local broadcast service and to create an equitable distribution of broadcast services among the various regions of the country. The legislation created the *non-duplication rules* (or *syndicated exclusivity*) that prevented cable systems from showing a *syndicated* program (reruns of a show that formerly aired in primetime on a network) from a distant station if a local station in the market was airing the same program. In addition, the FCC ruled that new cable systems had to have at least 20 channels and two-way capacity, allowing signals to travel both to and from the audience, eventually making cable Internet service possible. The syndicated exclusivity rules were dropped in 1980 after causing eight years of problems for cable operators wishing to include more program options in their line-ups.

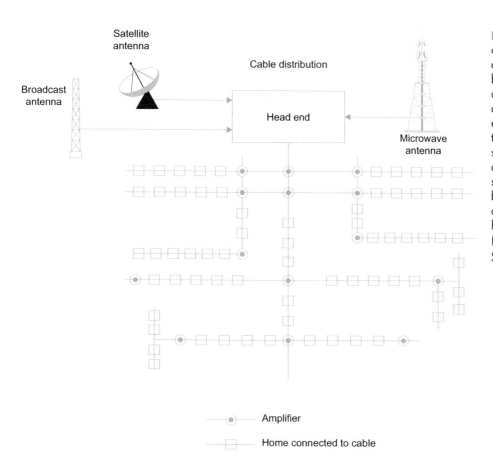

Diagram 3.3 Cable distribution to homes in a city. Signals are received by satellite, microwave, and broadcast antennas and combined at the head end. From the head end, the signal is sent through a system of wires that have amplifiers (i.e. the round symbols in the drawing) to boost the signal strength and continue the signal to houses connected to cable (i.e. the square symbols).
Source: Franklin, 2004

FYI: 'RENTING A CITIZEN' FOR CABLE

Some cable companies tried to influence city officials to choose them as local providers with attractive stock deals. For instance, in 1982, a Denver councilwoman received a terrific stock deal from one of the cable franchise applicants and observed that the only person she knew who did not own cable stock was the coach of her son's little league football team. (She later discovered that he did not live in the city of Denver.)

And in Milwaukee in 1982, an applicant for the cable franchise retained two local political consulting firms: One to help pitch its service as the city's provider and the other to prevent the first firm from helping any of the other applicants win the franchise license (Hazlett, 1989). Using the promise of stock in a company in return for granting an operating license was quite common.

Upheaval and Education

The Tumultuous 1960s

In 1961, recently appointed FCC Chairman Newton Minow stated at the National Association of Broadcasters convention that television programming was a "vast wasteland." Critics latched onto that remark as accurately depicting the quality of the programming offered by most television stations. Although few would dispute that television has traditionally offered many hours of intellectually light programs, it has and still does serve a function beyond pure entertainment. For example, television allowed Americans to witness history during the tumultuous decade of the 1960s. This journalistic function both solidified the importance of television in American society and gave it real credibility as a provider of valuable information.

Some of the many events covered by television during the 1960s included the presidential debates between candidates John F. Kennedy and Richard M. Nixon; the 1962 Cuban missile crisis; the 1963 assassination of President John F. Kennedy and the subsequent killing of his accused assassin, Lee Harvey Oswald; the 1968 assassinations of Martin

Luther King Jr. and Senator and presidential candidate Robert F. Kennedy; and the first walk on the moon in 1969. In addition, continuing coverage of the escalating war in Vietnam, domestic unrest regarding racial discrimination and women's rights, and violent anti-war and anti-government demonstrations during the 1968 Democratic Convention in Chicago gave the public a view of historical events in an up-close-and-personal way never before possible.

Viewers were deeply affected by what they saw on television – for example, live footage of rioting in the streets and the horrors of war direct from the battlefield. In addition, the proliferation of fictional violent action shows on television led many to wonder whether the media were somehow encouraging people to behave in violent ways. The U.S. government responded by creating a research commission, the Commission on the Causes and Effects of Violence.

The results from the committee, published in early 1972, stated that violence on television might influence some individuals to behave violently. Although the findings fell short of pointing to television as the *cause* of increased violence in society, they fueled the efforts of citizen action groups like the Action for Children's Television (ACT), which sought to focus congressional attention on the content of television and its effect on children.

ZOOM IN 3.4

Action for Children's Television (ACT) was a national grassroots organization founded in 1968 by Peggy Charren, who worked in the creative arts field and with reading programs for children. The goal of the organization was to raise the quality and diversity in television programming for children and adolescents. In addition, the group sought to limit the number of commercials directed at young audiences. The organization had thousands of members across the United States and had a major impact on the content and scheduling of children's television programs and advertising. Thanks to the efforts of ACT, the U.S. Congress passed the Children's Television Act of 1990, which required stations to air at least three hours of educational programs for children, and to make television more child-friendly. After achieving their main goals, ACT disbanded in 1992.

Educational Television Goes Public

During the television freeze that followed World War II, the FCC was lobbied both by commercial broadcasters and the Joint Committee on Educational Television (JCET) regarding non-commercial television stations. The commercial broadcasters tried to prevent the FCC from reserving television channels for non-commercial television stations, while the JCET lobbied for non-commercial television use. As part of the *Sixth Report and Order*, which ended the station-licensing freeze, the FCC increased the number of channels reserved for non-commercial stations from 10% to 35% of the available station allocations (242 channels – 80 VHF and 162 UHF). Since then, the FCC has increased the number of channels dedicated for non-commercial stations.

In 1959, non-commercial television producers formed the National Educational Television (NET) network to operate as a cooperative, sharing venture among stations that sent prerecorded programs by mail on Kinescope film and later on videotape. After one station aired a program, it sent it to the next station, and so on. This inexpensive and low-tech network, which became known as a **bicycle network**, allowed stations in different locations to air the same programs, but not at the same times because they did not have the films at the same times.

In 1967, the **Carnegie Commission on Educational Television (CET)** – a group composed of leaders in politics, business corporations, the arts, and education – published a report about non-commercial television that recommended that the government establish a corporation for public television.

As a result of the Carnegie report and recommendations, the **Public Broadcasting Act of 1967** was passed by Congress. The Act provided for the establishment of a non-profit organization to acquire, distribute, and promote educational programming for both radio and television. The **Corporation for Public Broadcasting (CPB)**, was formed a year later. The CPB's responsibility was to create and provide financial support for a network of educational radio and television stations. The CPB formed **National Public Radio (NPR)** and the **Public Broadcasting Service (PBS)**. Radio and television stations that affiliate with NPR or PBS receive networked programs to air, or they can produce their own local programs. Both NPR and PBS receive government funding and public donations to support their stations and program producers. As non-profit entities, the networks are not beholden to advertisers or the government. PBS stations began airing network programs in 1969, including *Sesame Street*. NPR hit the airwaves in 1970.

NPR and PBS are unlike the commercial networks because they do not sell advertising time, instead they are funded by a variety of national, regional, and local sources. For PBS, audience donations account for about 25% of the funding. State governments provide about 18%, and CPB (along with federal grants and contracts) adds about 16%. Businesses add another 16% through underwriting (company mentions on shows), state universities and colleges donate more than 6%, foundations and miscellaneous sources provide an additional 21%. PBS's programming philosophy is oriented to providing programs of cultural and educational interest.

Although NPR and PBS stations do not sell advertising, they do seek **underwriting**, which is similar to advertising, but without a call to action. In other words, an underwriting announcement may describe a product or service in factual terms but may not urge the audience to purchase or call or take any action. A phrase like, "call 888 555-1212 now for a 20% discount," is not allowed. These restrictions are put in place by the FCC, because donations to non-profit NPR and PBS stations are tax deductible, thus an underwriter should not profit from both the tax deductions and increased sales through blatant promotional messages.

Because the president of the United States appoints the CPB board members and executives, the organization sometimes gets embroiled in politics. For example, President Richard Nixon vetoed a funding bill for CPB because he did not like that it allowed PBS to air information programs that showed his administration in an unfavorable light. Nixon forced the resignation of several CPB officials, who were replaced by people with a more favorable view of the Nixon administration. Although Nixon's strong hold on PBS was somewhat unusual in the history of CPB, it shows that though public radio and television are supposed to be non-partisan, independent entities, they are not protected from partisan politics and politicians.

Satellite Delivery

The idea of man-made satellites orbiting the earth goes back hundreds of years to Sir Isaac Newton. Scientists and science fiction writers of the 20th century imagined using satellites for military and communication purposes. Satellites finally became a reality when the USSR launched the first man-made satellite into outer space on October 4, 1957. Named Sputnik, the satellite had many uses including monitoring radio signal distribution. The U.S. entered the satellite age when it launched

Explorer 1, on January 31, 1958. Explorer 1's main mission was to measure the level of radiation in the earth's orbit. The idea of a communication satellite that could capture a signal sent up from earth and echo it back to another part of the planet became a reality on August 12, 1960, with the U.S. launch of Echo 1. The notion of using satellites to distribute television programs was still 12 years away, with the advent of Home Box Office, the first satellite-delivered television network.

ZOOM IN 3.5

The idea of using satellites for communication purposes was first publicized in a 1945 article appearing in *Wireless World*, written by novelist and scientist Arthur C. Clarke. He theorized that three satellites in geostationary orbit (22,300 miles above the equator) could be used to relay information to the entire globe (Clarke, 1945).

HBO

In 1972 the television industry was changed by **Home Box Office (HBO)**. Initiated by Time, Inc.'s Sterling Manhattan Cable in New York City, HBO was the first service to deliver television programming via satellite. HBO placed a transponder on an existing satellite, Westar 1. HBO then uplinked a program signal to the transponder. The signal was then downlinked to HBO's satellite dish receivers. Not only did HBO bring subscribers live sporting events and full-length movies, but it also specialized in programming that was not offered, or offered minimally, by the networks, and it later created its own original programming.

But HBO did not start out as a satellite network, it originally delivered shows via microwave transmission, which requires a system of towers to relay the signal. Once HBO began delivering and charging for programs it delivered via satellite, the FCC stepped in to try to regulate content and other business matters. It took some legal wrangling (*HBO v. FCC*, 1977) before HBO was relatively free of the FCC. Rules regarding programs that could be shown on cable and HBO were relaxed, thus opening the door for other cable companies to offer satellite-delivered (premium) channels to their customers, for an extra fee. HBO, thus, became the first company to deliver non-broadcast, original programming via satellite.

HBO is known as the first **pay-cable channel** because it delivers its programs via satellite to a cable service that in turn delivers the shows to its cable subscribers. HBO was also called a **premium channel** because cable subscribers had to pay extra for it. Other cable-only satellite networks soon followed, such as ESPN (1979) and CNN (1980).

HBO filled the void for new and varied programming and ushered in a new world of television viewing. When satellite distribution of video programs to cable companies became both technically and economically feasible, the cable industry grew rapidly.

Satellite distribution services like HBO ushered in a new way to deliver programs and brought big-screen movies to in-home viewers.

Cable Systems and Satellites

Satellite delivery changed the use and purpose of cable television. Television signals delivered by a satellite can reach many cable company receivers throughout the country at once.

Three different types of satellites are used for television and other communications:

1. Geosynchronous satellites, like the one used by HBO, are parked in an orbit that is 22,300 miles above the earth's equator. These stationary satellites provide direct broadcast satellite video and audio to cable companies, television stations, and homes with receiving dishes.
2. Middle-earth orbiters are satellites that travel in a lower orbit – beneath 22,300 miles but more than 1,000 miles above the earth. These non-stationary satellites are used for voice and data transmission and also for global positioning system (GPS) devices.
3. Low-orbit satellites are non-stationary and travel from 100 to 1,000 miles above the earth. They are used for personal communication services, such as mobile phones, Internet access, and video conferencing.

After HBO became popular the cable industry quickly realized that it could make a fortune by expanding its offerings from simple retransmission of broadcast shows to creating new cable networks and new cable programs. Premium channels, such as HBO rival Showtime, and the first advertiser-supported networks like USA Network, ESPN, CNN, MTV, and BET began popping up in the late 1970s to early 1980s. Different from premium channels, advertising-supported cable channels were not billed separately but were offered as part of a package of programs that included

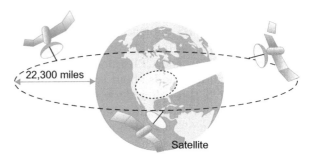

Diagram 3.4 Geosynchronous satellites orbit the earth 22,300 miles above the equator. At that distance from the earth, the satellite stays in the same location relative to the earth. A narrow-beam signal from the satellite creates a large signal pattern or 'footprint' in which the television signal can be received with a small receiving dish.

the broadcast networks. But all of these cable channels work the same way – they are uplinked to satellite from a programming source and then downlinked to a cable service that distributes them to subscribers' homes.

Once viewers took their first look at the new cable channels, they wanted more of them. And viewers who lived in areas that were not wired for cable service petitioned their local governments to get cable service in their communities. Cable systems were expensive to set up and miles and miles of cable wire had to be laid to connect each home to the system. In addition, existing cable systems had to add expensive satellite receiving dishes to downlink the new cable-only channels. But demand was strong. In 1978, there were only 829 **satellite-enabled** cable systems in the U.S. and within two years, that number had spiked to 2,500 such systems.

Satellites and Superstations

Satellite delivery and cable networks were turning viewers' attention away from broadcast television. In response, the broadcast industry started its own hybrid satellite/broadcasting system. Independent stations WGN (Chicago), WTBS (Atlanta), and WOR (New York) started uplinking their signals to satellite. Their signals were then downlinked by cable systems for delivery to their subscribers. Programming from these 'superstations' could be seen in distant cities across the country. Viewers in San Diego could watch a Chicago Cubs game or see a local New York weathercast on the superstations.

FYI: CABLE TERMS

Cable company: A company (e.g. Comcast, Verizon) that constructs the wire system that brings multiple television and audio channels to homes and businesses in a community. A cable company provides the infrastructure that sends television to cable subscribed homes.

Cable system: The cable system refers to the actual wiring through which television and audio signals travel to homes and businesses in a cable-wired community.

Cable channel: A program service that provides television shows to a cable company that it, in turn, distributes to its subscribers. Cable channels (e.g. ESPN, CNN, MTV) are formed for the purpose of creating and delivering shows over cable, but not for broadcasting over the air. No FCC license is required to start and operate a cable channel.

Cable network: This term is used interchangeably with 'cable channel.' ESPN is referred to as both a cable channel and a cable network.

MSO: This term stands for Multiple-System Operator, where one company owns more than one cable system.

Cable systems were required to carry signals from the big three networks (ABC, CBS, NBC) and to obtain those signals from the station in the market closest to the cable system, a requirement that became known as the **antileapfrogging rule** because cable was not allowed to skip over closer stations in favor of those in distant markets. Importing signals from independent stations was similarly limited until the tide turned from favoring more cable regulation to a deregulation sentiment in 1976. When the antileapfrogging rule was eliminated, cable systems could import a wider variety of signals, including signals from certain local stations carried via satellite to cable systems nationwide, called superstations.

Satellite Master Antenna Television (SMATV)

SMATV is an acronym for **satellite master antenna television**, which is also known as **private cable**. Essentially, an SMATV system is similar to a very small cable system. SMATV serves one or several adjacent buildings in a part of a city. Apartment buildings and housing complexes, large hotels and resorts, and hospitals commonly use SMATV systems as an alternative to each unit having to subscribe to cable separately. An SMATV head end is usually placed on the roof of a building. The head end receives the satellite signals. The signals are then delivered to each unit within the building.

Multichannel Multipoint Distribution Systems (MMDS)

Multichannel multipoint distribution systems provide channels that are similar to those available from a cable system but use a microwave signal to send the channels to individual homes. MMDS

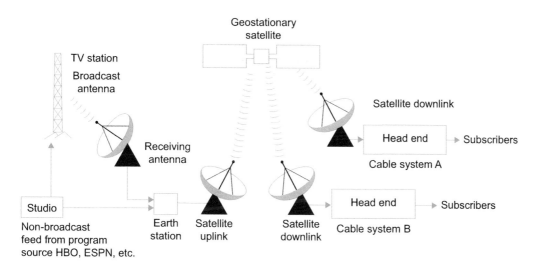

Diagram 3.5 This diagram shows how distant broadcast stations and non-broadcast channels such as HBO, ESPN, and CNN are sent to cable systems.

is used in parts of communities that are not wired for cable. MMDS does not require the stringing of cable but rather delivers television channels via a microwave signal. Subscribers need only a small antenna on their rooftop and a receiver near their television set to pick up MMDS. MMDS is best used in cities with high housing density and flat terrain. MMDS systems were successful, at least for some time and still exist in some locations, although traditional cable and direct broadcast satellite are far more common.

Consumers Get Their Own Satellite Television Dishes

The first satellite receiving dishes were referred to as *TVROs* (TV Receive Only). These dishes measured 8 to 15 feet across. A dish was costly, and because of its size it needed to be installed in a large open space and thus they were useful only to cable companies. The size of a receiving dish was directly related to the power of the satellite transponders and the frequency at which the signal was sent. The C band – the space in the electromagnetic spectrum used by communication satellites – required a large dish to best receive the signals.

Although television receive-only dishes were marketed heavily to the public in the late 1970s and 1980s, they were not appealing to the everyday television viewer. At about 8 feet by 15 feet, the dishes were cumbersome and unsightly. At first, when a TVRO could pick up premium channels (e.g. HBO, Showtime) and some adult entertainment for free, viewers thought it was well worth it to have one. But the cable networks were not happy about what they considered program pilfering – watching premium cable shows for free through a backyard TVRO – so they blocked premium reception by scrambling the signal. TVRO owners were upset that they could no longer receive premium channels and potential purchasers decided it was no longer worth it to buy a TVRO.

But all that changed in 1994 with DirecTV, the first multichannel delivery service that provided television directly to subscribers. A subsidiary of General Motors owned by Hughes Electronics, DirecTV works as a **direct broadcast satellite system (DBS)**. Subscribers could now downlink satellite television programs to a small dish provided by DirecTV instead of buying their own TVRO or subscribing to a cable company. And program packages

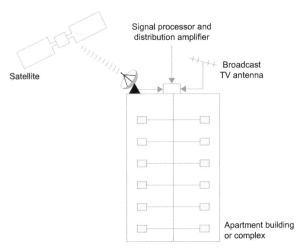

Diagram 3.6 This diagram shows how a satellite master antenna system works. A satellite sends program information and is received by a satellite receiving dish on the top of an apartment building. A broadcast antenna receives broadcast signals from the local broadcasters. These signals, channels of television, are then programmed to channels that are sent to individual apartments. The system is designed to give uniformly good signals to all subscribers in the building and avoid having individual apartments install multiple antennas or receiving dishes on the rooftop.

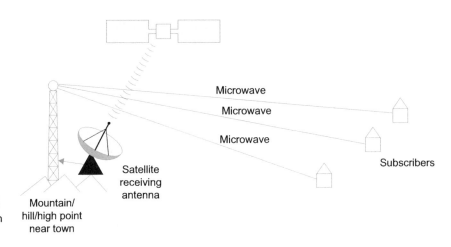

Diagram 3.7 MMDS program content is sent from a geostationary satellite to a receiving dish. The signal is then sent from the dish to an MMDS transmitter that uses microwaves to send the signal to subscribers in town. Each subscriber needed a microwave antenna and signal processor to receive the television channels.

Figure 3.13 This picture shows both an older C-band satellite receiving dish and a newer DBS dish mounted below it on the same mast. The DBS dish is about one tenth the size of the C-band dish above it.
Source: Photo courtesy of iStock #2687905

offered hundreds of channels of high-quality television, including premium cable networks. DirecTV was without competition for two years, when the DISH Network entered the marketplace.

Until 1999, DBS was not allowed to provide local broadcast stations to individual subscribers' homes. This limitation forced subscribers who wanted local programming, such as evening newscasts, either to subscribe to cable or to capture over-the-air signals with an antenna in addition to their DBS service. For many consumers, the hassle of switching back and forth between DBS and an over-the-air antenna, or paying for both DBS and cable was not acceptable. The local channel limitation was lifted in 1999, when the Satellite Home Viewers Act (SHVA) mandated that local broadcast signals be carried on DBS.

As DBS technology improved, so did satellite dish receivers. Not only did dishes become smaller (18 inches in diameter), but they became capable of receiving high-quality digital transmission. Local neighborhood associations and neighbors themselves no longer had a reason to complain, as dishes could be discreetly placed on a rooftop, on a windowsill, or in a backyard.

DirecTV and the DISH Network are still the two major players in the DBS business although they proposed to the FCC in March 2002 that they merge and create one national provider of DBS service. The FCC rejected the merger idea, preferring to maintain two competing providers to give the audience a choice of services.

Increased Choice and Competition

The television broadcast industry and viewers were shaped by major changes in television technology starting in about the 1970s. The number of cable channels proliferated and videocassette recorder (VCR) sales exploded. Viewers were no longer tied to television viewing schedules but could now 'time shift' by recording programs for viewing at a later time. Color television sets had become affordable for the lower and middle classes, as manufacturers lowered sales prices because of the economy of scale – the more sets that were produced, the less expensive it became to manufacture each one. Sony, Panasonic, and Toshiba began importing high-quality, competitively priced sets from Japan, further increasing the competition among manufacturers, and giving viewers more choices at even lower prices. Along with the soaring number of set purchases, sometimes two or more for one home, came a demand for high-quality, innovative programs and an increase in the amount of time spent viewing. And best of all for the networks, advertising revenue jumped.

The networks' high times started diminishing in the 1980s. As more and more cities were wired for cable, the broadcast networks' 90% share of the viewing audience became smaller as people started watching more cable shows. Interesting programs on new cable channels such as CNN (news), MTV (music videos), ESPN (live sports), and pay channels such as HBO and Showtime (movies) lured viewers away from the typical broadcast fare of comedies and dramas. Video stores that rented out and sold programs and movies on videocassette tapes popped up all over the country, giving viewers yet another viewing option that further eroded the network audience. Cable television, video rentals, and VCR time-shifting all contributed to broadcast losing about one-third of its viewers. Along with the smaller share of viewers came a loss in advertising revenue.

As viewers slipped away, so did the networks' power over their affiliates and advertisers. Affiliated stations took revenue matters into their own hands by pre-empting network shows for ones that they produced or obtained from syndicators on which they could sell advertising and keep revenue at the station level. Affiliates even questioned whether payments for clearance were high enough and whether long-term affiliate agreements were financially wise.

Despite these challenges, many executives of the big three networks (ABC, NBC, CBS) decided that their strategies were working fine and did not require overhauling. New ideas clashed with old, and the direction of the television industry became a source of tension. Generally, the big three networks managed to change very little with the times, leading some industry critics to refer to them as the "three blind mice" (Auletta, 1991). A fourth network, Fox, successfully launched in the mid-1980s, after the Rupert Murdoch-owned News Corporation purchased a group of stations previously owned by Metromedia, along with a 50% share of the 20th Century Fox film studio, making it possible to produce programs and air them on stations around the country. Promising to deliver hipper and edgier content than the big three networks, Fox debuted with *Married...with Children* and *The Tracy Ullman Show*. It introduced viewers to *The Simpsons*, and later brought on *The X-Files*, and with these shows captured young adult viewers and became known as the cool network.

SEE IT NOW

The advent of digital technology made more television channels possible, reducing the impact of 'spectrum scarcity' that was such a major limitation on the broadcasting industry since its inception. The very existence of the FCC was premised on equitably distributing available channels to allow diversity of media ownership and programming options, as well as maximizing media companies' service to the public interest. Because digital technology is more efficient, allowing more content to be distributed to the public with better picture and sound quality, it made sense for the FCC to facilitate its development and expansion, leading to the present-day circumstance in which thousands of media options are available to a great majority of the population, filling increasingly narrow niches, and leading to what is frequently described as audience fragmentation.

Digital Television

In 1996, the FCC appointed a task force to study the conversion of all television stations to digital broadcasting. On the task force's recommendation, the FCC ordered all television stations to upgrade to digital broadcasting by 2006. Station owners were up in arms, not so much because they opposed digital but because of the millions of dollars it would cost them to comply. Consumers were also concerned because conversion meant they had to buy new sets that were capable of picking up digital signals. After much brouhaha from the industry and viewers, the FCC capitulated by postponing the conversion until February 17, 2009. But continued pressure led to the Obama administration and Congress persuading the FCC to postpone the digital conversion one more time, to June 12, 2009.

Digital television (DTV) transmits television programs in high-definition television (HDTV), which is a widescreen, high-resolution format. It also uses standard definition (SDTV), similar to an analog television picture, but with better color reproduction and less interference. Traditional analog television images were made up of 525 lines of video (top to bottom) but the HDTV picture is made up of

1,080 lines, more than twice the picture resolution. The HDTV picture has a wider *aspect ratio* (the relationship between screen width and height), yielding a 16:9 picture (16 units wide, 9 units high) that more closely resembles a widescreen movie picture than the 4:3 picture of analog television, which originated in television's earliest days with screens that were nearly round and could not contain the types of wide visual compositions typically considered most pleasing to the eye. The HDTV standard for the configuration and delivery of higher quality video and audio adopted by the Advanced Television Systems Committee is known as ATSC A/53 or simply ATSC.

In making the move from analog to digital, the FCC forced all television stations to make large investments in new digital equipment. This switch to digital did not offer immediate financial rewards, as stations did not have a way to generate any more income with a digital picture than they did with analog-only broadcasting. One advantage to the digital system, however, is that stations have enough room in the 6 MHz channel to send out more than one program at a time. For example, a local station might have its main programming on its first channel, that is, a network feed, such as NBC. A second channel could carry weather and news, and a third could show foreign-language programs. Some stations carry programming from networks that emerged specifically for such secondary channels, including MeTV, owned by Weigel Broadcasting, and Laff, a network owned by Scripps, which airs on many ABC-owned stations' secondary channels. Digital conversion also meant that broadcasters had to give up the analog frequencies that they had been using since the 1940s. The FCC collected billions of dollars by auctioning these frequencies to telecommunications companies.

FYI: ANALOG VERSUS DIGITAL

Duplicating an analog signal is like pouring water from one jar to another, there are always a few drops left in the old jar. When an analog signal is duplicated, some of the signal gets lost in transit. Duplicating a digital signal is less like pouring water from one jar to another and more like pouring marbles from one container to another. In this case, the new container will be full of marbles, and the other container will be completely empty. Digital transfer allows the entire sampled signal to be duplicated, and the copies are identical to the original. Although the process of digital sampling misses some of the original signal (like the content represented by the spaces between the marbles), the process is very reliable and efficient.

Before the official switch to digital, the audience was slow to embrace the new digital broadcast channels because they appeared on different channels than the original station channel. Many viewers simply did not bother with hunting down the digital version of a show they were watching. Moreover, on older sets the digital version of a show was not perceptibly different from the analog version.

Viewers were skeptical about the benefits of digital television and reluctant to buy a new digital set. Before the digital conversion, only about 10% to 20% of viewers watched 'over-the-air' broadcasting, with the remaining 80% to 90% subscribing to either cable or DBS.

Digital conversion initially affected the small segment of the audience that watched television over the air, as both cable systems and DBS continued to provide analog signals for those subscribers with analog sets. Households that did not subscribe to cable or DBS could purchase set top converter boxes that received the digital signal over the air and converted it to an analog signal that was compatible with their older television sets. The government offered 34 million coupons worth $40 toward the purchase of these set-top boxes. However, once digital television was up and running and cable and satellite systems added high-definition channels in addition to their standard definition offerings, viewers were impressed with the picture quality and abandoned their analog sets.

Analog recording

Digital recording

Diagram 3.8 When content is recorded in analog mode, the recording is an approximation of the entire sound or picture (or both). The recording process is slightly distorted, and some information is lost. When the analog recording is played back, more distortion occurs, and more information is lost. When content is digitally recorded, the recording is sampled and encoded digitally. The process of sampling is recording (or measuring) the content at regular, very rapid intervals. The more samples of the content, the more accurate the recording and playback.

Television Sets and Viewing Habits Change

The distinction between television sets and computers has blurred in the modern era, with computers able to provide audio and video content as well or better than televisions, but televisions, in turn, are capable of streaming data directly from the Internet. The missing link was the ability to receive live television from broadcast network, cable, or satellite television on a computer monitor without special hardware or software, but this changed rapidly in the second decade of the 21st century. This blurring of distinction encouraged some of the big names in the computer industry – Apple, Microsoft, Dell, and Samsung – to market computers and television sets to satisfy users' entertainment needs and it also changed how audiences use their computers and televisions and where they place them in their homes.

TV-on-DVD also changed television viewing. The sales of television programs on DVD generated $1.5 billion in 2003. After sales peaked in 2004, DVDs have seen a decline in sales over the past decade, mainly because online streaming makes it unnecessary to acquire a physical product in order to watch prerecorded video content. Many viewers previously were willing to buy a whole season of a television series on DVD to avoid annoying commercials, to get a better-quality picture, and to set their own viewing times, but sales slowed somewhat because download sites – both legal and illegal – made buying the physical DVD less necessary. And on many small devices like tablets and small laptops that do not have internal DVD drives, shows and movies cannot be downloaded, only streamed. The movie industry has encouraged an increase in streaming movies by releasing digital versions of new movies online (pay per view) one to three weeks before they become available on DVD. As a response to COVID-19, which has been keeping people away from movie theaters, in 2021 some studio began releasing their new movies to streaming platforms (e.g. HBO MAX) on the same day that the movie is released to movie theaters.

Internet Delivery of Television Programs

Stations are now using the Web to connect with viewers through the use of online video, social networking, and other features. Stations host blogs about local news issues, up-to-the-minute traffic

ZOOM IN 3.6

The Advanced Television Systems Committee (ATSC) is an international non-profit organization that helps develop voluntary standards for digital television. For more information about the new ATSC 3.0 standard, go to: www.digitaltrends.com/home-theater/atsc-3-0-ota-broadcast-standard-4k-dolby-atmos/
 Or see the differences between analog and digital: www.youtube.com/watch?v=M_nTmRtAD98

and weather coverage, community chat rooms, in-depth coverage of stories, and – most importantly for station revenues – advertising space. News personalities offer email contact with the audience as well as a presence on Facebook and Twitter. Some stations post information and content from audience emails directly into local news shows.

Cable and telecommunication companies are reaching out to their users by encouraging them to subscribe to 'wired' programming services that are viewable on most mobile devices. For example, Comcast, the largest cable company in the United States, uses a system named Xfinity, which allows subscribers to access television shows and movies from computers, tablets, and smartphones, as well as download many television shows and movies to mobile devices. Verizon Wireless has expanded its telephone service and now also offers television shows via its FiOS cable system as well as its FiOS Mobile app that lets subscribers stream television programs and movies to their smartphone or tablet.

Contemporary Cable

The popularity of online streaming of video content has resulted in a change for cable providers in which some subscribers use traditional cable television. This does not necessarily result in a decline in revenue for the companies, however, because they

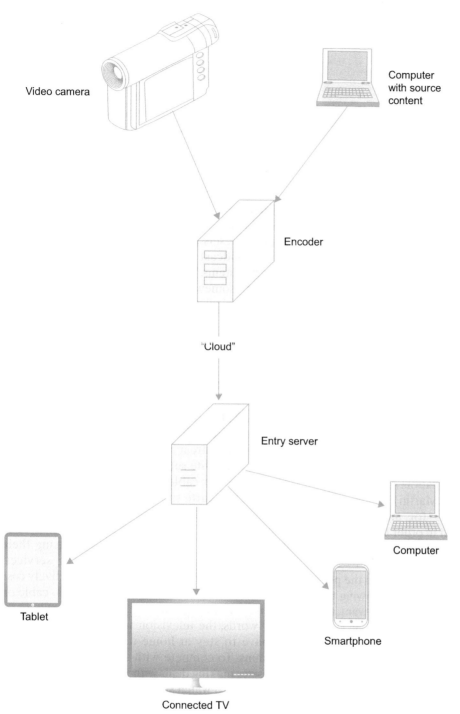

Diagram 3.9 Material from a camera, computer, or audio source is fed to a computer that encodes the material to a digital signal stream. The stream is sent to the 'cloud' and is retrieved by an entry server. The entry server decodes the signal and sends it to receiving devices (tablets, smartphones, television sets, and computers) that play the audio and video signals.

still provide Internet service even to some so-called 'cord cutters.' In 2013, cable and satellite delivery systems lost 250,000 subscribers, the first decline ever. Cable subscriptions peaked in 2001, when there were almost 67 million subscribers, but by 2010 the number of subscribers dropped to 63 million and by 2014 to just under 50 million. Obviously, the trend in cord cutting has continued. Viewers are watching television programs via DBS or other delivery methods such as online streaming. More than half of cable subscribers in 2014 said they would cancel their cable television subscriptions if they had a choice of another cable service. In addition, almost three-quarters believe that "cable companies are predatory in their practices and take advantage of the consumers' lack of choice" (Sherman, 2014). The

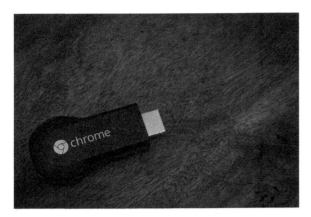

Figure 3.14 This small USB device, the Google Chromecast, allows a video signal retrieved on a mobile device to be sent to a large television via the Internet.

negative attitude toward cable is manifested in consumers 'cutting the cord' by canceling their cable and satellite television services. Despite the decline in the number of subscribers, cable is still a profitable business. Although there are fewer cable television subscribers, the financial loss in subscriptions is made up for by increased prices for television and Internet service. Cable television is not coming to an end, but many feel that the cable model is broken. Between 1995 and 2005, cable companies doubled the size of the 'bundles' that were marketed to subscribers, and the charges to the customers rose at a rate three times the rate of inflation. According to a former FCC chairman (Kevin Martin), "the average cable subscriber was paying for more than 85 channels that she didn't watch in order to obtain the approximately 16 channels that she does" (Popper, 2015, April 22).

Except for VOD (video on demand) and the possibility of a more 'à la carte' approach allowing the subscribers to select only the channels they want, the cable industry is consolidating and holding on to its ISP (Internet service provider) business, the cable industry seems to be technologically stagnant. The phenomenon known as 'cord cutting' has hurt both the cable and DBS industries, as more people have shunned large monthly fees in favor of streaming services such as Netflix and Hulu+ on their smart televisions, capable of Internet streaming.

SEE IT LATER

The television industry has changed immensely since 2010. Old delivery systems have given way to 'now media' technology that brings the programs and movies we want to see, when we want to see them to our television sets, tablets, smartphones, and laptops.

Subscriptions to cable and satellite systems have declined, while subscriptions to streaming services such as Netflix and Hulu continue to increase. Television sets have changed dramatically. Analog sets are no longer manufactured. Flat-panel smart television sets that come with built-in connections to the Internet have decreased in price, with a typical 47-in. diagonal HD set selling for about $300 or less. Increasingly, television sets are used as monitors that receive inputs from a variety of sources: DVD players, streaming devices such as Apple TV and Roku, tablets like the iPad, and laptop computers. In addition, sets are all capable of receiving signals from broadcast television, cable, and satellite delivery systems. With all of these inputs, viewers are becoming somewhat 'agnostic' about how they get their television. In other words, the audience does not care who or what delivers programs as long as they get it where and when they want it.

Television and the Internet

Essentially, the television industry is in flux. Both broadcast networks and premium cable services have launched streaming services that do not require a cable or satellite subscription. Instead of paying a monthly fee for cable or satellite services, viewers prefer an à la carte subscription model. Netflix and Hulu now offer an enormous selection of programs for a monthly or by-the-program fee, and they and other streaming services are now producing their own original content. These streaming services, known as *over the top* or OTT, can collectively cost more than a single monthly subscription to cable or satellite, but they provide mobility to subscribers. In other words, the television viewing audience does not have to stay at home to watch their favorite programs. To compete with streaming, cable companies are making deals with networks to allow their subscribers to watch shows for free via the Internet simply by entering their subscriber ID and password. The new subscriber model is available anywhere the viewers can get an Internet connection and puts cable companies in competition with streaming services for viewers who prefer to watch on the go. Simply put, streaming has disrupted the traditional television business in a way that is similar to how the Internet disrupted the music and print industries, but cable is trying to find a way to compete as 'cord cutting,' continues, though at a slower rate than previously.

New Television Sets

Television sets are changing as well. A few years ago, 3D sets were the big fad, but audiences did not like wearing 3D glasses to watch a show. Now 3D sets are scarce in retail showrooms and have been replaced by new sets with curved instead of flat screens. Purportedly, a curved screen creates a viewing experience that draws the viewer into the program, but one that comes at a significantly higher cost than flat-panel sets. As of 2021, curved screens have all but disappeared from retails stores. The second innovation, UHD (ultra high definition) television, holds more promise for the consumer market.

Until 2009, analog NTSC (National Television Standards Committee) television used 525 lines; now digital ATSC (Advanced Television Systems Committee) television can use up to 1,080 lines per frame. Now we are seeing ATSC 3.0, which includes '4K' video – four times the resolution of ATSC, along with better sound, high dynamic range for better clarity, wider color gamut, and higher frame rates. The term 4K refers to the 3,840 or roughly 4,000 pixels width. ATSC 3.0 standard also allows for an eventual update to 8K video.

It is expected that in the near future, HD sets will give way to UHD sets. HD sets are capable of reproducing a television picture with 1920 (wide) × 1080 (high) pixels (picture elements or dots of color that make up the picture). UHD sets are capable of reproducing a television picture with at least 3840 × 2160 pixels, yielding about four times as much picture content per frame. The UHD picture yields a clearer, more realistic picture, which is similar in resolution to 4K but based on a different standard. The change in resolution is technically similar to the change from SD (standard definition 480 lines of resolution) to HD (1,080 lines of resolution). Although some UHD sets are still sold at premium prices, the prices for most UHD sets have already dropped. Consumer adoption of UHD sets depends in part upon the availability of UHD programs. Currently the production companies that produce television programming are slowly changing from HD production to UHD production. Netflix and other content providers are streaming some of its shows in 4K.

Even as 4K becomes mainstream, there is already a buzz about 8K television sets. These sets double a 4K picture's pixel width and height and yield an amazingly realistic television image. Although these sets are demonstrated at technology and industry conventions, they will most likely be used only in movie theaters because they require a screen size that is too large for most homes. In addition, there is no commercially produced 8K content currently available to general audiences.

The change in resolution does not just change television sets but also affects cameras, smartphones, tablets, and all monitors. In other words, anything that records or plays back a television image will have to change to accommodate 4K or 8K content.

Organic Sets

Organic is not just a type of food; it also refers to a type of television set. *OLED* sets are constructed with **organic light emitting diodes** that produce an amazingly vivid image with very strong contrast levels. Although smaller OLED sets are affordable, larger sized ones (40+ inches) are very expensive and represent a small percentage of all television sets sold.

Broadcast Television

The broadcast network audience share will continue to decrease in the years to come, because the audience will have more choices. The networks may find that their profitability depends increasingly on delivering programs that other services cannot provide or that they, the networks, can best provide. For example, the networks can deliver live programs, like news and sports, and can operate profitably by offering reality shows that do not require large payments to stars, writers, and independent producers.

Each of the broadcast networks and most of the cable networks have their own Web site that streams video. NBC, ABC, and Fox created Hulu to attract viewers to clips and full episodes of shows. Hulu also offers two subscription rates, one with some commercials inserted into the programs and another slightly more costly option that eliminates commercials within programs.

While it is true that broadcast networks have entered the streaming market as well, they still have technological and human resources to produce live programming more readily than other streaming services. The networks will continue to be challenged by technological changes and other delivery systems. As a result, they will continue to vertically integrate by buying program-production houses and program-syndication companies to gain control over programming sources and outlets. They will also utilize new programming services on other delivery systems (e.g.

cable, satellite, and the Internet) and even new technologies to keep their audience share large enough to attract advertisers. CBS, the network that carried the 2011 Super Bowl, offered a live stream of the broadcast for the first time.

Although it does not now pose a threat to large television content producers, user-generated television content, such as YouTube videos, looms on the horizon. Over two billion people watch YouTube videos each month and hundreds of hours of user-made videos are uploaded to the site every minute. YouTube has the potential to keep drawing users away from television and streaming services, especially those who just have time to watch a short video.

Digital tools for the production of high-quality television, once the exclusive domain of 'big media,' are now becoming available to ordinary people. Personal technology of this type may change the future production of television programs.

Local broadcast stations are now producing video content not only for their broadcast newscasts but also for online streaming. A video package produced for a nighttime newscast is often archived as is or edited to a different length and repurposed for later use on the station Web site, and shared with other stations owned by the same broadcast groups.

Television stations compete not only with traditional media, other television stations in their market, cable, satellite, and commercial streaming services but also with the streaming content provided by non-professional audience members.

The future of broadcast television might depend upon its ability to provide live local news and big sporting events from the professional sports world. Since people do not want to watch time-shifted sports, broadcast television's (and the cable and satellite providers') ability to show live sports, at least for the time being, sets it apart from competing systems such as Netflix and Amazon Prime. Although some have predicted the end of broadcast television, its total demise is not in sight, although an overhaul of its structure is overdue.

Cable Television

Cable channels and networks are trying their best to reach young audiences. They are putting prime-time shows on later at night or perhaps repeating them at a later time to reach those that use other entertainment options during primetime.

Cable systems will keep trying to reverse the 'cutting the cord' trend and convince a new generation of 'cable nevers,' to subscribe. Cable will need to heavily promote its offerings to young people who, after leaving their parents' homes, have never subscribed to cable television because their viewing needs are met by Internet delivery. But mostly, cable needs to change its brand image from the 'old' way to get television, to a 'now' way to get broadcast and cable shows and movies on demand, on any device, with just a couple of button pushes on the remote control, and without bothering with Internet connections, passwords, or clunky interfaces.

Satellite Television (DBS)

Satellite television delivery faces the challenges of providing a unique service and making itself user friendly. Satellite delivery to homes has a marketing advantage by offering many more channels of high-quality video than cable, but it does not offer high-speed broadband at a price and speed competitive with cable.

Consumers would like to get their television in an à la carte style, meaning that consumers would rather pay per channel than pay for a bundle of television channels, many of which they do not watch. In this way, the satellite television industry is at least as vulnerable as the cable industry to 'cord cutting.' The difference between the cable and satellite industry relationship to consumers is that the cable industry has become the main provider of Internet connections. While any home connected to cable television can also be easily connected to cable's ISP service, this is not true for satellite subscribers. Internet service through a satellite service is primarily provided in remote areas, where cable television cannot provide service. Until 2015, the satellite industry's lack of interactivity put it at a disadvantage relative to cable as a multichannel provider, because interactivity will continue to be an essential part of the television viewing experience. In July of 2015, AT&T acquired DirecTV, which opened up possibilities for DirecTV to offer its customers broadband access through its parent company.

SUMMARY

Early experimenters in television tried two methods to obtain pictures: A mechanical scanning system and an electronic scanning system. The electronic system was eventually adopted as the standard. Although many inventors were involved in the development of electronic television, two of the most important were Vladimir K. Zworykin and Philo T. Farnsworth.

The FCC authorized commercial television broadcasting in 1941, but the industry did not catch on until after World War II, when the materials needed for television equipment manufacturing became available. After the war, television grew so rapidly that the FCC could not keep up with license applications or technical issues. In 1948, the FCC put a freeze on all television license applications that lasted until 1952. During the freeze, the FCC considered the allocation of spectrum space to stations, designed the UHF band, and dealt with the issues of VHF and UHF stations in the same markets, color television, and educational channels.

At first, television was a live medium. In the 1950s, many high-quality dramatic programs were written for live theater-type performances to attract educated audience members who could afford the cost of a television set. Program production changed when videotape became available and programs were no longer produced live in the studio. The television audience expanded when the price of a television set dropped and programs were adapted to appeal to a mass audience. The early days of television had some challenges, including blacklisting people working in the industry who supposedly had communist sympathies.

Stations with network affiliations did well because of network programming. Independent stations had to resort to older programs from syndication as well as sports and locally produced programs, especially children's programs and cooking shows.

The importance of television became more evident in the 1960s, when events such as the assassinations of John F. Kennedy, Martin Luther King Jr., and Senator Robert F. Kennedy were covered extensively by the networks. That decade also saw a growth in local news coverage and national news coverage of the Vietnam War, violence in the streets, racial tension, and the first man on the moon. In 1967, Congress approved the Public Broadcasting Act, signaling the birth of a network dedicated to non-commercial television broadcasting.

Cable television started out in the late 1940s as a small-time system to provide rural audiences with television programming. In the 1970s, cable television grew rapidly, many cities, large and small, became wired, and cable systems began to use communication satellites to receive programming from all over the country and developed its own programming through national cable channels, pay channels, local origination, and pay per view.

Throughout the 1980s, VCRs gave the audience the ability to rent movies and record programs off the air. The addition of cable networks such as CNN, MTV, and ESPN began to change the dominance of the broadcast networks. Broadcast deregulation was a guiding principle for the FCC, and many rules were modified or removed entirely. Television technology improved, and electronic news gathering using portable video equipment became common in stations across the country. A new major commercial network, Fox, began operating. Overall, the television audience size grew, but the networks' share of the audience declined because viewers had more choices, primarily cable channels.

In the mid-1990s, direct broadcast satellite began providing many high-quality channels and direct competition to local cable companies. The passage of the Telecommunications Act of 1996 changed many of the rules regarding ownership of broadcast television stations and allowed television group owners to acquire as many stations as they wanted. This relaxation of ownership rules resulted in fewer groups owning more stations and thus consolidating the television station business. In the last half of the 1990s, the Internet became popular and signaled huge changes in media use as people began to spend more time online and less time in front of their television sets.

At the beginning of the new century, consolidation of station ownership began to raise public awareness about the lack of diversity in ownership and localism in programming. The role of the networks continues to change from one of a delivery system for independently produced programs to one that delivers network-owned and -produced programs. The convergence of computers and television will continue drawing corporations such as Apple, Microsoft, Dell, and Samsung into the television business. Digital broadcasting replaced analog broadcasting in 2009 and ushered in a new era of high-quality video and sound and different delivery methods, including multiple signals and Web delivery of content to the audience. Television stations are now using 'now' media to provide additional viewing opportunities and social media to connect with the audience.

An important trend in television delivery systems is vertical integration. Large media companies now own networks, television stations, newspapers, magazines, cable channels, book publishing companies, movie studios, syndication companies, and television program-producing companies. Television networks are producing content for other services to distribute. This practice contributes to the profitability of the parent companies and signals a strategy change by the networks. The television networks are no longer just in the business of exhibiting their content.

They are owners of content regardless of the manner of distribution.

The audience of the 2010s consists of many who get television through streaming video from the Internet. This audience has increasingly cut the cord by dropping cable and satellite television subscriptions and relies upon subscription services like Netflix and Amazon for television viewing. Many content providers (e.g. HBO and ESPN) will continue to seek direct relationships with the audience and will skip the multichannel program services. This model of television delivery will be a major force in shaping the future of television.

Television sets have changed dramatically in the past ten years. The old, bulky cathode ray television sets have been replaced by slim, light, large television sets that have HD or UHD resolution that yields amazingly clear pictures. These digital sets, once very expensive, have now become affordable and dependable entertainment centers in homes. A variety of inputs brings television programs, movies, social media, and audio music services to the audience with the touch of the remote control.

'Now media' technology continues to present different viewing opportunities to the audience that will force the networks and multichannel television program services (cable companies, satellite companies, and telecommunication companies) to continually adjust their delivery and revenue models to stay financially viable in the competition for viewers and advertisers.

BIBLIOGRAPHY

Auletta, K. (1991). *Three blind mice: How the TV networks lost their way*. New York: Random House.

Bellamy, R., & Walker, J. (1996). *Television and the remote control: Grazing on a vast wasteland*. New York: Guilford Press.

Clarke, A. (1945, October). Extraterrestrial relays: Can rocket stations give world-wide radio coverage? *Wireless World*, 305–308.

Faulk, J. (1964). *Fear on trial*. New York: Simon & Schuster.

Fotrell, Q. (2014, September 3). *Cable companies should be afraid of this trend*. California: Market Watch, Inc.

Franklin, C. (2004). How cable television works. Retrieved from: http://electronics.howstuffworks.com/cable-tv.htm [April 1, 2016]

Halberstam, D. (1993). *The fifties*. New York: Ballantine Books.

Hazlett, T. (1989). Wiring the constitution for cable. *Regulation: The Cato Review of business and government*, *12(1)*. Retrieved from: www.cato.org/pubs/regulation/regv12n1-hazlett.html

Hendrik, G. (1987). *The selected letters of Mark Van Doren*. London, LA: Louisiana State University Press.

Higgins, J. (2004, January 5). The shape of things to come: TiVo's got a gun. *Broadcasting & Cable*, p. 33.

Hilliard, R., & Keith, M. (2001). *The broadcast century and beyond: A biography of American broadcasting*. Boston, MA: Allyn & Bacon.

Hutchinson, T. (1950). *Here is television: Your window to the world*. New York: Hastings House.

Jarman, M. (2008, August 30). Intel bets on web TV concept. *The Arizona Republic*.

Jessell, H. (2002, August 5). Sink or swim: With a set-top box of their own, TV broadcasters can take their place in the digital future. *Broadcasting & Cable*, pp. 14–16.

Local television in 2018: Evolving with shifting audience demos. *Nielsen*. Retrieved from: www.nielsen.com/us/en/newswire/2013/local-television-in-2018-evolving-with-shifting-audience-demos/

Mitchell, C. (1970). *Cavalcade of broadcasting*. Chicago, IL: Follett Books.

Perez, S. (2009, August 6). Livestation brings TV to the iPhone. *New York Times*. Retrieved from: www.readwriteweb.com/archives/livestation_brings_live_tv_to_the_ihone.php

Popper, B. (2015, April 22). The great unbundling: Cable TV as we know it is dying. *The Verge*. Retrieved from: www.theverge.com/2015/4/22/8466845/cable-tv-unbundling

Ritchie, M. (1994). *Please stand by: A prehistory of television*. Woodstock, NY: Overlook Press.

Schatzkin, P. (2003). *The boy who invented television*. Burtonsville, MD: Team Com Books.

Schwartz, E. (2000, September/October). Who really invented television? *Technology Review*. Retrieved from: www.technologyreview.com/Biztech/12186/?a5f

Sherman, E. (2014, June 9). Most consumers ready to dump their cable companies. *Money Watch*. Retrieved from: www.cbsnews/news/most-consumers-ready-to-dump-their-cable-companies/

Sterling, C., & Kittross, J. (2002). *Stay tuned: A history of American broadcasting*. Mahwah, NJ: Erlbaum.

The future of television: The weak staff. (2014, December 9). *The Week*. Retrieved from: http://theweek.com/articles/441909/future-television

Walker, J., & Ferguson, D. (1998). *The broadcast television industry*. Boston, MA: Allyn & Bacon.

Chapter four

Radio and Television Programs and Programming

In the early days of radio, amateur operators just put something on the air – often on the spur of the moment. Maybe someone would drop by a station to sing or play an instrument or to talk about some topic to whoever was listening. Some years later, radio station licensing required the transmission of *scheduled* programming. Listeners crowded around their radio to listen to sports, news, music, dramas, and church services. By today's standards, the level of static and poor audio quality would make radio unlistenable, but to yesterday's audience, radio was magic.

This chapter covers the development of radio programming, starting with station formatting, how stations obtain music, how songs get on the air, and who decides how often they are played. The chapter then moves on to television programming. It describes the different types of television programs and explains how program ideas are developed, how programs make it to the air, and how program-scheduling strategies are used to keep the audience tuned to a channel.

SEE IT THEN: RADIO

Types of Programs

In the old days of radio, programming was much more varied than it is now. Music was mostly in the form of live performances, and a station could play a variety of music. Fans could listen to fiction and non-fiction programs, whether dramas or comedies, and live performances of plays, comedy acts, and opera. In contrast, most contemporary stations are formatted such that they play one type of music or they are news, talk, or sports formatted.

Music

In one of the first live radio performances, opera singer Enrico Caruso sang 'O Sole Mio' from the Metropolitan Opera in New York in 1910. Lee de Forest, the disputed inventor of the audion tube, which amplifies radio sound, masterminded the performance to promote radio.

ZOOM IN 4.1

Hear Caruso as he sang live from the Met almost 100 years ago at www.youtube.com/watch?v=u1QJwHWvgP8

By the late 1920s, music was the main source of radio programming. For example, NBC Blue and NBC Red network-affiliated stations in New York devoted about three-quarters of programming to music. Because the quality of phonograph records was very poor, most of the music was broadcast live in a studio or other concert venue. The studios were often decorated with potted plants, such as palm trees, to make them seem like real concert halls or ballrooms; thus, radio programming of that era is often referred to as **potted palm music**.

Early model radio transmitting equipment was bulky, heavy, and hard to move around, so on-location broadcasts were troublesome and thus rare. Station KDKA broadcast the 1921 Dempsey–Carpentier boxing match with a temporary transmitter set up in New Jersey. It was fortunate for KDKA that Carpentier was knocked out in the fourth round, because shortly after the fight ended, the station's transmitter melted into one big heap of metal.

Industry developments in the early 1930s set the standard for radio programming for the next 20 years. Portable transmitting equipment made possible coverage of live sports and other types of on-location events, and creative writers developed new and sophisticated types of radio programs. By the late 1940s, only about 40% of the programming was music oriented. Comedies, soap operas, dramas, quiz shows, and children's programs filled the airwaves and were a prominent form of in-home entertainment until the late 1950s.

Early programs were not always scheduled at regular times; the audience rarely knew what was going to be playing when they turned on the radio. Station executives realized that to capture and maintain a steady listenership, programs needed to air on a regular schedule. Eventually, newspapers ran radio program listings, and special radio program guides were printed so listeners knew when it was time to gather around their sets to listen to favorite shows, such as *The Green Hornet*, *The Jack Benny Program*, and *The Lone Ranger*.

Dramas
Among the most popular radio programs were daytime serial dramas, which came to be known as *soap operas* because laundry soap manufacturers were the primary sponsors. However, the shows could have just as easily been called *cereal operas*, because the first serial drama, *Betty and Bob*, was sponsored by General Mills. These 15-minute continuing-storyline dramas appealed largely to females and dominated the daytime airwaves. In the evening, a more diverse audience tuned in to a different type of dramatic programming, the episodic drama, which resolved the story within a single episode.

ZOOM IN 4.2

Every Sunday from 1930 to 1954, the Mutual network aired one of the most popular radio programs ever, *The Shadow*. The program enthralled listeners with its famous opening line, "Who knows what evil lurks in the hearts of men; the Shadow knows," which was followed by a sinister laugh ('Famous Weekly Shows,' 1994–2002). Listen to audio clips of *The Shadow* on YouTube. Just enter the search term: '*The Shadow* radio shows.'

Comedies
Comedy programs were an early hit on the radio and were especially popular during the depression. Radios were inexpensive and provided in-home entertainment at a time when people could not afford a theater ticket. Radio comedies took listeners' minds off the depression and their financial worries.

The comedy show *Amos 'n' Andy* made its radio debut in 1928 and was the first nationwide hit on American radio. Avid fans would stop what they were doing and crowd around a radio set to laugh at the latest antics of their two favorite characters. *Amos 'n' Andy* was also one of the most controversial programs ever to air on radio. The title characters were "derived largely from the stereotypic caricatures of African-Americans" and played by White actors who "mimicked so-called Negro dialect" (*Amos 'n' Andy Show*, 2003). Freeman Gosden and Charles Correll, the White creators and voices of the program, "chose black characters because blackface comics could tell funnier stories than whiteface comics." Even so, the program quickly came under fire from the Black community and NAACP (National Association for the Advancement of Colored People). Eventually, the characters grew beyond being caricatures. In fact, NBC claimed the program was just as popular among Black listeners as among White listeners. The program moved to CBS television in 1951 but new episodes were only made for two more years.

It was not until 1947 with the debut of *Beulah* that an African American starred in a radio sitcom. Well-known actress Hattie McDaniel took over the part of Beulah from a White male actor. Beulah was a stereotypical simple-minded but warm, caring, funny maid who outwitted the family for whom she worked.

Figure 4.1 Charles J. Correll and Freeman F. Gosden wore blackface as characters Amos and Andy.
Source: Photo courtesy of RKO Pictures/Photofest.
© RKO Radio Pictures

ZOOM IN 4.3

Listen to clips of popular old radio shows on Old Time Radio https://archive.org/details/oldtimeradio

Watch television programs such as *Amos 'n' Andy*, *Bob Hope*, *Fibber McGee and Molly*, *The Red Skelton Show*, *Abbott and Costello*, and *The Adventures of Ozzie and Harriet* on YouTube. Just enter in the name of the program.

When McDaniel took the lead, the program was renamed *The Beulah Show* and remained on radio until 1954. The show also aired on television from 1950 to 1953 with the role of Beulah played by various African American actresses, including Ethel Waters.

Comedians such as Jack Benny, George Burns, Gracie Allen, and Bob Hope all got their start in radio between the late 1920s and mid-1940s and later made a successful transition to television. *The Burns and Allen Show* hit the radio airwaves in 1932. The show featured George Burns as a curmudgeonly husband and straight man to his wife, the very funny Gracie Allen (they were married for 38 years in real life). After a successful 18-year run, the program moved to television, where it aired for eight seasons and 291 episodes.

Back to the Music

Radio thrived throughout the 1930s and 1940s and became the primary source of entertainment. People could not imagine anything better than having entertainment delivered into their living rooms. But then came television – a new-fangled device that combined sound with images. Television quickly won the hearts and eyes of the public, who proudly displayed their new television sets and pushed their radio receivers into a dark corner of their living room. In 1946, only about 8,000 U.S. households had a television set, but just five years later, some 10 million households were enjoying the small screen.

Television programs were in huge demand but there were not enough shows to fill the airtime. Industry executives convinced radio show producers to adapt their programs to television. As radio soaps, comedies, quiz shows, and dramas transitioned to television, radio found itself scrambling to replace these programs or face its own empty airtime. Considering that the loss of network radio programs also meant the loss of advertising revenue, radio executives knew they had to do whatever they could to save radio from dying off altogether.

The easiest way to fill radio airtime was with music. Several technological innovations helped remake radio into a music-dominated medium, namely, music was now recordable on vinyl records and reel-to-reel tape, which vastly improved the overall sound quality and lessened the need for live performances. Further, thanks to the invention of the transistor, radio sets could be made much smaller. Pocket-sized transistor radios hit the consumer market in the mid-1950s and forever changed music listening habits as radio was now portable. Radio listenership rose even higher as automobile manufacturers offered optional in-dash radio receivers. By turning back to music, radio distinguished itself from television and reemerged as a medium in its own right.

But radio station managers knew music itself was not enough to draw and maintain an audience. It had to be the right kind of music, the kind that appealed to the local broadcast market. Stations saw value in developing their own identities and attracting particular listeners through **music formatting**. Different stations in the same broadcast area began playing different types of music. Maybe one station played classical, another blues and jazz, another big band, and so on. Realizing that music fans were spending a great deal of money on records, enterprising radio programmers soon got into the habit of checking record sales to predict what songs and what performers would go over well on their station and keep within the format. Station formatting caught on and is still the mainstay of radio programming today.

Figure 4.2 Radio's portability, as well as its ability to provide soothing background noise, lends it to listening while doing other things.
Source: Photo courtesy of Arnoud R. Beem

Rock 'n' Roll Despite the success of music formatting, stations needed something new to draw listeners back to radio – and rock 'n' roll saved the day. Young people went crazy over this new sound, and parents went crazy trying to keep their teens away from this wild new music. The more teens were told they could not listen, the more they wanted to listen, and rock 'n' roll stations flourished.

About the same time that rock 'n' roll steamrolled its way onto the music scene, radio stations were experimenting with disc jockeys (DJs) whose job it was to play the tunes and establish rapport with listeners. Cleveland's Alan Freed was the first DJ to introduce teenagers to this exciting new sound and to play rhythm-and-blues and Black versions of early rock to his mostly White audience. He is probably the most influential DJ of all time and is best known for dubbing the new music 'rock 'n' roll,' a term that had been used in many blues songs as a euphemism for having sex.

The new sound combined the rhythm-and-blues sound of Memphis with the country beat of Nashville and thus became the first "integrationist music" (Campbell, 2000, p. 72). Rock 'n' roll was embraced by many different types of listeners and thus provided a way to break away from the "racial, sexual, regional, and class taboos" of the 1950s (Campbell, 2000, p. 76). The music united Blacks and Whites, rich and poor, men and women, and Northerners and Southerners.

Some parents, teachers, and legislators called for banning rock 'n' roll, and some communities even held record-burning bonfires. Teenagers and rock 'n' roll enthusiasts were fiercely pitted against those who tried to keep the new tunes off the airwaves and out of the record stores. Newspapers ran front-page stories featuring the conflict between anti-rock crusaders and pro-rock activists. The resulting publicity sparked a renewed interest in radio, and as ratings shot up, protests against rock 'n' roll subsided. Rock 'n' roll became the music that defined the baby-boom generation and continues as a popular station format today.

Capitalizing on the appeal of rock 'n' roll, radio programmers Todd Storz and Bill Stewart created the first **Top 40 format** for a station in Omaha, Nebraska. Storz and Stewart listened to the music that other stations were playing and discovered that at any one time, the number of different hit songs was about 40 – hence the name Top 40. Storz and Stewart's new format featured up-and-coming hit songs and current hits.

The Top 40 format caught on with many stations across the country and appealed especially to young people, who wanted to keep up with the latest and most popular songs. But Top 40 music playlists were largely based on local or regional preferences and decency standards. Before cable music entertainment programs, pop culture magazines, the Internet, and other music sources, there were few ways for teens on the East Coast to know what West Coasters were listening to – that is, until Casey Kasem aired *Billboard* magazine's top national records of the week on his *American Top 40* program. At its peak, the program aired on more than 500 radio stations across the country, reaching millions of young music fans. After 39 years on the air, Kasem recorded his last show in July 2009.

Rock 'n' roll was boosted, ironically, by television. American teenagers were wild for *American Bandstand* (1957–1989), a Saturday-afternoon dance show that featured artists playing the latest hits. Originally broadcast from a studio in Philadelphia, operations moved to Los Angeles in 1963 at the height of the show's popularity. As many as 20 million viewers learned the latest dance moves and were sometimes the first to see new bands and singers. Host Dick Clark helped launch classic rock

Figure 4.3 Television host Dick Clark presides over the set of his show *American Bandstand* as teenagers dance to Top 40 popular music, circa 1957 in Philadelphia.
Source: Photo courtesy of Michael Ochs Archives/Getty Images

'n' roll artists like Jerry Lee Lewis, The Beach Boys, Madonna, Steely Dan, The Jackson 5, and KISS. *American Bandstand* largely promoted mainstream rock 'n' roll to White, middle-class teens. There was not a dance show counterpart for African Americans and the rhythm-and-blues (R&B) scene until Don Cornelius introduced *Soul Train* in 1971. For the next 12 years, *Soul Train* help popularize R&B, soul, rock, and hip-hop artists such as Paula Abdul, Mary J. Blige, James Brown, the Supremes, Sly and the Family Stone, Smokey Robinson, Elton John, and David Bowie. *American Bandstand* and *Soul Train* fans set the demand for certain songs and artists. Rock 'n' roll was initially thought of as a flash-in-the-pan musical style that would be popular only among a small number of teenagers. However, it proved to be the magic formula that rekindled radio listenership.

SEE IT NOW: RADIO

Types of Programs

Commercial radio of today is more standardized than before and stations stick to a preset format. Most stations play some genre of music, or they are news/talk formatted, while others are sports oriented. Non-commercial college or independent stations are freer to broadcast programs and music of local interest, while non-commercial stations that are affiliated with National Public Radio play a mix of network and local programming.

Music

Since the late 1950s, it has been standard for stations to establish an identifying format. A station chooses

Figure 4.4a & 4.4b Fats Domino and Chuck Berry were among the Black performers who helped usher in rock 'n' roll.
Source: Photo courtesy of Warner Bros./Photofest.
© Warner Bros. Courtesy Photofest

and changes a format based on budget, local audience characteristics and size, the number and strength of competing stations, and potential advertising revenue. Radio stations generally foot the bill for their own local programming. The stations pay for the talent, production equipment, physical space, and copyright fees (which go to music licensing agencies), but most of the music itself is provided free of charge by recording companies. It is very common for recording companies to send demo discs and downloads of new releases to stations across the country. This arrangement works well for stations that play current music but not so well for stations that play classical music or other old tunes. There is little financial incentive for recording companies to send out free CDs of 200-year-old classical pieces, which will draw small audiences, or 30-year-old classic rock 'n' roll that listeners are not going to rush out to buy, so stations that specialize in older or more specialized music often have to purchase their own music libraries or depend on donated music.

ZOOM IN 4.4

Learn more about different types of music and station formats at these Web sites:

- Guide to Radio Station Formats: http://newsgeneration.com/broadcast-resources/guide-to-radio-station-formats/
- Radio Station World: http://radiostationworld.com/

Given the cost of producing enough programming to fill airtime 24/7, some radio stations subscribe to network and syndicated programs, such as the popular *Delilah*. Network programs are usually all inclusive, meaning that music, news, commercials, on-air talent, and other program elements are supplied to a station. The station merely puts on the air 24 hours of packaged programming.

Syndicated programs, on the other hand, are individual shows that a station airs in addition to its own original programs. For instance, a local station may broadcast its own music until noon, air a syndicated talk show until 3 p.m., and then return to its own programming for the rest of the day.

Network and syndicated programs are commonly delivered via satellite and the Internet from radio networks and syndicators. Stations pay cash and/or negotiate for commercial time in exchange for network and syndicated programming. Some stations might cut a deal in which they get programming free of charge, but the network or syndicator reaps revenue by selling commercial time within a show to national advertisers.

FYI: TOP RADIO STATION FORMATS (2019) BY AUDIENCE SHARE

1. Country (13.2%)
2. News/Talk (12.0%) (commercial and non-commercial)
3. Adult Contemporary (AC) (8.6%)
4. News/Talk Commercial (8.3)
5. Pop CHR (7.3%)
6. Classic Rock (6.1%)
7. Classic Hits (5.8%)
8. Hot AC (4.7%)
9. Urban AC (4.1%)
10. Contemporary Christian (3.9%)
11. Urban Contemporary (3.8%)
12. All Sports (3.7%)
13. Rhythmic CHR (2.7%)
14. Mexican Regional (2.5%)
15. Alternative (2.0%)
16. Adult Hits + 80s Hits (2.0%)
17. Active Rock (2.0%)
18. Album-Oriented Rock (AOR) + Mainstream Rock (1.7%)
19. Spanish Contemporary/Spanish AC (1.6%)
20. Classical Music (1.5%)

Source: Audio Today, 2019

ZOOM IN 4.5

Learn about some different radio networks at these sites:

- Premiere Radio Networks: www.premrad.com
- Westwood One: www.westwoodone.com

News and Information

About two-thirds of 'all-news' and 'news/talk' stations in the United States are on the AM dial. In 1961, the Federal Communications Commission (FCC) opened up spectrum space for FM, which has a sound quality superior to AM, and authorized stereo broadcast FM. Almost all stations remained on the AM spectrum, largely because most radios were manufactured with only an AM tuner and could not receive stereo FM. As listener demand for stereo FM increased, more stations began moving to the richer-sounding FM dial, and manufacturers included an FM dial on most radio sets.

By the early 1970s, only about 30% of listeners tuned exclusively to FM stations, but 20 years later, three-quarters of radio listeners preferred FM, and now about 86% of radio listeners prefer the FM band. Music-formatted AM stations could not come close to the sound quality emanating from their FM competitors, so they either migrated to the FM band or stayed on the AM band but changed to a talk or news format for which audio quality was not so important.

News, talk, and sports programs are generally very expensive to produce, so it is usually only the big-market stations that create their own programs, while others rely on syndicators to provide shows hosted by big names such as Sean Hannity and Steve Harvey. The news/information format falls into three overlapping categories:

1. **All-news** stations primarily air national, regional, local news, weather, traffic, and special-interest feature stories. The reports are usually scheduled throughout the day. The news cycle may occasionally be interrupted for a special in-depth report or talk show or another program that is not part of the usual schedule. Instead of employing beat reporters, stations air news that they obtain from wire services or other news sources.
2. The **sports/talk** format is similar to the news/talk format but focuses mainly on sports issues and news and live sporting events. Sports talk usually includes call-in programs and may emphasize local sports, especially if a professional or big-name college team is located within the broadcast area.
3. The **news/talk** format consists of a combination of call-in talk shows and short newscasts that usually come on once an hour or between segments. Most talk programs air during regularly scheduled times and usually last from one to four hours. News/talk shows are often politically oriented, mostly conservative.

Non-commercial Radio

Non-commercial stations have long existed as educational radio with the aim of providing traditional school subjects and scientific and social information to children and adults. Many educational institutions held broadcast licenses for educational purposes, however each station had to produce its own programs.

Non-commercial independent radio stations are usually owned and operated by colleges and universities, religious institutions, and municipalities. Most non-commercial stations rely on their own music library of donated CDs and free downloads, but they also produce their own community-affairs and news shows.

To support the continued creation and distribution of high-quality educational non-commercial content, Congress established the **Corporation for Public Broadcasting (CPB)** in 1968. The CPB, in turn, set up the National Public Radio (NPR) network in 1970. NPR's mission "is to work in partnership with its member stations to create a more informed public – one challenged and invigorated by a deeper understanding and appreciation of events, ideas and cultures" (What Is NPR? 2019). NPR distributes both its own and independently produced programs to its member stations and its programs air on slightly more than 1,000 public stations, attracting 27.4 million over-the-air listeners each week. NPR has won many programming awards and is considered one of the most trusted news sources. NPR delivers such favorites as *Morning Edition* and *All Things Considered*.

Most non-commercial stations, whether or not affiliated with PBS, broadcast at the lower end of the FM dial, between 88 and 92 Mhz. Non-commercial stations are not as rigidly formatted as commercial stations. Listeners are often treated to a variety of music, such as classical, jazz, alternative rock, and blues. Additionally, news, talk, and sports programs may be thrown into the mix. The eclectic nature of non-commercial stations draws listeners who are tired of the same old rotated music offered by most commercial stations. Well-known bands and artists such as Nirvana, U2, The Cure, Elvis Costello, Nine Inch Nails, B-52s and R.E.M. owe much of their success to college radio.

How Radio Programs Are Scheduled

Airtime is a valuable commodity, and so every second needs to be scheduled. In the early days of radio, songs that made it to the airwaves and how often they were played was largely left to the DJ's discretion. After the **payola scandal** in the late 1950s, when DJs were caught taking money from record companies in exchange for playing and promoting new releases, stations shifted the programming function from the DJs to a **Program Director**, who is responsible for everything that goes out over the air.

The **program director** sets up program clocks (also known as **hot clocks** and **program wheels**), which are basically minute-by-minute schedules of music, news, weather, commercials, and other on-air offerings for each hour of the day. For example, the first five minutes of the noon-to-1:00 p.m. hour may be programmed with news, followed by two minutes of commercials, followed by a 15-second station promotion, followed by a 10-minute music sweep (music that is uninterrupted by commercials). Each hour may vary, depending on audience needs and time of day. For instance, morning drive-time music sweeps may be cut by 30 seconds for traffic updates, and an extra five minutes may be set aside for news during the 6:00 p.m. hour.

The director also comes up with a playlist of music, which is basically a roster of songs and artists featured on the station, and sets the **rotation**, which is the frequency and times of day the songs are scheduled. Favorite songs and current hits are usually rotated frequently, and older and less popular numbers may have a less frequent rotation. Most stations try to avoid playing different songs by the same artist too closely together or playing too many slow-tempo or similar-sounding selections in a row.

ZOOM IN 4.6

- Learn more about NPR at www.npr.org
- Tune in to your local NPR member station. Listen to at least one news program, one music program, and one other program of your choosing. How do NPR programs differ from those on commercial radio stations?
- Listen to your college radio station. How is it different from the commercial stations in your area?

ZOOM IN 4.7

BEST-SELLING ARTISTS THROUGH TIME (1969–2019)

Check out this YouTube video that takes you through a moving timeline of the best-selling artists through time 1969–2019: www.youtube.com/watch?v=a3w8l8boc_I

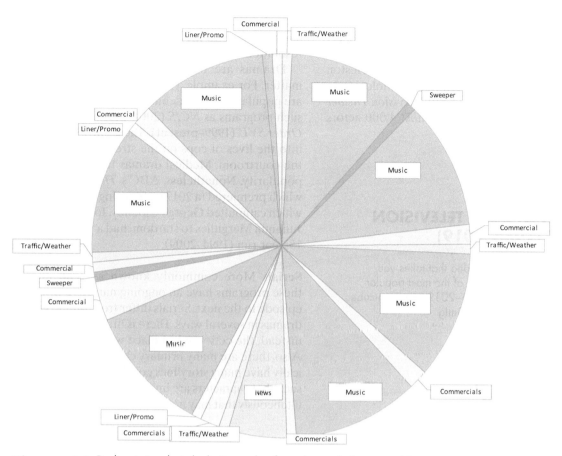

Diagram 4.1 Radio station hot clock. Example of one hour of afternoon drive time.

SEE IT THEN, SEE IT NOW: TELEVISION PROGRAMS

Types of Television Programs

In the early days of television, executives were faced with the challenge of what to put on the air. After all, people were not going to buy television sets if there was nothing to watch. Program executives smartly adapted existing radio programs for television. Instead of just using sound stages where radio actors spoke into microphones, sets were built and scenes were acted out in front of a camera. Many radio programs soon had televised counterparts, and in the process loyal fans followed the programs from one medium to the other.

There are many **genres** (or types) of television programs, each of which is distinguished by its structure and content. Television programs can be categorized as either narrative or non-narrative. **Narrative programs** weave a story around the lives of fictional characters played by actors. Dramatic programs and situation comedies, such as *The Good Doctor*, *Young Sheldon*, *Mixed-ish*, and *The Connors* present fictional stories that are scripted and acted out.

Even though it can be argued that every television program tells a story, **non-narrative programs** tell stories that are real and do not come from fictional scripts. Game shows, talk shows, reality shows, sports, and news are all types of non-narrative programs. Non-narrative programs present real situations and real people (e.g. game show contestants, news anchors, sports stars), not actors.

Narrative Programs

Anthologies In the early days of television, New York City was the central location for network headquarters and television studios, and thus programs were influenced by Broadway plays. Live productions of serious Broadway dramas were common fare on television in the late 1940s and early 1950s. These anthologies were hard-hitting plays and other works of literature that were adapted for television. They were often cast with young talent from radio and local theater, some of whom went on to become major theatrical and television stars. Robert Redford, Joanne Woodward, Angela Lansbury, Sidney Poitier, Paul Newman, and Charlton Heston all got their starts acting in televised anthologies. In its 11-year run on television, *Kraft Television Theater* produced 650 plays, featuring almost 4,000 actors and actresses.

ZOOM IN 4.8

MOST POPULAR TELEVISION SERIES (1986–2019)

Check out this YouTube video that takes you through a moving timeline of the most popular television series from 1986–2019: www.youtube.com/watch?v=7DemM7UGmlg

ZOOM IN 4.9

To watch old television programs of various genres, click on http://oldmovietime.com/tv/

Anthologies helped boost the fledgling television industry by attracting deep-pocket sponsors and well-educated and affluent viewers who were most likely to buy expensive television sets. Anthologies were enormously popular into the late 1950s. They lost their appeal toward the end of the decade when expanded production capabilities led to on-location, action-oriented shows that drew viewers away from anthologies, and where the viewers went, so did the sponsors.

Dramas A dramatic series presents viewers with a narrative that is usually resolved at the end of each episode; in other words, the story does not continue from one episode to the next. A drama typically features a recurring cast of primary characters who find themselves involved in some sort of situation, often facing a dilemma that gets worked out as the action peaks and the episode comes to a climax and resolution.

Dramas are often subcategorized by subject matter. For example, police and courtroom dramas are popular today, as demonstrated by the ratings of such programs as *NCIS* (2003–present) and *Law & Order SVU* (1999–present). These shows give a look into the lives of cops on the streets and lawyers in the courtroom. Medical dramas fade in and out of popularity. Nonetheless, ABC's *The Good Doctor*, which premiered in 2017 is still going strong, and *ER*, which catapulted George Clooney, John Stamos, and Julianna Margulies to stardom, had a very successful 15-year run (1994–2009).

Serials More commonly known as **soap operas**, these programs have an ongoing narrative from one episode to the next. Serials differ from other types of dramas in several ways. There is little physical action; instead, the action takes place within the dialogue. Also, there are many primary characters. Serials typically have many storylines going on at the same time, such that characters are involved in several plots simultaneously that may not be resolved for years, if at all.

Figure 4.5 Ryan O'Neal, Mia Farrow, and Barbara Parkins in *Peyton Place*.
Source: Photo courtesy of ABC/Photofest. © ABC

Figure 4.6 Scene from *Game of Thrones*.
Source: AA Film Archive / Alamy Stock Photo

When it finally seems like a resolution is at hand (say, a marriage and a happy life), a twist in the story leads to more uncertainty (did she unknowingly marry her long-lost brother?) and to a new, continuing storyline. *General Hospital* (1963–present) and *The Young and the Restless* (1973–present) are popular soap operas of today. *Guiding Light*, however, was the king of soap operas and holds the distinction of being the longest-running scripted (non-animated) program in radio and television history. The show made its radio debut on January 25, 1937, and began airing on television in 1952. It was still broadcast on radio until 1956, when it moved exclusively to television until it was canceled in 2009.

Although most soap operas air during the daytime and are targeted primarily to women, the prime-time hours (8 p.m.–11 p.m. EST and PST; 7 p.m.–10 p.m. CST and MST) have also seen their share of soaps (however, to attract male viewers, the networks are careful not to call them 'soaps'). Serial dramas were particularly popular on primetime in the 1980s, when viewers were treated to a peek into the fictional lives of the rich on such programs as *Dallas* (1978–1991), *Dynasty* (1981–1989), and *Falcon Crest* (1981–1990), and they seem to have resurged with *Desperate Housewives* (2004–2012), *Mad Men* (2007–2015), and most notably *Grey's Anatomy* (2005–present).

FYI: MOST VIEWED NETWORK TELEVISION SHOWS IN 2018/2019

Sunday Night Football (NBC) 19.2 million
Big Bang Theory (CBS) 17.4 million
NCIS (CBS) 15.9 million
Game of Thrones (HBO) 15.3 million
Young Sheldon (CBS) 14.6 million
NFL Thursday Night Football (Fox/NFL) 14.4 million
This is Us (NBC) 13.8 million
Blue Bloods (CBS) 12.8 million
FBI (CBS) 12.8 million
The Good Doctor (ABC) 12.6 million

Source: Statista

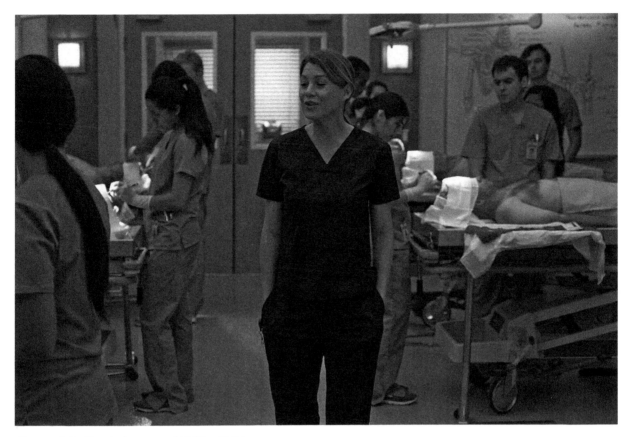

Figure 4.7 *Grey's Anatomy*, 'Walking Tall.'
Source: Photo courtesy of Richard Cartwright/ABC via Getty Images

Probably one of the most-watched serial dramas of all time was the fantasy *Game of Thrones* (2011–2019). Set in the fictional Seven Kingdoms of Westeros and the continent of Essos, the show featured medieval political intrigue, including dark magic and dragons. The dark tone, good vs. evil theme, and multiple storylines kept the audience rapt. The show garnered many Emmy and Golden Globe Awards.

Telenovela ('tele' meaning television and 'novella' meaning a literary work, or in some languages meaning 'romance') is a type of short-run serial that originated in Latin American countries. Hugely popular telenovelas captivate viewers in Mexico, Central and South America, Spain, and even Russia and China. In the United States, the emotionally charged, passionate telenovelas are typically aired on the Spanish channels, Univision and Telemundo.

Through its MyNetworkTV, Fox Television brought telenovelas to the United States in 2006, but with little success. Ultimately, the channel aired six telenovelas (*Desire, Fashion House, Wicked Wicked Games, Watch Over Me, American Heiress,* and *Saints & Sinners*), each showing five nights per week for 13 weeks. Despite the genre's popularity in other countries, MyNetworkTV telenovelas did not draw an enthusiastic audience in the U.S. and were dropped after only one year.

FYI: 71ST ANNUAL EMMY AWARD WINNERS (2019)

Outstanding Comedy Series: *Fleabag*
Outstanding Lead Actor in a Comedy Series: Bill Hader as Barry *(Barry)*
Outstanding Lead Actress in a Comedy Series: Phoebe Waller-Bridge as Fleabag *(Fleabag)*
Outstanding Drama Series: *Game of Thrones*
Outstanding Lead Actor in a Drama Series: Billy Porter as Pray Tell *(Pose)*
Outstanding Lead Actress in a Drama Series: Jodie Comer as Villanelle *(Killing Eve)*

Source: Academy of Television Arts & Sciences

Situation Comedies Situation comedies (sitcoms) are usually half-hour programs that present a humorous narrative that is resolved at the end of each episode. Sitcoms feature a cast of recurring characters who find themselves caught up in some crazy predicament. Situation comedies are perfect for television, because they can be shot in-studio on a typical three-sided stage that is decorated to look like a simple apartment or home. Most early sitcoms were family oriented, and the comic aspect was noted in the dialogue between the characters rather than in visual gags. However, Lucille Ball excelled at physical comedy and changed the face of television by insisting on using three cameras to film *I Love Lucy* (1951–1957). If a physical gag or antic failed, the scene could be reshot or edited for maximum effect. Multiple-camera filming paved the way for other comedies that featured more physical comedy than comedic dialogue.

ZOOM IN 4.10

- Connect to YouTube to watch clips and episodes of *I Love Lucy*.
- Listen to the *I Love Lucy* introduction music at: www.youtube.com/watch?v=-t4ql-r406Q

Situation comedies are often criticized for stretching the limits of what is considered funny. The airwaves of the 1970s and 1980s were filled with social-consciousness sitcoms. *All in the Family* (1971–1979) and *Maude* (1972–1978) lightheartedly poked fun at serious issues like the Vietnam War, abortion, feminism, and racism. These shows boldly took on controversial subjects, artfully and humorously addressed sensitive topics, and elevated the collective consciousness to greater acceptance and understanding of people of all types.

Missing from the networks' line-up since the cancellation of *Beulah* in 1953 were sitcoms featuring African Americans. When *Julia* premiered in 1968, it was hailed as the first starring television role that depicted an African American woman as an intelligent, educated, sensible person. The character Julia, played by Diahann Carroll, was a widowed single mother who worked as a nurse. The groundbreaking show's three-year run, however, was fraught with controversy, as critics claimed that Julia's middle-class life as a single mother was unrealistic.

Julia led the way to other African American-centered sitcoms, such as *Sanford and Son* (1972–1977), *The Jeffersons* (1975–1985), *What's Happening!!* (1976–1979), *Diff'rent Strokes* (1978–1986), and *Benson* (1979–1986). The 1980s brought *Webster* (1983–1989) and *The Cosby Show* (1984–1992). Over the next three decades, about 60 African American sitcoms premiered, including the popular *Family Matters* (1989–1998), *Fresh Prince of Bel-Air* (1990–1996), *Moesha* (1996–2001), *The Bernie Mac Show* (2001–2006), *That's So Raven* (2003–2007), *Everybody Hates Chris* (2005–2009), and *Black-ish* (2014–present). The concern that such shows are overly stereotypical continues, as does discomfort that some Black characters, such as on *The Cleveland Show* (2009–2013), were voiced by White actors.

Facing formidable competition from comedy cable channels and humorous online videos, contemporary network sitcoms, such as *Modern Family* (2009–2020) and *The Big Bang Theory* (2007–2019) continue toppling long-standing television taboos in the guise of humor. Modern sitcoms jab at drug addiction, drunkenness, casual sex, vomit, and teenage pregnancy, among other gross and sensitive topics. Humor often makes it easier for viewers to confront what they dislike most, but laughing at ourselves and at our foibles is a healthy way to release tension.

Movies and Miniseries

Theatrical movies, albeit mostly low-budget ones such as westerns, made their way onto television screens in the mid-1950s, but only after American movie studios took a cue from their British counterparts that profits could be gleaned by renting films to television stations. For example, eccentric millionaire and RKO studio owner Howard Hughes sold older films to General Tire & Rubber Company to air on its New York television station. NBC's *Saturday Night at the Movies* debuted in 1961 and was quickly followed by ABC's *Sunday Night at the Movies*. By the mid-1960s, television schedules across the country were filled with theatrical films, which drew very large audiences. When Alfred Hitchcock's *The Birds* hit the television airwaves in 1968, five years after its theatrical release, it reaped 40% of the viewing audience. Another movie classic, *Gone with the Wind*, captured half of all viewers for two nights in 1976.

Figure 4.8 Jackie Gleason, Art Carney, Audrey Meadows, and Joyce Randolph (from left to right) on the set of *The Honeymooners*.
Source: Photo courtesy of CBS/Photofest. © CBS

In the 1970s, the overabundance of theatrical movies on television led to a new program format, **made-for-television movies**. Different from theatrical releases, which are produced for showing in movie theaters, made-for-television movies are written, produced, and edited to accommodate commercial breaks. Although many made-for-television movies are low-budget productions, some – such as *Brian's Song* (1971), *Women in Chains* (1972), *Burning Bed* (1984), and *The Waltons' Thanksgiving Story* (1973) – drew larger audiences than some hit theatrical films.

The made-for-television movie spawned the television **miniseries**, which is a multipart, made-for-television movie that airs as several episodes rather than in one installment. Miniseries often tackle controversial issues. Successful miniseries from the 1970s and 1980s included *Lonesome Dove*, *Holocaust*, *Shogun*, *The Winds of War*, and, most notably, the 12-hour *Roots* (1977), which won nine Emmys and one Golden Globe award. Just over half of U.S. viewers tuned in for the final episode of *Roots*, and 85% watched at least some part of the program.

Contemporary miniseries are usually scheduled to air on consecutive nights or over successive weeks. As such, they are known as short-form and long-form productions, respectively. After the riveting 30-hour *War and Remembrance* (1988–1989), which cost more than $100 million to produce, ratings for such long-form miniseries plummeted. However, FX's *The People vs. O.J. Simpson: American Crime Story* (2016), was a spellbinding 10-episode series that won nine Emmy awards. The networks have

Figure 4.9 The cast of *I Love Lucy* (from left to right): Lucille Ball, Vivian Vance, Desi Arnaz, and William Frawley.
Source: Photo courtesy of CBS/Photofest. © CBS

generally turned to shorter 4- to 7-hour miniseries, such as *Dune* (2000), *Generation Kill* (2008), and *The Loudest Voice* (2019).

These days, made-for-television movies and theatrical films abound on television. Films are so much in demand that the movie industry adapted its release schedule to accommodate later television viewing. A theatrical film is first released to the movie theaters and then pulled when its box office receipts drop. Sometimes a film will be held for a second release, but usually within about six months of the final theatrical run, the movie is made available on streaming services and distributed for DVD rental and later for sale. Sometimes a film will be shown on pay per view and later licensed to pay-cable networks, such as HBO and The Movie Channel. After a film has made the rounds on the cable channels, it may be shown by a network-affiliated or independent station.

Non-narrative Programs

Variety Shows Radio variety shows were largely limited to audio acts such as singing and comedy and thus were one of the first types of programs to transition to television. Visual acts such as juggling, dancing, and magic tricks are much better suited for television than for radio. The audience loved these programs that often showcased new talent. Programs such as *The Ed Sullivan Show* (1948–1971), *The Red Skelton Show* (1951–1971), *The Jackie Gleason Show* (1952–1971), and *The Carol Burnett Show* (1967–1978) featured singing and dancing, stand-up comedy, and comedic skits, along with other lighthearted entertainment.

Variety shows often provided viewers the first glimpse at new talent such as Humphrey Bogart, Bob Hope, and Lena Horne. Elvis Presley's 1956

Figure 4.10 Modern Family.
Source: Photo by Eric McCandless/ABC via Getty Images

appearance on *The Ed Sullivan Show* (1948–1971) made headlines when producers refused to show his gyrating hips and only aired close-ups from the waist up. Nevertheless, Elvis's appearance attracted 83% of the viewing audience, the largest share in television history. On February 9, 1964, *The Ed Sullivan Show* hosted The Beatles' American television debut in front of a screaming studio audience of mostly teenage girls.

The Smothers Brothers Comedy Hour (1967–1969) was probably the most controversial variety show of all time. It was wildly popular among young adults, who loved how it mocked politics, government, church, family, and almost everything and everyone. Network censors often butted heads with the show's stars over the irreverent nature of the material, and many viewers were offended by the content. The show was canceled by CBS but later taken up by ABC and still later by NBC. The antiwar, left-wing gags ultimately ran their course, and the program just faded away after a brief reprise in 1975. The *Carol Burnett Show* (1967–1978) premiered the same year as the *Smothers Brothers* but had a longer and less controversial run. A parody scene of *Gone with the Wind*, in which Burnett as Scarlett O'Hara majestically walks down a staircase wearing a dress made from a window drape that includes the curtain rod, is known as one of the funniest moments in television.

Variety shows have generally fallen off the programming schedules. *All That* (1994–2005) and *The Amanda Show* (1999–2002) are examples of modern variety/comedy sketch programs that had somewhat successful runs. *The Maya Rudolph Show* aired in May of 2014 as a one-time special with the hopes of becoming a regularly scheduled program. The show's concept is a throwback to older skit comedy/song-and-dance shows. Although the show was well received, it has yet to be put on as a recurring show. *Saturday Night Live* is the most well-known, longest-running, non-prime-time variety/comedy/skit program. On the air since 1975, it has become a cultural icon. Grizzled *SNL* viewers contend that the show reached its peak in the early years, while Millennials think it is more hilarious now than ever before.

Game and Quiz Shows Studio-based game and quiz shows have always been very popular with viewers.

Figure 4.11 The Beatles made their U.S. debut on *The Ed Sullivan Show* in New York on February 9, 1964.
Source: Photo courtesy of CBS/Photofest. © CBS

Old shows like *What's My Line?* (1950–1967), *I've Got a Secret* (1952–1967), and *To Tell the Truth* (1956–1968) were all variations on simple ideas that invited the audience to get involved and play along. Shows like *Beat the Clock* (1950–1961) and *Truth or Consequences* (1950–1988) centered on contestants performing outrageous stunts for money and prizes.

At first, quiz show contestants competed for small prizes. Then big-money quiz shows, such as *The $64,000 Question*, became enormously popular after their introduction to prime-time television in the 1950s. When the highly rated quiz show *Dotto* was suddenly canceled in 1958, curiosity grew as rumors flew that the show was rigged. It turns out that producers were giving the questions to contestants in advance of the airing. The investigation kicked off television's first major scandal by revealing that cheating was prevalent on many of the quiz shows. *Dotto's* downfall spurred disgruntled contestants from other shows to tell of similar cheating.

Charles Van Doren, a young, handsome, witty faculty member at Columbia University, was selected as an ideal contestant for NBC's *Twenty-One*. Van Doren's winning streak lasted for 15 weeks. In early 1957, he became a media celebrity, was featured on the cover of *Time* magazine, and was offered a job on the *Today* show. All was well in Van Doren's life until defeated contestant Herb Stempel came forward, claiming that the show was rigged and that he had been forced to deliberately lose to Van Doren. Stempel divulged that the program's producers gave the answers to the favored contestants and coached them on how to create suspense by acting nervous and uncertain about their responses.

After rounds of televised congressional hearings, Van Doren and many other contestants, producers, writers, and others who worked behind the scenes were indicted by a federal grand jury for complicity in the deception. Most defendants received suspended sentences, but Van Doren, who was probably the

Figure 4.12 *Twenty-One* contestant Charles Van Doren answering questions from inside the isolation booth.
Source: Photo by Hulton Archive/Getty Images

most beloved and well-known contestant, suffered greatly. He lost his teaching position at Columbia, as well as his job on the *Today* show. Van Doren passed away in relative obscurity in 2019.

Although it took years for the public to regain its trust in game shows, the power of a good contest eventually overcame the skepticism. Quiz and game shows, such as *Joker's Wild* (1969–1974) and *Concentration* (1958–1978, 1987–1991), have been long-time mainstays of daytime television. Today, the most popular game shows, *Jeopardy!* (1964–present) and *Wheel of Fortune* (1975–present), air during the prime-time access hour. After the quiz show scandals, it took until the 1999 debut of *Who Wants to Be a Millionaire?* before viewers could again enjoy a quiz show during the prime-time hours.

Television viewers love the competitive nature of game shows. It is fun to watch contestants compete for prizes and money. Many game shows are formatted so the television audience can play along. Most viewers have 'bought a vowel' while watching *Wheel of Fortune* or guessed the cost of a refrigerator on *The Price Is Right*. Viewers root for their favorite contestants and develop a feeling of kinship with them, especially when they keep winning, like 2004 *Jeopardy!* champ Ken Jennings, who holds the record for the longest winning streak in television game show history. Viewers sat on the edge of their chairs as they rooted for Jennings as he amassed $2.5 million in prize money across 74 consecutive wins. Viewers were captivated again in 2019 as it looked like James Holzhauer was going to break Jennings' record. Although Holzhauer edged out Jennings in total winnings ($2.7 million), his 32-game winning streak was less than half of Jennings.

Reality Shows Big ratings, audience demand, and low production costs make reality shows a network executive's dream. Such unscripted reality shows account for about 25% of all prime-time broadcast

ZOOM IN 4.11

- Listen to an NPR report about the history of quiz shows and hear short audio clips from quiz programs at: www.npr.org/templates/story/story.php?storyId=93749534

- Learn more about the quiz show scandal at www.pbs.org/wgbh/americanexperience/films/quizshow/

Figure 4.13 The set of *Jeopardy!* hosted by Alex Trebeck.
Source: Photo by Ben Hider/Getty Images

programming. Reality shows generally follow the same format week after week. Will they make the deadline? Which house will they buy? Will they get married? How much weight will they lose? Will she be cured? Even though the answers are usually apparent and predictable, reality shows pull in the audience.

But all that glitters is not gold – at least not for the contestants. Reality show participants do not receive union protection, as do most television actors. Contestants complain of physical isolation, grueling work, sleep deprivation, bad food, and an endless supply of alcohol. Additionally, participants may not be thoroughly vetted. On VH1's *Megan Wants a Millionaire* (2009), one of the contestants went on the lam after being suspected of murdering his ex-wife. After the man was found dead of an apparent suicide, negative publicity forced the network to cancel the show even though it had only been on the air for three weeks.

Reality Situation. *Candid Camera* (1948–1979, 1991–1992, 1996–2004, 2014), which was the first reality-type program, premiered in 1948 under its original radio title, *Candid Microphone* (1947–1948, 1950). The show featured hidden-camera footage of unsuspecting people who were unwittingly involved in some sort of hoax or funny situation that was contrived as part of the program. After catching people's reactions to outlandish and bizarre situations, host Allen Funt would jump out and yell, "Smile, you're on *Candid Camera!*"

America's Funniest Home Videos (1989–present) breaks the typical reality format. Instead of professional camera operators taping antics, ordinary people tape each other doing silly things and send in their recordings with the hopes of having them shown on television and winning a cash prize.

Reality Family. On reality family shows, the camera literally follows family members through their daily lives. The first such program was *An American Family*, which aired on PBS in 1973. It followed a year in the real life of the Loud family from Santa Barbara, California. The program documented the parents' marital discord (which later led to divorce) and the lives of their five teenagers. The audience watched as son Lance struggled with his sexual identity. He became the first openly gay person to appear on television. *An American Family* reflected the changing lifestyles of the times and stood in stark contrast to unrealistic, idealized, fictional family programs, such as *The Brady Bunch* (1969–1974). In 2011, HBO aired

FYI: SURVIVING AN AVALANCHE – HOW *AMERICA'S FUNNIEST HOME VIDEOS* ARE SELECTED

"When *America's Funniest Home Videos* premiered in 1990, it hit with a bang. Within two weeks we were receiving about 5,000 tapes a week so we quickly had to develop a system to handle that avalanche. After each VHS cassette was assigned a unique number and logged into a database, it then went to the screeners. To find the 60 or more clips needed for each half-hour episode the ten screeners had to sift through at least 4,000 per week. Their selections were then scored on a scale of one to ten, with all five and above passing up to me for my review. The screeners had a very high 'kill ratio' but I usually watched at least 80 tapes a day, picking and packaging the best and tossing the rest. I then worked with the voiceover writers and sound effects editors to ensure that each clip got the best possible comedy icing on the cake."

"After the show's Executive Producer, Vin Di Bona, approved the playback reel of clips we took them into the studio and taped the show before a live audience. Each playback reel had at least 15 more clips than needed for each show. That way we could watch the reaction of the studio audience and toss out the tapes that weren't working before editing the final version of the show for air. This basic system, plus talent, dedication, and a unique sense of comedy has made AFHV an international mega-hit for over 20 years."

Source: Steve Paskay, Co-Executive Producer, *America's Funniest Home Videos* (1990–1995)

Figure 4.14 The Osbournes.
Source: Photo courtesy of MTV/Photofest. © MTV

its *Cinema Verite*, a drama about the making of *An American Family*.

A contemporary version of *An American Family* was MTV's *The Osbournes* (2002–2005), which portrayed the everyday family life (sans oldest daughter Aimee) of Ozzy Osbourne, former member of heavy-metal group Black Sabbath. The cameras were even there when Ozzy's wife, Sharon, underwent treatment for colon cancer. Attracting an average of six million viewers per week, *The Osbournes* was one of the most-watched shows ever on MTV. Whether they were throwing a baked ham at the noisy neighbors, potty training their dogs, or turning on yard sprinklers to repel nosy fans, the Osbournes created a whole new program genre: The reality sitcom.

Keeping Up with the Kardashians (2007–2021) is arguably the most popular family reality show of all time. The show documents the lives of the Kardashian-Jenner family, focusing on the five sisters – Kourtney, Kim, and Khloe Kardashian, and Kendall and Kylie Jenner, and parents Kris Kardashian and Caitlyn Jenner.

Reality Crime and Medicine. COPS (1989–2020) was one of the longest-running reality programs. The basic premise is simple: A camera crew follows police on their beats and tapes their confrontations with suspects. Other reality programs, such as *America's Most Wanted* (1988–2012), depict real-life crime situations but were scripted and played by actors reenacting the real-life situation. In *Cold Justice* (2013–2015), a former prosecutor and a former crime scene investigator travel around the country helping local law enforcement solve cold cases. The show has led to many arrests and convictions.

Emergency medical shows like *Rescue 911* (1989–1996) and *Trauma: Life in the ER* (1997–2002), kept viewers on their seats wondering if a life would be saved or if a victim would succumb to his or her injuries. Each episode of *Dr. G. Medical Examiner* (2004–2012) showed real-life medical examiner, Dr. Jan Garavaglia, conducting an autopsy to piece together the chain of events that led to a person's death.

Reality Dating. The forerunner to reality dating programs was *The Dating Game*, which first appeared in 1965 and aired on and off for 35 years. A bachelor or bachelorette questioned three hopefuls who were hidden from view and then selected one for a date. Although most contestants were ordinary people, occasionally stardom's knowns and still unknowns, such as Farrah Fawcett, Steve Martin, Burt Reynolds, Arnold Schwarzenegger, Tom Selleck, Ron Howard, Sally Field, and Michael Jackson, would appear on the show hoping to win a date.

Other dating shows, such as *Elimidate* (2001–2006) and *The Fifth Wheel* (2001–2004) followed couples as they went on dates and got to know each other. Other dating programs have a different spin on love and relationships. On *Parental Control* (2005–2010), parents interviewed potential dates and chose the most likely person to steal their son or daughter away from their current girlfriend or boyfriend. *Dating in the Dark* (2009–2010) literally took the idea of blind dating and made it into a show. Three single men and three single women would live together, but they were not allowed to see each other. They could only talk and get to know each other in the dark. The point was to take personal appearance out of the dating equation. On *The Bachelor/Bachelorette*, the bachelor or bachelorette chooses from among 25 suitors for a spouse. Each episode ends with potential mates being eliminated. In the final episode of the season, two potential spouses remain – one is 'broken up with' and the other gets a red rose and a marriage proposal. The U.S. version of the British *Love Island*, premiered on CBS in 2019. A group of contestants live together in an island villa, and must couple up or they are booted from the island. But perhaps the most risqué dating show was *Dating Naked* (2014–2016). Billed as a social experiment, the show focused on two daters who each went on three naked dates, including with each other. The contestants frolic around a tropical island, enjoying beach and poolside dates in the nude. However, when shown on television, the daters' genitals were blurred, as were the women's breasts – only the buttocks of both genders were shown.

Reality Game. The newest reality rage is the so-called reality game show, such as *Survivor* (2000–present), *The Amazing Race* (2001–present), *Dancing with the Stars* (2005–present), *America's Got Talent* (2006–present), *American Idol* (2002–2016), and *America's Next Top Model* (2003–2015). In these shows, contestants do seemingly impossible tasks to be declared the winner. On *Survivor*, a group is sent to a remote location and left to deal with nature, all while each contestant tries to position himself/herself as the most valuable person. Each week the castaways vote one person off the island until only the winner remains. *American Idol* was a kind of an updated *Gong Show* (1976–1989) but with an interactive twist, in that viewers voted for the winners either by phone, text, app, Facebook, or Google. *Idol* centered on a panel of judges who verbally jabbed at the contestants after they have sung their hearts out and tried their best to win. In the program's first season, about 110 million votes were cast, and by the tenth season votes were coming in to the count of 750 million.

Reality Home Improvement. Another new reality format revolves around home and gardening. Such programs, which are mostly on cable networks, include *Design on a Dime* (2003–present), redecorating for under $1,000; *House Hunters* (1999–present), in which a realtor shows buyers who are relocating nationally or internationally a selection of homes that are for sale; *Yard Crashers* (2008–present), in which a team of landscapers transforms homeowners' yards into amazing outdoor spaces; and *Love it or List it* (2008–present), in which homeowners can choose to 'love' their renovated house or 'list' it and move into a new house.

Sports Televising live sporting events in the early days presented several technological challenges. Directors had to figure out how to get heavy, bulky cameras and other equipment to the site of an event; then, once there, they had to strategize how to position the cameras to capture all the action. Boxing, wrestling, bowling, and roller-derby matches were easy to cover because they were played in relatively small arenas and required little camera movement. The first televised baseball game was between collegiate rivals Columbia and Princeton on May 17, 1939. The action was covered by only one camera that was positioned on the third baseline.

As production capabilities increased and cameras became more portable, television moved to covering more action-packed sports. Multiple cameras could be set up to follow basketball and football games and even to track golf balls as they flew long distances across fairways.

The 1960 Winter Olympics in Squaw Valley, California, was the first Olympics televised in the

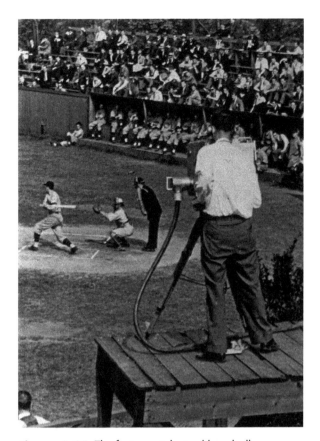

Figure 4.15 The first-ever televised baseball game was played on May 17, 1939, between Princeton and Columbia.
Source: Photo courtesy of Getty Images Sport Collection

United States (the 1936 Summer Games in Berlin was televised in Germany). Hosted by Walter Cronkite, viewers for the first time witnessed the outdoor excitement and challenges of the games. *Wide World of Sports* (1961–1998) was a groundbreaking Saturday-afternoon program that spanned "the globe to bring you the constant variety of sport. The thrill of victory, and the agony of defeat. The human drama of athletic competition." Gymnastics, track and field, skating, rodeo, skiing, surfing, badminton, and demolition derbies were just a few of the sports featured on the show.

Live sports and sports shows are now found on many cable and broadcast networks 24/7. Sports fans rabidly devour shows like *College Gameday*, *Baseball Tonight*, *Sports Center*, *NFL Live*, and *NBA Fastbreak*.

Late-Night Talk Late-night talk shows have been the rage since 1954, when *The Tonight Show Starring Steve Allen* hit the airwaves. Jack Parr, Johnny Carson, Jay Leno, Conan O'Brien, and Jimmy Fallon have all been at the helm of *The Tonight Show*. Whether *The Tonight Show*, *Jimmy Kimmel Live* (2003–present), *A Late Show with Stephen Colbert* (David Letterman retired in 2015 after hosting the show for 22 years), or *Late Night with Seth Myers* (2014–present), these programs follow the same basic format. The shows open with a short comic routine, followed by a skit or funny bit about a current issue or event, and lighthearted banter or serious talk with the nightly celebrity guests. The shows often wrap up with a band dropping a new album. Regardless of the comedic content and ironic nature, late-night talk shows have a strong influence on public opinion.

News and Public Affairs With the entry of the United States into World War II, there was renewed interest in and growth of radio news. **Edward R. Murrow** was perhaps the best-known journalist of that era. He broadcast dramatic, vivid, live accounts from London as it was being bombed. After the war, Murrow hosted *Hear It Now* (1950–1951), a weekly

FYI: MOST-WATCHED PROGRAMS IN U.S. TELEVISION HISTORY (SPORTS)

1 Super Bowl XLIX (Feb. 2015) Patriots/Seahawks	114.4 million
2 Super Bowl XLVIII (Feb. 2014) Seahawks/Broncos	112.2 million
3 Super Bowl 50 (Feb. 2016) Panthers/Broncos	111.9 million
4 Super Bowl LI (Feb. 2017) Patriots/Falcons	111.3 million
5 Super Bowl XLVI (Feb. 2012) Giants/Patriots	111.3 million

Sources: Bibel, 2015; Nielsen, 2017

Figure 4.16 *A Late Show with Stephen Colbert.* Colbert interviewing Dr. Sanjay Gupta during the pandemic. Source: Getty Images

news digest program, which led to *See It Now* (1951–1958), the first nationwide televised news show. *See It Now* was also the first investigative journalism program, a style that has since been imitated by contemporary programs such as *60 Minutes* (1968–present) and *Dateline NBC* (1992–present). Murrow focused not only on the major events of the day but also specifically on the everyday people involved in those events. Many Americans hailed Murrow for his heroic role in bringing an end in 1954 to Senator Joseph McCarthy's rampage about imagined communists. At a time when both legislators and the media were too timid to challenge McCarthy, Murrow stood up for the people who were being persecuted by claiming that McCarthy had gone too far with his accusations.

Making its debut in 1947, *Meet the Press* was the first television interview show and is still the longest-running network program. Program guests discuss politics and current events and treat subjects in a serious manner. *Meet the Press* was the precursor to public-affairs programs such as PBS's *Washington Week* (1967–present) and *PBS NewsHour* (1975–present), which had gone through several title changes since its debut.

News did not become a prominent part of broadcast television until about the 1960s. News events were shot on film and then had to be developed and edited, a time-consuming process that could prevent stories from being ready in time for the next newscast. Beginning in the late 1970s and becoming

ZOOM IN 4.12

- To learn more about Edward R. Murrow and hear audio clips of his news reports and read about this program, *See It Now*, click on: www.otr.com/murrow.html

- To see the original broadcast of Murrow taking on McCarthy go to www.youtube.com/watch?v=vEvEmkMNYHY

- To learn more about Edward R. Murrow and the McCarthy era, watch the movie, *Goodbye and Good Luck*.

widespread in the 1980s, film gave way to battery-powered portable video cameras and quick editing procedures. Stories shot in the field were processed more speedily and on the air within a short period of time. As portable video cameras became common at networks and television stations in the 1980s, news gathering and reporting became more timely and sophisticated.

Generally, a television news show is classified as (1) a local newscast, (2) a broadcast network newscast, or (3) a cable network newscast.

Local Newscast. A local newscast is produced by a television station, which is usually a network affiliate. Local news typically features community news and events. Depending on the size of the market and the community, most local stations have a small staff of reporters (who sometimes double as anchors) who gather the news and write the stories. It is economically unfeasible for each station to send reporters all over the world, so local stations also receive news stories from the wire services, such as the Associated Press (AP) and Reuters, and from their affiliated networks, which transmit written copy and video footage via satellite.

Broadcast Network News. Network news specializes in national and international rather than local news. Prime-time broadcast news programs, as known today, were slow to arrive on television. NBC and CBS expanded their 15-minute newscasts to 30 minutes in 1963, and ABC followed in 1967. In 1969, the networks aired newscasts six days per week and finally every day of the week by 1970. It was Walter Cronkite who brought respect and credibility to television news. On the air from 1962 to 1981, Cronkite's commanding presence, avuncular nature, and stoic and professional delivery earned him the title of the most trusted news anchor. Although he rarely expressed an opinion, when he did, it had a profound impact. For example, when Cronkite called the Vietnam War a stalemate, President Johnson fretted, "If I've lost Cronkite, I've lost Middle America." Cronkite, who died in 2009, set high standards for those who followed, such as David Muir (ABC), Lester Holt (NBC), and Norah O'Donnell (CBS).

Broadcast networks also produce morning and prime-time news programs. *Today* (NBC, 1952–present), *Good Morning America* (ABC, 1975–present), and *CBS This Morning* (CBS 1998–present) wake up viewers with lighthearted information and entertainment fare, sprinkled with a few minutes of hard news. The networks turn to more serious and concentrated reporting of important issues in their prime-time news programs. Shows such as *20/20* (ABC, 1978–present) and *48 Hours* (CBS, 1988–present) bring in-depth coverage of current issues and events.

Cable News Networks. Competition for broadcast network news arrived in 1980, when Ted Turner started Cable News Network (CNN). His critics dubbed it the Chicken Noodle Network and mocked, "It'll never work. No one wants 24 hours of news."

Figure 4.17 *The Today Show.*
Source: People. Nathan Congleton/NBC/NBCU Photobank/Getty

But Turner proved them all wrong. There was and still is an appetite for 24/7 news.

Using newer satellite technology, on-the-scene CNN reporters could shoot live video with portable cameras and uplink it to a satellite that would send it to the studio. During the 1991 Gulf War, CNN reporters relied heavily on satellite delivery to show the world what was happening with up-to-the-minute accounts. The broadcast networks had a hard time competing with CNN because they did not have nearly the same number of news bureaus and reporters in the field, and their on-air news time was limited to several half-hour segments throughout the day. CNN's reputation as a premier source for news was solidified.

CNN's success as a 24-hour cable news network demonstrates the need for an all-day newscast. Viewers are no longer content to wait for the 6:00 or 11:00 local or network newscast to learn about an event that happened hours earlier. They want to see the action when it happens – "All the news, all the time." Cable networks are perfect for filling that need. Today, CNN continues to be one of the most widely watched and relied-upon news sources around the world. Fox News, MSNBC (a Microsoft and NBC partnership), Newsy, ESPN (sports news), and CNBC (financial news) all specialize in around-the-clock international and national news, in-depth analysis, and special news programs.

The viewing public casts a critical eye toward television news. Only about one-half of viewers think MSNBC, Fox News, and CNN are very believable. CNN edges out its competitors, with 58% of viewers claiming that it can be trusted, compared to 50% for MSNBC and 49% for Fox News. Local television news shows fare a bit better. About 65% of viewers state that local television news is somewhat or very credible. The lack of trust in cable news is moving many viewers from television to the Internet as their primary source of news.

The expansion of broadcast network news and the emergence of cable news have cut into newspaper readership. Newspapers and broadcast and cable television executives are concerned that they cannot seem to capture young adults. Only about 16% of those between the ages of 18 and 24 read a newspaper daily compared to 50% of those older than 65. Despite the availability of around-the-clock television news, Millennials know less and care less about news and public affairs than any other generation in the past 50 years. This disinterest among young people brings new challenges to the television news industry to lure this group of viewers to the screen.

FYI: MOST-WATCHED TELECASTS IN U.S. TELEVISION HISTORY (NON-SPORT)

1	Apollo 11 moon walk (July 1969)	125–150 million (500 million around the world)
2	Nixon Resignation Speech (August 1974)	110 million
3	M*A*S*H finale (Feb. 1983)	106.0 million
4	*Roots* (miniseries) Part IV (Jan. 1977)	100.0 million
5	Police pursuit of O.J. Simpson (June 1994)	95 million

Children's Shows

Most children's shows in the early days were variety-type programs that featured clowns, puppets, and animals. *Kukla, Fran, and Ollie* (1948–1957) focused on real-life Fran and her puppet friends. Like many shows of its time, it aired live, but unlike other shows, the dialogue was unscripted. *The Howdy Doody Show* (1947–1960), hosted by Buffalo Bob Smith, featured Howdy Doody, an all-American boy puppet with 48 freckles (one for every state in the union at the time). Narrative programs such as *Zorro* (1957–1959) and *The Lone Ranger* (1949–1957) also were very popular with children, as were science fiction/adventure shows like *Captain Video and His Video Rangers* (1949–1950).

ZOOM IN 4.13

Watch this six-minute clip of the last episode of *Howdy Doody* and find out Clarabelle the Clown's secret: www.youtube.com/watch?v=IJ6ybvlsb4s

Captain Kangaroo (1955–1984), one of the longest-running children's programs of its time, prided itself on being slow paced and calming. Captain Kangaroo

and his sidekick, Mr. Green Jeans, taught children about friendship, sharing, getting along with others, and being kind to animals.

The longest-running and probably the most popular and highly regarded children's show in television history is *Sesame Street* (PBS 1969–2015; HBO/PBS 2015–present; new episodes first air on HBO then are rebroadcast on PBS). The program set the standard for contemporary educational and entertainment children's shows. *Sesame Street* teaches children about words, spelling, math, social skills, logic skills, problem solving, hygiene, healthy eating, pet care, and many other subjects. *Sesame Street* shows that learning is fun by using puppets, like the beloved Muppets, animation, and games. The show has almost 200 Emmy awards, more than any other television program.

Other PBS children's programs include *Odd Squad* (2014–present), *Curious George* (2006–2015), *Arthur* (1996–present), *Postcards from Buster* (2004–2008, 2012), and *Mr. Rogers Neighborhood* (1971–2001). Nickelodeon also airs many children's shows, such as *Sponge Bob Square Pants* (1999–present) and *Dora the Explorer* (2000–2015). The Disney Channel's Playhouse Disney features programs like *Mickey Mouse Clubhouse* (2006–2016) and *Imagination Movers* (2008–2013).

Non-commercial Television

In addition to establishing National Public Radio, the Corporation for Public Broadcasting also created the **Public Broadcasting Service** (PBS) in 1969. PBS is devoted to airing "television's best

Figure 4.18 Buffalo Bob Smith poses with Howdy Doody and Clarabell the Clown on the set of the *Howdy Doody Show* to celebrate its 10th anniversary, 1957.
Source: Photo courtesy of NBC/Photofest. © NBC

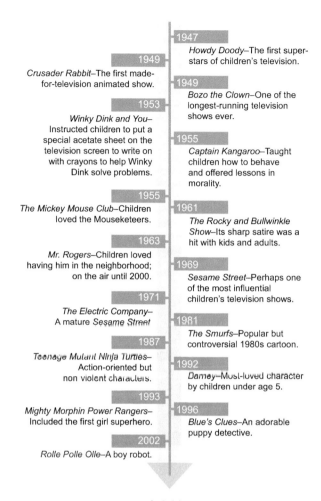

Diagram 4.2 History of children's shows.
Source: McGinn, 2002

children's, cultural, educational, history, nature, news, public affairs, science and skills programs" (About PBS, 2019). PBS has about 350 member stations that are watched by just over 100 million viewers every month. PBS is watched about eight hours per month by about 73.2% of all U.S. television-owning families.

In addition to its children's shows, PBS is known for airing quality educational and entertainment programs. PBS program milestones include *Masterpiece Theater* (1971–present) the longest-running prime-time drama series, *NOVA* (1974–present), *Great Performances* (1972–present), *American Playhouse* (1982–1993), and Ken Burns' documentaries including *Civil War* (1990), *Baseball* (1994), *Jazz* (2001), *The Vietnam War* (2018), and *Country Music* (2019).

Some contemporary PBS programs include *Antiques Roadshow* (1997–present), a traveling show that offers appraisals of antiques and collectibles,

ZOOM IN 4.14

- Learn more about the Public Broadcasting System (PBS) at www.pbs.org
- Watch a PBS program and consider how it differs from a broadcast or cable program of the same genre.

Frontline (1983–present), an investigative documentary program that delves into the issues of today, and *Nature* (1982–present).

The Production and Distribution of Shows

The process of getting a program on the air usually starts with a **treatment**, a description of the program and its characters. Next, the writer or producer pitches the treatment to a network or production company. Even though treatments are accepted from anyone, those submitted by established writers and producers usually get the most serious consideration. The treatment is basically a detailed description of the program and the characters as well as outlines of some episodes. If network executives think the project is worthwhile, they will go forward with a **pilot**, a sample episode that introduces the program and characters to viewers. Pilots are often spectacular productions that draw in viewers. Unfortunately, subsequent episodes are not always nearly as grand as the pilot, and disappointed viewers might stop watching the show.

About 2,000 to 4,000 program proposals are submitted each year to networks and production companies. Of those, maybe 100 are filmed as pilots, between ten and 20 actually make it to the air, and about five stay on for more than a single season.

Producers

Networks and television stations acquire program broadcast rights from independent producers. Most major production companies – such as Columbia TriStar, Warner Bros., and 20th Century Fox – started as movie production houses and have been around for years. Some of the major production companies own or have financial interest in some television networks and may produce programs for them. Major production houses have brought such programs as *According to Jim* (ABC Studios) and *Cold Case* (CBS Paramount Networks). In contrast to the huge conglomerates with ties to the networks,

Figure 4.19 Several of the *Sesame Street* characters.
Source: Photo courtesy of PBS/Photofest. © PBS

there are small independent production houses. For example, Bochco Productions produced *NYPD Blue*, *Philly*, and *Hill Street Blues*, and Cherry Productions kick started *Desperate Housewives*.

Independent producers also distribute their programs on cable networks such as HBO and Showtime, and along with pay-per-view and video-on-demand channels, they offer movies, sports, and special events.

When a network or production company produces a television program, it pays the costs. Producing just one episode can cost millions of dollars, but rarely is an episode sold for more than the cost of production. Producers therefore create programs with an eye toward being on the air long enough for either the profitable network prime-time airing and off-net syndication, or first-run syndication.

Broadcast Networks and Station Affiliates

Television stations air programs that they get from their affiliated network as well as from shows they produce themselves, such as daily newscasts, local public-affairs shows, weekend travel programs, weather and traffic reports, live sporting events, and other programs of interest to the local broadcast community.

Broadcast networks produce and pay the production cost for their own programs, and they pay independent producers for the rights to air programs on their affiliated stations. Specifically, a network buys the rights to a program, with the producer maintaining ownership. A broadcast network then pays its affiliated stations (through a contractual relationship) to air the programs the network produced or obtained from an independent producer. The amount of compensation depends on the affiliate/network contract, the popularity of a program, and other factors including station and size of the viewing audience. In turn, the network makes money by selling commercial time to advertisers. In some cases, such as for a very popular program or game or event, a station might pay the network for the privilege of airing the program, a move known as **reverse compensation**.

CAREER TRACKS: JENNIFER MCCLURE-METZ, FREELANCE TELEVISION EXECUTIVE PRODUCER

WHAT IS YOUR JOB?

I am a freelance television executive producer. I have been working in the television industry since right after graduating from Southern Illinois University-Carbondale in 1997 with my Bachelor of Arts degree in Radio/TV. I have worked on many unscripted shows throughout my career including Amazing Race, Survivor, I Survived a Japanese Gameshow, The Voice, Sunday Best, Freakshow, Married at First Sight, Growing Up Chrisley, and The Biggest Loser.

WHAT LED TO YOUR PRESENT JOB?

Growing up, I was inspired by television, the way it could transport and entertain me. I was interested in theater, drama, writing, music, and creating narratives. I think all of those interests came together in television, particularly unscripted television, where you have a lot of freedom to be creative and experiment. Things aren't as regimented as scripted television or film and you can dabble in a lot of different areas, rather than focus on only one small part of the big picture. I like the idea of being able to touch many parts of a show that I work on. I worked my way up in television, starting as a tape logger for a wildlife series. Always the first to arrive and the last to leave, my work ethic was and continues to be what leads me to success. I am always striving for better and am extremely passionate about whatever show I'm doing. Every single show teaches me a lesson, something that I can add to my playbook and I have to be open to receive that gift of constantly learning, so that I can identify the lesson learned from each experience. I always had the goal of being an Executive Producer and continued to take steps towards that goal throughout my career.

Figure 4.20 Jennifer McClure-Metz.

WHAT ARE YOUR PRIMARY RESPONSIBILITIES?

I am a show runner, which means that I am in charge of the crew – hiring and firing the team, working with networks and production companies to create and refine the creative vision of a show pre-production through post-production. I have to be a jack of all trades and a detective of sorts, if there's something that isn't working I have to figure out why and fix it. Although I work in unscripted television, there is a creative vision that must be written, on some shows it's a logline or creative plan, on others it's a full outline of the story. On some shows, I write host dialogue, or create challenges or activities. I run field production, ensuring safety of the cast and crew, stick to budgets and schedules and drive creative to ensure we have the most compelling content we can. I conduct interviews and dig to find real, relatable and emotional stories. I work in edit bays, working with editors to create a visual story and work with composers to create the musical style of a show. I supervise the finishing processes of the show (color correction and sound mix) to ensure the show looks and sounds the best it can. I also manage a ton of personalities within the crew, production company, and at the network. I ensure everyone's vision can be aligned.

WHAT ADVICE WOULD YOU HAVE FOR STUDENTS WHO MIGHT WANT A JOB LIKE YOURS?

You have to love it. You have to learn as many skills as possible. But also focus on the specific portion of television and focus on getting jobs around that goal. I also suggest that you find a mentor, someone who is doing what you want to do. Remember you can change your mentor at any time. I have had several over my career and continue to feel thankful that I listened to what they were willing to teach. Television production is hard, it's long hours and dedication. If you want to be successful, you have to have a thick skin and be tenacious. If you love it, are willing to put in extreme hours and feel passionate about it, you will likely be successful.

Hoping to promote competition among competing television stations, the FCC enacted the **prime-time access rule** that was implemented in the 1971 television season. FCC's goal was to encourage more local programming at the station level and to give independent producers and stations one hour to show their programs before the start of primetime. That is why even today the hour before primetime is dominated by game shows (e.g. *Jeopardy!*) and other low-budget programs (e.g. *Entertainment Tonight*).

Syndicators and the Fin/Syn Rule Television syndicators are companies that distribute television programs to television stations, cable television networks, and other media outlets. Television affiliates and independent stations depend on syndicated material to fill out their broadcast schedules. There are two basic types of broadcast syndicated programs: First-run and off-network.

First-run syndicated programs are made for non-prime-time network airing. First-run syndicated shows mostly fill the morning, afternoon, and early-evening time slots. Examples of **first-run syndicated programs** include *Jeopardy!*, *Wheel of Fortune*, and *Judge Judy* (syndicated by CBS Television Distribution), and *The Ellen Degeneres Show* (Telepictures Productions Inc.).

Off-network syndicated programs are those that once aired, or are still shown, as regularly scheduled programs on one of the broadcast networks. For example, even though new episodes of *Seinfeld* (1989–1998) and *The Big Bang Theory* (2007–2019) are no longer being produced, many television stations buy the rights to old episodes from a syndicator. The stations air the show at whatever time of day it will draw the largest number of viewers in the market.

The off-network syndication market was the brainchild of Lucille Ball and Desi Arnaz, stars of *I Love Lucy* (1951–1957), who were the first in the industry to envision the financial advantages of rerunning episodes. The major drawback to starting syndication was that many live shows were not recorded at all, and others were preserved on Kinescope, a pre-videotape process that filmed the show from the video monitor itself and resulted in very poor-quality playback. To their credit, Ball and Arnaz insisted that *I Love Lucy* episodes be shot and preserved on high-quality film rather than Kinescope. Almost 70 years later, *I Love Lucy* is still a syndication favorite that keeps the audience in stitches.

Prime-time network programs usually air once a week, but once they have been syndicated to a station, they often run five days a week (a practice known as **stripping**). To have enough episodes for syndication, prime-time programs need to stay on the air for at least 65 episodes, so they can be shown Monday through Friday for 13 weeks. However, most stations will not pick up a syndicated program unless there are at least 100 to 150 episodes, with 130 being the ideal number (26 weeks of Monday-through-Friday airings).

Television stations pay for syndicated first-run and off-network shows. With a **cash purchase**, the station simply pays the syndicator for the right to air the program, and the station, in turn, sells commercial time. In a **straight barter agreement**, the station gets a program for free, but the syndicator gets to sell a portion of the available commercial time, leaving only a small amount of time for the station to sell. A **cash-plus-barter arrangement** works almost the same way as a straight barter, but the station pays a small amount of cash in exchange for more commercial time.

The FCC in 1970 enacted **financial interest and syndication rules (fin/syn)**, which allowed a broadcast network to produce only 3.5 hours' worth of its weekly prime-time programs; the rest had to come from outside production companies. In mandating fin/syn, the FCC hoped that independent producers would bring more diversity to prime-time programming and break what it considered the networks' monopoly over the production and distribution of television programs.

As competition from cable channels eroded broadcast network audiences, the FCC relaxed and then rescinded fin/syn in 1995. As a result, the competition between the networks and independent producers has heated up. Today, about one-half of all network programs are produced by network-owned or -affiliated production companies. Independent producers complain that their programs are less likely to make it to the airwaves than network-produced shows, and they claim that because the networks stand to make quite a bit of money from syndication, they are more likely to keep their own poorly rated programs on the air and instead cancel independently produced programs that have higher ratings.

The networks contend, however, that this ongoing competition has led to higher-quality programs. Moreover, they counter charges of keeping low-rated programs on the air with examples of networks canceling their own programs, such as when ABC canceled *Ellen* (1998). Nevertheless, networks are not as quick to cancel shows as they were in the past several decades. In earlier days of television, programs were given a chance to grab an audience. Hit shows like *Cheers* (1982–1993), *Seinfeld* (1989–1998), *The Office* (2005–2013), and *The Big Bang Theory* (2007–2019) all struggled in their first season to attract viewers. But savvy network executives gave the shows a second and even a third season to rise to the top. The reason shows are canceled is often a mystery to viewers. Eva Longoria-produced *Grand Hotel* (2019) was canceled after one season despite debuting with good reviews and a respectable number of viewers. Even shows such as *Grey's Anatomy* (2005–present) and *The Simpsons* (1989–present) both remain on the air despite an audience loss of about 70% from their peaks.

FYI: TOP 10 SYNDICATED PROGRAMS: WEEK OF MARCH 20–29, 2018 (MON–FRI) (FIRST-RUN AND OFF-NETWORK SYNDICATION)

Family Feud (first-run syndication)
Weekend Adventure (first-run syndication)
Judge Judy (first-run syndication)
Jeopardy! (first-run syndication)
Wheel of Fortune (first-run syndication)
Big Bang Theory (off-network syndication)
Law & Order SVU (off-network syndication)
Dr. Phil Show (first-run syndication)
Modern Family (off-network syndication)
Inside Edition (first-run syndication)
Two and a Half Men (off-network syndication)

Source: National Media Spots, 2018

Primetime Broadcast television singles out primetime as its most-watched three hours of programming. Most prime-time programs are designed to appeal across the demographic (e.g. age, education, income) spectrum. Situation comedies and dramas are the most common forms of prime-time entertainment. Primetime also includes theatrical and made-for-television movies, one-time specials (such as the Emmy, Oscar, and Grammy Award shows), and made-for-television miniseries.

Non-Primetime Non-prime-time programs are generally less expensive to produce and contain five to seven more commercial minutes than prime-time programs. Granted, non-prime-time commercials do not sell for nearly as much as those aired during primetime, but the additional minutes coupled with lower production costs maximize revenue.

Daytime and late-night programs are different from their prime-time counterparts and are driven by audience size and composition. Fewer viewers tune in during non-prime-time hours and the daytime audience is less diverse and made up mainly of children, stay-at-home parents, senior citizens, students, and shift workers. Children's shows, after-school specials, soap operas, talk shows, and game shows make up most non-prime-time day shows, while the later hours are dominated by entertainment/talk, sports, comedies, and movies.

The big three networks (ABC, CBS, NBC) dominated broadcast television until about the late 1990s when new broadcast networks (Fox, WB, UPN, PAX,) hit the screen (The PAX name was changed to Independent Television in 2005 and then to ION Television in 2007). Other network additions include MeTV (Memorable Entertainment Television), which was created in 2003 to show classic television programs from the 1950s–1990s, The CW in 2006 (joint venture by WB, UPN, and CBS), and MyNetworkTV, a Fox-owned network and syndication service that airs repeats of old programs such as *Bones* and *Law & Order: SVU*. Additionally, the Spanish-language networks, Univision and Telemundo, are formidable competitors for the Hispanic audience.

Cable Companies and Networks

Cable channels typically show programs that are either local originated, independent or network produced, or supplied by syndicators. Cable companies, such as Comcast, are often required to produce and provide local origination and local-access public/education/government (PEG) channels as part of their franchise agreements with cities and local communities. The cable company pays for the production of these channels. Cable services also provide **public-access** channels, which are produced and paid for by local citizens, schools, community groups and government entities. Public-access programs run the gamut from city council meetings, PTA meetings, public roundtables, and religious sermons to some guy playing a guitar in a local coffeehouse. The 1992 film comedy *Wayne's World*, starring Mike Myers and Dana Carvey, is about two loser guys with a public-access show who make it to the big time.

For other programs, cable companies pay syndicators and cable networks distribution fees based on their numbers of subscribers. Cable companies, in turn, generate revenue by collecting subscriber fees and by selling commercial time. Some cable network programs come with presold commercial time. In other cases, cable companies barter with cable networks for more commercial time, which they sell locally. Premium cable networks, such as HBO, require a fee-splitting financial relationship.

In exchange for carrying a premium cable network, a cable company agrees to give the network about half of the fees it collects from its customers, who subscribe to the premium network.

There are hundreds of cable networks that focus on specific topics, such as golfing, home and garden, or history. In contrast to the broadcast networks, which target a large, mass audience, the cable networks target smaller niche or specialty audiences. Thus, a cable network such as the Food Network appeals to viewers interested in cooking and so attracts a much smaller audience than an ABC sitcom, which might appeal to millions of viewers. Cable network shows have become so engaging that they now get about 60% of the audience share, compared to a 40% share for ABC, CBS, NBC, and Fox combined.

Premium channels like HBO, and Cinemax, differ from standard cable channels by the fee structure: An extra charge is added to an existing cable subscription. Premium channels specialize in showing movies, miniseries, and special pay-per-view sporting events such as boxing. Some premium channels also produce their own shows, like HBO's popular *Game of Thrones* (2011–present) and Showtime's *Dexter* (2006–2013).

Streaming Services/Over-the-Top Content

Over-the-top content (OTT) is a relatively new term that refers to streaming audio and video programming over the Internet, bypassing or going 'over the top' of cable and satellite providers, in favor of third-party content 'now media' such as Hulu, Netflix, Amazon Prime Video, Disney+, and HBO Now.

Netflix, which started out as a DVD sales and rental business, and other streaming services are generally digital archives where people can watch syndicated programs from old broadcast and cable networks at any time they wish. Old favorites such as *Family Guy*, *Grey's Anatomy*, and *South Park*, newscasts, and movies are available for viewing for a monthly fee.

Streaming services pay the producer or syndicator for the rights to stream a show. The licensing agreement sets the streaming terms, such as a particular number of episodes or all of the program's episodes for a limited time or for a number of years. The terms are renegotiated at the end of the contract period. Sometimes show licenses are not renewed as in the case of *The Office*, which is leaving Netflix after 2020 and going to NBC's new Peacock streaming service. Peacock has also signed for the rights to stream *Parks and Recreation*. The hit comedy series *Seinfeld* has been available on Hulu since 2015, but Hulu's rights expire in 2021 and renewal will be competitive as Sony Pictures Television, which handles the program's distribution, is calling for bids from other streaming services.

Hulu, Netflix, and Amazon Prime are also producing their own shows, which are available for viewing on their services. In 2012, Hulu's *Battleground* was its first foray into program production, Netflix stepped in with *House of Cards, Hemlock*, and *Orange Is the New Black* in 2013, and it produced new episodes of the canceled Fox sitcom *Arrested Development*. In 2015 Amazon Prime's first original shows were *Bosch, Hand of God*, and *The Man in the High Castle*, and Tina Fey is planning to produce an original series for NBC's new Peacock streaming service that started in April 2020.

Streaming services' growing influence in content production was demonstrated in 2019 when Netflix dominated the Golden Globe nominations and was competitive with HBO for Emmy nominations and awards. Amazon's original programs, *The Marvelous Mrs. Maisel* and *Transparent* both received critical acclaim in 2019.

Television has now entered an era in which it is easier to stay on the air than to get on the air. In 2009, there were 211 scripted (non-news, non-sports) television series on broadcast and cable networks. Five years later that number jumped to 376 programs including those on streaming services. In 2019, there were 532 programs across broadcast, cable, and streaming. And the number of shows will continue to rise as the streaming services themselves produce more programs.

The successes of Netflix and Amazon have compelled traditional television networks to follow their lead and stream their own programs. CBS did just that when it notably live streamed for free the Super Bowl in 2015 and 2016 on cbssports.com. About 5% of those who planned to watch the game in 2016 said they would do so online, compared to 57% via cable, 28% on satellite, and 9% with an antenna. Other broadcast and cable networks now offer their programs directly from their Web site, and as with most streaming services, they charge users a monthly or by program subscription fee for an array of programs that can be watched any time, from almost anywhere, without having to adhere to a program schedule.

CAREER TRACKS: JAY RENFROE, TELEVISION PRODUCTION CO-OWNER, WRITER, AND PRODUCER

WHAT IS YOUR JOB? WHAT DO YOU DO?

I own a Los Angeles-based television production company. I create and produce reality and scripted television shows for the broadcast networks (Fox, NBC, ABC, CBS, CW), several cable channels (Discovery, A&E, Lifetime, TLC, Animal Planet, Oxygen, Travel Channel, Turner), and streaming services such as Amazon, HBO Max and Quibi. In the reality world, I am an executive producer, and in the scripted world, I am both an executive producer and a writer.

HOW LONG HAVE YOU BEEN DOING THIS JOB?

Been in the business for 35 years. I've been a partner in Renegade Entertainment for over 20 years now.

Figure 4.21 Jay Renfroe.

WHAT WAS YOUR FIRST JOB IN ELECTRONIC MEDIA?

I wrote/directed commercials and industrials in Atlanta.

WHAT LED YOU TO THE JOB?

Studied TV/film production in college, did some sports production, had a comedy troupe in Atlanta, produced stand-up comedy for HBO/Showtime, wrote a play in Los Angeles, which led to writing the screenplay for TriStar based on the play, which led to creating and writing a sitcom for CBS, which led to forming my own company with an old college friend from Florida State University. We're still partners.

WHAT ADVICE WOULD YOU HAVE FOR STUDENTS WHO MIGHT WANT A JOB LIKE YOURS?

Do everything. I was a camera operator, editor, lighting director, theatrical director, writer. Learn every job, and you'll be a better producer. The more material you can create and produce, the more confidence you'll have in yourself as a leader. Learn your own voice and how to communicate your vision to everybody involved in the production. Execution is everything.

Non-commercial Stations

Most non-commercial independent stations produce their own shows or obtain and pay for programming from other entities, such as independent producers and syndicators. PBS-affiliated stations obtain their programs from independent producers, for such shows as *Downton Abbey* (2010–2015) or from member stations. PBS stations such as WGBH (Boston), KQED (San Francisco), and WETA (Washington, DC), produce programs for airing on their own station or on stations throughout the PBS network. For example, *PBS News Hour* and *Washington Week* are produced by station WETA and *Nova* by WGBH, but all three programs are shown on PBS stations nationally.

The PBS network acts only as a distributor in that it transmits programs to its member stations via satellite. Thus, the PBS network/member station programming relationship is opposite the relationship between the commercial networks and affiliates. That is, PBS member stations pay the network for programs and generate revenue through membership drives and from federal, state, and local government funding.

FYI: TOP 10 CABLE ENTERTAINMENT NETWORKS BY TOTAL VIEWERS (2018)

USA	1.48 million
HGTV	1.45 million
TBS	1.32 million
TNT	1.26 million
History	1.21 million
Hallmark	1.18 million
Discovery	1.14 million
Investigation Discovery	1.10 million
TLC	1.03 million
A&E	1.03 million

Source: de Moraes, 2018

How Programs Are Scheduled

Seasons

Broadcast networks historically introduced new programs and new episodes of programs in 'seasons.' The big introduction began in September after the Labor Day weekend, which coincided with when the car companies would bring out the year's new models. Not surprisingly, the automakers were some of the biggest spenders among television advertisers. Television seasons usually adhered to a 30- to 39-episode year with summer reruns.

Television seasons shortened a bit starting in the late 1960s. Most shows aired 24 to 26 episodes from September through March with reruns from April through August. Since the 1980s, competition from the growing number of cable channels and a stronger emphasis on Nielsen ratings extended the fall season until May, which is one of the 'sweep' months in which Nielsen collects audience viewing data. Programs still typically consist of 20 to 26 episodes but take a hiatus during the winter holidays.

Although September is still the official new season kick-off month, the full season is sometimes split into two separate units of 10 to 13 weeks or in bursts of 8-, 10-, and 13-week episodes. AMC's acclaimed *Mad Men* is a perfect example of a 'season-free' program. Making its debut in July 2007 instead of the usual September, its first two seasons and the fourth season ran 13 episodes from July through October, its third season ran August to November, and then the fifth season shifted to March to June, with season six airing April through June. To heighten the drama surrounding the long-anticipated final episode, *Mad Men's* seventh season was split into two seven-episode 'seasons' airing in April and May of 2014 and 2015.

Successful shows are 'picked up' for the next season and new episodes shot, while less popular ones are replaced by other programs, sometimes in mid-season. *Batman* (1966–1968) was the first mid-season replacement (in January), but it turned out to be such a big hit that the networks took to the idea and began replacing their ratings duds with new shows before the season ended.

More recently, the first episode is released online days, weeks, and even months before its actual television premiere. An early online preview of Starz's *Outlander* drew almost one million viewers. The Disney Channel now regularly debuts almost all of its shows online, on its official Disney+ streaming app. In the autumn of 2014, NBC's *A to Z* (premiered 10/2/14), ABC's *Selfie* (premiered 9/30/14), and Fox's *Red Band Society* (premiered 9/17/14) were posted on Hulu and their networks' respective Web sites before the series started on television. But the strategy of online previewing is only as good as the program, and in the case of the three shows above, all were canceled within weeks of their television premiere, but they could still be seen on the Internet.

Programming Strategies

The big three television networks use various programming strategies to attract viewers to their channels. In the pre-cable-network, pre-remote-control days of television, these strategies worked more effectively because there were only three networks, and viewers were more comfortable sticking with one channel than getting up and walking to their sets to change to another. A network's goal is to control audience flow, or the progression of viewership from one program to another, and to keep viewers from changing to another channel.

- **Tentpoling:** A popular program is scheduled between two new or poorly rated programs. The theory is that viewers will tune to the channel early in anticipation of watching the highly rated

program and thus see at least part of the less popular preceding show.
- **Hammocking:** A new or poorly rated show is scheduled between two successful shows. After watching the first program, the network hopes that viewers will stay tuned to the same channel and watch the new or poorly rated show while waiting for the next program to begin.
- **Leading in:** The idea is to grab viewers' attention with a very strong program, anticipating that they will watch that popular show and then stay tuned to the next program on the same channel.
- **Leading out:** A poorly rated program is scheduled after a popular show with the hope that the audience will stay tuned to the same channel.
- **Bridging:** A program is slotted to go over the starting time of a show on a competing network. For example, the season finale of a reality show could be scheduled from 8:00 p.m. to 9:30 p.m. to compete with another network's reality program that is scheduled to start at 9:00 p.m.
- **Blocking:** A network schedules a succession of similar programs over a block of time – for example, four half-hour situation comedies scheduled over two hours.
- **Seamless programming:** One program directly follows another, without a commercial break or beginning or ending credits. Some programs use a split-screen technique, in which program credits and closing materials blend in with the start of the following program.
- **Counterprogramming:** One type of program, such as a drama, is scheduled against another type of a program, say, a sitcom, on another network.
- **Head-to-head programming:** This strategy is the opposite of counterprogramming; that is, two popular shows of the same genre are pitted against each other.
- **Stunting:** A special program, such as an important sporting event or a holiday show, is scheduled against a highly rated, regularly scheduled show on another network.
- **Repetition:** Used mostly by cable networks. Repetition involves scheduling a program such as a movie to air several times during the week or even during the day.
- **Stripping:** Normally used for syndicated programs, stripping occurs when a program is shown at the same time five days a week.

Getting Viewers to Watch

Even the most aggressive programming strategies by the broadcast and cable industries have not been enough to keep users away from streaming services. Viewers who have dropped their cable service to watch television programs solely online are known as '**cord cutters**,' and those who have never subscribed are called '**cord nevers**.' About 17% of U.S. viewers have cut the cable cord (a percentage that has grown from 2% in 2010), joining the 12% (mostly Millennials) who have never subscribed to cable or satellite. But yet 15% of cable customers say they will never cut the cord.

Although traditional 'linear' scheduled viewing on a television set still dominates, about 28% of American adults and six in ten of those aged 18–29 primarily access television through streaming services. The difference in viewing habits by age is striking – 70% of 50–64 year olds mainly watch television via cable compared to 10% who do so via streaming. While streaming has caught on with younger viewers, 59% of U.S. adults of all ages still prefer cable or satellite.

SEE IT LATER: RADIO AND TELEVISION

Marketing Television Programs

The National Association of Television Program Executives (NATPE) was first organized in 1963 as a way to bring together television program executives faced with a rapidly changing television industry. NATPE is now more than 4,000 members strong and holds an annual convention at which writers, program creators, and producers market their shows to syndicators, stations, networks, and other program distributors. The NATPE convention is considered the primary venue for buying and selling television programming. Industry executives schmooze and wheel and deal for licensing and distribution rights.

ZOOM IN 4.15

Learn more about the **National Association of Television Program Executives (NATPE)** at www.natpe.com

Cable Television and Streaming Services

Perhaps the biggest change in television programming is the rise of award-winning cable- and streaming services-originated programs. Until about the late 1990s, broadcast network shows received almost all Emmy Award nominations and wins. HBO's *The Larry Sanders Show* (1992–1998) was one of the first cable-originated shows to win multiple Emmy awards. But the popularity of HBO's *The Sopranos* (1999–2007) cemented the reality that cable-originated shows could compete head to head with broadcast programs. In its eight-year run *The Sopranos* won 21 Emmy awards and five Golden Globes. Until 2008, the only winners of an Emmy for top dramatic series were broadcast network and HBO programs. Then came AMC's *Mad Men* (2007–2015), which took the highly coveted honor of best dramatic series at the 2008, 2009, 2010, and 2011 Emmy Award ceremonies. *Breaking Bad* (AMC, 2008–2013) racked up 58 Emmy nominations and 16 wins, including the top prize for Outstanding Dramatic Series in 2013 and 2014. Cable networks are basking in the glory of many hit shows. Programs such as PBS's *Downton Abbey* (2010–2016) give the big four broadcast networks fierce competition for accolades and viewers. The wildly successful shows have catapulted Netflix and other streaming services into the world of television program production. Cable networks and streaming companies are discovering that they can be as successful, if not more successful, than broadcast networks in drawing a large following of viewers. What is ahead for television is a shift from older viewers who watch broadcast and cable programs to younger viewers taking to streaming services. In the 2013–2014 television season, the typical broadcast/cable viewer was 44.4 years of age, by 2018 the average viewer was about 54 years old. Moreover, hit broadcast programs attract older viewers: The median age for *The Good Doctor* is 58.5, *Young Sheldon* 57.4, *The Voice* 57.3, and *Grey's Anatomy* 53.1.

SUMMARY

Early radio was quite the magical medium. For the first time, music floated into homes. People could enjoy a concert from the comfort of their living room instead of having to go to a theater. Dramas, comedies, soap operas, quiz shows, music, and many other types of programs filled the radio airwaves.

Television emerged in the late 1940s and took its place as a mass medium in 1948, when the numbers of sets, stations, and audience members all grew by 4,000%. Radio listeners gravitated to television especially as radio programs transitioned to the screen. Anthologies and dramas were popular in the 1950s, but as new production techniques and more portable cameras made outside location scenes possible, the audience's taste moved to more realistic and action-packed shows.

As television gained in popularity, radio's audience shrank. To survive as a medium, radio needed to find a way to live alongside television, and it did so by focusing on music instead of programs. Stations played one or two types of music, or formats, be it rock 'n' roll, classical, middle-of-the-road, easy listening, jazz, or country. Station formatting depends on the market and the competition.

With the growth of cable television, and now streaming services, the broadcast networks' share of the audience declined. Looking for ways to maintain their audience, television programmers devised programming strategies to control audience flow from one program to another.

'Now television' in the form of streaming services is poised to become the primary way to watch television. By subscribing to online streaming, television viewers gain control of their own viewing schedules and become their own programmers. Broadcast and cable networks' best-laid plans are thwarted as consumers find ways to avoid being tied to programming schedules and serving as a captive audience for advertisers. Television viewers have much to look forward to in the coming years as new programs, new delivery systems, and new ways of watching 'television' emerge.

BIBLIOGRAPHY

71st Primetime Emmy nominees and winners. (2019, September). *Television Academy of Arts & Sciences*. Retrieved from: www.emmys.com/news/awards-news/71st-emmy-award-winners-live

About Meet the Press. (2003). *MSNBC*. Retrieved from: www.msnbc.com/news/102219.asp

About PBS. Corporate Facts. (2019). *PBS.org*. Retrieved from: www.pbs.org/aboutpbs/aboutpbs_corp.html

About Premiere. (2015, n.d.). *Premiere Networks*. Retrieved from: www.premiereradio.com/pages/corporate/about.html

Ahrens, F. (2006, August 20). Pausing the panic. *The Washington Post*, p. F1.

An American family. (2002). *PBS*. Retrieved from: www.pbs.org/lanceloud/american

Arango, T. (2009). Broadcast TV faces struggle to stay viable. *The New York Times*, pp. A1, A15.

Barnes, B. (2019, December 1). The streaming era has finally arrived. *The New York Times*, pp. F3–F4.

Beers, B. (2019, May 6). How Netflix pays for TV show licensing. *Investopedia*. Retrieved from: www.investopedia.com/articles/investing/062515/how-netflix-pays-movie-and-tv-show-licensing.asp

Berr, J. (2014, November 20). Has the cord-cutter threat to pay TV been exaggerated? *CBS Money Watch*. Retrieved from: www.cbsnews.com/news/has-the-cord-cutters-threat-to-pay-tv-been-exaggerated/

Best streaming video services of 2020. (2020, January 9). *Tom's Guide*. Retrieved from: www.tomsguide.com/us/best-streaming-video-services,review-2625.html

Bibel, S. (2015, February 2). Super Bowl XLIX is most-watched show in U.S. television history with 114.4 million viewers. *TVByTheNumbers*. Retrieved from: http://tvbythenumbers.zap2it.com/2015/02/02/super-bowl-xlix-is-most-watched-show-in-u-s-television-history/358523/

Broadcasting and cable yearbook. (2010). Newton, MA: Reed Elsevier.

Brooks, T., & Marsh, E. (1979). *The complete directory to prime time network TV shows*. New York: Ballantine Books.

Butler, J. G. (2001). *Television: Critical methods and applications*. Cincinnati, OH: Thomson Learning.

Campbell, R. (2000). *Media and culture*. Boston, MA: St. Martin's Press.

Captain Kangaroo. (2003). *The Fifties Web*. Retrieved from: www.fiftiesweb.com/tv/captain-kangaroo.htm

Carter, B. (2006, January 25). With focus on youth, 2 small TV networks unite. *The New York Times*. Retrieved from: www.nytimes.com/2006/01/25/business/25network.html [August 30, 2009]

Carter, B. (2008, September, 23). A television season that lasts all year. *The New York Times*, p. B3.

Cord cutters alert: 5% of TV broadband users watch all their TV online. (2013, August 1). *Gigaom Research*. Retrieved from: https://gigaom.com/2013/08/01/five-percent-cord-cutters/

de Moraes, L. (2018, December 19). Cable Network 2018 Ratings Rankings. *Deadline*. Retrieved from: https://deadline.com/2018/12/cable-tv-network-ratings-2018-lists-fox-news-usa-network-1202523019/

Digital: News sources for Americans by platform. (2012, September 27). *The Pew Research Center*. Retrieved from: www.journalism.org/media-indicators/where-americans-get-news/

Dizard, W., Jr. (2000). *Old media, new media*. New York: Longman.

Eastman, S. T., & Ferguson, D. A. (2002). *Broadcast/cable/web programming* (6th ed.). Belmont, CA: Wadsworth.

Enger, J. (2009, August 25). Share your *Guiding Light* memories. *The New York Times*. Retrieved from: https://artsbeat.blogs.nytimes.com/2009/08/18/share-your-guiding-light-memories/

Famous weekly shows. (1994–2002). *Old Time Radio*. Retrieved from: www.oldtime.com/weekly

Farhi, P. (2009, May 17). Click, change. *The Washington Post*, pp. E1–E8.

For '*SNL*,' doubts follow a banner year. (2009, September 8). *The Washington Post*, pp. C1, C6.

Fung, B. (2015, December 22). Survey: Many cord-cutters don't have broadband. *The Washington Post*, p. A12.

Garvin, G. (2009, July 9). When Casey Kasem started counting down, America rocked together. *The Washington Post*, p. C10.

Gertner, J. (2008, January). A clicker is born. *The New York Times Magazine*, pp. 34–35.

Golden years. (1994–2002). *Old Time Radio*. Retrieved from: www.old-time.com/golden_age/index.html

Graham, J. (2019, March 24). In defense of cable. *USA Today*. Retrieved from: www.usatoday.com/story/tech/talkingtech/2019/03/24/who-needs-cut-cord-when-cable-just-works-we-love-our-remote/3261793002/

Guiding Light. (2009). *Wikipedia*. Retrieved from: en.wikipedia.org/wiki/Guiding Light [August 26, 2009]

Gundersen, E. (2002, November 22). Uncovering the real Osbournes. *USA Today*, p. E1.

Hare, B. (2014, May 20). 'The Maya Rudolph Show:' What's the verdict? *CNN.com*. Retrieved from: www.cnn.com/2014/05/20/showbiz/tv/maya-rudolph-show-whats-the-verdict/index.html

Heimlich, R. (2012, October 11). Number of Americans who read a daily newspaper continues to decline. *Pew Research Center*. Retrieved from: www.pewresearch.org/fact-tank/2012/10/11/number-of-americans-who-read-print-newspapers-continues-decline/

Hilliard, R., & Keith, M. (2001). *The broadcast century and beyond* (3rd ed.). Boston, MA: Focal Press.

History of the Batman. (2010). *Batman-on-Film.com*. Retrieved from: www.batman-on-film.com/historyofthebatman_batman66.html [February 6, 2010]

How do public broadcasters obtain programming? (n.d.). *Corporation for Public Broadcasting*. Retrieved from: www.cpb.org/aboutpb/faq/programming.html

Hsu, T., & Lee, E. (2020, January 17). NBC to offer free option for streaming (with ads). *The New York Times*, pp. B1, B3.

Ingram, B. (2003). Captain Kangaroo. *TV Party*. Retrieved from: www.tvparty.com/lostterrytoons.html

Internet use by age. (2017, January 11). *Pew Research Center*. Retrieved from: www.pewresearch.org/internet/chart/internet-use-by-age/

Java, J. (1985). *Cult TV*. New York: St. Martin's Press.

Johnson, P. (2002). Fox News enjoys new view – From the top. *USA Today*, pp. 1A–2A.

Johnson, T. J., & Kaye, B. K. (2002). Webelievabilty: A path model examining how convenience and reliance on the web predict online credibility. *Journalism & Mass Communication Quarterly*, 79(3), 619–642.

Just two in five Americans read a newspaper daily. (2010, January 14). *Marketing Trends*, Retrieved from: www.marketingcharts.com/industries/media-and-entertainment-11646

Kang, C. (2014, September 5). TV is increasingly for old people. *The Washington Post*. Retrieved from: www.washingtonpost.com/news/business/wp/2014/09/05/tv-is-increasingly-for-old-people/

Kaye, B. K., & Johnson, T. J. (2003). From here to obscurity: The Internet and media substitution theory. *Journal of the American Society for Information Science and Technology, 54*(3), 260–273.

Koblin, J. (2015, November 19). The ax falls, but slowly. *The New York Times*, pp. A1, A4.

Lafayette, J. (2018, March 16). TV viewers getting older fast, analyst says. *Broadcasting & Cable*. Retrieved from: www.broadcastingcable.com/news/tv-viewers-getting-older-fast-analyst-says-133656

Lim, D. (2011, April 17). Reality TV originals in drama's lens. *The New York Times*, Television, pp. 24–25.

Lovely, S. (2019, February 17). Hulu vs. Netflix: By the numbers. *Cordcutting*. Retrieved from: https://cordcutting.com/compare/hulu-vs-netflix/

Maheshwari, S., & Koblin, J. (2018, May 13). Why traditional TV is in trouble. *The New York Times*. Retrieved from: www.nytimes.com/2018/05/13/business/media/television-advertising.html?login=email&auth=login-email

Massey, K. B., & Baran, S. J. (1996). *Television criticism*. Dubuque, IA: Kendall/Hunt.

Matsa, K. E. (2014, January 28). Local TV audiences bounce back. *Pew Research Center*. Retrieved from: www.pewresearch.org/fact-tank/2014/01/28/local-tv-audiences-bounce-back/

Masterpieces and milestones. (1999, November 1). *Variety*, p. 84.

McGinn, D. (2002, November 11). Guilt free TV. *Newsweek*, pp. 53–59.

Millennials on Millennials. TV and digital news consumption. (2018). *The Nielsen Company*. Retrieved from: www.nielsen.com/wp-content/uploads/sites/3/2019/04/millennials-on-millennials-news-consumption-report.pdf

Most viewed TV shows in the United States in the 2018/2019 season. (2020). *Statista*, Retrieved from: www.statista.com/statistics/321390/most-viewed-tv-shows-usa/

National Syndication Broadcast TV Show Ratings. (2018). *National Media Spots*. Retrieved from: www.nationalmediaspots.com/national-syndicated-broadcast-tv-show-ratings.php

Newspaper Fact Sheet. (2019, July 9). *Pew Research Center*. Retrieved from: www.journalism.org/fact-sheet/newspapers/

Newspapers: Daily readership by age. (2016, June 10). *Pew Research Center*. Retrieved from: www.journalism.org/chart/5802/

Nielsen. (2017, February 6). Super Bowl LI draws 111.3 million viewers. Retrieved from: www.nielsen.com/us/en/insights/article/2017/super-bowl-li-draws-111-3-million-tv-viewers-190-8-million-social-media-interactions/

Nielsen TV ratings. (2010, February 10). *Nielsen*. Retrieved from: en-us.nielsen.com/rankings/insights/rankings/television

NPR fact sheet. (2019). *National Public Radio*. www.npr.org/about/press/NPR_Fact_Sheet.pdf

About NATPE. (2002). *National Association of Television Program Executives*. Retrieved from: www.natpe.com/about

Pareles, J. (2012, February 1). A smooth operator in the name of soul. *The New York Times*, pp. C1, C8.

Parsons, P. R., & Frieden, R. M. (1998). *The cable and satellite television industries*. Boston, MA: Allyn & Bacon.

Pew Internet. (2017). *About 6 in 10 young adults in U.S. primarily use online streaming to watch TV*. Retrieved from: www.pewresearch.org/fact-tank/2017/09/13/about-6-in-10-young-adults-in-u-s-primarily-use-online-streaming-to-watch-tv/

Peyser, M., & Smith, S. M. (2003, May 26). Idol worship. *Newsweek*, pp. 53–58.

Poindexter, P. (2012). Too busy for news: Unlimited time for social media. In *Millennials, news, and social media: Is news engagement a thing of the past?* (pp. 53–69). New York: Peter Lang Publishing.

Powers, L. (2011, February 7). 10 most watched TV shows ever. *The Hollywood Reporter*. Retrieved from: www.hollywoodreporter.com/blogs/live-feed/10-watched-tv-shows-97180

Premiere Radio Networks. (2009, August 26). *Premiere Radio Networks.com*. Retrieved from: www.premiereradio.com/category/view/talk.html [August 26, 2009]

Public Broadcasting Service. (2008, December 2). *History. Radio*. Retrieved from: www.pbs.org/wnet/makeemlaugh/comedys-evolution/history-radio/35/

Research: 30m+ Americans 'cord nevers.' (2019, April 10). *Advanced Television*. Retrieved from: https://advanced-television.com/2019/04/10/research-30m-americans-cord-nevers/

Sandomir. R. (2009, July 19). Amid blizzard, Cronkite helped make sports history. *The New York Times*, Sports, pp. 1, 8.

Seipp, C. (2002). Online uprising. *American Journalism Review, 24*(5), 42.

Sesame Street Awards. (2014). *IMDb*. Retrieved from: www.imdb.com/title/tt0063951/awards

Shieber, J. (2019, December 17). Placement is coming for TV and movies and Ryff has raised cash to put it there. *TechCrunch*. Retrieved from: https://techcrunch.com/2019/12/17/virtual-product-placement-is-coming-for-tv-and-movies-and-ryff-has-raised-cash-to-put-it-there/

Stanley, A. (2008, September 21). Sitcoms' burden: Too few taboos. *The New York Times*, pp. MT 1, 6.

State of the media: Audio Today. (2019). *Nielsen. Insider Radio*. Retrieved from: www.nielsen.com/wp-content/uploads/sites/3/2019/06/audio-today-2019.pdf

Stelter, B. (2009, August 1). Voices from above silence a cable TV feud. *The New York Times*, pp. A1, A3.

Stelter, B. (2009, August 24). Reality TV star, a killing suspect, is found dead. *The New York Times*. Retrieved from: www.nytimes.com/2009/08/24/arts/television/24real.html [August 30, 2009]

Stelter, B. (2013, September 21). Emmys highlight a changing TV industry. *The New York Times*, pp. C1, C4.

Stelter, B. (2020, January 9). In ten years, the number of scripted shows on American TV has more than doubled. *CNN*. Retrieved from: www.cnn.com/2020/01/09/media/scripted-television-series-tca-reliable-sources/index.html

Stelter, B., & Carter, B. (2009, November 20). A daytime network franchise bets on her future with cable. *The New York Times*, pp. A1, A3.

Sterling, C., & Kittross, J. (2002). *Stay tuned: A history of American broadcasting*. Mahwah, NJ: Erlbaum.

U.S. cord cutters, 2017–2022. (2018, July 1). *eMarketer*. Retrieved from: www.emarketer.com/chart/220822/us-cord-cutters-2017-2022-millions-change-of-population

United States population. (2020). World Population Review. Retrieved from: http://worldpopulationreview.com/countries/united-states-population/

Vanderbilt, T. (2013, April). The new rules of the hyper-social, data-driven, actor-friendly, super-seductive platinum age of television. *Wired*, pp. 90–103.

Walden, A. (2015, August 26). Who reads what? And where? *Forbes*. Retrieved from: www.forbes.com/sites/alanwalden/2015/08/26/96/#5667d1655938

Walker, J. R., & Ferguson, D. A. (1998). *The broadcast television industry*. Boston, MA: Allyn & Bacon.

Weber, B. (2012, April 19). TV Emperor of rock 'n' roll and New Year's Eve dies at 82. *The New York Times*, pp. A1, A24.

What is NPR? (2019). *National Public Radio*. Retrieved from: www.npr.org/about/

Why your favorite long-running TV shows are exiting Netflix. (2019, August 13). *The Strait Times*. Retrieved from: www.straitstimes.com/lifestyle/entertainment/why-your-favourite-long-running-tv-shows-are-exiting-netflix

Wolbe, T. (2014, February 13). Can we save AM radio? *The Verge*. Retrieved from: www.theverge.com/2014/2/13/5401834/can-we-save-am-radio

Wolk, D. (2005, November 17). College radio. Slate.com Retrieved from: www.slate.com/id/2130587/ [August 26, 2009]

Wyatt, E. (2009, August 2). TV contestants: Tired, tipsy and pushed to the brink. *The New York Times*, pp. A1, A14.

Young adults are heavy users of streaming services. (2017, September 13). *Pew Research Center*. Retrieved from: www.pewresearch.org/fact-tank/2017/09/13/about-6-in-10-young-adults-in-u-s-primarily-use-online-streaming-to-watch-tv/ft_17-09-13_streaming-1/

Zietchik, S. (2018, September, 29). Why TV shows hang on long past their prime. *The Washington Post*, A1, A10.

Zietchik, S., & Timberg, C. (2019, April). Cord cutters' dreams dashed as streaming sector splinters. *The Washington Post*, A1, A5.

Chapter five

Interconnected by the Internet

Although many people believe that the Internet is a 1990s invention, it was actually envisioned in the early 1960s. Over the last 50 or so years, the Internet has gone from having the technological capability of sending one letter of the alphabet at a time from one computer to another to a vast system in which trillions of messages are sent around the world every day. The Internet ushered in the age of 'now media.'

This chapter covers the Internet and the various Internet resources: Email, electronic mailing lists, chat rooms, instant messaging, newsgroups, Web sites, blogs, microblogs, social networking sites, video sharing sites, and mobile apps. This chapter focuses primarily on the use of online resources and how radio, television, and print have adapted to a 'now' digital world. Understanding how a technology developed and made its way into everyday life is the first step to guiding where it is going.

SEE IT THEN

The Internet

In the early 1960s, scientists approached the U.S. government with a formal proposal for creating a decentralized communications network that could be used in the event of a nuclear attack. By 1970, **ARPAnet (Advanced Research Projects Agency Network)** was created to advance computer interconnections.

The interconnections established by ARPAnet soon caught the attention of other U.S. agencies, which saw the promise of using an electronic network for sharing information among research facilities and schools. While disco music was hitting the airwaves, **Vinton Cerf**, later known as '**the father of the Internet**,' and researchers at Stanford University and UCLA were developing packet-switching technologies and transmission protocols that are the foundations of the Internet. In the 1980s, the National Science Foundation designed a prototype network that became the basis for the Internet. At the same time, a group of scientists in the **European Laboratory of Particle Physics (CERN. Conseil Européen pour la Recherche Nucléaire)** was developing a system for worldwide interconnectivity that was later dubbed the World Wide Web. **Tim Berners-Lee** headed the project and was dubbed '**the father of the World Wide Web**.'

ZOOM IN 5.1

See Tim Berners-Lee on YouTube talking about how the Internet was created:
www.youtube.com/watch?v=sSqZ_hJu9zA

Simply stated, the **Internet** is a worldwide network of computers. Millions of people around the globe upload and download information to and from the Internet every day. The Internet is a digital venue that connects users to each other and to vast amounts of information through an **Internet service provider (ISP)**, such as Earthlink, a cable company like Comcast, a phone company like Verizon,

Figure 5.1 Media adoption rates.
Source: Visual Capitalist

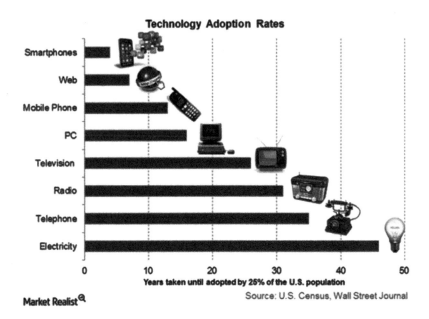

or a public wired or wireless connection. Like radio and television, the Internet is a mass medium in the respect that it reaches a broad range of users.

Before any medium can be considered a **mass medium**, it needs to be adopted by a critical mass of users. In the modern world, about 50 million users seems to be the benchmark. The Internet became a mass medium at an unprecedented speed. Radio broadcasting (which began in an era with a smaller population base) took 38 years to reach the magic 50 million mark, and television took 13 years. The Internet surpassed 50 million regular U.S. users sometime in late 1997 or early 1998, only about five years after the World Wide Web made it user friendly.

FYI: THE WORD 'INTERNET'

The word *Internet* is made up of the prefix *inter*, meaning 'between or among each other,' and the suffix *net*, short for *network*, which is 'an interconnecting pattern or system.' An *inter-network*, or *internet* (small *i*), refers to any 'network of networks' or 'network of computers,' whereas the *Internet* with a capital *I* is the specific name of the computer network that provides the World Wide Web and other interactive components (Herring, 2015; Krol, 1995).

How the Internet Works

The Internet operates as a **packet-switched network**. It takes bundles of data and breaks them up into small packets or chunks that travel through the network independently. Smaller bundles of data move more quickly and efficiently through the network than larger bundles. It is kind of like moving a home entertainment system from one apartment to another. The DVD player might be packed in the car and the large-screen television in the truck. The home entertainment system is still a complete unit, but when transporting, it is more convenient to move each part separately and then reassemble all of the components at the new place. The Internet works in almost the same way, except it disassembles bits of data rather than a home entertainment system and reassembles the bits into a whole unit at its destination point.

The speed of online delivery is constrained by **bandwidth**, which is the amount of data that can be sent all at once. Think of bandwidth as a water faucet or a pipe. The circumference of the faucet or pipe determines the amount of water that can flow through it and the speed at which it flows. Similarly, bandwidth determines the speed of information flow and thus affects how quickly content appears on the screen. Web designers might choose to reduce the amount of content to increase speed. Bandwidth is becoming less of a concern, however, now that fast broadband (100 megabits per second or faster) and wireless connections are commonplace, though still only about six in ten Americans live in areas where broadband is available.

The bits of data that make up online information flow through interconnected computers from their points of origin to the destination. The sender's computer is the origination point, known as the **client**. The message bits leave the client computer and travel in separate packets to a **server**.

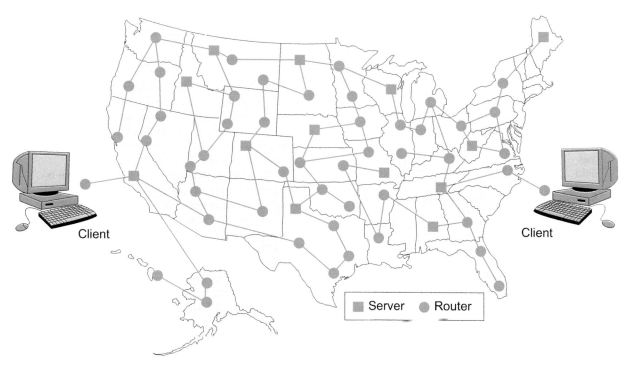

Diagram 5.1 How the Internet works.
Source: Kaye and Medoff, 2001

A server is basically a powerful computer that provides continuous access to the Internet. From there, packets move through routers. A **router** is a computer that links smaller networks and sorts each packet of data until the entire message is reassembled, and then it transmits the electronic packets either to other routers or directly to the addressee's server. The server holds the entire message until an individual directs his or her client computer to pick it up.

Servers and routers deliver online messages through a system called **transmission control protocols/Internet protocols (TCP/IP)**, which define how computers electronically transfer information to each other on the Internet. TCP is the set of rules that governs how smaller packets are reassembled into an IP file until all of the data bits are together. Routers follow IP rules for reassembling data packets and data addressing so information gets to its final destination. Each computer has its own numerical **IP address** (which the user usually does not see) to which routers send the information. An IP address usually consists of between eight and 12 numbers and may look something like *166.233.2.44*.

Because IP addresses are rather cumbersome and difficult to remember, an alternate addressing system was devised. The Domain Name System (DNS) basically assigns a text-based name, **uniform resource locator (URL)**, to a numerical IP address using the following structure: *username@host.subdomain*. For example, in MaryC@anyUniv.edu, the user name, *MaryC*, identifies the person who was issued Internet access. The @ literally means 'at,' and *host.subdomain* is the user's location. In this example, *anyUniv* represents a fictitious university. The top-level domain, which is always the last element of an address, indicates the host's type of organization. In this example, the top-level domain, *.edu*, indicates that *anyUniv* is an educational institution.

The Internet Corporation for Assigned Names and Numbers (ICANN) assigns top-level domains. The first top-level domains created for the Internet were .com (commercial), .org (organization), .net (network), .int (international), .edu (education), .gov (government), and .mil (military). Over the years, other domain names were added such as .biz (business), .museum (museum), and .coop (cooperatives), and domains for countries and regions such as .ec (Ecuador), .it (Italy), .pl (Poland), .uk (United Kingdom), and .eu (European Union). But the largest expansion came after 2013 when ICANN began accepting applications for custom domain names. By early 2015, there were just over 700 top-level domain names and by 2019 about 1,500.

> **FYI: THE FOUR MOST USED TOP-LEVEL DOMAINS (EXCLUDING COUNTRY CODES) (2019)**
>
> .com
> .org
> .net
> .info
>
> Sources: Domain Numbers, 2019; Usage Stats 2019

The World Wide Web

For many years, the Internet was the domain of scientists and researchers. It came into widespread use only in 1993 with the advent of the easy-to-use World Wide Web. Prior to the creation of the Web, Internet content (black-and-white pages of text) could be retrieved only through a series of complicated steps and commands. The process was difficult, time-consuming, and required an in-depth knowledge of Internet protocols. As such, the Internet was of limited use. In fact, it was largely unnoticed by the public until 1993 when undergraduate Marc Andreessen and a team of University of Illinois students developed **Mosaic**, the first **Web browser**. Different from the Internet in which data were retrieved by entering cumbersome commands, Mosaic was based on a system of clickable, intuitive hyperlinks. Mosaic caught the attention of Jim Clark, founder of Silicon Graphics, who lured Andreessen to California's Silicon Valley to enhance and improve the browser. With Clark's financial backing and Andreessen's know-how, Netscape Navigator was born. This enhanced version of Mosaic made Andreessen one of the first new-technology, under-30-year-old millionaires. Once Netscape Navigator hit the market, the popularity of Mosaic plummeted. Since the advent of Mosaic and Netscape Navigator, other browsers, such as Netscape Communicator, have come and gone, and new and improved ones, such as Firefox and Chrome, have taken their place.

Navigating the World Wide Web

Web browsers are based on **Hypertext Markup Language (HTML)**. Hypertext is "non-linear text, or text that does not flow sequentially from start to finish" (Pavlik, 1996, p. 134). The beauty of hypertext is that it allows non-linear or non-sequential

> **FYI: HYPERTEXT MARKUP LANGUAGE (HTML)**
>
> HTML is the World Wide Web programming language that basically guides an entire document or site. HTML tells browsers how to display online text and graphics, how to link pages, and how to link within a page. HTML also designates font style, size, and color.
>
> Specialized commands or tags determine a document's layout and style. For example, to center a document's title – say, How to Plant a Containerized Rose in Houston – and display it as a large headline font, the tags <HTML><HEAD><TITLE> are inserted before and after the title, respectively:
>
> <center><H1><I>How to Plant a Containerized Rose in Houston </TITLE></HEAD></HEAD>
>
> The first set of commands within the brackets tells the browser to display the text centered and in headline bold font. The set of bracketed commands containing a slash tells the browser to stop displaying the text in the designated style.
>
> The HTML source code for most Web pages can be viewed by users. In Firefox, click on the Tools pull-down box in the browser's tool bar and then click on the Web Developer option. Then click on Page Source.

> **ZOOM IN 5.2**
>
> To see Web pages from the old days, go to WayBack Machine, http://archive.org/web/web.php, a service that brings up Web sites as they were in the past. The archives go as far back as 1996.

movement among and within documents. Hypertext is what lets users skip all around a Web site, in any order they please, and jump from the beginning of a page to the end and then to the middle, simply by pointing and clicking on hot buttons, links, and icons.

The importance of Web browsers is that they transformed the Internet from a collection of text-only documents to a collection of Web information displayed and stored in text, graphics, and audio and video formats. Web browsers also improved old Internet data storage and organization through the creation of Web sites, in which related documents and information could be housed in one area. Because of Web browsers, accessing information is as easy as clicking a mouse.

SEE IT NOW

Internet Resources

The Internet is a vast connection of trillions of bits of information and it is the gateway to many types of digital communication like one-to-one or one-to-many 'conversations.' Email, electronic mailing lists, chat rooms, instant messaging, and newsgroups are some of the more common ways people interact with others through Internet technology. Web sites, blogs, microblogs, social network sites, video sites, and apps are venues for posting information and interacting with the public. Each of these Internet resources is described below in more detail.

Electronic Mail

Electronic mail, or email, is one of the earliest Internet resources and one of the most widely used applications. The first known email was sent from UCLA to Stanford University on October 29, 1969, when researchers attempted to send the word 'login.' They managed to send the letter L and then waited for telephone confirmation that it had made it to Stanford. They then sent the letter O and waited and waited until it arrived. Then they sent the letter G, but due to a computer malfunction, it never arrived. Just as the letter S was the first successful transatlantic radio signal, the letters L and O made up the first successful email message.

Email has come a long way since that first attempt. Email has become an easy and quick way to communicate and share documents one-to-one or to a group of known individuals. Almost 300 billion email messages fly through the global cyberspace every day from more than six billion accounts. The number of personal emails is decreasing as consumers turn to social networking, instant messaging, texting, chatting, and other forms of digital communication.

FYI: HOME INTERNET ACCESS

Percentage of U.S. households with access

Year	%
2019	85.0%
2014	84.0%
2009	80.0%
2005	58.5%
2001	51.0%

Source: Leichtman Research Group, 2019

The rise of email and the creation of online marketing and billing, business-to-business email, and new systems for collecting and verifying online signatures have reduced reliance on the United States Postal Service (USPS). The volume of mail handled by the USPS has been declining in recent years. From 2000 to 2010, volume declined from 208 billion pieces of mail to 170.9 billion, and to 146.4 billion in 2018 (much of it due to online bill paying and the Great Recession when less business mail was sent). There are now more emails sent each day than letters and packages through the USPS each year. A delivery agreement between USPS and Amazon doubled shipping and package delivery to 6.2 billion pieces from 2010 to 2018, and though delivery only accounts for about 5% of postal volume it brings in 30% of its revenue.

Unfortunately, the coronavirus has taken its toll on the USPS. Even though package volume jumped by 53% during the first few months of the outbreak as the homebound bought more goods online, it is not enough to make up for the 30% plunge in revenue from the steep drop off of business mail and third-class advertisements.

Electronic Mailing Lists

Electronic mailing lists are similar to email, in that messages are sent to electronic mailboxes for later retrieval. The difference is that email messages are addressed to individual recipients, whereas electronic mailing list messages are addressed to the list and then forwarded only to the electronic mailboxes of the list's subscribers. Electronic mailing lists are often referred to generically as listservs; however, LISTSERV is the brand name of an automatic mailing list server that was first developed in 1986.

Electronic mailing lists connect people with similar interests. Most lists are topic specific, which means subscribers 'converse' about specific subjects, like college football, gardening, computers, dog breeding, and television shows. Mailing lists are commonly used in the workplace to connect personnel. Many clubs, organizations, special-interest groups, classes, and media use electronic mailing lists as a means of communicating among their members. Electronic mailing lists can be open to anyone, or can be restricted to subscribers who sometimes need permission to join.

Chat Rooms

A chat room is another type of interpersonal communication. Chat participants exchange live, real-time

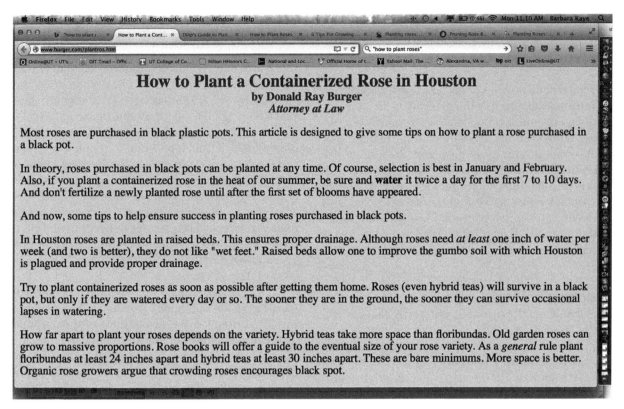

Figure 5.2 Simple How to Plant Roses Web site www.burger.com/plantros.htm.
Source: Photo courtesy of Donald Burger

messages. It is almost like talking on the telephone in that a conversation is going on, but instead of talking, messages are typed back and forth. Chat is very useful for carrying on real-time, immediate-response conversations, such as for online tech or customer product support.

Instant Messaging

Instant messaging (IM) is another way to carry on real-time typed conversations. Different from chat-room conversation, which can occur among anonymous individuals, IM takes place among people who know each other. IM is basically a private chat room that alerts users when friends and family are online and available for conversation. Because users are synchronously linked to people they know, IM has a more personal feel than a chat room and is more immediate than email. Some IM software boasts video capabilities to create a more realistic face-to-face setting.

'Instant messaging' differs from 'text messaging' in that IM is an Internet computer-to-computer connection, whereas texts are sent to and from cell phones and other handheld devices. Instant messaging services were once the domain of America Online (AOL) and Microsoft, but Skype, Facebook, and Google have since added IM applications.

Newsgroups

Similar to electronic mailing lists, newsgroups bring together people with similar interests. Web-based newsgroups are discussion and information exchange forums on specific topics, but unlike electronic mailing lists, participants are not required to subscribe, and messages are not delivered to individual electronic mailboxes. Instead, newsgroups archive messages that users access at their convenience. Think of a newsgroup as a bulletin board in a hallway outside of a classroom. Flyers are posted on the bulletin board and left hanging for people to sift through and read.

Web Sites

Most people think of the Internet and the Web as synonymous, but the Web is the part of the Internet that brings graphics, sound, and video to the screen. A **Web site** is a collection of related information, or **Web pages**. Some Web sites carry news, others

HTML Source Code for the website "How to Plant a Containerized Rose in Houston"
<HTML><HEAD> <TITLE>How to Plant a Containerized Rose in Houston by Donald Burger, Houston, TX</TITLE>

</HEAD></HEAD><BODY bgcolor="#FFE4E1"><font size=5 color=brown>
**How to Plant a Containerized Rose in Houston</font>
<font size=3 color=brown>by Donald Ray Burger
Attorney at Law** </font>

<p>Most roses are purchased in black plastic pots. This article is designed to give some tips on how to plant a rose purchased in a black pot.

<P> In theory, roses purchased in black pots can be planted at any time. Of course, selection is best in January and February. Also, if you plant a containerized rose in the heat of our summer, be sure and water it twice a day for the first 7 to 10 days. And don't fertilize a newly planted rose until after the first set of blooms have appeared.

<P> And now, some tips to help ensure success in planting roses purchased in black pots.

<P> In Houston roses are planted in raised beds. This ensures proper drainage. Although roses need at least one inch of water per week (and two is better), they do not like "wet feet." Raised beds allow one to improve the gumbo soil with which Houston is plagued and provide proper drainage.

<P> Try to plant containerized roses as soon as possible after getting them home. Roses (even hybrid teas) will survive in a black pot, but only if they are watered every day or so. The sooner they are in the ground, the sooner they can survive occasional lapses in watering.

<P>How far apart to plant your roses depends on the variety. Hybrid teas take more space than floribundas. Old garden roses can grow to massive proportions. Rose books will offer a guide to the eventual size of your rose variety. As a general rule plant floribundas at least 24 inches apart and hybrid teas at least 30 inches apart. These are bare minimums. More space is better. Organic rose growers argue that crowding roses encourages black spot.

Figure 5.3 HTML source code for the "How to Plant a Containerized Rose in Houston" Web page. Source: Photo courtesy of Donald Burger

are online stores, or archives of related information about most any topic. As astonishing as it seems, there are an estimated 1.2 to 1.7 billion Web sites, up from one in 1991, ten in 1992 and 130 in 1993. Moreover, each site contains multiple pages, totaling up to an astounding 15 trillion pages of content. As Tim Berners-Lee pointed out during an interview on PBS's *Nightly Business Report*, there are more Web pages than there are neurons in a person's brain (about 50–100 billion). Further, Berners-Lee claims that more is known about how the human brain works than about how the Web works in terms of its social and cultural impact.

ZOOM IN 5.3

MOST POPULAR WEB SITES THROUGH TIME

Check out this YouTube video that takes you through a moving timeline of the most popular Web sites from 1996 to 2019: www.youtube.com/watch?v=2Uj1A9AguFs

Web sites are accessed through a Web browser. All Web browsers operate similarly, yet each has its own unique features. The competition among browsers has always been stiff, and market share fluctuates. In the mid-1990s, Netscape Navigator led the U.S. market, but by 2003 Internet Explorer (IE) commanded a 90% share of users, which left Netscape with only about a 7% share and Mozilla and other browsers fighting for the remaining market.

Two decades into the new millennium, the two pioneering browsers, Mosaic and Netscape Navigator, are now defunct, having been replaced with more sophisticated ones such as Firefox and Google Chrome, which is the most widely used browser (67.6% of users), with Firefox a distant second capturing about 9% of Internet users.

America Online was once a popular way to access online content, but it was more than just a browser. America Online was a proprietary online content provider. Sometimes the term 'walled garden' was used to describe America Online's business model because only subscribers could access certain services. In 1997, almost half of all U.S. homes with Internet access had a subscription to America Online. In 2000, America Online merged with Time Warner, and several years later its name changed to AOL. In 2015, AOL was acquired by Verizon Communications. Through the years, AOL shed its model as a proprietary service and became an Internet portal, similar to Yahoo!. AOL is now a news and information site and offers free access to the Internet, email accounts, and other online resources.

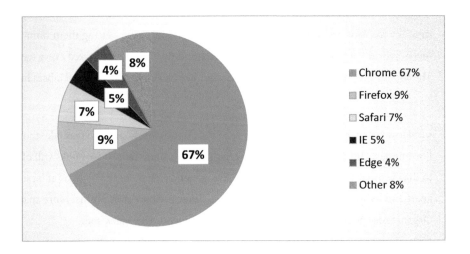

Diagram 5.2 Most used browsers by market share.
Source: Browser Market Share, 2020

ZOOM IN 5.4

BROWSER WARS THROUGH TIME (1993–2020)

Check out this YouTube video that takes you through a moving timeline of the most popular browsers from 1993 to 2020: www.youtube.com/watch?v=W4wWdmfOibY

Figure 5.4 Anyone can write a blog.
Source: Photo courtesy of iStockphoto. © AlexValent, image #3220999

Blogs

Blogs (originally dubbed Weblogs) are a type of Web site that allows users to interact directly with the blog host (blogger). It is hard to know exactly when the first blog came online, but the term was coined in 1997. Blogs have been part of the Internet landscape since the late 1990s, but they became popular shortly after September 11, 2001. These diary-type sites were an ideal venue for the outpouring of grief and anger that followed the terrorist attacks on the United States. Blogs are now exceedingly popular, with an estimated 500 million worldwide.

Blogs are free-flowing journals of self-expression in which bloggers post news items, spout their opinions, criticize and laud public policy, opine about what is happening in the online and offline worlds, and connect visitors to essential readings. Blog users, in turn, add their own comments and links. Blogs are places where everyday people and the digital and political elite meet to exchange ideas and discuss war, peace, the economy, politics, celebrities, and a myriad of other topics without the interference of traditional media. As such, bloggers are part of a tech-savvy crowd who sometimes scoop the media giants and provide more insight into current events than the traditional journalists.

Moving beyond the typical written blogs are videoblogs, also known as **vlogs**. Vlogs are kind of mini-video documentaries. Rants and raves are presented as full-motion video. Anyone equipped with a digital video camera and special software can produce his or her own vlog. Vlogs have become popular alternatives to blogs, especially on social media sites.

Microblogs

Microblogs are similar to blogs but the messages are much shorter. Twitter is the most popular microblog. Twitter is considered microblog because of the limited length of a tweet – 280 characters. But microblog or not, there is nothing micro about the size of some Twitter users' networks, which reach hundreds of thousands or even millions of followers.

Twitter took only a few years to infiltrate the social network world. Twitter has moved beyond connecting small groups of friends to being a way to reach millions of followers at once. Celebrities, politicians, and companies use Twitter to promote themselves, their ideas, and their products to their followers. The downside of Twitter is that once a message is out there, it cannot be taken back. One wrong word or misstatement can and probably will come back to haunt the author. Some users have lost

CAREER TRACKS: GLENN REYNOLDS, BLOGGER, INSTAPUNDIT.COM

WHAT IS YOUR JOB?

I'm Professor of Law at the University of Tennessee. I also write a twice-monthly column for *USA Today*. Before coming to Tennessee, I went to Yale Law School, clerked for a federal judge on the U.S. Court of Appeals for the Sixth Circuit, and practiced law at the Washington, D.C. office of a Wall Street law firm.

WHAT GAVE YOU THE IDEA OF STARTING YOUR BLOG?

In 2000/2001, I was operating some music Web sites and reading some of the very first blogs – Kausfiles, Andrew Sullivan, Virginia Postrel, etc. – when I ran across the Blogger.com site, which Ev Williams was then running from his basement, I believe. I had been using a program called Dreamweaver, which was very labor intensive. With Blogger you just logged on to a Web site, typed in your post, and hit 'publish.' It was so easy, I decided to give blogging a try. Also, I was teaching Internet law and wanted to keep up with new things. I didn't expect it to last so long, or to have such a large audience.

Figure 5.5 Glenn Reynolds.

DESCRIBE YOUR BLOG?

Today, Instapundit gets around 600,000+ page views on a typical weekday. It consists mostly of short posts on things that interest me. Often, it's just a link with a brief description, sometimes I'll make a few pithy observations, and sometimes I'll produce a lengthy post incorporating reader comments. It just depends. I tend to put politics, war, and the 'heavier' stuff in the morning, with science, technology, and pop culture in the afternoon, though that's more of a guideline than a rule. Basically, I post whatever I find interesting. A few years ago, I went from a solo blogging format to a group blog with several other bloggers ranging from John Tierney of *The New York Times* to fellow law professors David Bernstein and Gail Heriot. I think I still produce more than half the total content, though.

HOW IS PJ MEDIA CONNECTED TO INSTAPUNDIT?

PJ Media is a company I started with a couple of other bloggers back in 2005. It was sold to Salem in 2019, but my relationship – essentially with PJ as an advertising agent – has stayed the same. They also have editors who will keep an eye on my blog when I'm busy and fix typos, etc.

WHAT DO YOU DO ON A DAILY BASIS TO KEEP YOUR BLOG GOING?

On a typical day, the blog will have about 30–50 posts. Since about 15 years ago, I've been able to write posts in advance and set a time when they will self-publish, a technological improvement that has made my blogging life much easier. I usually set up a skeleton for the day the night before, so that if I get busy there will still be new stuff appearing every hour. Then I add breaking news and other new stuff that comes up throughout the day as I have time. I get a lot of my links from my readers, and a lot from Twitter, where I follow a lot of people from different circles. Sometimes I travel to events – conferences, protest rallies, etc. – and report on them via my blog, posting photos and videos that I shoot and edit myself.

WHAT ADVICE DO YOU HAVE FOR STUDENTS WHO MIGHT WANT TO START A BLOG?

Don't start a blog if you can't handle criticism. On the Internet, everyone's a critic. Fortunately, I'm pretty thick skinned. Focus on stuff that interests you, not on copying blogs that you like. The Internet audience is vast, so if you're good at blogging even on a niche target you can build up a big audience. Cultivate a personal relationship with your readers – that's what distinguishes a blog from a Big Media outlet. And don't be afraid to mix up your interests: The joke about my blog is 'come for the politics, stay for the nanotechnology.' Specialization is good, but too much specialization is for insects.

Also, bear in mind that your blog is a marketing tool for your skills. I was writing the occasional op-ed for newspapers before I started blogging (many law professors do that sort of thing), but I've gotten a lot of higher-level writing opportunities because editors know my general take on things, know that I can write, and know that I can write fast. Plus, when I publish a piece and then post a link on my blog, it brings traffic of its own. In today's click-conscious environment, that's a big advantage that comes with having your own audience.

their jobs or have been socially ostracized because of something they wrote on Twitter. Although proponents love the collection of voices, critics claim that most tweets are nothing but useless babble that wastes time.

Social Networking Sites

Millions of online users are drawn to social networking sites (SNS) as a means of keeping in touch with friends and family and building a network of new 'friends' based on shared interests and other commonalities, such as politics, religion, hobbies, and activities. Social media information is called **user-generated content (UGC)**, written by ordinary people, to differentiate it from news and information penned by journalists or other experts.

Social network sites have been in existence since the late 1990s, but hit their stride in the late-2000s. But of all SNS, about 200 have emerged as the most popular, with Facebook leading the pack with almost 70% of all Americans having an account. The growing popularity of social network sites is signaling a shift in how consumers are using the Internet. Almost eight in ten online users have a profile on at least one of the estimated thousands of networking sites, and a typical user has seven social media accounts.

Video Sharing Sites

There are countless sites on which users can post their own videos. People put up all types of videos that range from amateur content of funny cat capers, to graduation and wedding celebrations, to travel logs, to instructional aids, musical performance, to professionally produced movies and

ZOOM IN 5.5

Go to any search service, find a list of blogs, newsgroups, and social network sites, and click on some you are unfamiliar with. Join some you find interesting.
www.ebizmba.com/articles/blogs
www.bloggingfusion.com

television shows, and any other kind of video imaginable.

Unfortunately, some user-generated videos are nothing more than a pernicious attempt to alter reality. These include 'deep fake' videos in which digital files are manipulated in such a way that what seems real is actually doctored through voice-cloning and face-recognition technologies. A 2018 parody of a deep fake video created by actor/director Jordan Peele, shows former president Barack Obama using an obscenity when referring to his successor Donald Trump. Peele's point is that it is almost impossible for the average viewer to notice that a video is a 'deep fake.' The Obama video circulated among millions of online users – no telling how many of them thought it was real. Deep fake ends an era in which videos were once regarded as trustworthy, as 'seeing is no longer believing.'

Of all video sharing sites, YouTube is the most popular. Created in 2005 and acquired by Google in October 2006, slightly more than two billion unique users each watch more than 30 billion hours of videos and short clips each month. Every minute, 500 hours of video are uploaded to the site for open-access viewing.

FYI: TOP YOUTUBE CHANNELS (MARCH 2020)

Rank	YouTube channel	Description	Number of subscribers March 2020
1	T-series	Bollywood music, dance	130 million
2	PewDiePie	Variety, comedy	103 million
3	Cocomelon	Nursery rhymes, animated videos	74.2 million
4	SET India (Sony Entertainment Television)	Music, dance, comedy specials	66.1 million
5	5-Minute Crafts	Crafts and how-to	64.7 million
6	Canal KondZilla	Music videos	55.9 million
7	WWE	Wrestling, entertainment	55.1 million
8	Justin Bieber	Music	51.9 million
9	Z music	Music videos	50.9 million
10	Dude Perfect	Comedy sports group	49.5 million

Source: Larsen, 2020

Mobile Apps

In today's on-the-go world, consumers rely heavily on their smartphones and tablets to access the Internet. Users have embraced portable, location-based digital devices to receive phone calls, text messages, alerts, news, and many other types of communication. The small screen size of early hand-held devices made it hard to see Web sites or to use functions such as calendars or calculators. To solve this problem, smartphone and tablet manufacturers and independent developers began creating 'applications' (apps), which are computer software programs especially designed for viewing and interacting with online content on devices with small screens. Apps 'reside' on a mobile device's opening screen and are accessed with a quick tap on the icon. There are millions of apps to choose among – most Web sites have an accompanying app.

The World Online

Going online is often the first activity of the morning, and these days almost all daily activities revolve around the Internet. Completing routine tasks, such as making airline, hotel, or dinner reservations, contacting friends and family, keeping up with the latest news, finding recipes, shopping, and playing games, are conducted online. In developed countries, it is rare to find someone who does not use the Internet.

Using the Internet is no longer an in-home or in-office activity. With wireless connectivity and mobile smartphones and tablets, people access the Internet from almost anywhere. Café and coffee houses are filled with customers using the Internet while having a snack.

FYI: MOST POPULAR ONLINE ACTIVITIES (2017)

	Percentage of users
Sending or reading email	90.8%
Text messaging or instant messaging	90.2%
Using social media	74.4%
Watching video	69.5%
Shopping, making reservations, or using other consumer services	68.5%
Using financial services (billing paying, investing)	65.9%
Streaming or downloading music, radio, podcasts	52.6%
Online video calls or conferences	46.6%
Working remotely	22.6%
Job searching	20.8%
Taking classes/job training	19.1%
Interacting with household equipment (Alexa, Echo)	11.0%

Source: Statista, 2017

Internet Users

The Internet has come a long way since 1993, when there were about 14 million users (0.3%

of the world population). The number of users increased tenfold in the first ten years of its existence, and by mid-2016 there were about 3.5 billion users (48.6% global penetration), and by 2019, Internet users numbered 4.5 billion (58.8% global penetration).

U.S. figures for 2019 show 312 million (94.5% of the population), with most going online every day. The U.S., however, does not have the highest Internet penetration rate; that honor goes to Kuwait and Qatar both with 99.6%. Other countries that exceed the U.S. in percentage of Internet users are Falkland Islands (99.3%), Bermuda (99.2%), and Iceland (99.0%).

Estimates of the number of hours users spend online vary widely, but most individuals spend about three to five hours per day, which constitutes about 40% of their daily media use. Generally, online users tend to be younger than age 65, highly educated, and more affluent than the U.S. population in general, and they tend to live in urban and suburban areas.

FYI: HOW MANY ONLINE?

The number of Internet users and percentage of population by region (June 2019).

World regions	Internet users	% of population
North America	327.5 million	89.4%
Europe	727 million	87.7%
Latin America/ Caribbean	453.7 million	68.9%
Oceania/ Australia	28.6 million	68.4%
Middle East	175.5 million	67.9%
Asia	2.30 billion	54.2%
Africa	522.8 million	39.6%

Source: World Internet Users, 2019

FYI: DEMOGRAPHICS OF INTERNET USERS (2019) COMPARED TO 2000

	2000	2019
	Percentage	Percentage
Age		
18–29	70%	100%
30–49	61%	97%
50–64	46%	88%
65+	14%	73%
Gender		
Male	79.4%	90%
Female	77.6%	91%
Education		
Less than high school	53.7%	71%
High school graduate	69.7%	84%
Some college/ associate degree	82.4%	95%
Bachelor's degree or higher	91.5%	98%

Sources: Pew Research Center, 2019; Share of Adults, 2020a, 2020b

FYI: MILLENNIALS AND TECHNOLOGY USE (2019)

100% use the Internet
93% own a smartphone
86% use social network sites
78% have home broadband service
53% own a tablet computer

Source: Vogels, 2019

A Wireless World

To encourage settlement of the American West in the mid-1880s, prominent newspaper editor Horace Greeley urged, "Go west, young man, go west." If he was alive today, he would probably exhort us to venture into an even newer territory with "Go wireless, young man, go wireless."

Wireless technology **(Wi-Fi)** first became accessible to consumers when Apple introduced its AirPort Base Station in 1999. Base station technology gave way to wireless routers that send signals to any computer via a wireless card or transmission receiver that needed to be connected to a computer.

Most computers now come equipped with built-in wireless receiving capabilities. The ease and convenience of Wi-Fi has made it the preferred mode of connecting to the Internet. Wi-Fi makes it possible to pick up Internet signals from almost anywhere. These signals cannot be seen or felt, but in most restaurants, hotel lobbies, airports, universities, and other public places, Internet signals are bouncing off the walls and traveling through the walls.

Like a cell phone, Wi-Fi operates as a kind of radio. The 4G (fourth-generation) wireless technologies significantly improve data transmission speed and range over 3G wireless, and newer 5G promises yet greater speed and capabilities. The rollout for 5G is spotty at best, especially in countries, such as India, that are still building a 4G infrastructure. Switzerland is the global leader in 5G installation, followed by Kuwait and South Korea. 5G coverage is forecast to hit about 50% of U.S. mobile connections by 2025. Other wireless technologies, such as Bluetooth, are meant for very short-range connections, such as between a computer and a printer.

Wireless is not perfect. The speed of data transmission fluctuates and the signals are sometimes blocked by thick walls and other solid objects. The best wireless signal is obtained in a large, unobstructed space, like in an office with an open floor plan. Despite these obstacles, once wireless has been set up, it is quick and easy to use and compatible with most computers and mobile devices.

FYI: FIFTH GENERATION – 5G

The term 5G stands for 'fifth-generation' wireless service. The first generation (1981) was comprised of early analog cell phones and services. The second generation (1992) provided digitized phones and services. The third generation (2001) provided more efficient delivery and multimedia capabilities. Smartphones depend on 3G capabilities. The fourth generation (4G) (2012) provides mobile broadband Internet access to smartphones and other mobile devices through long-term evolution (LTE) technology that has even faster connection speeds. The fifth generation (5G), provides still faster service, quicker response rate, like the ability to download a full-length movie in just a few seconds, and enhanced capabilities.

The World Wide Web and the Mass Media

The Web is altering existing media use habits and the lifestyles of millions of users who have grown to rely on it as a source of entertainment, information, and social connections. The Web is distinguished from other mass media by its inclusion of text, graphics, video, and sound into one unique medium. Whereas newspapers are print only, radio audio only, and television audio and video, these properties converge through Internet technology. Convergence is the "coming together of all forms of mediated communication in an electronic, digital form, driven by computers" (Pavlik, 1996, p. 132), and the "merging of communications and information capabilities over an integrated electronic network" (Dizard, 2000, p. 14).

The Internet and Radio

As online technology improved, radio stations saw the potential for increasing their audience and revenue by delivering their content online. Almost all of the nation's 15,500 broadcast radio stations now have some sort of Internet presence. Online radio offers so much more to its audience than broadcast stations, which are limited by signal range and audio-only output. Internet radio delivers audio, text, graphics, and video to satisfy a range of listener needs.

The rise of Internet radio somewhat mirrors the development of over-the-air radio. In the early days of radio, amateur (ham) operators used specialized crystal sets to transmit signals and voice to a limited number of listeners (usually other ham operators) who had receiving sets. Transmitting and receiving sets were difficult to operate and the reception was poor and full of static, leaving only a small audience of technologically advanced listeners who knew how to operate them.

In the 1990s, the technologically savvy again took the lead, but this time, they paved the way for **cybercasting**. In its early years, the Internet was accessed through a modem, which allowed digital signals to travel from computer to computer via telephone lines. Just as primitive radios and static-filled programs once kept the general public from experiencing the airwaves, bandwidth limitations and slow computer and modem speeds kept many radio fans from listening via the Internet. Early online radio was hard to access, and it took an excruciatingly long time to download a song. For example, using a 14.4-Kbps modem, *Geek of the Week* – a 15-minute audio-only program – took almost two hours to download, which was considered very fast in the mid-1990s. Moreover, the playback was tinny and would often fade in and out. Few Internet users had computers powerful enough to handle audio, and few stations were streaming live content. Small- to medium-market and college radio stations generally led the way to the Internet. As their success stories quickly spread throughout the radio industry, other stations eagerly set up their own music sites. But in the late 1990s, online audio was still so undeveloped that only 5–13% of Web users were spending more time listening to radio online than to over-the-air stations.

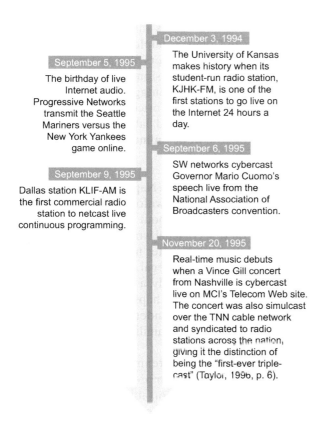

Diagram 5.3 Highlights of early Internet radio.

RealNetworks, the provider of **RealAudio** products, was the first application to bring real-time audio on demand over the Internet. Since the introduction of RealAudio in 1994, thousands of radio stations have made the leap from broadcast to online audiocast. By using RealAudio technology, AM and FM commercial radio, public radio, and college stations captured large audiences and moved from simply providing prerecorded audio clips to transmitting real-time audio in continuous streams. **Streaming** technology pushes data through the Internet in a continuous flow, so audio and video selections can be played before the entire file has finished downloading.

Since RealAudio's introduction, several other companies have developed audio-on-demand applications and protocols for increasing bandwidth for even faster streaming. Computer software and audio technology have become less expensive, easier to use, and better able to deliver high-quality sound.

Music Pirating In 1999, online music was transformed when college student Shawn Fanning created **Napster**, the first software for finding, downloading, and swapping **MP3 (Moving Picture Experts Group, Audio Layer 3)** music files online. Young adults' love for music made the MP3 format and MP3 player the hottest trend since the transistor radio hit the shelves in the 1950s. At first, music file-sharing sites seemed like a good idea, but they quickly ran into all kinds of copyright problems and found themselves knee deep in lawsuits. Although it has always been legal for music lovers to record their personal, store-bought LPs, tapes, and CDs to a portable player, sharing copies with others who have not paid for the music is considered piracy and copyright infringement.

In the early 2000s, the recording industry launched a vigorous campaign against so called 'music pirates,' who shared or downloaded music files online for free. In 2003, the **RIAA (Recording Industry Association of America)** sued 261 music lovers, targeting excessive pilferers. In some cases, the RIAA held individuals liable for millions of dollars in lost revenue, sometimes equaling up to $150,000 per song. After making its point, the RIAA worked out settlements in the $3,000 to $5,000 range and instituted the **Clean Slate Amnesty** program for those who wanted to avoid litigation by issuing a written promise to purge their computers of all files and never download music without paying again.

Since the first round of lawsuits, RIAA has filed another 30,000 or so. In the summer of 2009, a 32-year-old Minnesota mother of four and a Boston University graduate student were both fined for illegally downloading music. The woman was fined $80,000 for each of 24 songs she illegally downloaded, for a total judgment of $1.92 million – for songs that only cost 99 cents each; the graduate student was fined $675,000 for illegally downloading and sharing 800 songs between 1999 and 2007. Both fines were later settled for lesser amounts.

The RIAA is serious about getting its message out: "Importing a free song is the same as shoplifting a disc from a record store" (Levy, 2003b, p. 39). The RIAA's legal and educational efforts have paid off; just over half of those who regularly downloaded music illegally claimed that the crackdown has made them less likely to continue to pirate music. Prior to the initial lawsuits, only about 35% of people knew file sharing was illegal; now about double that percentage is aware of what constitutes illegal downloading. Despite efforts to protect copyright, just over one-third of online music consumers still download music illegally. Although some online users might not think there is much harm in illegally downloading a song here and there, the scope of the problem shows otherwise – illegal downloads cost the U.S. economy $12.5 billion in 2012, rising to over $20 billion by 2015.

File sharing and downloading, both legal and illegal, spawned a new way to obtain and listen to music and other audio files. Cybercasting, either by live streaming or by downloading, is a great way for little-known artists who have a hard time getting airplay on traditional stations to gain exposure. Additionally, listeners can easily sample new music, and links are often provided to sites for downloading and purchasing songs and CDs.

Radio Sites Although some radio station Web sites retransmit portions of their over-the-air programs, other audio sites produce programs solely for online use and are not affiliated with any broadcast station. The advantage of both over-the-air radio and online radio is that they can be listened to while working or engaging in other activities. But online radio has a distinct advantage over AM/FM radio – users can listen to stations from all over the world, and they can choose from a huge selection of music genres. If the radio stations in a local area play a limited range of music, listeners can go online to hear their favorite artists. Internet users may not be turning their backs on radio as such, but rather might be abandoning the old over-the-air delivery for online delivery, which provides clearer audio and a wider array of music.

Radio station Web sites are promotional by nature, and although free streaming is a great way for consumers who do not have access to a radio or who live outside the signal area to listen to their favorite station, early online radio hit a snag. After scores of complaints from record labels, artists, and others in the music business, the Library of Congress implemented royalty fees, which required Webcasters to pay for simultaneous Internet retransmission. The fees vary from per-song costs to a percentage of gross revenues, depending on the online pricing model. Whatever the exact amount charged, labels and artists claim this fee is too low, while online music providers say it is too high.

Audio Streaming Services Napster was at the center of the illegal music download imbroglio. After building a clientele of about 80 million users but facing several years of legal wrangling, Napster went offline in July 2001. Napster made its comeback in late 2003, this time as a legal site. In 2011 Napster was acquired by Rhapsody, a music streaming service with a monthly fee. Rhapsody, in fierce competition with Spotify and Pandora, rebranded itself in 2016 as Napster, a more familiar name in the music streaming business.

The audience for streaming Internet radio grows as sound quality improves and as listeners embrace newer, more portable playback devices such as laptops, smartphones, and tablets, and customizable services such as iTunes, Pandora, and Spotify. They offer an array of musical choices well beyond local over-the-air radio, plus music lovers get to select only what they want to hear, and they can see video of their favorite artists and learn more about them and their music. In 2018, about 162 million subscribers obtained music from any of the hundreds of licensed digital music services, with Apple Music, Spotify, Pandora, Sound Cloud, Google Play, iHeartRadio, and Amazon Music leading the pack.

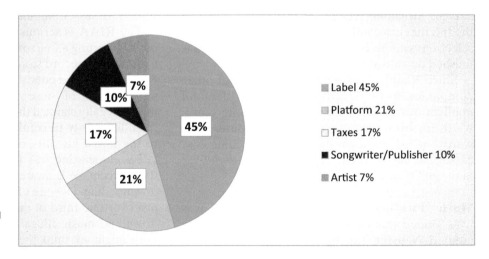

Diagram 5.4 An artist's share of streaming revenue.
Source: TechDirt, 2018

FYI: TOP DIGITAL SONG DOWNLOAD SALES 2019

Artist, Song

Lil Nas X, "Old Town Road"
Lady Gaga & Bradley Cooper, "Shallow"
Post Malone & Swae Lee, "Sunflower"
Halsey, "Without Me"
Lizzo, "Truth Hurts"
Lauren Daigle, "You Say"
Billie Eilish, "Bad Guy"
Post Malone, "Wow"
Panic! At the Disco, "High Hopes"
Jonas Brothers, "Sucker"

Source: Billboard, 2020

FYI: U.S. AUDIO LISTENING: 32 HOURS PER WEEK (2019)

HOW PEOPLE LISTEN:

Smartphone: 43.5%
PC/Laptop: 36.6%
Television: 23.0%
Tablet/e-reader: 20.0%
Stand-alone radio: 11.0%
Specialized headphones: 9.4%
Satellite radio: 8.7%
Portable combination music player/boom box: 6.4%
Voice-controlled devices: 6.1%
Wearables: 5.4%

Source: Nielsen, 2019

Radio vs. Streaming Radio stations strive to balance listenership so over-the-air does not keep listeners from the online site and streaming does not take away from traditional broadcasts. Local news and information is a strong draw to broadcast radio, and it is the main reason half of U.S. adults tune in. Stations understand that over-the-air is listened to more for non-music programming, whereas audio streaming is more for music, especially for learning about new tunes.

Stations have learned how to maximize listenership by teaming over-the-air delivery with online delivery. Most broadcast station Web sites stream their music as it is being played on air. It is good news for stations that, despite strong growth in streaming listenership, over-the-air AM/FM listeners remain loyal, except perhaps for adults aged 18 to 34 who are more likely to listen to online audio than over-the-air transmissions.

Radio stations and their online sites are adapting to the existence of streaming services; after all, they both play a part in increasing attention to music. Online listening continues to trend upward. In 2010 there were 93 million Internet radio listeners (30% of the U.S. population). By 2015 that number had risen to almost 170 million users (52%), and by 2019 to 209 million listeners (67%).

Total U.S. music revenue in 2018 grew to $9.8 billion, with streaming services generating $7.3 billion; about 75% of total music revenues. Physical sales accounted for 12% of revenue, digital downloads 11%, and synchronization 2%.

Podcasting Podcasts are audio-only programs produced by radio and television stations and networks, talk show hosts, celebrities, professors, and anyone who has something interesting to say about topics such as movies, music, popular culture, cooking, and sports. Many corporate media Web sites – such as npr.org, espn.com, and abc.go.com – host news podcasts. There are about 850,000 unique podcasts with 54 million episodes in over 100 languages, of which about 78% are recorded in English.

One of the first times that the term 'podcasting' was mentioned was in an issue of the British newspaper *The Guardian* in 2004. The term is a combination of the acronym 'personal on demand' (POD) and the word 'broadcasting.' According to Jason Van Orden, author of *Promoting Your Podcast*, a podcast is "a digital recording of a radio broadcast or similar programme, made available on the Internet for downloading to a personal audio player" (Van Orden, 2008).

The purpose of a podcast is not just to transfer readable information to audio format but to provide users with updated and sometimes live online audio. Different from a typical audio file, a podcast is created through **RSS (Real Simple Syndication)** feed, a standardized format used to publish frequently updated materials, such as news headlines and blog entries. Podcasts are subscribed to through a software program, called a podcatcher, that syncs podcasts to a computer-connected MP3 player, and they can be archived on a computer or burned onto a CD.

About 50% of the U.S. population has listened to a podcast. With about 330 million people in the United States, that percentage translates into

165 million listeners – an incredibly high number considering that podcasting first appeared online only in 2004. Slightly more males (56%) than females (44%) listen to podcasts, and the average age of a listener is 39. Further, podcast listeners have a higher-than-average annual income, about $87,000.

One of the most critically acclaimed podcasts to date first appeared in October 2014. *Serial* delves into police investigation of the death in 1999 of an 18-year-old woman in Baltimore, MD. The victim's boyfriend (Adnan Syed) was found guilty and sentenced to life plus 30 years despite his pleas of innocence. Each of the 12 *Serial* episodes dramatically unravels the story of the woman's strangulation and painstakingly retraces each step of the investigation, including scrutinizing witnesses' testimony and analyzing court documents. The gripping audio documentary becomes especially compelling as the boyfriend's guilt comes into question – was he wrongly convicted? *Serial* had caught the ears of about 40 million listeners, including lawyers for Project Innocence, who retested DNA samples found on the victim's body. Based on the evidence presented on *Serial*, a circuit court judge overturned Syed's conviction, but in 2019 it was reinstated by the Maryland Supreme Court.

The Internet and Television

Many ardent fans have long hailed television as the ultimate form of entertainment and news. Although cable was established early in the life of television, it did not take hold with viewers until the 1980s. Yet despite the popularity of watching television, the viewing public is always looking for new forms of entertainment. New video technologies gave rise to newer means of program delivery, which slowly enabled television stations and networks to show their programs online.

Television Embraces the Internet Just as early radio shows, such as *Guiding Light*, *You Bet Your Life*, and *The Lone Ranger*, migrated from radio to television, the television industry knew it had to embrace the Internet. But, television had the added burden of delivering quality video along with audio. With limited bandwidth, and slow modem speeds, the best video was choppy and pretty much unwatchable. Only after **RealVideo**, video compression software, came out in 1997 could the television industry seriously consider moving programs online. Doing so was a slow process and by the close of the 1990s, very few videos were available online, and the quality and slow loading speed kept users from bothering to watch.

FYI: TOP 10 PODCAST SITES 2019 (U.S. UNIQUE MONTHLY AUDIENCE)

1. NPR
2. iHeartRadio
3. Wondery
4. The New York Times
5. PRX
6. Barstool Sports
7. ESPN
8. WYNC Studios
9. Kast Media
10. NBC News

Source: Podcast Industry Audience Rankings 2019, 2020

Diagram 5.5 Highlights of early Internet television.

Despite its flaws, the Web was being hailed as the 'television of the future,' and was touted as the up-and-coming substitute for television. The physical similarities between a television and a computer monitor, coupled with the promise that online content would soon be as plentiful and exciting as that on television, led people to believe that the Web would soon replace television. The Internet caught on largely because it is used the same way television is watched – sitting in front of a screen watching something. Switching from Web site to Web site is, in some ways, similar to changing television channels – browsers' 'back' and 'forward' buttons are similar to up and down arrow keys on a television remote-control device. Even the lingo of Web browsing is borrowed from television. Commonly used terms such as 'surfing' and 'cruising' are used to describe traversing from one Web site to another and are also used to describe television channel-switching behavior.

Even though about 18–37% of Web users were watching less television in favor of the Internet, until about the mid-2000s television was a bit protected from Web pillaging for several reasons. Online technology could not deliver the same clear video and audio/video syncing as television, not all television programs were available online in their entirety, and television-quality original Web programs were scarce. Even the short episodes (**Webisodes**) of online-only programming failed to draw large numbers of viewers.

Television executives and online users crossed into the 21st century with the hopes that the Internet would soon be technologically capable of receiving television programs and other types of video content. It took well over a decade of technological improvements for online video to finally be almost as clear as television.

Almost all television networks and stations now host a Web site on which they post their shows. New smart television sets with built-in Internet, and links to Netflix and YouTube, a Web browser, and connections to DVD players, DVRs, and computers are bringing down the barriers that once separated a television set from the Internet and are, instead, setting up a symbiotic relationship that makes it unnecessary to have both a traditional television set and a separate computer with Internet access. On some televisions, while a program is shown in one window, computer functions such as email, are open in another window on the same screen. Television and the Internet are functionally combined in many ways and inseparable in the minds of many viewers.

FYI: TOP MULTIPLATFORM PROPERTIES 2019

Property	Unique visitors
Google Sites	255.9 million
Facebook	215.7 million
Microsoft Sites	208.7 million
Verizon Media	208.7 million
Amazon Sites	206.1 million
Comcast NBC Universal	184.5 million
CBS Interactive	166.5 million
Apple Inc.	156.1 million
Walt Disney Company	155.2 million
Hearst	154.6 million

Source: comScore, 2020

Watching Online Millions of viewers are watching television online every day. The top video sites by time spent are Netflix, YouTube, HBO Now, and Hulu. Viewers are most likely to watch full-length movies on Netflix rather than on the other services, and thus they spend about two hours per day using the service, compared to between 3.5 hours watching traditional television. Online viewing still has not caught up to television in terms of time spent, but the gap is quickly narrowing as younger viewers turn exclusively to streaming.

The glory days of broadband viewing might, however, be waning. The time when consumers could watch what they wanted, when they wanted, for far less money than paying for a cable subscription that included channels they never watch, is giving way to expensive services that offer less content for more money. For consumers to watch what they want to watch, they often must subscribe to a variety of streaming services, each with its own monthly bill, which can add up to more than the cost of cable, but with fewer program offerings. The average American spends about $1,300 a year for digital media. Not to mention of course, just more passwords to have to remember. Newer services, such as Disney+, are slowly pulling their shows and movies from services like Netflix into their own service, adding yet another subscription and another monthly bill. Searching for a particular program or movie could mean connecting to several services before finding the one that carries the show. As content is siloed among the services, offerings will be more limited unless consumers add new subscriptions, but the fatiguing and time-consuming nature of searching among many services keeps viewers to around two to three subscriptions. Even though Internet technology is the catalyst for changing the culture of television viewing, people love television and, so far, it has proven resilient against the lure of online delivery.

Second Screening The clarity of digitally delivered television programs shown on services such as Netflix and Hulu has changed the way viewers watch television. Viewers have combined their two favorite pastimes, television viewing and Web surfing, into a new activity – **second screening**. Second screening is when viewers who are watching a television program on a 'first screen,' such as a television or a computer, use a 'second screen,' a computer or mobile device, to look up program information or chat, text, or tweet with others who are also watching the show. Most viewers tend to second screen mostly by texting with friends in real life, but connecting with unknown others on social media is common too. Second screening connects television viewers who are watching the same content at the same time. Programs viewed through streaming services are second screened more often than those shown on broadcast and cable television. Presumably, viewers who watch programs online are tech-savvy and online connected, thus second screening is merely an extension of viewing.

Second screening is very popular, with 70% of television viewers doing so regularly, mostly on their smartphones and laptops, followed by desktops and tablets. As with other viewing trends, young people are leading the way – almost eight in ten viewers aged 18 to 34 second screen, while only about one-quarter of 50+ viewers do so.

The Internet and News

For many people, the Internet is the first stop for finding out about daily events. News junkies eat up reports posted on traditional media Web sites and social media. Online news reports are richly presented in audio, text, video, and graphic formats. Whereas radio is bound to audio, television to audio and video, and newspapers to text and graphics, on the Web traditional media cross over into other forms of presentation. The characteristics that distinguish radio and television news presentation from one another blur on the Web.

The Web has become increasingly important to both television networks and affiliate stations as an alternative means of distributing around-the-clock and up-to-the minute information. Television stations work with Internet companies to help develop and maintain their online news presence. The Web abounds with news sites that are hosted by reliable and known sources, such as CNN, NBC, *The New York Times*, and *The Los Angeles Times*. For example, stories that appear in *The Washington Post* also appear on its online site, news videos aired on *NBC Nightly News* also appear on its Web site. Web site news stories are generally produced by a network or an independent producer or written by a credentialed journalist, and are fact-checked and edited for accuracy. Media Web sites also often post stories, video, and interactive content that has not been printed or aired. News sites like *HuffPost*, *BuzzFeed*, and *Salon* are just a few of many respectable news sites that started as online-only news without a traditionally delivered counterpart.

The Web is an ideal venue for reporting the news in that it is free from the constraints of time and space. Most news organizations gather more information than they have the time to air or the space to print, but they can post an almost unlimited amount of information online. Online news is sometimes written as a summary with a link to an in-depth version as well as to related stories. Late-breaking news can be added almost instantaneously, and stories can be updated and amended as needed. For example, TMZ.com was the first news outlet to inform the public about Kobe Bryant's death in 2020 and Michael Jackson's death in 2009. TMZ even posted the Los Angeles coroner's office report of Jackson's death six minutes before anyone else. Critics claim that the posting was premature and TMZ just got lucky that Jackson did indeed die, but the organization countered that it has an inside track because of its large network of reliable sources.

Online media are having a difficult time distinguishing themselves from their competitors; therefore, media sites are turning to brand awareness as a strategy for getting their site selected over another, such as cbssports.com over espn.com. The media aim to transfer their strong brand names, such as NBC and CNN, to the online environment. Marketers speculate that early users of the Internet were not brand sensitive but rather tried out many different sites before returning to the ones they liked the best. Users who are more Web tentative tend to be brand loyal and stick to known sites. In other words, someone who regularly watches CNN on television may be more likely to access cnn.com than another online news site. But sticking with what one already likes and setting preferences for certain types of content, keeps users in a **filter bubble** of comfortable information, and unaware of the full range of stories and sources.

Moreover, attending to UGC, such as on social media, which are not always source checked or edited, could leave some users, especially novices, believing news that is not true. Such UGC now accounts for more online information than that produced by journalists, writers, and other professionals. Internet users must grapple with the amount of credence to give to what they see and read online. Thus, it is a good idea to sift through cyber information very carefully and to double-check the veracity of online information. Users should be cautious of accepting conjecture as truth and of using Web sources as substitutes for academic texts, books, and other media that check the facts for accuracy before publication.

Figure 5.7 The Internet keeps us connected to the latest news and events.
Source: Photo courtesy of iStockphoto. © sjlocke, image #9652079

Figure 5.6 With the Internet, you can watch online video anywhere.
Source: Photo courtesy of iStockphoto. © adamdodd, image #4481014

Millennials tend to get most of their news from social media. Although adults of all ages rely on various sources of news, mainly television and newspapers (online and traditionally delivered), and the Internet, Millennials are the least interested in mainstream media, or in news in general.

Privacy, Surveillance and Trust

Privacy is particularly relevant in the Internet era. The right to privacy of personal information such as how people spend their money and their time is a big concern to online users. Generally, people do not want personal information such as their phone numbers, addresses, credit card numbers, and medical information released without their consent.

Companies and other organizations track online surfing by placing cookies (tracking software) on a Web site visitor's computer. Cookies reveal what sites have been visited, monitor online purchases and product preferences, and collect geographic and demographic information. Moreover, because the Internet is commonly used for commerce between individuals, such as eBay auction transactions, as well as between individuals and commercial entities, private information is sometimes shared (or sold) to third parties. For example, credit card numbers intentionally given to one person or company could be stolen and used surreptitiously by someone else.

Data are collected even when users are doing the most mundane online activities, such as watching a video. Even common Web browsers are watching what users watch. Doing a search on Google means being tracked by third-party monitors, which are employed by about 90% of all Web sites. And even though a company might claim they do not 'sell' personal data, they sure do 'share' it. The importance of data collection is underscored by its financial impact on Internet commerce. Online companies utilize consumer data to send ads to targeted consumers, which in turn leads to the sale of products and services. Without the ability to collect data, the cycle would break and revenue would drop precipitously, as at least one-half of the Internet's economic value stems from consumer data.

Figure 5.8 In 1993, cartoonist Peter Steiner created for *The New Yorker* magazine a one-panel cartoon now known as "On the Internet Nobody Knows You're a Dog." The cartoon illustrated that Internet users at the time had almost total anonymity. The cartoon became the most reproduced in the magazine's history. The lower cartoon above, created by cartoonist Tom Toles, parodies the original to show how much the Internet had changed in seven years from "nobody knowing you're a dog" to "everybody knowing you're a dog," even your breed, hobbies, friends, online activities, and other personal information (Cavna, 2013).
Source: Toles © 2000 *The Buffalo News*. Reprinted with permission of Universal Press Syndicate. All rights reserved

Many online users go along with monitoring, thinking that because they have 'nothing to hide' there will not be any personal consequences. Browser history, however, could be used as a reason to be fired from a job if the boss does not like what employees are reading or writing online. A primary browser should be chosen carefully; some allow stronger privacy settings than others. Firefox lets users select 'private browsing' mode, and Safari applies 'intelligent tracking protection,' which screens out some cookies. Google Chrome, however, offers little protection against online data collectors. Users should learn how to set a browser's privacy settings. Another way to minimize surveillance perhaps is to switch phones – Androids collect about ten times as much personal data as iphones. And, for those who can afford the monthly fee, premium services, such as Apple News Plus, promise high-quality content that is ad-free, so less tracking takes place.

There is even a trade-off for using mobile apps. Users, wittingly or unwittingly, give apps access to their online activities and other personal information through preference settings, and algorithms track users' movements. Apps know where users have been, what they did, and where they are going – sometimes updating a user's activities thousands of times a day. Location information can reveal the intimate details of someone's life, like a doctor visited, who a user is dating, clubs belonged to, sporting activities participated in or watched, all of which could be accessed by an employer, insurer, or other person or entity with a vested interest in the user. Although many online users object to government scrutiny, they do not seem to mind that Google, Yahoo!, Facebook, and other Internet giants collect and store millions of bits of private information for commercial and marketing purposes.

Not only do online users need to worry about their private information being stolen, they also have to be on the lookout for their image or photo being used without permission. Individuals relinquish much of their right to privacy when they gain celebrity or are elected to public office.

Private citizens are entitled to more protection. People have tried to sue when videos are posted online of them at large public parties and gatherings, but the courts have held that individuals give up some of their right to privacy when they are in public places. Although the video producers profit (e.g. *Girls Gone Wild*), the individuals shown in the videos do not receive compensation nor, in most cases, did they grant permission to use their images. The wide availability of video cameras has made many people aware of the fact that their public behavior may become everyone else's entertainment.

ZOOM IN 5.6

VIDEO GONE WILD

Watch out when posting video on YouTube. If you will be embarrassed if others see it, do not do it. Look what happened to the two Domino's Pizza workers who foolishly sneezed on and otherwise tainted food they were preparing for delivery and captured it all on video. The workers claim they were just fooling around, but once the video hit YouTube, disgusted viewers spread the word all over the Internet. In the end, the workers were fired and the pair was arrested. Additionally, Domino's reputation was harmed, and it filed civil charges against the former employees.

See the original video at www.youtube.com/watch?v=oMO_uysMOXU

Baby Boomers and GenXers are probably the last two generations to know what it means to have private information kept private, whereas Millennials have grown up with their lives as an open book. But after years of privacy breaches, hijacked identities, and embarrassing social media photos, a security uprising is brewing. No matter what generation they are from, online users are embracing 'disappearing' message apps, like Snapchat, and they are pressuring social media such as Facebook to allow them greater control over what information is public and what is for friends only. Users are diligently setting up encryption tools, creating stronger passwords, enabling additional authentication tools, and just being more careful about what they post about themselves and what they upload to the cloud. Even though 86% of online users have taken steps to secure their privacy, 60% of them believe that online anonymity is an impossible dream. Moreover, the more people disclose about themselves online the more they desire privacy.

Some privacy advocates have endorsed encryption technology, which scrambles information and makes it unintelligible to third parties. While this seems like a simple enough way to guarantee privacy, the U.S. government has restricted the availability of powerful encryption programs. In fact, the government believes that it should have access to personal data to protect citizens against terrorist plots and that encryption technology hinders its information-gathering efforts. A consequence of breach of privacy and over-surveillance is a lack of trust in the Internet in general. Only about one-third of users trust what they read on general Web sites, 56% deem established media sites as credible, and about two-thirds trust government sites. Only about 13% believe information trusted by strangers. Clearly, online users can discern between 'official' sites and general use sites, and find the official sites most believable. With all the misinformation floating around cyberspace it is hard to know what is true and what is phony.

Some relief is on the way, at least for Europeans. The European Court of Justice ruled in 2014 that people have a right 'to be forgotten' by the Internet. The Court asserted that search engines like Google must erase reputation-harming search results generated by inputting someone's name. The ruling limits the Internet's ability to archive personal information indefinitely, especially if it is incorrect or harmful, and it places the right to privacy above the public's right to know. The ruling does not directly require that information be taken off a site; it applies only to the search results and the links that direct users to the information.

Figure 5.9 College students are the most avid of computer and Internet users.

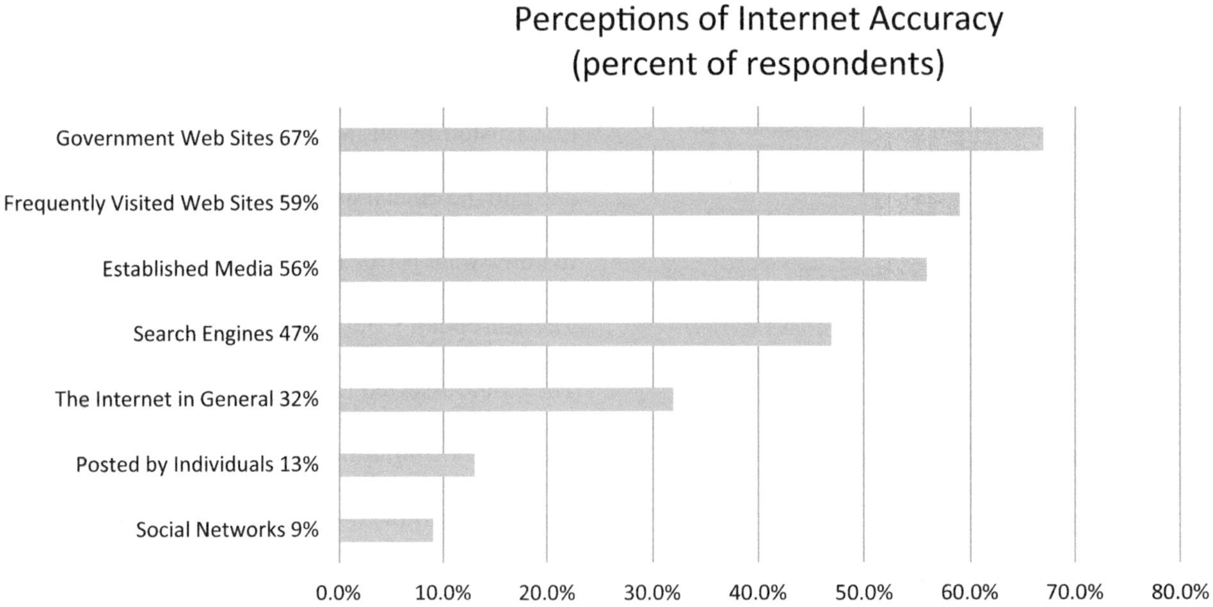

Perceptions of Internet Accuracy
(percent of respondents)

- Government Web Sites 67%
- Frequently Visited Web Sites 59%
- Established Media 56%
- Search Engines 47%
- The Internet in General 32%
- Posted by Individuals 13%
- Social Networks 9%

Diagram 5.6 Credibility of online information.
Source: Center for Digital Future, 2018

Legal experts claim that such a ruling in the U.S. would be regarded as censorship and clash with the First Amendment. In today's world, it is impossible to remove or correct information from sites that are not in a user's direct control. Anyone can say anything, post false data, or paint a misleading picture, but there is almost nothing a victim can do about it. Unfortunately, it is doubtful that the U.S. will adopt a European type of a 'right to be forgotten.'

SEE IT LATER

A Changing World

The Internet is changing so rapidly that what is written here in the 'See It Later' section of this chapter may very well be more appropriate for the 'See It Now' section by the time the book is published. It is not that the publishing process is slow – it is more that digital technology and digital culture move at lightning speed.

The traditional media are struggling to keep up with the Internet. As radio listeners and television viewers come to prefer online delivery, the media must offer online content. But just putting up content is not enough; the media have to figure out how to survive in the digital age.

Radio has survived broadcast television, cable television, records, cassettes, CDs, and music videos, but can it survive the Internet? Even though the audience for over-the-air radio is still large, broadcasters are not doing very well financially. As advertisers are moving online, so must broadcasters. The future is broadcasting to an audience of one. Radio must take full advantage of digital technology that allows users to be their own programmers. MP3 players and streaming should not be viewed as a radio station's enemies but rather as ways to reach an audience. Indeed, those who listen online also tune in to over-the-air radio, which indicates that radio listeners are not abandoning radio but are merely turning to its online counterparts.

Nearly three-quarters of all radio listening occurs in vehicles, but CDs, MP3 players, and satellite radio compete for drivers' ears as well. And now, Web radio is calling shotgun. Almost 60% of drivers still listen to AM/FM radio, 15% to CDs or their own music collection through iPod/MP3 players, 14% to satellite radio, online radio commands the attention of only about 10% or drivers, with the remaining not listening at all. Listening to in-car Web-streamed audio is possible with Bluetooth, which links a smartphone to the car sound system, by connected the two with an auxiliary cable, or by buying a special in-car version of Amazon Echo that by voice commands a smartphone to play a tune, "Alexa, play 'This is America.'"

It used to be that watching television was a family or group activity during which everyone would squeeze together on the sofa and sit back and make comments about the show. As the number of multiple-television households grew and with the proliferation of cable channels, television viewing became more of a solo activity and more individualized. But still, watching programs when they aired gave viewers something to talk about the next day. The Internet and other video delivery systems are contributing to a culture in which a shared television viewing experience may become an activity of the past. Perhaps that is why second screening has taken off as a new way to watch television. Second screening is a way for viewers to once again enjoy the experience of shared viewing.

Since Mosaic hit the market in 1993, there has been much talk about the potential for viewing television online. Now the talk is over. Network television must find a way to retain an audience and turn a profit. Offering full episodes containing commercials and making them available for downloading on mobile devices are positive steps in attracting an audience and revenue. Many networks have started offering full-length programs for streaming to cable and satellite subscribers. Perhaps offering more behind-the-scenes clips, unedited news and interviews, and making it easier for viewers to help create news stories would make networks, stations, and online sites more engaging and interesting. Perhaps instead of thinking of television as an imperiled medium, it should be thought of as one with new and exciting ways of delivering content.

The younger Web-savvy generation enjoys interacting online and feeling like they are part of the action. Blogs and vlogs are instrumental in creating the interactive culture that marks the online world. Users are no longer content to be mere receivers of news – they want to create news and report on events. During the 2009 protests in Iran challenging the election results, a young protester, Neda Agha-Soltan, was killed. She became a symbol of unrest as millions of viewers watched in horror as her death was captured on cell phones and video cameras and posted on YouTube as it happened. The day Michael Jackson died, several witnesses videoed the EMTs putting him into the ambulance and his arrival at the hospital before the major news outlets were on the scene. And because news travels in seconds across

Figure 5.10 The Internet is a combination of television, film, radio, newspaper, magazines, and other media.
Source: Photo courtesy of iStockphoto. © Petrovich9, image #6030817

the Internet, thousands of mourners gathered at the hospital within minutes of Jackson's death. These are examples of how individuals became reporters rather than just viewers. The difficulty for the television networks is verifying the user-generated content – anyone can shoot a fake video. Television news organizations need to strategize how to deal with these types of situations and set information policies regarding privacy, surveillance, ethical standards, verifying sources, and protecting anonymity.

As laptops once freed us from our desktop computers, mobile connectivity by cell phone and other handheld devices holds promise to unfetter us from our laptops. Mobile devices are quickly becoming the primary way to connect to the Internet. Their portability and low cost make them the optimal way to stay socially in touch and linked to online media and the real world.

Net Neutrality

Net neutrality and what it means to the Internet's future is one of the hottest issues of the day. Internet neutrality is basically the idea that the "Internet should be an impartial conduit for the information that flows through it" (Scola, 2014, p. B2), and everyone should have equal and unfettered access to online content ensured by regulations that high-speed Internet service providers cannot charge extra or slow or block the flow of certain types of content.

Online content providers such as Netflix, Google, and Yahoo! use huge amounts of broadband data to efficiently and effectively deliver content, especially video and graphics. But the ISPs, cable companies, and cell phone carriers that provide broadband access to the content companies are demanding that they be allowed to increase charges for certain amounts and types of access. As an executive for SBC Communications so eloquently proclaimed, "For a Google, or a Yahoo, or a Vonage to expect to use these pipes for free is nuts!" (Scola, 2014, p. B2). On the other hand, content providers and consumer groups protest that such a fee increase, and the right to slow bandwidth for some content would hinder access and go against the principle of a free Internet.

In 2005, the Federal Communications Commission (FCC) issued the Internet Policy Statement, which listed four principles to keep the Internet 'open.' According to the 2005 policy:

1. Consumers can access all lawful Internet content.
2. Consumers can use applications and services that they choose as long as they are lawful.
3. Consumers can connect devices to the Internet as long as the devices do not harm the network.
4. There should be free and open competition among network providers, application and service providers, and content providers.

The net neutrality controversy centers on the last point. The fear is that some Internet providers may block some Internet applications or content, especially from their competitors, thus creating a two-tiered system of Internet content haves and have-nots, depending on how much they can afford to pay. For example, if a couple of young entrepreneurs want to start a Web-based video service that requires high-speed bandwidth, under net neutrality they could simply set up their site and be in business. But without neutrality rules, an ISP could charge them the same amount of money for high-speed-quality service that they charge large established sites, thus competitively disadvantaging the newcomers. But opponents claim that ISPs should have the ability to provide tiered services, such as giving companies the option of paying more for high-quality data transmission at higher speeds than their competitors.

In late February 2015, the FCC voted to regulate the Internet as a utility, as are telephone companies and other utilities, so they cannot filter or interrupt online content. Common carrier status would allow the FCC to uphold and enforce its net neutrality rules. Tom Wheeler, former FCC commission chairman, justified the agency's stance as protecting innovators and consumers and preserving "the Internet's role as a core of free expression and democratic principles" (Ruiz & Lohr, 2015, p. B1).

ZOOM IN 5.7

NUMBER OF USERS OF STREAMING SERVICES THROUGH TIME (2005–2019)

Check out this YouTube video that takes you through a moving timeline of the number of users of streaming services, 2005 to 2019: www.youtube.com/watch?v=MbKZQUZEabM

Opponents of the FCC utility ruling have not given up and are looking into legal ways to curb it, even hoping that the Supreme Court will intervene. With President Trump's appointment of Ajit Pai, an opponent of net neutrality, as the chair of the FCC, some net neutrality rulings have been reversed and Internet services reclassified. The battle continues as individual states try to pass their own neutrality laws. California was the first to pass a neutrality Act, but it is being challenged by the U.S. Department of Justice. There is much at stake, as shown by the almost four million public comments sent so far to the FCC. Clearly, the controversy continues and could take years to settle.

SUMMARY

The history of the Internet is longer than most people think. It dates back to the 1960s, when scientists were experimenting with a new way to share information and keep connected in times of crisis. In its early days, the Internet was largely limited to communicating military, academic, and scientific research and was accessed by using complicated commands. In 1993, the first Web browser, Mosaic, came onto the scene, and the Internet quickly caught the public's attention. Since then, it has become the most quickly adopted new medium in history. But in the year 2020, the Internet is no longer considered 'new', rather, as coined by this book, it is a 'now medium.' The Internet is 'now' in the sense that it delivers information and entertainment in unprecedented ways, and those ways are ever changing. A technology or medium that is 'new' today, is 'old' tomorrow, but while we are using it – it is 'now.'

Without a doubt, the Internet has changed our lives tremendously. We no longer have to passively sit and absorb whatever news and information the traditional media want to send our way, but we can select what we want to know. Online technologies have sprouted new ways of gaining news and information and having an individual voice. Despite these unique qualities, the Internet also has many of the same properties as the traditional print and broadcast media. It delivers audio, video, text, and graphics in one package.

Many tech-savvy Internet users have begun the quest for access to information without intrusion by advertisers and big data mining companies. For now, the answer seems to be monthly fees for proprietary services that guarantee privacy and anonymity. The early Internet promised an egalitarian world in which all users, regardless of socioeconomic status, education, place of residence, and so on, would have equal access to information. Unfortunately, instead the Internet has divided the world. Not everyone has access, not everyone has access to the same information, and those who can afford it get to connect to high-quality, ad- and tracking-free sites.

Not only has the Internet changed the way we receive and provide information, but it is also altering our traditional media use behaviors. Millions of people around the world log onto the Internet on a regular basis. For some, the Internet will always be a supplement to radio and television, but for others, the Internet may become the medium they turn to first for news, information, and entertainment.

Online radio delivery has attracted many users who prefer clicking a button to hear their favorite audio over tuning in to an over-the-air station. For audio providers, online radio is relatively easy to set up and inexpensive to maintain. Cyber radio is the ideal medium for those who want to reach a global audience. People have been enamored with television and the act of watching television since the 1940s, which means it will be hard, if not impossible, to tear them away from it. Instead of abandoning television, new program delivery systems provide new ways to watch – on a computer, laptop, tablet, or smartphone. Television content still rules; the only difference is what type of screen it is watched on.

Clearly, the Internet is quickly catching up with radio and television when it comes to news delivery. In many ways, the Internet has surpassed radio and television when it comes to providing in-depth news. The Web is not constrained by time and space as are the traditional media. News is posted immediately and updated continuously, as the situation warrants. Moreover, Weblogs and vlogs redistribute news and information from established media into the hands of everyday people.

Broadcast radio and television networks and stations are competing among and between themselves and with the Internet, often with their own online counterparts. Television and radio must compete with the Web for a fragmented audience and precious advertising dollars. The traditional media are concentrating their efforts on designing Web sites that draw viewers away from their online competitors. But at the same time, they have to be sure that they do not lure viewers to the Web at the expense of their over-the-air fare. Even though research differs on whether the Internet is taking time away from radio and television, even a short amount of time spent online is time taken away from the "old-line media" (Dizard, 2000).

The Internet is still a relatively new medium, and so no one knows for sure what form it will end up taking as it keeps changing and adapting to technological innovations and diverse social and cultural needs. But what we do know is that it has had an enormous impact on the radio and television industries. If the prognosticators are right, the Web will eventually merge with radio and television, offering both conventional radio and television fare and Web-based content through a single device. But new net neutrality laws could change everything, from how much we have to pay for content, the speed and quality of content, and the type of content we receive to the Internet business model in general.

BIBLIOGRAPHY

126 amazing social media statistics and facts. (2019, September 30). *Brandwatch*, Retrieved from: www.brandwatch.com/blog/amazing-social-media-statistics-and-facts/#section-2

The 16th annual study of the impact of digital technology on Americans. (2018). *Center for the Digital Future*. Retrieved from: https://digitalcenter.org/wp-content/uploads/2018/12/2018-Digital-Future-Report.pdf

2020 Podcast Stats and Facts. (2020, January 28). *Podcasts Insights*. Retrieved from: www.podcastinsights.com/podcast-statistics/

Adgate, B. (2019, November 18). Podcasting is going mainstream. *Forbes*. Retrieved from: www.forbes.com/sites/bradadgate/2019/11/18/podcasting-is-going-mainstream/#3ed0d23f1699

Ali, A. (2018, November 13). The second screen viewing trend that's reshaping the market. *Curatti*. Retrieved from: https://curatti.com/second-screen-viewing-trend/

AOL. (2009). *Wikipedia*. Retrieved from: en.wikipedia.org/wiki/American_Online

Aratani, L. (2018, November 8). Altered video of CNN reporter Jim Acosta heralds a future filled with 'deep fakes.' *Forbes*. Retrieved from: www.forbes.com/sites/laurenaratani/2018/11/08/altered-video-of-cnn-reporter-jim-acosta-heralds-a-future-filled-with-deep-fakes/#70dc09133f6c

As Web sites go, which are the click magnets? (2014, September 28). *ComScore, The Washington Post Magazine*, p. 7.

Audio and podcasting fact sheet. (2019, July 9). *Pew Research Center*. Retrieved from: www.journalism.org/fact-sheet/audio-and-podcasting/

Audio by the numbers: The state of the news media. (2012). *Pew Research Center*. Retrieved from: www.stateofthemedia.org/2012/audio-how-far-will-digital-go/audio-by-the-numbers/

Audio today 2019: How America listens. (2019). Nielsen, Retrieved from: www.nielsen.com/wp-content/uploads/sites/3/2019/06/audio-today-2019.pdf

Average number of social media accounts per Internet user from 2013 to 2018. (2019, July 22). *Statista*. Retrieved from: www.statista.com/statistics/788084/number-of-social-media-accounts/

Berniker, M. (1995, October 30). RealAudio software boosts live sound, music onto the Web. *Broadcasting & Cable*, p. 67.

Billboard. (2020). Digital song: Year-end sales 2019. Retrieved from: www.billboard.com/charts/year-end/digital-songs

Blood, R. (2001, September 7). Weblogs: A history and perspective. Inkblotsmag.com. Retrieved from: rebeccablood.net/essays/weblog_history.html

Bogage, J. (2020, April 16). A 2006 law is partly responsible for the Postal Service's woes. *The Washington Post*, A18.

Bromley, R. V., & Bowles, D. (1995). The impact of Internet use on traditional news media. *Newspaper Research Journal, 16*(2), 14–27.

Browser market share. (2020, February). *Net Market Share*. Retrieved from: www.netmarketshare.com/browser-market-share.aspx?

Byers, K. (2019, January 2). How many blogs are there? *Growth Badger*. Retrieved from: https://growthbadger.com/blog-stats/

Campbell, F. B. (2014, December 26). The slow death of 'do not track.' *The New York Times*, Op Ed, 4.

Castillo, Michelle (2018, July 17). Netflix only takes up about 8 percent of the time you spend watching video, but the company wants to change that. *CNBC.com*. Retrieved from: www.cnbc.com/2018/07/17/netflix-small-portion-of-overall-watch-time-and-competition-is-stiff.html

Cavna, M. (2013, July 31). Nobody knows you're a dog. *The Washington Post*. Retrieved from: www.washingtonpost.com/blogs/comic-riffs/post/nobody-knows-youre-a-dog-as-iconic-internet-cartoon-turns-20-creator-peter-steiner-knows-the-joke-rings-as-relevant-as-ever/2013/07/31/73372600-f98d-11e2-8e84-c56731a202fb_blog.html

Center for the Digital Future. (2018). Surveying the digital future. *University of Southern California*.

Chadwick, A., O'Loughlin, B., & Vaccari, C. (2017). Why people dual screen political debates and why it matters for democratic engagement. *Journal of Broadcasting & Electronic Media, 61*(2), 220–239.

Clark, T. (2019, March 13). Netflix says its subscribers watch an average of 2 hours a day. *Business Insider*, Retrieved from: www.businessinsider.com/netflix-viewing-compared-to-average-tv-viewing-nielsen-chart-2019-3

Clover, J. (2019, February 28). Streaming music contributed 75% of total U.S. music industry revenues for 2018. *MacRumors*, Retrieved from: www.macrumors.com/2019/02/28/streaming-services-music-industry-revenues-2018/

Cohen, N. (2009, July 13). How the media wrestle with the Web. *The New York Times*. Retrieved from: www.nytimes.com/2009/07/13/technology/internet/13link.html [July 14, 2009]

comScore. (2019, December). Top U.S. online video content properties ranked by unique video viewers. Retrieved from: www.comscore.com/Insights/Rankings#tab_video_top_properties/

comScore. (2020). ComScore March 2019 top video multi-platform website properties. Retrieved from: https://seattleorganicseo.com/comscore-march-2019-top-50-multi-platform-website-properties-desktop-and-mobile/2016?ns_campaign=comscore_general&ns_source=social&ns_mchannel=social_post&ns_linkname=link_name&ns_fee=0

Cook, S. (2019, March 28). 60+ video and music streaming statistics. *Comparitech*. Retrieved from: www.comparitech.com/tv-streaming/streaming-statistics/

Digital audio usage trends: A highly engaged listenership. (2011). *Parks Associates*. Retrieved from: www.iab.net/media/file/TargetSpotInc_DigitalAudioUsageTrends_WhitePaper2011.pdf

Distribution of music industry revenue in the United States in 2017 and 2018. (2019). *Statista*. Retrieved from: www.statista.com/statistics/186304/revenue-distribution-in-the-us-music-industry/

Dizard, W. (2000). *Old media, new media*. New York: Longman.

Email Statistics Report, 2015–2019. (2015, March). *The Radicati Group, Inc*. Retrieved from: http://radicati.com/wp/wp-content/uploads/2015/02/Email-Statistics-Report-2015-2019-Executive-Summary.pdf

Fahri, P. (2009). TMZ earns second look with scoop on Jackson. *The Washington Post*, pp. C1, C12.

Federal Communications Commission. (2019, December 31). *Broadcast Station Totals*. Retrieved from: https://docs.fcc.gov/public/attachments/DOC-361678A1.pdf

First U.S. Web page made its debut 10 years ago. (2001, December 13). *Knoxville News-Sentinel*, p. C1.

Fitzpatrick, A. (2013, February 26). Custom top-level domains will be live this year. *Mashable*. Retrieved from: http://mashable.com/2013/02/26/top-level-domains-2013/#

Flatow, I. (2009, October 30). Happy birthday, Internet. Transcript. NPR.com. Retrieved from: www.npr.org/templates/story/story.php?storyId5114319703. [November 4, 2009]

Fowler, G. A. (2002, November 18). The best way ... Find a blog. *Wall Street Journal*, p. R8.

Fowler, G. A. (2019, June 30). Google's Chrome browser has become surveillance software. *The Washington Post*, G1, G5.

Frankel, D. (2019, May 31). Smart TV penetration in U.S. now up to 32%. *MultiChannel News*. Retrieved from: www.multichannel.com/news/32-percent-of-us-tvs-are-smart-tvs

Frauenfelder, M., & Kelly, K. (2000). Blogging. *Whole Earth*, 52–54.

Friend, E. (2009, June 18). Woman fined to tune of $1.9 million for illegal downloads. CNN.com. Retrieved from: www.cnn.com/2009/CRIME/06/18/minnesota.music.download.fine/index.html [October 25, 2009]

Fung, B. (2015, March 18). GPO: FCC's IG is looking into net neutrality. *The Washington Post*, p. A13.

Gamerman, E. (2014, November 13). 'Serial' podcast catches fire. *The Wall Street Journal*. Retrieved from: www.wsj.com/articles/serial-podcast-catches-fire-1415921853

Gil de Zúñiga, H., García-Perdomo, V., & McGregor, S. (2015). What is second screening? Exploring motivations of second screen use and its effect on online political participation. *Journal of Communication, 65*, 793–815.

Global spam volume as a percentage of total email traffic. (2019, December 4). *Statista.* Retrieved from: www.statista.com/statistics/420391/spam-email-traffic-share/

Gross, G. (2015, January 20). Republican Net neutrality bill allows 'reasonable' network management. *InfoWorld.* Retrieved from: www.infoworld.com/article/2871945/government/republican-net-neutrality-bill-allows-reasonable-network-management.html

Heeter, C., D'Allessio, D., Greenberg, B., & McVoy, S. D. (1988). Cableviewing behaviors: An electronic assessment. In C. Heeter & B. Greenberg (Eds.), *Cableviewing* (pp. 51–63). Norwood, NJ: Ablex.

Helft, M. (2009). Google to add captions, improving YouTube Videos. *The New York Times,* p. B4.

Herring, S. (2015, October 19). Should you be capitalizing the word 'Internet'? *Wired,* Retrieved from: www.wired.com/2015/10/should-you-be-capitalizing-the-word-internet/

Herman, J. (2019, November 17). Big techs big stalemate. *The New York Times Magazine,* pp. 38–45.

Hesse, M. (2009). Web series are coming into a prime time of their own. *The New York Times,* pp. E1, E12.

Hotz, R. L. (2009). A neuron's obsession hints at biology of thought. *The Wall Street Journal,* p. A14.

How many websites are there around the world? (2020, January 9). *Mill for Business.* Retrieved from: www.millforbusiness.com/how-many-websites-are-there/

How old media can survive in a new world. (2005, May 23). *The Wall Street Journal.* Retrieved from: online.wsj.com [September 29, 2009]

Human Brain. (2009). *Wikipedia.* Retrieved from: en.wikipedia.org/wiki/Human_brain [October 25, 2009]

ICANN. Registry Listing. (2010). *Internet Corporation for Assigned Names and Numbers.* Retrieved from: www.icann.org/en/registries/listing.html [January 27, 2010]

Indvik, L. (2011, March 25). Internet piracy is on the decline. *USA Today.* Retrieved from: http://content.usatoday.com/communities/technologylive/post/2011/03/us-internet-piracy-is-on-the-decline/1#.VtnbHiNN_xY

Interactive Advertising Bureau. (2017, May). The changing TV experience: 2017. Retrieved from: www.iab.com/wp-content/uploads/2017/05/The-Changing-TV-Experience-2017.pdf

Internet 2008 in numbers. (2009, January 22). *Royal. Pingdom.com.* Retrieved from: www.bizreport.com/2009/06/americans_greatly_increasing_time_spent_online_1.html [October, 25, 2009]

Internet 2013 in numbers. (2014. January 14). *Dubai Chronicle.* Retrieved from: www.dubaichronicle.com/2014/01/31/internet-2013-numbers/

Internet and American life. (2000). *Pew Research Center.* Retrieved from: www.pewinternet.org/reports [January 14, 2002]

Internet history. (1996). *Silverlink LLC.* Retrieved from: www.olympic.net/poke/IIP/history.html

Internet radio's audience turns marketers' heads. (2013, February 6). *eMarketer.* Retrieved from: www.emarketer.com/Article/Internet-Radios-Audience-Turns-Marketer-Heads/1009652

Internet users. (2015, January 19). *Internet Live Stats.* Retrieved from: www.internetlivestats.com/internet-users/

Internet usage stats. (2019). *Internet World Stats.* Retrieved from: https://internetworldstats.com/stats.htm

Irvine, M. (2010, February 3). Is blogging a slog? Some young people think so. *Excite News.* Retrieved from: apnews.excite.com/article/20100203 [February 4, 2010]

Itzkoff, D. (2009, August). Graduate student fined in music download case. *The New York Times,* p. A3.

Johnson, T. J., & Kaye, B. K. (2002). Webelievability: A path model examining how convenience and reliance on the Web predict online credibility. *Journalism & Mass Communication Quarterly, 79*(3), 619–642.

Johnson, T. J., & Kaye, B. K. (2004). Wag the blog: How reliance on traditional media and the internet influence perceptions of credibility of weblogs among blog users. *Journalism & Mass Communication Quarterly, 81*(3), 622–642.

Johnson, T. J., & Kaye, B. K. (2011). Can you teach a new blog old tricks? How blog users judge credibility of different types of blogs for information about the Iraq War. In B. K. Curtis (Ed.), *Psychology of trust* (pp. 1–25). New York: NOVA Publishers.

Judge cues Napster's death music. (2002, September 3). *Wired News.* Retrieved from: www.wirednews.com

Kaye, B. K. (1998). Uses and gratifications of the World Wide Web: From couch potato to web potato. *New Jersey Journal of Communication, 6*(1), 21–40.

Kaye, B. K., & Johnson, T. J. (2002). Online and in the know: Uses and gratifications of the Web for political information. *Journal of Broadcasting & Electronic Media, 46*(1), 54–71.

Kaye, B. K., & Johnson, T. J. (2003). From here to obscurity: The Internet and media substitution theory. *Journal of the American Society for Information Science and Technology, 54*(3), 260–273.

Kaye, B. K., & Johnson, T. J. (2004). Blog day afternoon: Weblogs as a source of information about the war on Iraq. In R. D. Berenger (Ed.), *Global media go to war* (pp. 293–303). Spokane, WA: Marquette.

Kaye, B. K., & Medoff, N. J. (2001). *The World Wide Web: A mass communication perspective*. Mountain View, CA: Mayfield.

Knight, K. (2009, June 26). Americans greatly increasing time spent online. *BizReport.com*. Retrieved from: bizreport.com/2009/06/americans_greatly_increasing_time_spent_online_1.html [October 25, 2009]

Krol, E. (1995). *The whole Internet*. Sebastopol, CA: O'Reilly & Associates.

Lafayette, J. (2019, March 19). Traditional TV still sinking in stream of digital video. *TV Technology*. Retrieved from: www.tvtechnology.com/news/traditional-tv-still-sinking-in-stream-of-digital-video

Larsen, L. (2020, April 17). The most-subscribed to YouTube channels. *Digital Trends*. Retrieved from: www.digitaltrends.com/web/biggest-youtube-channels/?itm_medium=editors

Leichtman Research Group. (2019). 85% of U.S. households get an Internet service at home. Retrieved from: www.leichtmanresearch.com/85-of-u-s-households-get-an-internet-service-at-home/

Levy, S. (2003a, March 29). Blogger's delight. *MSNBC*. Retrieved from: www.msnbc.com

Levy, S. (2003b, September 22). Courthouse rock? *Newsweek*, p. 52.

Li, D. K. (2019, March 8). 'Serial' podcast's Adnan Syed has murder conviction reinstated by Maryland court. *NBC News*. Retrieved from: www.nbcnews.com/news/us-news/serial-podcast-s-adnan-syed-has-murder-conviction-reinstated-maryland-n981101

Listserv. (2003). *Webopedia*. Retrieved from: www.webopedia.com/TERM/L/Listserv.html

Luckerson, V. (2013, February 28). Revenue up, piracy down: Has the music industry turned a corner? *Time*. Retrieved from: http://business.time.com/2013/02/28/revenue-up-piracy-down-has-the-music-industry-finally-turned-a-corner/

Marc Andreessen, co-founder of Netscape. (1997). *Jones Telecommunications and Multimedia Encyclopedia Homepage*. Retrieved from: https://web.archive.org/web/19980202013525/www.digitalcentury.com/encyclo/update/andreess.htm

McGregor, S. C., Mourão, R. R., Neto, I., Straubhaar, J., & Angeluci, A. (2017). Second screening as convergence in Brazil and the U.S. *Journal of Broadcasting & Electronic Media*, 61, 163–181.

McKetta, I. (2019, October 1). How 5G is changing the global mobile landscape. *Speedtest*. Retrieved from: www.speedtest.net/insights/blog/5g-changing-global-mobile-landscape-2019/

Mitova, T. (2019, August 30). How many emails are sent per day? *Review 42*. Retrieved from: https://review42.com/how-many-emails-are-sent-per-day/

Most popular music streaming services in the United States as of March 2018. (2018). *Statista*. Retrieved from: www.statista.com/statistics/798125/most-popular-us-music-streaming-services-ranked-by-audience/

Murphy, K. (2014, October 4). We want privacy, but can't stop sharing. *The New York Times*, Op Ed 4.

Newman, N., Fletcher, R., Kalogeropoulos, A. L., Levy, D. A. L., & Nielsen, R. K. (2018). *Reuters Institute digital news report 2018*. Retrieved from: http://media.digitalnewsreport.org/wp-content/uploads/2018/06/digital-news-report-2018.pdf

Nielsen. (2019). Devices used for music listening in a typical week. Retrieved from: www.nielsen.com/us/en/insights/article/2017/time-with-tunes-how-technology-is-driving-music-consumption/

Nocera, J. (2009, August 7). It's time to stay the courier. *The New York Times*. Retrieved from: www.nytimes.com/2009/08/08/business/08nocera.html [October 26, 2009]

Number of digital music users in the U.S. from 2016 to 2022. (2019, May 20). *Statista*. Retrieved from: www.statista.com/statistics/455716/digital-music-users-format-digital-market-outlook-usa/

Number of sent and received emails per day worldwide, 2017–2024. (2021, January 27). *Statista*. Retrieved from: www.statista.com/statistics/456500/daily-number-of-e-mails-worldwide/

Our mission. (2015). *Facebook newsroom*. Retrieved from: http://newsroom.fb.com/company-info/

Palser, B. (2002). Journalistic blogging. *American Journalism Review*, 24(6), 58.

Parks, M. (2015, January 2). He came, Reddit conquered, now what? *The New York Times*, pp. C1, C2.

Patterson, T. (2009, May 8). Is the future of TV on the Web? *CNN.com*. Retrieved from: www.cnn.com/2008/SHOWBIZ/TV/05/01/tv.future/index.html [October 28, 2009]

Pavlik, J. V. (1996). *New media technology: Cultural and commercial perspectives*. Boston, MA: Allyn & Bacon.

Pegoraro, R. (2009, May 31). Web radio hits the road. *The New York Times*, p. G2.

Pew Research Center. (2019). Internet use by age. Retrieved from: www.pewresearch.org/internet/chart/internet-use-by-age/

Plummer, M. (2019, January 22). How to spend way less time on email every day. *Harvard Business Review*. Retrieved from: www.statista.com/statistics/456500/daily-number-of-e-mails-worldwide/

Podcast. (2009). *Wikipedia*. Retrieved from: en.wikipedia.org/wiki/Podcasting [October 25, 2009]

Podcast industry audience rankings 2019. (2020). *PodTrac*. Retrieved from: http://analytics.podtrac.com/industry-rankings/

Postal service stretched thin. (2010, February 18). *The Washington Post*, p. B3.

Progressive networks launches Real Audio Player Plus. (1996, August 18). *Real Audio*. Retrieved from: www.realaudio.com/prognet/pr/playerplus.html

Quittner, J. (1995, May 1). Radio free cyberspace. *Time*, p. 91.

Radio facts and figures. (2014). *News Generation*. Retrieved from: www.newsgeneration.com/broadcast-resources/radio-facts-and-figures/

Rafter, M. (1996b, January 24). RealAudio fulfills Web's online sound promise. *St. Louis Post Dispatch*, p. 13B.

Rainie, L. (2008, December 14). The next future of the Internet. *Pew Internet & American Life Project*. Retrieved from: www.pewinternet.org/Commentary/2008/December/The-Next-Future-of-the-Internet.aspx#

Rein, L., & Bogage, J. (2020, April 25). Trump threatens to block virus aid to Postal Service. *The Washington Post*, A1, A18.

Richtel, M. (2001, November, 29). Free music is expected to surpass Napster. *The New York Times*. Retrieved from: www.nytimes.com/2001/11/29/technology/29MUSI.html [November 29, 2001]

Roberts, A. (2014, December 23). The 'Serial' podcast by the numbers. *CNN.com*. Retrieved from: www.cnn.com/2014/12/18/showbiz/feat-serial-podcast-btn/index.html

Roose, K. (2019, November 17). The rise of the luxury Internet. *The New York Times Magazine*, pp. 28–37.

Royalties. (2009). *Wikipedia*. Retrieved from: en.wikipedia.org/wiki/Royalty_fee#Music_royalties [October 25, 2009]

Ruiz, R. R., & Lohr, S. (2015, February 27). F.C.C. votes to regulate the Internet as a utility. *The New York Times*, pp. B1, B2.

Schellmann, H. (2018, October 15). Deepfake videos are getting real, and that's a problem. *The Wall Street Journal*. Retrieved from: www.wsj.com/articles/deepfake-videos-are-ruining-lives-is-democracy-next-1539595787

Schonfeld, E. (2012, January 8). How people watch TV online and off. *Tech Crunch*. Retrieved from: http://techcrunch.com/2012/01/08/how-people-watch-tv-online/

Scola, N. (2014, June 15). 5 myths about net neutrality. *The New York Times*, p. B2.

Scott, M. (2014, October 8). Tim Berners-Lee, Web creator, defends net neutrality. *The New York Times*. Retrieved from: http://bits.blogs.nytimes.com/2014/10/08/tim-berners-lee-web-creator-defends-net-neutrality/

Segailer, S. [Executive producer] (1999). *Nerds 2.0.1: A brief history of the Internet*. Television broadcast. Public Broadcasting Service.

Self, C. (2003, September 22). Admitted copyright offenders avoid suits. *Daily Beacon*, p. 1.

'Serial' season one. (2014). *Serial*. Retrieved from: http://serialpodcast.org/

Share of adults in the U.S. who use the Internet in 2019, by gender. (2020a). *Statista*. Retrieved from: www.statista.com/statistics/327130/internet-penetration-usa-gender/

Share of adults in the U.S. who use the Internet in 2019, by education. (2020b). *Statista*. Retrieved from: www.statista.com/statistics/327138/internet-penetration-usa-education/-gender/

The size of the world wide web. (2020, February 2). *World Wide Web Size*. Retrieved from: www.worldwidewebsize.com/

Smith, C. (2016, February 19). By the numbers: 60+ amazing Reddit statistics. *DMR*. Retrieved from: http://expandedramblings.com/index.php/reddit-stats/

Snapes, L., & Beaumont-Thomas, B. (2018, October 9). More than one-third of music consumers still pirate music. *The Guardian*. Retrieved from: www.theguardian.com/music/2018/oct/09/more-than-one-third-global-music-consumers-pirate-music

Social media fact sheet. (2019, June 12). *Pew Research Center*. Retrieved from: www.pewresearch.org/internet/fact-sheet/social-media/

Statista. (2017). *Most popular online activities of adult internet users in the United States*. Retrieved from: www.statista.com/statistics/183910/internet-activities-of-us-users/

Stelter, B. (2009, March 10). Serving up television without the TV set. *The New York Times*, pp. C1, C3.

Stelter, B. (2009, June 27). A web site scooped other media on the news. *The New York Times*, p. A10.

Stelter, B. (2009, November 11). TV news without the television. *The New York Times*, pp. B1, B10.

Stelter, B. (2011, September 26). Pew Media study shows reliance on many outlets. *The New York Times*, pp. B1, B10.

Stelter, B., & Helft, M. (2009, April 17). Deal brings TV shows and movies to YouTube. *The New York Times*, pp. B1, B2.

Stephens, H. (2017, May 8). Calculating the economic impact of counterfeiting and piracy. *HughStephensBlog*. Retrieved from: https://hughstephensblog.net/2017/05/08/calculating-the-cost-of-piracy-its-in-the-trillions/

Strauss, M. (2018, October 9). 38 percent of people still pirate music. *Pitchfork*. Retrieved from: https://pitchfork.com/news/38-percent-of-people-still-pirate-music-study-finds/

Streitfeld, D. (2014, May 14). European court lets users erase records on the Web. *The New York Times*, p. A1, A3.

Stross, R. (2009, February 8). Why television still shines in a world of screens. *The New York Times*, pp. D1, E4.

Taylor, C. (1996, September 23). Zapping onto the Internet. *MediaWeek*, p. 32.

TCP/IP green thumb: Effective implementation of TCP/IP networks. (1998). *LAN Magazine, 8*(6), 139.

TechDirt. (2018, August 18). Only 12% of music revenue goes to actual artist. Retrieved from: www.techdirt.com/articles/20180819/00051140461/only-12-music-revenue-goes-to-actual-artists.shtml

The average age of the podcast listener is ... (2019, November 14). *Podcast Business Journal*. Retrieved from: https://podcastbusinessjournal.com/the-average-age-of-the-podcast-listener-is/

Total number of web sites. (2020). *Internet Live Stats*. Retrieved from: www.internetlivestats.com/total-number-of-websites/

Top 25 countries with the highest Internet penetration. (2020). *Internet World Stats*. Retrieved from: www.internetworldstats.com/top25.htm

Twitter. (2009). *Wikipedia*. Retrieved from: en.wikipedia.org/wiki/Twitter [October 25, 2009]

United States: Number of Internet Users 2000–2019. (2020). *Statista*. Retrieved from: www.statista.com/statistics/276445/number-of-internet-users-in-the-united-states/

United States Postal Service. (2015). *Size and Scope*. Retrieved from: https://about.usps.com/who-we-are/postal-facts/size-scope.htm

United States Postal Service. (2015, November 15). *U.S. Postal Service Expects to Deliver more than 15 Billion Pieces of Holiday Mail and Packages this Year*. Retrieved from: https://about.usps.com/news/national-releases/2015/pr15_059.htm

United States Postal Service. (2019). *A decade of facts & figures*. Retrieved from: https://facts.usps.com/table-facts/

U.S. Census Bureau projects U.S. and World populations on New Year's Day. (2019, December 31). U.S. Census Bureau. Retrieved from: www.commerce.gov/news/blog/2019/12/us-census-bureau-projects-us-and-world-populations-new-years-day

Van Orden, J. (2008). *How to podcast*. Retrieved from: www.how-to-podcast-tutorial.com/what-is-a-podcast.htm [October 24, 2009]

Vogels, E. A. (2019, September 9). Millennials stand out for their technology use but older generations also embrace digital life. *Pew Research Center*. Retrieved from: www.pewresearch.org/fact-tank/2019/09/09/us-generations-technology-use/

Walker, R. (2009, October 18). The song decoders. *The New York Times Magazine*, pp. 49–53.

Weekly Internet radio listenership jumps from 33 to 42 million. (2009, April 16). *About.com: Radio*. Retrieved from: http://radio.about.com/b/2009/04/16/weekly-internet-radio-listenership-jumps-from-33-to-42-million.htm [October 26, 2009]

What is AudioNet? (1997, February 4). *AudioNet*. Retrieved from: www.audionet.com/about

Who music theft hurts. (2015). *Recording Industry Association of America*. Retrieved from: http://riaa.com/physicalpiracy.php?content_selector=piracy_details_online

Why we do what we do. (2015). *Recording Industry Association of America*. Retrieved from: http://riaa.com/physicalpiracy.php?content_selector=piracy-online-why-we-do-what-we-do

Wood, M. (2014, September 8). Younger generation still values privacy. *International The New York Times*, p. 15.

World Internet users. (2019). *Internet usage stats*. Retrieved from: www.internetworldstats.com/stats.htm

YouTube by the numbers. (2020, January 13). *Omnicore*. Retrieved from: www.omnicoreagency.com/youtube-statistics/

Zittrain, J. (2014, May 15). Don't force Google to forget. *The New York Times*, p. A25.

Chapter six

Social Media: Private Conversations in Public Places

Jason Stamm, Ph.D., University of Tennessee

Nowadays, it may seem impossible to be 'off the grid,' or disconnected from the rest of the world, at least digitally. Not only do people have the ability to post about their lives and interact with others on their laptop or desktop computers, but they can take their conversations through social media on the go with their smartphones, tablets, and other personal devices. Social media are 'now' in so many ways. Applications notify users to the latest social media posts from their friends and breaking news moments as they happen. Through blogs, Twitter, Facebook, Instagram, and most recently, TikTok, users selectively have the ability to share information about themselves and others and learn about the latest happenings in what has become a societal norm. Even through the COVID-19 pandemic that sheltered millions at home, social media could be counted on as a way to remain abreast of world and local news, as well as maintain relationships.

This chapter delves into the world of social media. Beginning with a historical overview of social media sites, it covers today's ever-evolving online social world and examines some of what is to come. This chapter is not so much about how to use social media as it is about the cultural and social shifts brought on by the ability to always be digitally connected through 'now media.'

SEE IT THEN

Historically, the communication process has been somewhat linear. There is a sender with a message designed to reach an audience or person. The rise of the information age has created a paradigm shift in communication. Instead of a linear process, communication is now a circular feedback loop. In other words, communication is no longer just one way from a communicator to a person or mass audience but instead is a two-way conversation that empowers the receiver to provide feedback.

Pre-World Wide Web Social Networking

It is not easy to pinpoint what was the first social media network. Several early Internet venues could be loosely considered types of social networks. **Multi-user dungeons (MUDs)** were pre-Web Internet games set in a fantasy world in which players took on fake identities and, using various keyboard commands, could slay a dragon or complete a quest. MUDs were played online as early as 1978 and could be considered the origin of interactive, social gaming. The **Bulletin Board System (BBS)** created in the late 1970s was a way for users to exchange messages. Users could post a message or read what others had posted. Early BBS allowed only one person to log on at a time. The Whole Earth 'Lectronic Link (WELL) went online in 1985, networking writers and readers of *The Whole Earth Review*, which was a magazine dedicated to technological innovation and making the world a better place through science. **LISTSERV** went online in 1986 as an automatic mailing list. Listservs are still used today to automatically email messages to everyone signed up to a group. Created in 1988, Internet Relay Chat (IRC) was the first form of one-to-one real-time chatting, and was the precursor to instant messaging as known today. These new interactive venues ushered in the era of consumers as content creators.

Post-World Wide Web Social Networking

The creation of browser technology (HTTP-hypertext markup language) changed the face of the Internet. Instead of being text-based, browsers enhanced the Internet with graphics and photos and later with audio and video. Pages of information were now linked through hypertext so users could bounce among them instead of having to read them in order like pages of a book.

New user-friendly and intuitive Web pages were prime for linking users together. *Six Degrees* is commonly thought of as the first Web-based social network. Named for the concept of 'six degrees of separation,' the site was conceptualized as a place for "meeting people you don't know through people you do know" (Adams, 2011). The site contained user profiles, friend lists, and an area for private messaging. At its peak, the site boasted nearly 3.5 million users but it could not sustain itself and went offline in 2001.

The year 1999 is known for the emergence of several key social media sites, such as sixdegrees.org, AsianAvenue, BlackPlanet, and LiveJournal, which in turn were the precursors to the second wave of social network sites (SNS), which includes Facebook and Instagram. Of all the new Web sites that popped up in the late 1990s and early 2000s, Friendster is considered the groundbreaking SNS that led to the ones of today. It was also the first network to attain more than one million users – within the first few months of its existence, three million users had signed on. Anyone could join Friendster and create a personal profile that consisted of such information as occupation, favorite music and television shows, location, and relationship status. Users would invite their friends into their networked circle and thus become friends of their friends plus their friends and so on. For its first two years, Friendster was the most popular social network, but by 2004, the one-year-old MySpace had overtaken it. Friendster never regained its status as a social network, and in 2011 it was repositioned as a social gaming site.

As Friendster declined, MySpace soared. By the end of 2007, MySpace claimed 100 million users worldwide, mostly teenagers and young adults. MySpace built its community around music and entertainment. Signifying the company's growth, News Corp acquired it in 2005 for $580 million. But after several high-profile events, MySpace quickly found itself embroiled in scandal and having to fend off skeptics who claimed that it was a place where child molesters and murderers lurked. When the Connecticut Attorney General claimed that children were being exposed to pornography on MySpace, the site began providing the names of registered sex offenders who used the site. With all eyes turned on MySpace as a "vortex of perversion" (Gillette, 2011), parents started making their children close out their accounts and shifted them to what they deemed safer online havens, such as the new Facebook. MySpace limped along despite the gradual decline of users. In February 2011 alone, traffic fell 44% from 95 million to 63 million users. That year, a group, including music artist Justin Timberlake, purchased the company for $35 million. In 2013, MySpace rebranded itself as "the new MySpace," aimed at 17- to 25-year-olds. The new site focuses more on music, games, movies, and the entertainment industry and less on social connections. MySpace is still alive but not doing so well. In 2019, a 'technical glitch' wiped out between 30 and 50 million tracks that were uploaded before 2016. Before the loss, MySpace was drawing about 29 million users but within three years that number has plummeted to only about two million.

Google+, introduced in 2011 as the social media arm of the formidable Google, was the "social layer across all of Google's services" (Bosker, 2012). The biggest criticism of Google+ was that once a user signed up it became the portal to all Google services. On the surface, this aggregation gave Google unprecedented access to users' personal and shopping information and Internet habits, and tapped into their friends' data. Privacy activists claimed that providing a social network was a mask for the primary purpose of Google+, which was selling in-depth consumer information to advertisers. Although Google+ attracted millions of active users per month, it struggled to gain a strong foothold. Google discontinued the service in April of 2019.

Twitter-owned Vine was a social media site built around looping video clips of ordinary life. It was basically a collection of moments linked together in a six-second video that was shared among followers. Immensely popular among Millennials (71% use), Vine videos showed others what was going on at events around the world. In October 2016, Twitter announced plans to shut down Vine, while the move officially took place in January 2017.

SEE IT NOW

Social Media Versus Social Networking Sites

It is ironic that social media sites have become the hubs of the real-life world – users go online to find

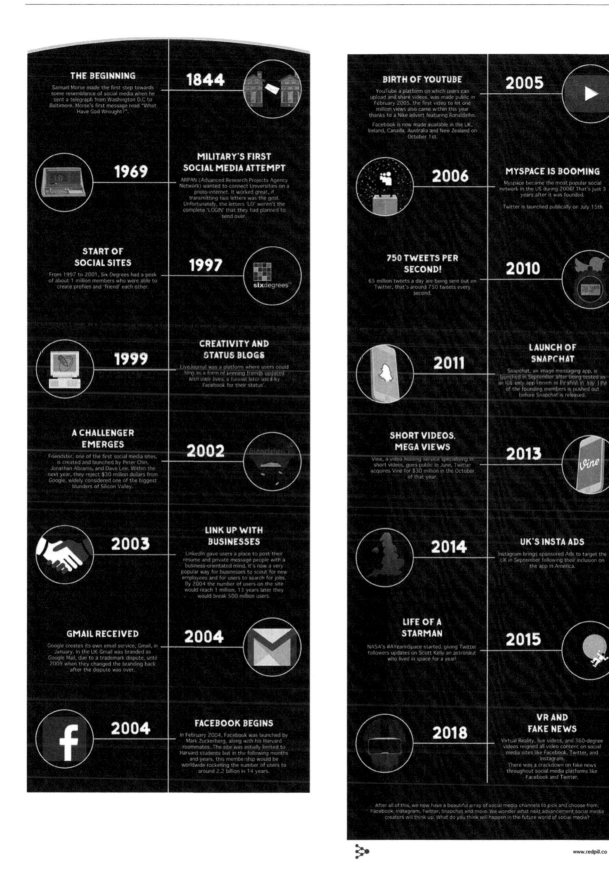

Figure 6.1 History of social media.
Source: Photo courtesy of Cendrine Marrouat (research), http://socialmediaslant.com; Karim Benyagoub (design)

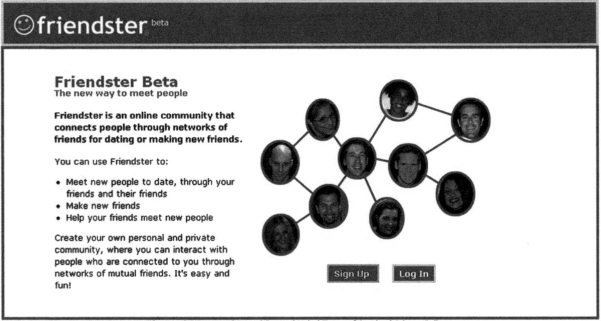

Figure 6.2 Friendster homepage.

Figure 6.3 Friendster toolbar.

out what is going on offline. It is not just teenagers and young adults but people of all ages who flock to social media. The use of social media has surged in recent years. The percentage of online U.S. adults with a social media profile soared from 8% in 2005 to 26% by the end of 2008, to 50% by 2011, to 62% in 2014, to 72% in 2019. But more astounding than these percentages is the sheer number of social media users in the world. Out of about 7.3 billion people on earth, about 3.5 to 3.8 billion use some form of social media.

The number of users tells only one part of the story; the amount of time spent with social media reveals a true obsession. Those 16 to 24 years of age socialize online about three hours per day compared to 2 hours and 37 minutes for 25- to 34-year-olds, and to 55- to 64-year-olds who use social media for 1 hour and 13 minutes per day.

Given that most people start using social media at about the age of ten and that global life expectancy is 72 years, The World Health Organization estimates that at about 2.5 hours of daily use, a typical social media user will spend about 3.5 million minutes using social media over a life time. Another way to look at this is that 6 years and 8 months out of a life time will be spent on social media – that is more time than spent doing housework, socializing face to face, shopping or eating and drinking. Collectively, at about 2.5 hours a day online, billions of social media users spent about 330 million years of human time socializing in front of a screen instead of face to face in 2019 alone.

But what are social media? The term 'social media' is often used loosely as a catch-all to describe any type of user-generated content (UGC). But for UGC to be considered social media, it must meet three primary standards:

> 1) UGC must be published either on a publicly accessible web site or on a social networking site accessible to a selected-group of people; 2) UGC needs to show a certain amount of creative effort, 3) and it needs to have been created outside of professional routines and practices.
>
> Kaplan & Haenlein, 2010, p. 61

Subcategorization of social media further complicates the understanding of it. Some social media are also classified as **social network sites**, which are:

> Web-based services that allow individuals to 1) construct a public or semi-public profile within a bounded system, 2) articulate a list of other users with whom they share a connection, and 3) view and traverse their list of connections and those made by others within the system.
>
> boyd & Ellison, 2007, p. 211

For simplicity, the remainder of this chapter uses the term 'social media' as a descriptor of any type of online site that is driven primarily by user-generated content. The term 'social network sites' refers specifically to sites that are social networks by design, such as Facebook.

FYI: WHO USES SOCIAL MEDIA (2019)

All Internet users	74%
Men	65%
Women	78%
18–29	90%
30–49	82%
50–64	69%
65+	40%
High school grad or less	64%
Some college	72%
College graduate	79%
Less than $30,000 per year	68%
$30,000–$49,999	70%
$50,000–$74,999	83%
$75,000+	78%

Source: Pew Research Center, Social media fact sheet, 2019

Top Social Networking Sites

The number of social networking sites is anyone's guess. Estimates range from 300 well-known ones to thousands of obscure specialty sites. What is known, however, is that the number of SNS is growing faster than they can be counted. Several social networking sites, however, far outrank others in terms of numbers of global users. At the top of the heap is Facebook, which has added one billion monthly users since 2016. The exact number of monthly users is elusive and changes from day to day, such that any number of users is an estimate.

FYI: ACTIVE SOCIAL NETWORK PENETRATION IN SELECTED COUNTRIES (2020)

Country	Share of Internet users on SNS
United Arab Emirates	99%
Taiwan	88%
South Korea	87%
Malaysia	81%
Singapore	79%
Hong Kong	78%
Argentina	76%
New Zealand	75%
Thailand	75%
Sweden	73%
Saudi Arabia	72%
China	72%
USA	70%
Worldwide	49%

Source: Statista: Active social network penetration in selected countries as of January 2020

SOCIAL MEDIA USE – GLOBAL BY TIME PER DAY

Country	Average hours/minutes per day
Philippines	4.11
Brazil	3.41
Indonesia	3.22
Argentina	3.20
Thailand	3.17
Mexico	3.12
United Arab Emirates	3.11
Malaysia	3.09
South Africa	3.04
Saudi Arabia	2.57
Turkey	2.50
Vietnam	2.40
India	2.30
Russia	2.26
Singapore	2.12
USA	2.08
Global Average	2.22

Source: Social, 2018

FYI: NUMBER OF U.S. USERS ON THE MOST POPULAR SOCIAL NETWORKING APPS

Facebook	169.7 million monthly visitors
YouTube	126 million
Instagram	121.2 million
Facebook Messenger	106.4 million
Twitter	81.4 million
Pinterest	66.8 million
Reddit	47.8 million
Snapchat	45.9 million
WhatsApp	25.6 million

Sources: Clement, 2019b; Most popular mobile social, 2019

Facebook (2004)

Facebook has good reason to brag about its 2.5 billion monthly global users. Facebook has grown into the largest social media portal on the planet. Yet, several countries ban Facebook: North Korea, Iran, and the tiny island nation of Nauru. Other countries, such as Pakistan, Syria, and Egypt, have at various times enacted temporary bans.

In the U.S., Facebook is the center of attention. It is the most popular social networking site and is used by about 70% of online adults and three-quarters of that group visit the site at least once per day. U.S. users spend 39 minutes per day taking care of and feeding furry and feathered friends, and yet 40 minutes on Facebook. Recent trends indicate that the over-65 crowd is the fastest-growing age group, having more than doubled since 2012, when 20% reported using the platform. In 2019, that figure jumped to 46%. In turn, use of Facebook by those aged 13 to 17 has decreased from 71% in 2014 to 51% in 2018. Facebook photos of young adults playing beer pong and drinking shots of tequila are being quickly replaced with ones of proud parents with their little soccer stars.

Facebook dominates global social media in terms of number of users and time spent on the site. In the U.S., it also captures the largest market share in terms of number of visits. Users are attracted to Facebook's interface, which makes it easy to quickly scroll through and scan photos, updates, and news and to keep up with friends just by clicking the 'like' button. But because the 'like' option often fell short of conveying the appropriate response, in 2016 Facebook added five new 'reactions': love, haha, wow, sad, angry. Now when someone posts news about Grandma's heart attack, a friend can click on 'sad' rather than 'like,' which could have been interpreted as being happy that Grandma is sick. Amid the coronavirus pandemic, in April of 2020, Facebook added the 'care' reaction, an emoji hugging a heart.

Mark Zuckerberg along with several other partners founded 'The Facebook' (as it was originally known) in 2004 while he was studying psychology at Harvard University. A keen computer programmer, Zuckerberg had already developed a number of social networking Web sites for fellow students, including *Coursematch*, which connected Harvard

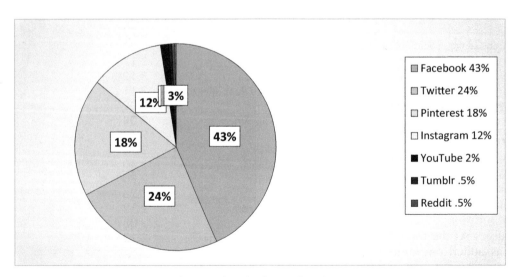

Diagram 6.1 Eight most popular social media by market share.
Source: Clement, 2020

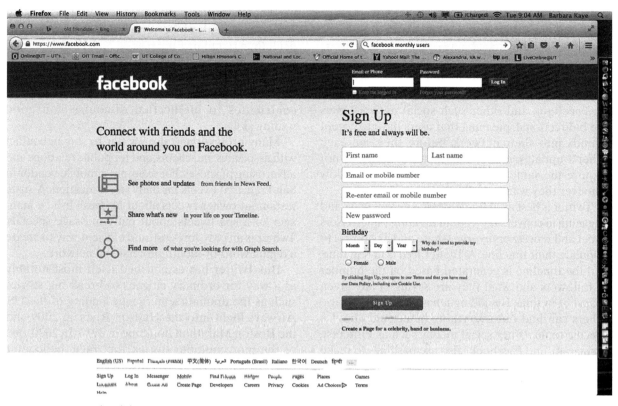

Figure 6.4 Facebook homepage.

students based on the courses they were taking, and *Facemash*, where students could rate their fellow students' attractiveness.

Facebook began as a way to connect college students. Only subscribers with an 'edu' email address could join. Within 24 hours of opening Facebook online, 1,200 Harvard students had signed up, and after one month, more than half of the undergraduate population had a profile. But registrants soon wanted to include non-student friends and family members and leaned on Facebook to open the site to the general public.

The name 'Facebook' was taken from the printed booklet that was distributed to freshmen to familiarize them with Harvard students and staff. Many other universities also distributed to potential employers printed 'facebooks' that contained profiles and resumés of graduating seniors.

There are so many people in the world with a Facebook account that dead users could one day outnumber the living. As creepy as it sounds, 30 million Facebook users died in the first eight years of its existence. Because it is so difficult to delete someone else's profile, it stays online in perpetuity. In some ways, knowing that a loved one's profile is still accessible is comforting. Faced with the possibility that by the year 2065 there could be more dead people than live ones on the site, the company now allows a profile to be turned into a memorial page on which friends can post comments and memories.

Twitter (2006)

Twitter is a microblogging social networking site (combination of a blog and a social network site) that distinguishes itself from other social media by limiting messages to 280 characters. Twitter was first created in 2006 and was a bit slow to catch on, but by 2009 it was ranked as one of the top 50 Web sites worldwide, with about 106 million registered users. By 2015, Twitter was claiming about 320 million monthly active users who fire off 500 million tweets per day, ten times more than five years earlier.

About 22% of online adults (19% of the U.S. adult population) use Twitter, but it has become more popular with middle-aged users. Still, Twitter is one of the top venues where the Millennial generation goes for news. The percentage is much lower for older individuals – only 19% of users are between the ages of 50 and 64. Early on, parents were concerned that it was unsafe for their children to tweet their locations or what they were doing, so only about 4% of Twitter users were under the age of 18, but these days one-third flock to Twitter.

Different from social network sites where users approve who gets to become a friend and only friends can see postings, Twitter is an open-access network that anyone can read. Whereas connections on Facebook and other such social network sites are bidirectional, meaning that users have to accept friends into their network before they see each other's updates, email, and chat, Twitter does not require the same reciprocation – users can follow whoever they want to follow.

Twitter is best used for delivering instant news and engaging in conversation about *trending* topics. Users tweet and *retweet* or *favorite* (share and like posts) to populate their timeline. A Twitter *feed* is in real time, and the timeline is generated based on the number of followers and what they are saying at any given snapshot in time. By adding a **hashtag** to a message, others can find out everything being said about a specific topic. Other social media, such as Pinterest, Instagram, and Facebook, also use hashtags to make it easy to search for topics of interest.

The Twitter network was aflutter when in early 2016 the company announced that it would allow tweets up to 10,000 characters in length. Only the first 140 characters would be displayed, but users could tap on a link to see full-length text. Some users worried that longer tweets would interrupt Twitter's sleek flow and would make it more like any ordinary blog with long babblings. Others, though, welcomed the idea of clicking on a short post for more information. After two months of anticipation, Twitter's chief executive made a brief, 81-character announcement on the *Today* program referring to the original character limit, "It's staying. It's a good constraint for us. It allows for of-the-moment brevity." But Twitter eventually capitulated, and in November 2017, the platform expanded the character limit but only up to 280.

So how did Twitter get its name? The definition of twitter is 'a short burst of inconsequential information' and 'a series of chirps from birds.' The name was fitting, and so the new platform became Twitter. Soon the 'chirps' of many 'tweeters' would be heard throughout the Twitterverse as the microblogging platform caught on with Internet users. As with texting, users have developed Twitter shorthand to stay below the maximum number of words. For example, IMHO means 'In my honest opinion,' and FYI means 'for your information,' ICYMI 'in case you missed it,' and SMH means 'shaking my head.'

Twitter has built a strong brand name and is used for a variety of purposes. For example, Twitter is used as a way to teach students to write concisely. Expressing an idea in fewer words is often more difficult than rambling for pages on end. Professors and teachers also encourage students to tweet questions and share information. On the other hand, it is difficult to understand how writing in tweet-speak contributes to intellectual discourse and good writing skills.

Many professionals use Twitter to stay in contact with associates and clients, and for public relations and advertising purposes. For example, a mobile vendor in San Francisco tweets his daily street location. A sushi restaurant owner tweets about his fresh fish. A manager of a spa tweets about daily massage specials. Twitter is an easy, low-cost, and efficient way to create a digital word-of-mouth professional network.

But Twitter has established itself most notably as a way for ordinary citizens to break big stories such as the dramatic emergency landing of the US. Airways flight into the Hudson River in 2009 and the Boston Marathon bombing in 2013. In 2020, one of the more notable stories, the tragic helicopter crash death of Kobe Bryant, was broken by TMZ via Twitter. Activists used Twitter to mobilize citizens and set up protests during the Arab Spring, a series of anti-government uprisings and protests throughout the Middle East in early 2011.

Pinterest (2009)

Many proud parents keep a pinboard in the kitchen or family room on which they display vacation photos and their children's hand-drawn artwork, or leave notes for family members. Pinterest is basically the online version of a pinboard in that it is used for sharing 'interesting' images, known as 'pins'; hence the name Pinterest. Pinterest is the perfect online venue for archiving important moments and sharing interests. Pinterest is more than a platform for storing visuals; it is also a 'self-expression engine,' on which users create montages of images that reflect a mood or vision.

Just like pinning a montage of photos on the kitchen pinboard, Pinterest users create their own boards that group pins of similar topics. For example, a user could have a vacation board, a recipe board, or a movie board. But Pinterest is a social medium because users can share someone else's pin – known as 'repinning.' 'Pins' can also be shared on Facebook and Twitter, and, like these social media, Pinterest users add friends to their accounts and everyone follows each other, and a user's main page displays a 'pin feed' that chronologically lists friends' Pinterest activities.

FYI: BARACK OBAMA AND DONALD J. TRUMP ON TWITTER

On May 18, 2015, then-President Barack Obama surprised the Twitterverse with his first tweet: "Hello Twitter! It's Barack. Really! Six years in, they're finally giving me my own account." Obama's @POTUS account is notable for being the fastest to achieve one million followers – in 4 hours and 52 minutes.

Donald Trump sent out his first tweet as president within an hour of being sworn into office, starting with "Today we are not merely transferring power from one Administration to another, or from one political party to another – but we are transferring ... power from Washington, D.C. and giving it back to you, the American people" (from @realDonaldTrump). Trump has since sent out about 16,000 tweets (as of May 2020), sometimes tweeting 100 times per day – his record is 142 tweets sent on December, 12, 2019. The National Archives has preserved every one of Trump's pre- and post-presidential tweets – 52,000 as of May 2020.

An analysis of just over 11,300 tweets sent during Trump's first 33 months in office shows that 5,889 (51.7%) attacked someone or something, and 2,026 (17.7%) praised himself. But whatever someone might think about Trump's tweets, he has learned how to use the power of Twitter to energize his base and anger his critics.

Both Obama and Trump have amassed tens of millions of Twitter followers. Trump's outrageous and conspiracy theory tweets, however, capture the public's eye and stir debate. Nevertheless, Trump has turned Twitter into a formidable communications tool.

Sources: Crunched, 2015; Cummings, 2020; Rogers, 2017; Shear et al., 2019

In the first year or so of its existence, Pinterest was largely the domain of female users. Males, however, have started gravitating to the site but still account for only about 30% of its audience. Pinterest's typical user is female and the site is used by 80% of all American mothers who report using the Internet. Moreover, 81% of females trust recommendations they read on Pinterest compared to Twitter (73%) and Facebook (67%).

Instagram (2010)

Instagram is the largest photo-sharing social media site. Since its inception in 2010, Instagram has added *direct messaging, tagging, hashtags,* and *video sharing.* From the very beginning, Instagram was a success. It amassed one million users within two months after its launch. Its wild success led Facebook to snap it up for a cool $1 billion in 2012.

Instagram is not just a place to store visuals, but it is a rich social network as well. The point is to share experiences by uploading photos and videos in real-time as an event is occurring. Similar to other social networks, account holders have a profile, news feed, and a network of 'followers.' Instagram is a mobile platform, which means that though visuals may be viewed, commented on, and liked on the Web, uploading must be done through the mobile application.

Instagram is used by 67% of all young adults aged 18 to 24. Of that group, 76% reported using the platform daily, while 60% say they do so multiple times each day. About 68% of all users are female, and nine in ten users are under the age of 35. Instagram users are very engaged with the site, spending about 4 hours and 16 minutes per month, as compared to 2 hours and 50 minutes on Twitter. The number of Instagram photos is astounding. As of January 2020, more than 50 billion photos had been shared, and 100 million photos were being posted daily. Although Instagram is for all types of visuals, over 35 million selfies have been posted.

YouTube (2005)

YouTube is the world's most popular online video site, with just over two billion monthly users watching one billion hours of video each day and uploading 500 hours of video every minute (720,000 hours of video per day). YouTube lets anyone upload short videos for private or public viewing. Although YouTube was not created as a social medium per se, but as a way to showcase amateur and professional videos, it has quickly become a place to network. Social connections are created through friend requests, channel subscriptions, and commenting on videos.

YouTube has emerged as a major venue for videos of political speeches. In the 2008 presidential campaign, videos of Barack Obama and John McCain were viewed more than three billion times, and 39% of voters watched at least one campaign-related video on YouTube. The 2012 Obama-versus-Romney contest yielded just about as many video views. For the 2016 election, Donald Trump benefitted greatly by YouTube as about eight out of ten partisan videos showed him in a favorable light, while only about 15% were positive about Hillary Clinton, and a large number contained conspiratorial content that damaged her reputation. YouTube has become a formidable force in setting and changing public opinion.

But YouTube is not all political. On YouTube, anyone can be a filmmaker, a teacher, an artist, or a performer. Videos run the gamut from professional instruction to amateur 'how-to' to humorous clips of silly pets and silly people to recordings of live events and even of a tornado as it bears down on a town. Viewers turn to YouTube for a variety of reasons. YouTube's comments section keeps users connected socially. YouTube is often used as a substitute for television – someone without cable or satellite can watch what he or she missed on air. Users think the videos are informative, funny, and inspiring, and they like to find out what other people think of the same videos.

YouTube is a global medium – most of its traffic comes from over 100 countries outside the U.S., and content is available in 80 languages. Its visibility and strength and its ability to bring nations together compelled Google to acquire YouTube in 2006 for $1.65 billion. YouTube has also gained footing with a younger demographic – 85% of Americans aged 13–17 use the platform, with the majority of users between the ages of 18–34.

Tumblr (2007)

Tumblr is a cross between a social media site, live streaming application, and microblog, on which users post short-form blog entries, music, and visuals (photos, graphics, video). Although there are no length restrictions, Tumblr encourages short updates. Tumblr's interface is easy to use. It categorizes posts by 'text,' 'photo,' 'quote,' 'link,' 'chat,' 'audio,' and 'video.'

As of August 2019, Yahoo!-owned Tumblr hosted about 475 million blogs. But the platform appears to be in decline. Yahoo! purchased Tumblr in 2013 for $1.1 billion. In August 2019, Automattic purchased the platform for just under $3 million.

Reddit (2005)

Reddit is yet another student success story. Thought up by two University of Virginia roommates in 2005, Reddit's success comes both from its content and from its fun-to-use, interactive interface. Reddit is a combination news and entertainment Web site, a social networking platform, and online BBS. The name 'Reddit' is a play on the words 'read it,' as in "I 'read it' on the Internet yesterday."

Registered users, known as 'redditors,' post links and content on the board. Community members then vote each submission 'up' if they like it or find it interesting or 'down' if they are not in favor of it. The goal is to send the best stories or submissions to the top of the main page. The homepage lists the 'hot topics,' but other tabs show 'hot,' 'new,' 'top,' and 'rising,' as well as posts made in every American state and various countries. Thousands of subreddits, which are topic-specific subcommunities, such as science, gaming, music, fitness, and gadgets and technology, make the site search friendly. The site gets its social media flair from users' comments. By the beginning of 2020, there were 1.7 billion comments on Reddit, with more than 430 million active monthly users. Moreover, 22% of Americans aged 18 to 29 use the platform, the largest of any adult demographic. Reddit's spiffy name plus the intuitive way it is organized give it enough online cachet that users spend an average of 16 minutes on each visit, more than Facebook (10 minutes), Google (8 minutes), and YouTube (7 minutes).

ZOOM IN 6.1

Visit this Web site and click on the short 'about Reddit' video: www.theatlantic.com/technology/archive/2013/09/what-is-reddit/279579/

LinkedIn (2003)

The digital age has changed the way people look for jobs. The old ways of reading newspaper employment listings and knocking on doors are long gone. Simply finding an appealing job description and submitting a resumé is not enough to land employment – now companies are looking for personalized approaches and recommendations from networks, and LinkedIn is the best way to make business connections. LinkedIn brings to life the adage, "It's not what you know but who you know."

LinkedIn is the largest online professional profiling platform. It boasts 675 million registrants across 200 countries and territories, with more than

70% of users outside the United States. It is designed for posting professional resumés, connecting with industry leaders, and endorsing others to enhance users' profiles. LinkedIn is an effective way for those early in their careers to introduce themselves to industry leaders. About 27% of U.S. adult Internet users have a profile on LinkedIn. The typical LinkedIn user completed some college or has a college degree, is between the ages of 30 and 64, and makes nearly $50,000 per year.

Once reserved only for those who had a foot in the door, LinkedIn aspires to give everyone an equal chance of being seen and heard and showing what they do best. Users' profiles are enhanced when others in their network 'endorse' them for their skills and knowledge. Most important – about 93% of professional recruiters look for potential employees on the site.

ZOOM IN 6.2

Learn more about LinkedIn and see the infographic of World Wide Web membership at: https://press.linkedin.com/about-linkedin

Other Social Networking Sites

Flickr (2004)

Flickr is an image-storing and -hosting site that is used for sharing photos and video among an online community of users. By sheer volume, Flickr's one million photo uploads per day pale in comparison to Facebook's 211 million per day. But SmugMug-owned Flickr is not so much about the act of sharing a photo as about the photo itself. On Facebook, once a photo slips down the news feed scroll, it is largely forgotten, whereas on Flickr, photos are showcased on "endlessly scrolling pages" (Roush, 2013) and photos can easily be sorted and categorized into albums or 'sets.' One of the coolest features is that geotagged photos are displayed on a world map.

Flickr's photo-editing tools attract a community of professional and serious amateur photographers and those who could use a terabyte of free storage. Like on any other social media site, Flickr users form social networks and comment on each other's photos and videos.

Snapchat (2011)

Snapchat is for sharing a special moment among friends. Once a photo is shared, the receiver has a set number of seconds to view it before it disappears. If someone tries to defeat the purpose of Snapchat by taking a screen shot of the photo, the sender is informed. Senders can personalize a photo by drawing on it or adding a caption. Snapchat is used by 24% of American adults, particularly those 18 to 24 years of age. In that demographic, 73% report using the platform.

Snapchat is a welcome antidote to a cyberworld where every movement, activity, and emotion is photographed, video recorded, or written and forever archived. Snapchat is a here-and-now app that builds social bonds through photos and messages, but users do not have to worry that an unflattering picture or a silly post will come back to haunt them years later. Snapchat's impermanence is its strongest draw.

TikTok (2016)

Originally available only in China, TikTok entered other world markets one year after its debut. Like Instagram and Snapchat, TikTok is a mobile platform. TikTok can be viewed through a desktop, tablet, or other device online, but posts are made only through iOS and Android devices.

Whereas Instagram and other social networking sites emphasize photos, TikTok emphasizes video, with short time limits. TikTok allows users to create looping lip-synched or music videos 3 to 15 seconds in length, or other looping videos 3 to 60 seconds in length. The application was downloaded 738 million times in 2019, and by the end of the year it boasted 200 million active monthly users. Perhaps more telling, 37% of all TikTok users were 10 to 19 years of age.

Blogs

Blogs are not social network sites per se. Blogs stand out from Facebook, Twitter, and other social network sites because conversation occurs between the blogger and fellow blog users who might be friends, acquaintances, or strangers, but users do not create networks or circles of 'friends.' Blogs are both conversational and informational but are not used primarily for social purposes. Although blog users express their opinions, on the major blogs of today there is very little exchange of photos or discussion about personal matters as on Facebook. Instead, most blogs tackle political and social issues. Blogs, then, are a social medium in that conversation occurs, but they are not used primarily for social networking but for the exchange of ideas and opinions. The basic or general-topic blog is an open forum, in which anyone can read and participate in discussions about a myriad of topics. Some general-information blogs tackle many issues and topics, whereas others focus on specific topics such as dog breeding or gardening.

Blogging has exploded on the cyberscene. A new blogger jumps on the bandwagon every 40 seconds. Although estimates vary widely on the number of blogs (depending on how 'blog' is defined), there were between 20,000 and 30,000 in the late 1990s, almost three million by the end of 2002, and 70 million blogs worldwide in 2007. By 2009, an astounding 133 million blogs inhabited cyberspace, and by 2015, that number had reached about 250 million, and had doubled by 2020.

Today's **blogosphere** (the collective world of blogs) is inhabited mostly by White, highly educated, high-income, conservative and libertarian males. But the blogosphere is becoming more diverse as new bloggers tackle multicultural and feminist issues and meet the needs of minority users.

ZOOM IN 6.3

TOP SOCIAL MEDIA SITES THROUGH TIME (2003–2019)

Check out this YouTube video that takes you through a moving timeline of the most popular social media sites from 2003–2019: www.youtube.com/watch?v=aOymOiQdNaE

FYI: MILLENNIALS AND MEDIA USE (% WHO SAY THEY …)

Own a smartphone	93
Own a tablet computer	55
Use social media	86
Use Facebook	84
Use the Internet	100
Use only Internet on smartphones	19
Subscribe to broadband Internet service	78
Play video games	60

Source: Millennials stand out for their technology use, 2019

Social and Cultural Benefits and Consequences of Social Media

Social media have drastically changed the way we live, love, and work. Whether for good or for bad, social media are here to stay for a very long time. Eventually, some other new social technology might replace the social networking of today, but even so, we will never go back to the old ways of making friends and maintaining relationships. Social media have even altered the way we do our jobs and our bosses' expectations, and it has affected the way we learn about current events. This chapter next explores several key ways that social media have altered our reality.

News and Information

Social media have ushered in the age of immediate information. News from across the globe shows up on our mobile feeds in an instant, 15-second TikTok videos grace our screens a few seconds after they are recorded, and the latest rumors are ours to judge and spread long before they are verified. Social media has created the need to know, now. Waiting until the next day to read a printed edition or even a couple of hours for a television newscast is passé; after all, by then, something more interesting might have happened.

The intersection of social media and news is the number-one online place for keeping up with the world. In 2018, 34% of Americans reported getting their news online, while 37% said they wanted to receive their local news online as well. But most telling of the current landscape is that more Americans report receiving their news from social media (20%) than newspapers (18%). Television is still the primary source (49%), though it has declined markedly since 2016 (57%). This trend is driven by immediate access to the Internet and a plethora of sources. One news story or one perspective is no longer satisfying; news-hungry consumers rely on many sources, both professional and amateur. Social media have become our eyes on the world.

Citizen Journalism Social media empower the ordinary person on the street to act as an observer and as a reporter. A mobile device is a tool that captures an event, and a social medium is the tool that spreads the word of the event. Anyone with both tools can become a roving reporter. Amateur videos were among the first to show the aftermath of the 2013 Boston Marathon bombings and the Black Lives Matter protests of 2020, while more recently, TikTok videos have been used by doctors to provide facts about the COVID-19 pandemic, and by other users to discuss anything to do with the coronavirus.

While there have been countless false bits of information circulating in cyberspace, at other times online news scooped the media. For example, Keith Urbahn, chief of staff to former Defense Secretary Donald H. Rumsfeld, tweeted the news of Osama bin Laden's killing about 20 minutes before the news media made the official announcement. While the news media were verifying the account, and putting their broadcasts together, Urbahn tweeted, "So I'm told by a reputable person, they have killed Osama bin Laden. Hot damn." That tweet became the first credible report of what happened on May 2, 2011, in Abbottabad, Pakistan.

Blogs were probably the first online medium on which the ordinary person could become a reporter or a columnist. Blogging redistributes traditional media's power "into the hands of many" (Reynolds, 2002). One of blogging's first coups came in 2002 when then-Senator Trent Lott glorified Senator Strom Thurmond's 1948 desegregationist campaign. The story was printed on page 6 of *The Washington Post* and omitted entirely by *The New York Times*. Only when bloggers homed in on the remark did the mainstream press run with the story. But it was pressure from the bloggers that eventually led to Lott's resignation.

The media consider themselves the watchdogs of the government, and now social media have taken on the role of watchdogs of both the government and the media. Like vultures, social media users hover over media reports, often criticizing and commenting on news stories and looking for errors even before the printed versions have hit the newsstands or the electronic versions have zipped through the airwaves. As one blogger commented, "This is the Internet and we can fact check your ass" (Reynolds, 2002).

Some media executives and journalists, concerned that blogs and social media sites spread misinformation and blur fact and opinion, have set up their own venues as a counterpoint. Almost all major news sites contain several blogs that are hosted by their own journalists and pundits. But because media-hosted blog content is edited and fact-checked and is sometimes merely expanded print or on-air stories that may reflect the views of the organization, blog purists do not consider them true blogs. Blog purists claim that real blogs are those that provide a space for uninhibited public deliberation and an open marketplace of ideas where everyone's views are considered seriously.

Citizen journalism on social media is also fraught with misinformation and disinformation. **Misinformation** means inadvertent mistakes, and **disinformation** is false information posted on purpose with the intent of causing harm. Unfortunately, all it takes is to plant the seed for a lie to be thought as the truth. Once false information is out there, it is very difficult to right the wrong. People do not always know what to believe, and they also want to believe what benefits them the most or feels right, a concept Stephen Colbert coined as 'truthiness.'

Never was 'truthiness' more apparent than on social media during the pandemic. Social media were fraught with false claims, what the World Health Organization refers to as an 'infodemic,' such as that Microsoft founder Bill Gates conjured up the virus in his lab, the coronavirus was no more dangerous than the common flu, the virus spreads though eye infections, drinking warm water every 15 minutes makes a person immune, and military helicopters spraying cities with disinfectant will cure the illness. Although these claims seem outlandish to most social media users, other people believe them to be true. With human lives at stake, such 'fake news' spread by citizen journalists is nothing to shrug off, but must be contained.

Political Efficacy and Grassroots Advocacy When blogs, and later Facebook, Twitter, and other social media, first started appearing in cyberspace, they were hailed as catalysts for democratic discussion and political understanding and unity. It was hoped that these sites would not only bring people together socially but would also strengthen **political self-efficacy**, the confidence to bring about governmental change. It does seem that 'get out the vote' messages are effective on social media, especially on Facebook. Those 'I voted' messages strongly influence others to go to the polls. When Facebook users see their friends voting or engaging in some other political activity, they are likely to do so as well, a phenomenon known as the 'social contagion effect.'

Facebook, Twitter, and other social media are credited with fueling revolutions and protests around the globe. From the Arab Spring uprisings to the occupation of Cairo's Tahrir Square that toppled President Hosni Mubarak to Hong Kong's Umbrella Revolution to the Occupy Wall Street sit-ins and the women's marches and science rallies in the U.S., social media amplified the voices of the discontented and empowered ordinary citizens to initiate change.

But despite the marches and demonstrations, once the initial exhilaration of the online push for democratic deliberation and worldwide attention wore off, protestors gradually returned to their everyday lives with little social or political change. Although there were some successes, social media have a way to go before becoming true instruments of social and cultural change. Rumors, lack of deep and thoughtful conversation, a mob mentality, 'echo chambers' in which discussion takes place only among like-minded individuals, and a general atmosphere of rudeness hinder the power of social media.

When Traditional Media Rely on Social Media When Andy Carvin was a social media strategist for NPR, he was tweeting hundreds of messages per day chronicling the latest news from items picked up on the Internet and turning himself into a one-man wire service. A browse through job listings using the search 'social media jobs' turns up positions such as 'social media manager,' 'social media specialist,' 'social media content creator,' and 'social media coordinator.' These positions involve anything from managing an organization's social media platforms and content to marketing, promotion, and HTML and IT work.

But Twitter's successes mask factors that keep it from emerging as a credible source of news. First, it is still used by a minority of the population; 22% have used it, but only about 10% do so in a typical day. Although Millennials account for the largest group of Twitter users, they are not big consumers of news – only about four in ten access news every day. Further, only about two in ten tweets can be considered news; the rest are celebrity or personal in nature, with television, film, sports, and music as the most common topics. Also, the 280-character count does not allow for in-depth exploration of hard-news issues. Anyone can tweet anything, true or false, so it is not surprising that Twitter news is regarded as fairly low in credibility.

Opinions about Twitter run the gamut from idiotic to its being a historical 'godsend.' When the Library of Congress announced in May 2010 that it had acquired the Twitter archive, historians were elated. The spontaneous nature of tweets contributes to their value as an authentic cultural record, but try convincing those who think tweets are 'culturally vacant.' The 'snarkosphere' lit up with complaints and cynical comments about why anyone would want an archive of tweets (Hesse, 2010, p. 1).

News and Information Consumption In his book *The Shallows: What the Internet Is Doing to Our Brain*, author Nicholas Carr (2011) contends that the Internet is a system of interruption that constantly pings users with morsels of information. The flashing of sights and sounds stimulates our brains like a drug: The more we get the more we need. Online hyperactivity has changed the way we process information. It has diluted our ability to read deeply and concentrate for long periods of time. We are in a constant state of distraction: Looking here, looking there; reading this sentence, reading that sentence; watching this part of this video, watching that part of that video; writing this text, writing that text; viewing this news clip, viewing that news clip. Although we believe that hyper information intake is making us smarter and better informed, it actually has the opposite effect. We know less, we understand less, and we think less.

Social media have added to the information frenzy. Social media information is delivered in rapid fragments rather than in thought-out pieces. As a result, we know 'less about more,' rather than 'more about less.' But many believe that is an acceptable trade-off. In his book *Smarter Than You Think*, author Clive Thompson (2013) claims that because the Internet is a collection of specks of information about a wide variety of subjects, it is making us smarter and better informed.

Social media are also blamed for dumbing down the news and precipitating a decline in journalistic substance. After all, a 280-character story is not in-depth news. And much of what is out there is unverified and written by anonymous sources. In sorting through the changes brought on by social media, scientists have discovered that Millennials consume less news and information than Baby Boomers, and they are more interested in entertainment and cultural news than 'hard' news or international events. Further, half as many Millennials as Boomers keep up with problems facing the country. In just a few generations, our brains have changed, and so has the way we learn and consume information.

Relationships

Many sociologists, psychologists, and other experts question whether social media have enriched our relationships or diluted them. There is little debate that the ways we socialize and interact with one another have changed dramatically, but the broader consequences and benefits of social media are just beginning to emerge.

FYI: SOCIAL MEDIA TRIVIA

- On average, people have 7.6 social media accounts.
- People report spending an average of 2 hours and 22 minutes per day on social media.
- Between October 2018 and October 2019 there were ten new social media users per second.
- Companies spent more than $90 billion in advertising in 2019 on social media.
- Those aged 55 to 64 are more than twice as likely to engage with branded content than those 28 or younger.

Sources: Brandwatch, Our Social Times

Loneliness The typical Facebook user has 338 friends, and 15% of account holders have more than 500 friends. Not all friends are 'known'; some are strangers and some are friends of friends. In a world of increasing personal isolation, social media friends have in some ways taken the place of in-person contact. The elderly living in small apartments on fixed incomes, Boomers busy with travel, work and family, and Millennials absorbed with college and careers have little time, money, and opportunity to get together with friends and family. Although it may seem contradictory, the modern ways of online socializing might actually be making us lonelier.

Americans live more solitary lives than ever before. In 1950, less than 10% of households contained only one person, but by 2020, that percentage had jumped to 28%. Although the quality, not the quantity, of relationships determines loneliness, proximity is a major determinant of with whom and how often we socialize.

Loneliness and being alone are not the same – loneliness is a psychological state, being alone a physical one, and sometimes a blissful one. A study conducted in 2020 by health insurer Cigna found that 61% of those surveyed reported feeling lonely, up 13% from the previous year. Perhaps surprisingly, however, the largest demographics to report loneliness are Generation Z (79%) and Millennials (71%), compared to Baby Boomers (50%). Moreover, in 1985, only 10% of Americans said they had no one to confide in, and by 2019, that figure had risen to 25%.

Now here comes Facebook, well timed as a panacea for loneliness. But at question is whether Facebook and other social media alleviate loneliness, or do they heighten it by making users keenly aware of their physical isolation from others? It is yet to be seen how the COVID-19 pandemic compounded feelings of loneliness, when people were physically isolated from each other and relied heavily on social media and online video conferencing to maintain communication with others.

The relationship between social media and loneliness is complex. On one hand, social media widen social circles, increase feelings of belonging, and keep people in close contact with distant friends and family. On the other hand, they make it easy to be physically isolated without feeling friendless. Social media create 'ambient awareness.' Photos, videos, and messages make people feel physically close though miles away.

Ambient Awareness 'Sharing' is the buzz word of social media, but just how much is 'good sharing' and how much is 'bad sharing'? Confessions of extramarital affairs, binge drinking, failing grades and drug use, revelations of hook-ups, car accidents, health problems, whining about mean bosses, college course assignments, bad hair days and bad days in general, and brags about new jobs, smart pets, marathon runs, expensive cars, and highly accomplished children, are the fabric of social media. Viewing social network sites is like being pummeled with annoying holiday letters every day of the year. We roll our eyes and mutter, "oh, brother," yet we cannot tear ourselves away. Social scientists call this type of incessant online contact '**ambient awareness**,' because it is so much like being with someone. We have become accustomed to being what sociologist Sherry Turkle calls 'alone together.' We can be in the same classroom, at the same dining table, or in the same living room as friends and family, yet we are each in our own digital world and thus together but not together. We mistake a 280-character conversation for intimacy and Facebook banter as close friendship. It may seem ironic that we use digital technology to forge closeness, yet in the long run it keeps people apart and makes us feel lonely. We feel uncomfortable being alone, thus, we are lonely. Further, the constant need for ambient intimacy is a form of narcissism in which people think that every one of their thoughts or actions is of profound interest to someone else. The need to share every little moment in life can be a bit too much ambient awareness. Almost four in ten Facebook users dislike it when their friends share too much information.

Like it or not, ambient awareness can be a satisfying substitute for getting together with someone. As we would in real life listen to a friend sharing his or her latest accomplishments and disappointments, through social media we listen online. In real life, though, we can control the conversation somewhat, but in social media life, we take what we are given.

Social Anxiety Social networking, whether through social media or texting, is ripe for misinterpretation and misunderstanding. When people speak face to face or even on the telephone, body language, facial expressions, and vocal intonations reveal what they are feeling. We know when someone is feeling happy, angry, sad, frustrated, or calm, but emotional cues are missing from social media, often resulting in barren conversation. Matter-of-fact messaging might be fine for business, commercial, or other formal relationships, but it is not so good for interpersonal ones.

Emoticons were created as a way to let others know what we are feeling. Emoticons represent moods and cue recipients on how they should respond. The word 'emoticon' is a cross between 'emotion' and 'icon.' Although emoticons, such as smiley faces, have long been used in casual and humorous writing, the earliest Internet versions began with an online bulletin board message written by Carnegie Mellon University professor Scott Fahlman:

19-Sep-82 11:44 Scott E Fahlman :-)
   From: Scott E Fahlman <Fahlman at Cmu-20c>
   I propose that the following character sequence for joke markers:
      :-)
   Read it sideways. Actually, it is probably more economical to mark things that are NOT jokes, given current trends. For this, use
      :-(

Although character sequences are still used to represent emotion, most software automatically replaces them with a corresponding icon. Thus typing :-) autocorrects to ☺.

Emoticons are different from **emoji**, which are characters that represent words. The word 'emoji' is a combination of the Japanese words, 'picture' and 'character.' Although some emoji represent moods and emotions, emoji are standardized icons that are used as shortcuts in place of a word.

Until there is a set of standardized emoji or software capable of translating them as users intend across all electronic devices, users are relying on emoticons to set the mood. Even grammatical punctuation has come to have meaning beyond merely signifying the end of a sentence or a clause. Even a seemingly innocuous text confirming dinner plans can turn into a nightmare if the proper punctuation is not used. For example, a text that reads, "Dinner still on. What time" could be interpreted as meaning that the sender does not want to go because the question mark was left off. But if more than one question mark was used, "Dinner still on?? What

Figure 6.5 Examples of emoticons.
Source: Photo courtesy of Shutterstock

Figure 6.6 Examples of emojis.

time??" then it could be thought that the sender is not sure that a dinner date was set in the first place. Inserting an exclamation mark, "Dinner still on! What time??" makes the sender seem excited, but two exclamation marks, "Dinner still on!! What time??" could make the sender seem overeager. In the dating scene, the inclusion or exclusion of a punctuation mark could make or break a relationship. Ending a text sentence with a period or using just the letter 'K' instead of 'OK' could signify that someone is angry or exasperated. Texting "I caaaannn't wait for toooonight" might make someone seem clingy. Knowing the new and ever-changing rules of punctuation is a must when texting. Improper or unacceptable punctuation causes much relationship angst and anxiety.

FYI: SOCIAL MEDIA AND LOVE

- People who use Twitter daily or very often are more likely than less frequent users to cheat, break up, or divorce.
- 81% think it is acceptable to respond to a text or email during a date as long as there is an explanation.
- 67% of committed couples share online passwords with each other.
- 17% of single people 21 to 34 have used their phones during sex.
- 40% of American couples met through a social networking site, doubling since 2009.

Sources: Kopf, 2019; Shelasky, 2014

FYI: EMAIL ETIQUETTE

How should you sign your email messages? Best? Fondly? Yours truly? Sincerely? Cheers? Love? XOXO? Kisses? Hugs? Apparently, the sign-off can take on a bigger meaning than you think. For instance, the wrong sign-off to a boyfriend or girlfriend could signify an emotional intensity or coolness. Or a too-casual or too-familiar sign-off could be a show of disrespect.

One survey found that the most frequently used professional sign-off is 'sincerely,' followed by a thank you of some kind, followed by no close at all. With personal emails, 'love' is the most popular sign-off.

Source: Best for Last, 2009

Crisis Communication

In the spring of 2020, social media became an even more important mode of communication, during the COVID-19 coronavirus pandemic. As cities, counties, states, and countries issued mandatory 'shelter in place' orders, social media took on increased significance and, as during any other time, its strengths and weaknesses were highlighted.

The pandemic and resulting quarantines and self-quarantines by residents around the globe left many sheltering at home. Businesses shuttered their stores, stocks tumbled to record lows, and 3.3 million Americans filed for unemployment during the first full week after the pandemic was announced in mid-March. That easily surpassed the previous weekly record of 695,000 in 1982. By May 2020, unemployment had soared to 14.7% and 30 million Americans had filed initial unemployment claims.

News organizations, many of them traditional forms, such as broadcast television stations and newspapers, continued to report, write, and broadcast news as they were considered essential work during the health crisis. But many Americans and others also turned to social media for news, first-hand accounts, and updated numbers and other facts, such as the growing number of positive tests and deaths associated with COVID-19.

Along with these vital pieces of information, people took to social media to thank first responders, to plead to abide by the shelter in place orders, and to raise money for various causes associated with the pandemic, such as with GoFundMe drives. Former NBA player Rex Chapman, for instance, started a COVID-19 relief fund and used Twitter to 'tag' celebrity friends, athletes, and coaches and challenge them to donate to the fund. Twitter itself also donated $1 million to journalism non-profit agencies to support reporters affected by the pandemic.

Due to social distancing measures that asked people to remain at least six feet away from others, social media followers were rallied to honk at a certain time to wish a child a happy birthday. In the United Kingdom, residents in a neighborhood were all asked via social media to open their windows to sing 'happy birthday' when a young girl walked out on her doorstep. 'Zoom,' 'Skype,' and Microsoft 'Teams' parties also became a method to digitally connect in real-time. Users 'went' to a Skype happy hour or Zoom birthday party, and 'met up' with friends for happy hour. Although not social media per se, Zoom, Skype and other such tools have become critical in bringing people together who otherwise cannot be physically united.

Work

Savvy employers are using social media as a way to tap into 'collective intelligence.' Coming up with new ideas and solutions to problems is no longer confined to brainstorming sessions in windowless conference rooms; the process is now open on collaborative platforms. Employees now share what they are working on, with whom they are meeting, how they are feeling, and what they are accomplishing. Thanks to social media, the work environment is much more transparent and less hierarchical.

Wikipedia, though not a social medium per se, is the ultimate collaborative product. Since its creation in 2001, the online encyclopedia, built entirely from user-generated content, has grown to 53 million articles, of which six million are written in English. About 17,000 new articles are added each month and fact-checked by 132,000 volunteer editors. Wikipedia has shown employers that collaboration often results in a better product than otherwise would have materialized from one person or from assigning several people to rigid tasks.

While social media have in many ways improved the workplace, they have also created more work and have changed expectations. For example, although email is not a 'social medium' in the strict sense of the term, it is the most common way colleagues connect one-on-one. Just over six in ten workers say that email is 'very important' to doing their job, 54% feel the same way about the Internet, and only 35% think so about their telephones. Many employees feel overburdened and overwhelmed with emails. Email boxes are clogged with tens of thousands of messages. Some employers require that all messages be saved, while others leave it to the employees' discretion.

Slack is one of several new applications that aim to replace interoffice email and foster collaboration. Slack is customizable and easy for departments and companies to maintain. Slack, for instance, defaults to open communication so anyone in the company can see what anyone has written, which could be unsettling to some employees. Whether such applications actually cut back on the relentless number of emails or just increase the barrage of messages remains to be seen.

Social networking sites are a big work distraction as well. Although about one-quarter of employees use social media for professional networking and for information to help solve work problems, many workers take a break to spend a few minutes checking Facebook or some other social medium. All too often, however, a few minutes turns into 30 minutes and more, leading some employers to block access to social media sites altogether.

Other pitfalls of social media result from who owns the data. Lawsuits are being heard in courtrooms across the country over who owns a Twitter account and thus the messages – the employee who tweets as part of his or her job requirement or the company. At question is whether an employee is required to cede the Twitter account and its followers to the company when he or she finds another job.

Other lawsuits and much furor have popped up concerning employees being fired for inappropriate remarks made on Twitter: A woman who dressed as a Boston Marathon bombing victim; a California Pizza Kitchen employee who complained about the new uniforms; a waiter who told about getting stiffed by an actress; a guy at a conference whose stupid joke to a friend was overheard by a woman sitting nearby who took a photo of him and tweeted his comment. Getting fired does not seem like the worst part of these stories – the horrible, stinging comments, death threats, and public ostracism sent some victims into hiding and emotional depression.

Privacy

Once any bit of information – whether true or false, trivial or important – goes online, it will stay there, forever. Drunk dialing is a dumb thing to do, but now drunk posting is the ultimate in stupidity. As the saying goes, "if you don't want the world to know, don't put it on the Internet," and hope that your friends do not do so either. Many Baby Boomers rambling around in their parents' attic or garage have stumbled across their dusty old diary. Reading the entries years later in private is embarrassing enough, but the thought of such information being online can make even the thick-skinned cringe. Boomers were probably the last generation to not have their formative years plastered all over cyberspace.

Online users are very concerned about the surveillance of their digital communications, especially about the invasion of privacy and the ability to retain and control confidentiality. Privacy is a hot button that has different meanings to different people. Some online users are fearful of government surveillance; others are more worried about commercial spying. Some say privacy breaches are acceptable as long as they are used to monitor terrorist activity; others claim their civil liberties are being violated when data are collected for any reason.

Apple Computer has taken a stand against breaches of privacy by refusing to unlock an encrypted cell phone belonging to Syed Farook, one of the killers of 14 people in San Bernardino, California, in December 2015. Suspecting that the phone held crucial evidence about the shooting, the FBI attempted to unlock the phone but did not know the password. The FBI asked Apple to disable the feature that erases data on the phone after entering an incorrect password on ten tries. Apple refused to do so, stating that it would set a dangerous precedent if asked to decrypt phones at the request of law enforcement or anyone. Apple maintains that only the person who owns the phone or who knows the password should be able to unlock it. The Justice Department stepped in and ordered Apple to create a way into the seized phone, but Apple still refused, and challenged the order. A federal magistrate judge ruled for the company and stated that the government had overstepped its authority. The Justice Department has appealed the latest ruling, and so the dispute continues. Dozens of technology companies have presented a united front against the government by filing legal briefs in support of Apple's position. Apple, however, has also been at the forefront of lawsuits objecting to its handling of privacy. In 2019, the company was sued by millions of iPhone users over its iOS virtual listening assistant, Siri, and how it recorded conversations without consent. As of spring 2020, that lawsuit remains unresolved.

Trying to control personal information is like trying to swim against a tsunami. It is impossible to control the wave; online users are swept in whatever direction privacy policies take them. Nine of ten online adults say they have no control over how personal information is collected or used. While almost six in ten online users are willing to share personal information with companies in return for free services, about the same number would like the government to do more to regulate the types of information advertisers can collect.

Users feel especially vulnerable on social media sites; only 15% of online adults believe social media are secure. Texting and emailing are deemed more secure than social media, but only about 40% of online users trust that personal information will not be released. While some users like to track their online reputation and profile, only about 60% have ever 'googled' themselves, and eight in ten acknowledge that it would be very difficult to remove or correct false information.

ZOOM IN 6.4

PRIVACY SETTINGS FOR SOCIAL MEDIA PLATFORMS

- Facebook: www.facebook.com/help/193677450678703
- Twitter: https://support.twitter.com/articles/14016-about-public-and-protected-tweets#
- Instagram: https://help.instagram.com/116024195217477/
- Pinterest: https://help.pinterest.com/en/articles/change-your-privacy-settings
- YouTube: https://support.google.com/youtube/answer/157177?hl=en
- LinkedIn: https://help.linkedin.com/app/answers/detail/a_id/66/~/managing-account-settings

FYI: SOCIAL MEDIA BEST PRACTICES

- Start with a clear plan. Do not send messages without knowing exactly why, when, and where they are going.
- Set goals. Know in advance what you want to achieve.
- Positively promote your cause or organization.
- Put your networking power into play with a smile.
- Create a user experience and dialogue. Listen and respond to your audience.
- Build relationships and a community.
- Be open and honest.
- Be respectful, professional, and non-argumentative.
- Post timely and up-to-date information.
- Prepare to lose control of your message. Others will edit, interpret, change, and repost it.
- Think before you act. An inappropriate or emotional outburst cannot be undone.
- Manage your reputation.

SEE IT LATER

The proliferation of social media has outpaced even the most generous projections. Some thought it was a fad. Facebook's beginning as a site for collegiate socializers was not the best predictor of what was to come. Today, how we communicate has changed. In fact, most scholars agree that this paradigm shift is cemented as 'normal.' We are hard pressed to find a situation in today's 'now' media world that is not affected by the proliferation of social media.

Social Media and Society

Companies, schools, civic groups, and individuals must take into account social media ethics, posting protocol, standards, and best practices. The push to 'share' has created a wave of litigation that will take years to sort out and settle. Slowly, we are beginning to understand that our social media profiles never go away.

'Influencers' are now what many on social media strive to be. These are an "independent third-party endorser who shape audience attitudes through blogs, tweets, and the use of other social media" (Freberg, Graham, McGaughey, & Freberg, 2011). Influencers shape others' purchasing decisions because of an achieved status of their experience, what they do, and who they know. Essentially, these are online socialites who have earned a sense of trust from other online users. These influencers have spread good news, such as raising awareness for Amyotrophic lateral sclerosis (ALS) through the ice bucket challenge that began in 2015. The dark side of influencers, however, was shown in the failed 2017 Fyre Festival. Influencers such as Kendall Jenner and model Bella Hadid were blamed for duping people to purchase tickets to what was ultimately a scam.

Whether one is a chief marketing officer managing a global brand, an editor in a local newsroom, or an ordinary person, properly managing and curating a social profile is essential. Growing up as digital natives, the Millennial demographic does not know a world without social media. However, even the most prolific posters, pinners, and tweeters need to know what they are doing and how it will stay with them forever.

FYI: ATTENTION SPAN

The average attention span for humans was 12 seconds in 2000 but had dropped to seven seconds by 2015. For a goldfish, it is nine seconds.
Source: Attention Span Statistics, 2015

Although some social media users complain about the invasion of privacy, they are still willing to share their lives with the world. Social media has brought on the belief that the need to know trumps the need for privacy, and this collective cry for transparency is unlikely to change in the near future. Ambient awareness is likely to intensify as new online tools make it easier to keep in touch and know what others are doing 24/7. Mobile devices and social media have given us a new sense of security we are unlikely to cede to independence.

There were a whopping 328 million sign ups across the various social media platforms from October 2018 to October 2019 with 3.725 billion users now in total. Pew Research's compilation of 'expert' opinion of the digital world between 2020 and 2070 includes that Internet will be as important as oxygen, with seamless connectivity a vital part of that, while artificial intelligence will be more widely used for routine activities, and the Internet will be controlled by a powerful elite in order to surveil and manipulate, as well as distract the masses.

SUMMARY

For most people, deleting their Facebook or Instagram account would be like banishing themselves to a desert island. For as much as we hate social media, we love them just as much. This chapter covered the rise of social media, which began in the late 1970s when the Internet was limited to text and traversed by using complicated commands. Early online socializing meant sending an email, reading a bulletin board, or joining a MUD. Social media as known today came about after the emergence of the Web and hypertext connections. The earliest social network sites, Six Degrees and Friendster, were slow to catch on and never really had much of an online impact. MySpace is credited for starting the social media movement. Teenagers and young adults went wild over the site. Media coverage, both positive and negative, fed the frenzy, but several high-profile incidents involving bullying, kidnapping, and other crimes turned users from MySpace to the newer and seemingly safer Facebook, which is currently the most popular social network site in the world.

Social media, as 'now media,' have changed the way news and information are delivered and consumed. Journalists and ordinary citizens share snippets of events with their friends and followers, who share the news items within their circles. The news cycle is now 24/7 with not day-to-day or hour-to hour updates but second-to-second news flashes. The word of an event goes out even before it is verified. Facts are checked after the news breaks, not before.

Critics complain that in-depth professional coverage has been taken over by shallow amateur accounts and, thanks to social media, Millennials are less knowledgeable than previous generations about national and world affairs and turn their attention instead to frivolous celebrity gossip. As social media change the way we think and consume information, the news media and educators will need to adapt to the next generation's learning style. Established professionals might balk at what they consider the dumbing down of the nation's intellect, but perhaps the swing between extremes will balance somewhere in the middle and social media will foster in-depth conversation and bolster knowledge of ourselves and of our world.

Although about 85% of online users believe their personal information is at high risk, they are willing to create a personal profile, post photos, and reveal intimate details of their lives. Peer pressure, work requirements, and the need to know what others are doing compel even the most reluctant to jump into the social media world. Further, social media users are willing to trade personal autonomy and independence for the feeling of security, to feel like they belong, and for the ability to peek in on their friends.

Social media have also changed the workplace. Employees are expected to have social media profiles on which they promote their employers' products and services. Yet employers are hypervigilant about what is posted about them and are merciless when it comes to an employee who inadvertently crosses the line with an unacceptable or offensive post.

Protocols, ethics, and standards are being developed to maximize outcomes and keep social media users out of trouble. The 'now' world of social media is complex. It is enlightening yet uninformative, serious yet frivolous, and egalitarian yet repressive. Despite the contradictions, social media are omnipresent and here to stay.

BIBLIOGRAPHY

126 Amazing Social Media Statistics and Facts. (2019, December 30). *Brandwatch*. Retrieved from: www.brandwatch.com/blog/amazing-social-media-statistics-and-facts/

24 Twitter acronyms and abbreviations. (2015). *AllAcronyms.com*. Retrieved from: www.allacronyms.com/twitter/topic

About Reddit. (2015). *Reddit*. Retrieved from: www.reddit.com/about/

Adams, D. (2011). The history of social media. *InstantShift.com*. Retrieved from: www.instantshift.com/2011/10/20/the-history-of-social-media/

Afshar, V. (2013, March 24). Five ways social media has forever changed the way we work. *Huffington Post*. Retrieved from: www.huffingtonpost.com/vala-afshar/social-media_b_2944407.html

Agarwal, A. (2015, May 27). 5 predictions for the future of social media. *INC*. Retrieved from: www.inc.com/ajagrawal/5-predictions-of-the-future-of-social-media.html

Arrington, M. (2009, January 22). Facebook now nearly twice the size of My Space worldwide. *TechCrunch*. Retrieved from: http://techcrunch.com/2009/01/22/facebook-now-nearly-twice-the-size-of-myspace-worldwide/

Attention span statistics. (2015). *Statistics Brain Research Institute*. Retrieved from: www.statisticbrain.com/attention-span-statistics/

Average time spent daily on social media (latest 2020 data). (2020). *BroadbandSearch*. Retrieved from: www.broadbandsearch.net/blog/average-daily-time-on-social-media

Baer, J. (2013). 11 shocking new social media statistics in America. *Convince & Convert*. Retrieved from: www.convinceandconvert.com/social-media-research/11-shocking-new-social-media-statistics-in-America

Bates, D. (2012, July 31). You've got (more) mail: The average worker now spends over a quarter of their day dealing with email. *DailyMail.com*. Retrieved from: www.dailymail.co.uk/sciencetech/article-2181680/Youve-got-mail-The-average-office-worker-spend-half-hours-writing-emails.html

Benner, K., & Goldstein, J. (2016, February 29). Apple wins ruling in New York iPhone hacking order. *The New York Times*. Retrieved from: www.nytimes.com/2016/03/01/technology/apple-wins-ruling-in-new-york-iphone-hacking-order.html

Benner, K., Lichtblau, E., & Wingfield, N. (2016, February 26). Apple goes to court, and the F.B.I. presses Congress to settle iPhone privacy fight. *The New York Times*, pp. B1, B7.

Bertrand, N. (2015, May 6). One of the world's smallest countries just banned Facebook. *Business Insider*. Retrieved from: www.businessinsider.com/one-of-the-worlds-smallest-countries-just-banned-Facebook

Best for last? (2009, August 3). *The New York Times*, pp. C1, C8.

Biggs, J. (2011, December 25). A dispute over who owns a Twitter account goes to court. *The New York Times*, pp. B1, B2.

Bosker, B. (2012, March 10). Vic Gundotra explains what Google+ is (but not why to use it). *Huffington Post*. Retrieved from: www.huffingtonpost.com/2012/03/10/vic-gundotra-google-plus_n_1336601.html

boyd, d. m., & Ellison, N. B. (2007). Social network sites: Definition, history, and scholarship. *Journal of Computer-Mediated Communication, 13(1)*, 210–230.

Bradford, K. T. (2010, December 8). What Tumblr is and how to use it: A practical guide. *Laptop*. Retrieved from: http://blog.laptopmag.com/tumblr-tips

Brandon, J. (2020, February, 26). Coronavirus misinformation is spreading on social media. Will Facebook and Twitter react? *Forbes*. Retrieved from: www.forbes.com/sites/johnbbrandon/2020/02/26/coronavirus-misinformation-is-spreading-on-social-media-will-facebook-and-twitter-react/#75dad95a785e

Brustein, J. (2014, July 23). Americans now spend more time on Facebook than they do their pets. *Bloomberg Business*. Retrieved from: www.bloomberg.com/bw/articles/2014–07–23/heres-how-much-time-people-spend-on-facebook-daily

Brusilovsky, D. (2009, July 13). Why teens aren't using Twitter: It doesn't feel safe. *The Washington Post*. Retrieved from: www.washingtonpost.com [July 14, 2009]

Carr, N. (2011). *The shallows: What the internet is doing to our Brains*. London: W.W. Norton & Co.

Chokshi, N. (2019, March 19). Myspace, once the king of social networks, lost years of data from heyday. *The New York Times*. Retrieved from: www.nytimes.com/2019/03/19/business/myspace-user-data.html?auth=login-email&login=email

Clement, J. (2019a, August 9). Hours of video uploaded to YouTube every minute 2007–2019 (2019, August 9). *Statista*. Retrieved from: www.statista.com/statistics/259477/hours-of-video-uploaded-to-youtube-every-minute/

Clement, J. (2019b, August 9). YouTube usage penetration in the U.S. by age group. *Statista*. Retrieved from: www.statista.com/statistics/296227/us-youtube-reach-age-gender/

Clement, J. (2020, March 25). Leading social media websites in the United States in February. *Statista*. Retrieved from: www.statista.com/statistics/265773/market-share-of-the-most-popular-social-media-websites-in-the-us/

Cole D., & Klein, B. (2020, May 10). The latest predictions from Trump officials on unemployment are dire. *CNN*. Retrieved from: www.cnn.com/2020/05/10/politics/unemployment-mnuchin-hassett-kudlow-economy-coronavirus/index.html

Collins, D. (2020, March 30). Coronavirus goes viral: It's essential we all fight back. *Wired*. Retrieved from: www.wired.co.uk/article/coronavirus-lies-social-media

Company info. (2015, March 15). *Facebook*. Retrieved from: https://newsroom.fb.com/company-info

Constine, J. (2016, January 5). Twitter may increase tweets to 10,000 characters, but hide all past 140. *TechCrunch*. Retrieved from: http://techcrunch.com/2016/01/05/information-density/

Corcoran, M. (2009, July 12). Death by cliff plunge, with a push from Twitter. *The New York Times*. Retrieved from: www.nytimes.com/2009/07/12/fashion/12hoax.html [July 29, 2009]

Costill, A. (2014, January 16). 30 things you absolutely need to know about Instagram. *Search Engine Journal*. Retrieved from: www.pewinternet.org/2015/01/09/demographics-of-key-social-networking-platforms-2/

Coyle, J. (2009, July 1). Is Twitter the news outlet for the 21st century? *The Washington Post*. Retrieved from: washingtonpost.com/wp-dyn/content/arti cle/2009/07/01 [July 14, 2009]

Cummings, W. (2020, January 23). Trump sets record for most tweets in a single day since he took office. *USA Today*. Retrieved from: www.usatoday.com/story/news/politics/2020/01/23/trump-record-most-tweets-since-taking-office/4551815002/

Crunched. (2015, August 23). *The Washington Post Magazine*, p. 8.

Dickinson, B. (2013, November 26). What's in a name? *Brand Marketing*. Retrieved from: www.paceco.com/whats-in-a-name-defining-the-millennial-generation/

Doyne, S. (2015, March 2). Does punctuation in text messages matter? *The New York Times*. Retrieved from: http://learning.blogs.nytimes.com/2015/03/02/does-punctuation-matter-in-text-messages/

Elmore, T. (2014, March 4). I'd rather lose my Ford than my phone. *Huffington Post*. Retrieved from: www.huffingtonpost.com/tim-elmore/id-rather-lose-my-ford-or-my-finger-than-my-phone_b_4896134.html

Experts optimistic about the next 50 years of digital life. (2019, October 19). *Pew Research*. Retrieved from: www.pewresearch.org/internet/2019/10/28/experts-optimistic-about-the-next-50-years-of-digital-life/

Facebook by the numbers: Stats, demographics and fun facts. (2020, April 22). *Omnicore*. Retrieved from: www.omnicoreagency.com/facebook-statistics/

Fallows, J. (2011, April). Learning to love the (shallow, divisive, unreliable) new media. *The Atlantic*, pp. 34–49.

Falman, S. E. (n.d.). Smiley lore. Carnegie School of Computer Science. Retrieved from: www.cs.cmu.edu/~sef/Orig-Smiley.htm

Farhi, P. (2011, May 3). Beat by a tweet: How the bin Laden story broke. *The Washington Post*, pp. C1, C7.

Freberg, K., Graham, K., McGaughey, K., & Freberg, L. A. (2011). Who are the social media influencers? A study of public perceptions of personality. *Public Relations Review, 37(1)*, 90–92.

Friedman, T. L. (2016, February 3). Social media: Destroyer or creator? *The New York Times*. Retrieved from: www.nytimes.com2016/02/03/opinion/social-media-destroyer-or-creator.html

Gillette, F. (2011, June 21). The rise and inglorious fall of Myspace. *Bloomberg Business*. Retrieved from: www.bloomberg.com/bw/stories/2011-06-21/the-rise-and-inglorious-fall-of-MySpace

Global social networks ranked by number of users 2020. (2021, January 28). *Statista*. Retrieved from: www.statista.com/statistics/272014/global-social-networks-ranked-by-number-of-users/

Harlow, S., & Johnson, T. J. (2011). Overthrowing the protest paradigm? How *The New York Times*, Global Voices and Twitter covered the Egyptian Revolution. *International Journal of Communication*, 5(feature), 1359–1374.

Hermida, A. (2010). Twittering the news: The emergence of ambient journalism. *Journalism Practice*, 4(3), 297–308.

Hern, A. (2019, February 1). Closure of Google+: Everything you need to know. *The Guardian*. Retrieved from: www.theguardian.com/technology/2019/feb/01/closure-google-plus-everything-you-need-to-know

Hesse, M. (2010, May 6). Twitter archive at Library of Congress could help redefine history's scope. *The Washington Post*. Retrieved from: www.washingtonpost.com/wp-dyn/content/article/2010/05/05/AR2010050505309.html

Hesse, M. (2011, January 13). Celebrating a decade of the age of Wikipedia. *The Washington Post*, pp. C1, C4.

Heussner, K. M. (2010, June 8). 'The Facebook effect': Inside Zuckerberg's coups, controversies. *ABC News*. Retrieved from: abcnews.go.com/Technology/Media/facebook-effect-inside-zuckerbergs-coups-controversies/story?id= 10853306

Hiscock, M. (2014, June 26). Dead Facebook users will soon outnumber the living. *The Loop*. Retrieved from: www.theloop.ca/dead-facebook-users-will-soon-outnumber-the-living/

The history of Instagram. (2015, February 3). *Social Experiment*. Retrieved from: http://socialexperiment.net/uncategorized/the-history-of-instagram/

Hughes, D. J., Rowe, M., Batey, M., & Lee, A. (2012). A tale of two sites: Twitter vs. Facebook and the personality predictors of social media usage. *Computers in Human Behavior*, 28(2), 561–569.

Internet usage of Millennials in the United States – Statistics & Facts. (2019, November 20). *Statista*. Retrieved from: www.statista.com/topics/2576/us-millennials-internet-usage-and-online-shopping/#dossierSummary__chapter2

Is social networking changing childhood? (2009). *Common Sense Media*. Retrieved from: www.commonsensemedia.org/teen-social-media [February 10, 2010]

Isaac, M. (2016, March 19). Twitter opts to keep its characters. *The New York Times*, p. B2.

Johnson, T. J., & Kaye, B. K. (2016). Some like it lots: The influence of interactivity and reliance on credibility. *Computers in Human Behavior*, 61, 136–145. doi: 10.1016/j.chb.2016.03.012.

Kang, C. (2014, August 23). Famous in a flash. *The Washington Post*, pp. A1, A4.

Kavulla, K. (2012, January 19). Pinterest: What it is, how you use it, and why you'll be addicted. *SheKnows.com*. Retrieved from: www.sheknows.com/living/articles/852875/pinterest-what-it-is-how-to-use-it-and-why-youll-be-addicted

Kaye, B. K., & Johnson, T. J. (2015). I only have eyes for YouTube: Motives for political use. *Journal of Social Media Studies*, 1(2), 91–104. doi:10.15340/2147336612841.

Kelly, J. (2020, May 8). U.S. unemployment is at its highest rate since the depression at 14.7% – with 20.5 million jobs lost in April. *Forbes*. Retrieved from: www.forbes.com/sites/jackkelly/2020/05/08/us-unemployment-is-at-its-highest-rate-since-the-great-depression-at-147-with-205-million-more-jobs-lost-in-april/#7aae72a1656d

Kemp, S. (2015, January). Time spent on social media. *Social Media Today*. Retrieved from: www.socialmediatoday.com/content/global-digital-social-media-stats-2015

Kemp, S. (2019, January, 31). Digital 2019: Global digital overview. *Data Reportal*. Retrieved from: https://datareportal.com/reports/digital-2019-global-digital-overview

Key findings about the online news landscape in America. (2019, September 11). *Pew Research*. Retrieved from: www.pewresearch.org/fact-tank/2019/09/11/key-findings-about-the-online-news-landscape-in-america/

Kinzie, S. (2009, June 26). Some professors' jitters over Twitter are easing: Discussions expand in and out. *The Washington Post*, p. B1.

Kopf, D. (2019, February 12). Around 40% of American couples now first meet online. *Quartz*. Retrieved from: https://qz.com/1546677/around-40-of-us-couples-now-first-meet-online/

Korkki, P. (2009). An outlet for creating and socializing. *The New York Times*, p. B2.

Krashinsky, S. (2012, November 2). How YouTube has transformed the 2012 presidential election. *The Globe and Mail*. Retrieved from: www.theglobeandmail.com/news/world/us-election/how-youtube-has-transformed-the-2012-presidential-election/article4871244/

Kushin, M. J., & Yamamoto, M. (2010). Did social media really matter? College students' use of online media and political decision making in the 2008 election. *Mass Communication & Society, 13*(5), 608–630.

Lampe, C., & Ellison, N. (2016, June 22). Social media and the workplace. *Pew Research Center*. Retrieved from: www.pewresearch.org/internet/2016/06/22/social-media-and-the-workplace/

Lampe, C., Ellison, N., & Steinfield, C. (2007). A Face(book) in the crowd: Social searching vs. social browsing. Paper presented to the proceedings of the 2006 20th Anniversary Conference of the Association for Computing Machinery. Banff, Vancouver, Canada.

Landry, T. (2014, September 8). How social media has changed us: The good and the bad. *Social Media Today*. Retrieved from: www.socialmediatoday.com/content/how-social-media-has-changed-us-good-and-bad

Lange, P. G. (2007). Publicly private and privately public: Social networking on YouTube. *Journal of Computer-Mediated Communication, 13*(1), 361–380.

Leetaru, K. (2019, February 26). Is Twitter really faster than the news? *Forbes*. Retrieved from: www.forbes.com/sites/kalevleetaru/2019/02/26/is-twitter-really-faster-than-the-news/?sh=638735401cf7

Lenhart, A. (2009a, January 14). Pew research center. Retrieved from: pewresearch.org/pubs/1079/social-networks-grow [September 28, 2009]

Lenhart, A. (2009b, January 14). Social networks grow: Friending Mom and Dad. *Pew Internet & American Life Project*. Retrieved from: pewresearch.org/pub/1079/social-networks-grow [March 20, 2009]

Lewis, P., & McCormick, E. (2018, February 7). How an ex-YouTube insider investigated its secret algorithm. *The Guardian*. Retrieved from: www.theguardian.com/technology/2018/feb/02/youtube-algorithm-election-clinton-trump-guillaume-chaslot

Lichtblau, E., & Benner, K. (2016, March 10). Apple and U.S. bitterly turn up volume in iPhone privacy fight. *The New York Times*. Retrieved from: www.nytimes.com/2016/03/11/technology/apple-iphone-fight-justice-department.html

Lohr, S. (2010, April 14). Library of Congress will save tweets. *The New York Times*. Retrieved from: www.nytimes.com/2010/04/15/technology/15twitter.html?_r=0

Loneliness and the workplace. (2020, n.d.) *Cigna*. Retrieved from: www.cigna.com/static/www-cigna-com/docs/about-us/newsroom/studies-and-reports/combatting-loneliness/cigna-2020-loneliness-fact sheet.pdf

Madden, M. (2014, November 12). Public perceptions of privacy and security in the post-Snowden era. *Pew Research Center*. Retrieved from: www.pewinternet.org/2014/11/12/public-privacy-perceptions/

Manjoo, F. (2016, February 25). Maintain privacy in an always-watch future. *The New York Times*, p. B1.

Marche, S. (2012, May). Is Facebook making us lonely? *The Atlantic*. Retrieved from: www.theatlantic.com/magazine/archive/2012/05/is-facebook-making-us-lonely/308930/

Markoff, J. (2012, September 13). Social networks can affect voter turnout, study says. *The New York Times*, p. A17.

McAlone, N. (2016, October 28). Twitter is shutting down Vine. *Business Insider*. Retrieved from: www.businessinsider.com/twitter-shutting-vine-down-2016-10

McCarthy, N. (2014, March 13). Millennials rack up 18 hours of media use per day. *Statista*. Retrieved from: www.statista.com/chart/2002/time-millennials-spend-interacting-with-media/

Millennials get news on Twitter, Facebook. (2014, March 25). *Yahoo! Finance*. Retrieved from: http://finance.yahoo.com/news/facebook-front-page-millennial-news-194700637.html

Millennials stand out for their technology use, but older generations also embrace digital life. (2019, September, 9). *Pew Research*. Retrieved from: www.pewresearch.org/fact-tank/2019/09/09/us-generations-technology-use/

Miller, C. C. (2009, August 26). Who's driving Twitter's popularity? Not teens. *The New York Times*, pp. B1, B2.

Miller, C. C. (2014, February 14). The plus in Google plus? It's mostly for Google. *The New York Times*. Retrieved from: www.nytimes.com/2014/02/15/technology/the-plus-in-google-plus-its-mostly-for-google.html?_r=0

Moreau, E. (2016, February 13). The top 25 social networking sites people are using. *About Tech*. Retrieved from: http://webtrends.about.com/od/socialnetworkingreviews/tp/Social-Networking-Sites.htm

Most popular mobile social networking apps by monthly users. (2019). *Statista*. Retrieved from: www.statista.com/statistics/248074/most-popular-us-social-networking-apps-ranked-by-audience/

Musgrove, M. (2009, June 17). Twitter is a player in Iran's drama. *The Washington Post*, p. A10.

Nakashima, E. (2016, February 17). Apple vows to resist FBI demands to crack iPhone linked to San Bernardino attacks. *The Washington Post*. Retrieved from: www.washingtonpost.com/world/national-security/us-wants-apple-to-help-unlock-iphone-used-by-san-bernardino-shooter/2016/02/16/69b903ee-d4d9-11e5-9823-02b905009f99_story.html

Neal, R. W. (2014, January 16). Facebook gets older: Demographic report shows 3 million teens left social networks in three years. *International Business Times*. Retrieved from: www.ibtimes.com/facebook-gets-older-demographic-report-shows-3-million-teens-left-social-network-3-years-1543092

Original bboard thread in which :-) was proposed. (1982, September 19). Retrieved from: www.cs.cmu.edu/~sef/Orig-Smiley.htm

Percentage of U.S. population with a social network profile 2008 to 2016. (n.d.). *Statista*. Retrieved from: www.statista.com/statistics/273476/percentage-of-us-population-with-a-social-network-profile/

Phillips, S. (2007, July 25). A brief history of Facebook. *The Guardian*. Retrieved from: www.theguardian.com/technology/2007jul/25/media.new media

Pinterest by the numbers: Stats, demographics & fun facts. (2020, February 11). *Omnicore*. Retrieved from: www.omnicoreagency.com/pinterest-statistics/

Plambeck, J. (2016, March 3). An industry lines up behind Apple. *The New York Times*. Retrieved from: www.nytimes.com/2016/03/04/technology/an-industry-lines-up-behind-apple.html

Poindexter, P. (2012). Why Millennials aren't into news. In *Millennials, news, and social media: Is news engagement a thing of the past?* (pp. 16–34). New York, NY: Peter Lang Publishing.

Purcell, K., & Ranie, L. (2014, December 30). Technology's impact on workers. *Pew Research Center*. Retrieved from: www.pewinternet.org/2014/12/30/technologys-impact-on-workers/

Ramirez, J. (2008, November 9). The YouTube election. *Newsweek*. Retrieved from: www.newsweek.com/youtube-election-85069

Reynolds, G. (2002, January 9). A technological reformation. *Tech Central Station*. Retrieved from: www.techcentralstation.com/1051

Rogers, J. (2017, January 20). Trump makes first tweet as president. *Fox News*. Retrieved from: www.foxnews.com/tech/trump-makes-first-tweets-as-president

Rosen, R. J. (2013, September 11). What is Reddit? *The Atlantic*. Retrieved from: www.theatlantic.com/technology/archive/2013/09/what-is-reddit/279579/

Roush, W. (2013, May 31). 11 reasons why Flickr, not Facebook, is the place to put your photos. *Xconomy*. Retrieved from: www.xconomy.com/national/2013/05/31/11-reasons-why-flickr-not-facebook-is-the-place-to-put-your-photos/2/

Rusli, E. M. (2012, April 9). Facebook buys Instagram for $1 billion. *The New York Times*. Retrieved from: http://dealbook.nytimes.com/2012/04/09/facebook-buys-instagram-for-1-billion/

Sadlier, A. (2019, April 19). 1 in 4 Americans feel they have no one to confide in. *New York Post*. Retrieved from: https://nypost.com/2019/04/30/1-in-4-americans-feel-they-have-no-one-to-confide-in/

Shapira, I. (2009). In a generation that friends and tweets, they don't. *The Washington Post*, pp. A11, A1.

Share of U.S. adults using social media, including Facebook, is mostly unchanged since 2018. (2019, April 10). *Pew Research*. Retrieved from: www.pewresearch.org/fact-tank/2019/04/10/share-of-u-s-adults-using-social-media-including-facebook-is-mostly-unchanged-since-2018/

Shelasky, A. (2014, September). The love-tech connection. *Self*, p. 60.

Shear, M. D., Haberman, M., Confessore, N., Yourish, K., Buchanan, L., & Collins, K. (2019, November, 2). How Trump reshaped the presidency in over 11,000 Tweets. *The New York Times*. Retrieved from: www.nytimes.com/interactive/2019/11/02/us/politics/trump-twitter-presidency.html

Shields, M. (2015, January 14). MySpace still reaches 50 million people each month. *The Wall Street Journal/CMO Today*. Retrieved from: blogs.wsj.com/cmo/2015/01/14/my-space-still-reaches-50-million-each-month/

Smith, A. (2014, February 3). 6 new facts about Facebook. *Pew Research Center*. Retrieved from: www.pewresearch.org/fact-tank/2014/02/03/6-new-facts-about-facebook/

Smith, A., & Brenner, J. (2012). Twitter use 2012. *Pew Research Center*. Retrieved from: www.pewinternet.org/Reports/2012/Twitter-Use-2012.aspx

Smith, C. (2016, March 6). By the numbers: 70 amazing Snapchat statistics. *DMR*. Retrieved from: http://expandedramblings.com/index.php/snapchat-statistics/

Smith, C. (2020, June 14). 80 amazing Reddit statistics and facts. *Digital Marketing*. Retrieved from: http://expandedramblings.com/index.php/reddit-stats/3/

Social. (2018). GlobalWebIndex's flagship report on the latest trends in social media. *GlobalWebIndex*. Retrieved from: www.globalwebindex.com/hubfs/Downloads/Social-H2-2018-report.pdf

Social media active users. (2016, January 28). *The Social Media Hat*. Retrieved from: www.thesocialmediahat.com/active-users

Social media fact sheet. (2019, June 12). *Pew Research Center*. Retrieved from: www.pewresearch.org/internet/fact-sheet/social-media/

Social networking eats up 3+ hours per day for the average American user. (2013, January 9). *MarketingCharts.com*. Retrieved from: http://marketingcharts.com/online/social-networking-eats-up-3-hours-per-day-for-the-average-American-user-26049

Social networking fact sheet. (2019, June 12). *Pew Research Center*. Retrieved from: www.pewinternet.org/fact-sheets/social-networking-fact-sheet

Social networking statistics. (2015, December 15). *Statistic Brain*. Retrieved from: www.statisticbrain.com/social-networking-statistics/

Stanley, A. (2009, February 28). What are you doing? Media Twitters can't stop typing. *The New York Times*, pp. C1, C6.

State of the media: The social media report. (2012). *Nielsen*. Retrieved from: https://postmediavancouversun.files.wordpress.com/2012/12/nielsen-social-media-report-20122.pdf

Statistics. (2015). YouTube. Retrieved from: www.youtube.com/yt/press/statistics.html

Steinmetz, K. (2014, September 19). In praise of emoticons. *Time*. Retrieved from: http://time.com/3341244/emoticon-birthday/

Stenovec, T. (2011, June 29). My Space history: A timeline of the social network's biggest moments. *Huffington Post*. Retrieved from: www.huffingtonpost.com/2011/06/29/myspace-history-timeline_n_887059.html

Sterling, G. (2013, May 21). Pew: 94% of teenagers use Facebook, have 425 friends, but Twitter and Instagram adoption way up. *Marketing Land*. Retrieved from: http://marketingland.com/pew-the-average-teenager-has-425-4-facebook-friends-44847

Tate, R. (2012, December 12). Social media is eating our lives (and Pinterest is chewing fastest). *Wired*. Retrieved from: www.wired.com/2012/12/social-spike/

Teens fact sheet. (2012). *Pew Research Center*. Retrieved from: www.pewinternet.org/fact-sheets/teens-fact-sheet/

'The word – truthiness.' (2005, October 17). *Comedy Central*. Retrieved from: www.cc.com/video-clips/63ite2/the-colbert-report-the-word---truthiness

Thompson, C. (2008, September 5). Brave new world of digital intimacy. *The New York Times*. Retrieved from: www.nytimes.com/2008/09/07/magazine/07awareness-t.html?pagewanted=1

Thompson, C. (2013). *Smarter than you think*. New York: Penguin Press.

TikTok – statistics & facts. (2020, March 17). *Statista*. Retrieved from: www.statista.com/topics/6077/tiktok/

Tsukayama, H. (2016, March 19). As it turns 10, Twitter to stick to 140 characters. *The Washington Post*, p. A16.

Tsukayama, H. (2016, February 24). Facebook officially expands beyond the 'like.' *The Washington Post*. Retrieved from: www.washingtonpost.com/news/theswitch/wp/2016/02/24/facebook-officially-expands-beyond-the like/

Turner, J. (2016, April 10). Are there really more mobile phone owners than toothbrush owners? *LinkedIn*. Retrieved from: www.linkedin.com/pulse/really-more-mobile-phone-owners-than-toothbrush-jamie-turner

Twitter usage. Company facts. (2015, March 31). *Twitter*. Retrieved from: https://about.twitter.com/company

U.S. Census Bureau releases 2018 families and living arrangements tables. (2018, November 14). *U.S. Census Bureau*. Retrieved from: www.census.gov/newsroom/press-releases/2018/families.html

Wagner, K. (2014, December 15). Instagram hits 300 million users, now larger than Twitter. *Recode.net*. Retrieved from: http://recode.net/2014/12/10/instagram-hits-300-million-users-now-larger-than-twitter/

Which of the following social networks have you used within the last two weeks? (2020, January 20). *Statista*. Retrieved from: www.statista.com/forecasts/1088802/usage-of-social-networks-in-the-usv

World's largest photography libraries. (2013, December). *Smithsonian*, p. 24.

Worthan, J. (2013, February 8). A growing app lets you see it, then you don't. *The New York Times*. Retrieved from: www.nytimes.com/2013/02/09/technology/snapchat-a-growing-app-lets-you-see-it-then-you-dont.html

Yeung, K. (2013, May 5). LinkedIn is 10 years old today: Here's the story of how it changed the way we work. *TNW News*. Retrieved from: http://thenextweb.com/insider/2013/05/05/linkedin-10-years-social-network/

YouTube about. (2020). *YouTube*. Retrieved from: www.youtube.com/about/press/

Chapter seven

Digital Devices: Up Close, Personal, and Customizable

Mary Beadle, Ph.D. (John Carroll University)

This chapter begins with a historical look at the telegraph, landline phones, and fax machines, the first one-to-one communication tools. The chapter then moves to contemporary personal communication devices, such as smartphones, tablets, and smart watches and other wearables, what they are and how they work. It then examines the 'now' world of customizable radio and television, in which listeners and viewers choose what they want to listen to or watch and design their own 'now media' uses. In addition to providing simple descriptions of these new technologies, this chapter discusses their cultural implications and how they affect media content and distribution and personal lifestyles.

SEE IT THEN

Point-to-point or person-to-person communication has been around since people first began speaking to each other but, except through writing letters, there was no other way for people to contact each other directly. After the invention of the printing press, authors, through a medium – books, newspapers, magazines – communicated to the public; the one-to-mass communication model. Later, radio, television, and film produced content for the mass audience. But what was missing was a way for one person to connect with another person over distance, at least until the telegraph and telephone.

Point-to-Point or One-to-One Communication

Landline Telephone

The invention of the **telegraph** in 1844 ushered in the era of electronic communication and, more importantly, an almost instantaneous way to communicate person-to-person (point-to-point). Now, instead of writing a letter and sending it by Pony Express, train, or other ground transportation, a person could write a short message to another person and send it long distance over the telegraph wires. A telegram would arrive at its destination (a telegraph office) and be decoded from Morse code to text within a few seconds. The message (telegram) would then be carried to the addressee, sometimes arriving the same day it was sent. Senders were charged by the word, so most telegrams were between 10–15 words. Telegraphy was very popular – at its peak in the late 1920s about 200 million telegrams were sent.

The telegraph was the precursor to telephone voice transmission and faxing. Alexander Graham Bell invented the telephone in 1876. By 1915, about 30% of U.S. households had at least one telephone installed, by 1945 one-half of all homes had a phone, by 1960 that percentage rose to three-quarters, and by the 1980s, landline penetration was at its highest, in about 95% of all homes. **'Landline' phones** are connected by copper wire to outside telephone transmission lines strung from pole-to-pole to a telephone company.

Telephone wires were not only used for voice communication but also to send text, photographs, and line drawings to distant locations, a process known as **facsimile** or telefax. The term 'fax,' as it is called today, literally means a copy or reproduction. Although the technology for 'printed telegraph' transmissions has been in existence since the late 1800s, it was not until the 1920s when scientists at AT&T improved the technology and in the mid-1940s when Western Union built a compact machine that 'faxing' became a tool for business communication. By the 1970s and 1980s, desktop versions of fax machines made their way to the consumer market and began appearing in offices and homes.

At about the same time, the telephone answering machine started being marketed to consumers for home use. The answering machine made it possible for callers and receivers to leave and retrieve voice messages, freeing people from having to sit at home waiting for a call, and changing interpersonal relationships. Some callers preferred to leave a message rather than talk to someone, and receivers would screen calls to know who was calling without having to answer. Outgoing messages became informative and entertaining: "If you're calling about the red Toyota, sorry, it's already been sold"; "If your call is about good news or money, leave a message. If not, send me a letter."

Cell Phones

The first prototype of a cell phone was developed in 1947 as a way to communicate from cars and when a landline phone was not available. The first cell phones were very expensive, and the Federal Communications Commission (FCC) allowed only 23 simultaneous conversations in the same geographical area. And because of the FCC's apathy in allocating airwave space for wireless one-to-one communication, the introduction of wireless phones to the public was further delayed. Then in 1978, AT&T and Bell Labs introduced a **cellular system** and held general consumer test trials in Chicago. Within a few years, other companies were also testing cellular systems and pressuring the FCC to authorize commercial cellular services.

The first portable or **mobile** telephone units were the size and weight of a brick. They were called **transportables** or **luggables**. Cell phones were commercially available starting in 1984, and by the mid-1990s they were gaining in popularity. Initially, the cost of a cell phone was too high for most consumers, but as demand increased, costs went down, and by 1990 there were 5.3 million cell phone subscribers.

FYI: FIRST MOBILE TELEPHONE CALL

The first mobile telephone call was made on April 3, 1973. Martin Cooper, an employee of Motorola, placed the first call using a prototype of what would become the company's DynaTAC cell phone. Cooper developed the phone, which weighed 2.5 pounds, was 9 inches long, 5 inches deep, and 1.75 inches wide. The cost was $3,995. Although no one remembers what words were said or the length of the first call, it was a historic step in the birth of the cell phone industry.

Source: Seltzer, 2013

ZOOM IN 7.1

Watch an ad for the DynaTAC cell phone from the 1980s: www.youtube.com/watch?v=0WUF3yjgGf4

Figure 7.1 Dr. Martin Cooper with the first portable handset.
Source: © Ted Soqui/Corbis

Personal Digital Assistants

The electronic personal digital assistant (PDA) freed people from the burden of lugging around a three-ring binder type of daily organizer. Apple's Newton was the first organizing and messaging handheld PDA. Introduced in 1993, Newton was an immediate hit; sales soared to 50,000 units in the first ten weeks the product was on the market. But users quickly became disillusioned with Newton's poor handwriting recognition, complexity of use, size, and expense. Palm Pilot debuted in 1996 as a competitor to Newton. "Palm" was a lightweight, small, easy-to-use organizer with enough memory to store thousands of addresses and notes. Palm's simple interface caught on with the public, and the name Palm was almost synonymous with PDA, even though several other manufacturers began making their own version of a PDA. Hewlett Packard acquired Palm but eventually discontinued its production.

A PDA was basically an electronic calendar that stored contact information (addresses, phone numbers, email addresses) and sent reminders for appointments and special events, and homework assignments. It also contained simple spreadsheets for tracking expenses and other numerical information. All those functions fit onto one small, palm-sized electronic device. PDAs were generally self-contained, meaning that stored information could not be readily shared. PDAs eventually gave way to multifunctional and Internet-connected smartphones and tablets.

SEE IT NOW

Point-to-Point or One-to-One Communication

Mobile Phones

The term 'mobile telephone' has given way to the more familiar 'cellular' or 'cell phone.' 'Mobile' describes how it is used, and 'cellular' describes how it works. A cell phone is really a type of two-way radio. Basically, a city or county is divided into smaller areas called cells, usually a few miles in radius. Each cell contains a low-powered radio transmitting/receiving tower that covers the cell area. Collectively, the cell towers provide coverage to an entire area, be it a city or county or some other region. The size of each cell varies according to geographic terrain, number of cell phone users, demand, and other criteria.

When a call is made, it is picked up by a cell tower and transmitted over an assigned radio frequency. When the caller travels out of the cell area while still talking, a switching mechanism transfers the call from that cell area to another cell area and to the corresponding radio frequency. Most of the time,

ZOOM IN 7.2

How cell phones work: www.youtube.com/watch?v=kxLcwlMYmr0

callers do not notice the switch, but sometimes a call might be dropped because of spotty coverage.

By mid-2002, more than 135 million Americans were cell phone customers – 18 times the number of users as in 1992. To keep up with the demand, in July 2002 the Bush administration allotted space in the radio spectrum for the use of wireless communication services. Within six years the number of cell phone users almost doubled to 262 million. The U.S. Commerce Department's National Telecommunications and Information Administration worked closely with the FCC and the cellular industry and made even more spectrum space available at the end of 2010 to continue to meet consumers' wireless voice and data communication needs. Since then the FCC has increased the available bandwidth seven times.

Today, 96% of Americans (about 317 million) own a cell phone of some kind, with 81% being a smartphone, up from just 35% in 2011. Along with mobile phones, Americans own a range of other information devices. Nearly three-quarters of U.S. adults now own a desktop or laptop computer, while roughly half now own a tablet computer and about the same percentage an e-reader. Further, about one in five American adults are "smartphone-only" Internet users, which means they do not have a traditional home broadband service. Reliance on smartphones for online access is especially common among younger adults, non-Whites and lower-income Americans.

Global statistics show a slightly different picture as about one-half of the 6.8 billion mobile users access the Internet only through a smartphone/mobile device. In many other countries, the broadband network is not as extensive as in the U.S. so in many areas the only way to access the Internet is through a mobile device.

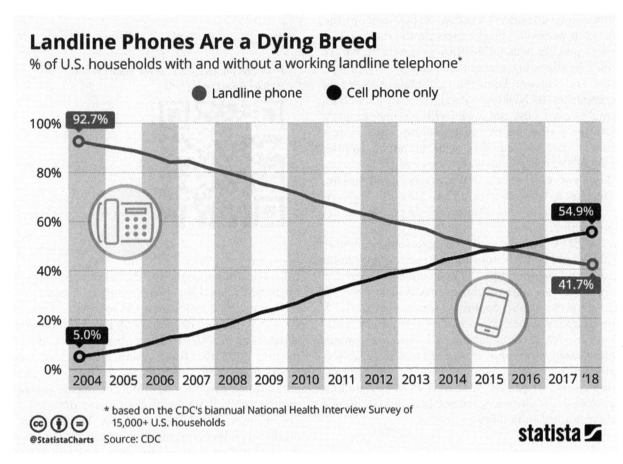

Figure 7.2 New landlines are a dying breed.
Source: Photo courtesy of The National Center for Health Statistics and Statista, 2019

Mobile Phones: Beneficial or Irksome?

In an episode of the old sitcom *Friends*, Ross was getting married (again), this time in England to Emily. As his friend Joey was about to escort Emily's mother down the aisle, her cell phone started to ring. She answered it and handed the phone to Joey. Much to Ross's annoyance, Joey not only took the call but held up the phone throughout the ceremony so a friend back in the States could listen to the vows.

Although this funny scene happened in a television sitcom, such use of mobile phones is common today. Stories about mobile phones ringing during speeches given by dignitaries and heads of state pervade the media. In fact, in response to these increasing interruptions, former President George W. Bush banned cell phones from his staff meetings. No event or location is sacred from mobile phones. They ring during funerals, in theaters, during concerts, and in classrooms, often despite regulations and urgings to turn off the ringer. And of course, talking or texting on a mobile phone while driving is particularly dangerous as well, even more risky than driving while intoxicated. Even using voice-activation systems, such as Siri, can be just as distracting as using a handheld cell phone.

Voice-to-voice calls, however, are not the only way cell phones are used to communicate. **Text messaging** has become a more popular means to get in touch. Users send or receive about 40 texts per day on average and check their phone hundreds to sometimes thousands of times per day. Six billion text messages are sent every day, and more than 2.2 trillion are sent each year in the United States. Globally, 8.6 trillion text messages are sent each year. Although there are many advantages to texting, critics claim that it is out of control. Some teenagers send and receive more than 3,000 text messages per month, which equates to about 100 texts per day. A 13-year-old from California racked up 14,528 texts in one month. Parents, teachers, and psychologists are concerned that adolescents often text late into

Figure 7.3 The mobile phone generation.
Source: Photo courtesy of iStockphoto. © elcor, image #3648770

FYI: DEMOGRAPHICS OF U.S. ADULTS WHO OWN SMARTPHONE/CELLPHONES

	Any cellphone	Smartphone	Cellphone, but not smartphone
Total	96%	81%	15%
Men	98%	84%	14%
Women	95%	79%	16%
Ages 18–29	99%	96%	4%
30–49	99%	92%	6%
50–64	95%	79%	17%
65+	91%	53%	39%
White	96%	82%	14%
Black	98%	80%	17%
Hispanic	96%	79%	17%
Less than high school graduate	92%	66%	25%
High school graduate	96%	72%	24%
Some college	96%	85%	11%
College graduate	98%	91%	7%
Less than $30,000	95%	71%	23%
$30,000–49,999	96%	78%	18%
$50,000–74,999	98%	90%	8%
$75,000+	100%	95%	5%
Urban	97%	83%	13%
Suburban	96%	83%	13%
Rural	95%	71%	24%

Source: Mobile Fact Sheet, Pew Research Center, 2019

the night, during school, and as an escape from homework and other activities.

In the 20 or so years since the cell phone made its way to the consumer market, it has had a strong social and cultural impact. In some respects, cell phone users have become tethered to their friends and family. Some people are so connected that they have to let someone know where they are every minute of the day. Moreover, families and friends come to expect constant and immediate contact and may become alarmed if there is no response. Despite the trivial uses of cell phones, the advantages can outweigh the disadvantages. Cell phones help users keep in touch about important matters and give comfort when physically isolated. They can make users feel popular, important, and independent yet needed. Moreover, they are convenient and easy to use and, in many cases, less costly than making long-distance calls over traditional landlines. They can also be credited with saving lives in emergencies.

Figure 7.4 Text message meeting.
Source: Photo courtesy of iStockphoto. © Michael DeLeon, image #930020

FYI: CELL PHONE ETIQUETTE

- *Respect.* When you are with others give them your complete and undivided attention. If a call is important, apologize and ask permission before accepting it.
- *Dining out.* No one wants to be ignored by a dining companion who is on the phone. Always silence and store your phone before being seated. Never put your cell phone on the table.
- *Use voicemail.* When you are in the company of others, let voicemail handle non-urgent calls.
- *Do not argue.* Nobody can hear the person on the other end. All they are aware of is a one-sided screaming match a few feet away.
- *Personal space.* When you must use your phone in public, try to keep at least 10 feet between you and others.
- *International protocol.* Cell phone etiquette varies from country to country.
- *Filter your language.* If you would not walk through a busy public place with a particular word or comment printed on your T-shirt, do not use it in cell phone conversations.
- *Quiet zone.* Do not talk in theaters, churches, libraries, classrooms, funerals, or other such quiet areas.
- *Lower your voice.* The average person talks three times louder on a cell phone than they do in a face-to-face conversation.
- *Be available.* Whether it is your turn in line for a service or purchase, do not make service people wait for you.
- *Hands-free calls while driving.* Refrain from holding a cell phone while driving.

Source: The do's and don'ts of cell phone etiquette, 2013

From Cell Phones and PDAs to Smartphones and Tablets

The cell phone, which used to be just a handy conversation device, morphed into smartphone with the combined functionality of a PDA, the calling ability of a cell phone, and the Internet capabilities of a laptop. Smartphones, in turn, have led the way to the modern-day tablet. Although smartphones and tablets offer many similar functionalities, they differ in several key ways. Cell calls can be made from a smartphone but not from a tablet, but video calls can be made from both. A tablet has more computing functionality than a smartphone and is considered a general-purpose mobile computer with a touch-screen panel and/or pop-up keyboard.

Contemporary tablets are somewhat different than earlier models. Early tablet computers required a stylus, whereas modern tablets use a finger-sensitive touch-screen, a pop-up keyboard, and an optional stylus. The iPad, the first touch-screen tablet, was released by Apple in 2010. Tablets and smartphones both rely on function-specific applications (apps). Mobile apps are to smartphones/mobile devices what software programs are to computers. There is an app for almost anything a user needs to know or needs to do. There are apps to read the weather forecast, to find a recipe or the closest Starbucks, to keep up with the stock market, to check flight times, and to engage in many other activities while on the go. The app craze began in 2008, and users were quick to download new apps as they came along. Today 90% of mobile time is spent on apps.

Since the iPad's launch, it has changed the market in terms of technology and consumer expectations, as Apple's iPad was the first tablet to achieve significant success among consumers. By 2018 Apple had sold about 425 million iPads worldwide. As of February 2019, 52% of American adults owned at least one tablet.

FYI: MOBILE APP STATISTICS

- Mobile apps are expected to generate $189 billion in revenue by 2020.
- The Apple App Store has 2.2 million apps available for download.
- Google Play Store has 2.8 million apps available for download.
- 21% of Millennials open an app 50+ times per day.
- 49% of people open an app 11+ times each day.
- 57% of all digital media usage comes from mobile apps.
- The average smartphone owner uses 30 apps each month.

Source: Blair, 2019

FYI: TOP 10 USES OF SMARTPHONES (2019) (PERCENT OF USERS)

1. Text – 88%
2. Email – 70%
3. Facebook – 62%
4. Camera – 61%
5. Reading news – 58%
6. Online shopping – 56%
7. Checking the weather – 54%
8. WhatsApp – 51%
9. Banking – 45%
10. Watching videos on YouTube – 42%

Source: C., Andy, 2017

ZOOM IN 7.3

How wireless technology works: www.wireless-technology-advisor.com/how-does-wireless-technology-work.html

How home networks work: www.cnet.com/how-to/home-networking-explained-part-1-heres-the-url-for-you/#!

FYI: CELL PHONE TURNS ON OVEN

Imagine if every time your cell phone rang, your oven turned on full blast. That is what happened to a New York man. At first, he was puzzled as to why his oven was automatically turning on. Then he noticed that whenever his cell phone rang while it was in the kitchen, the oven broiler would heat up and the clock would start blinking. It took quite a bit of testing to figure out that the cell phone was triggering the oven to turn on.

The explanation? Cell phone signals can create electromagnetic signals that often interfere with baby monitors, computer speakers, heart pacemakers, and other electronic devices. Apparently, the phone's signals caused the oven keypad to go haywire and turn on various functions.

Source: Dwyer, 2009

Phone Calls Over the Internet

Plain old telephone service (POTS) use is decreasing as more people use cell phones. Dealing yet another blow to standard landline telephone-to-telephone services is Internet telephony, which is a way to make phone calls over the Internet through systems known as **VoIP (voice over Internet protocol)** and **IP telephony**.

ZOOM IN 7.4

INTERNET PHONE CALLS

Click on this link to see how Internet phone calls work: www.explainthatstuff.com/how-voip-works.html

Internet telephony is a packet-switching system – similar to the Internet. Packet switching is much more efficient than the cellular phone system and the POTS circuit-switching system, in which circuits are open and dedicated to the call. With packet switching, the connection is kept open just long enough to send bits of data (a packet) back and forth between the caller and the receiver, so connection time is minimized and there is minimal load on the computer. In the transmission space taken up by one POTS call, about six Internet calls can be made. Basically, Internet telephony digitizes voice and compresses it so it will flow as a series of packets through the Internet's bandwidth.

Computer/Mobile Device to Telephone Free or low-cost phone calls from a computer or a mobile device to a telephone number – whether to a landline, cell phone, or smartphone. To make a free call using an Internet phone, software needs to be downloaded from an Internet phone provider Web site.

Computer/Mobile Device to Computer/Mobile Device This Internet method is probably the easiest way to connect, and most calls are free. All that is usually needed is an Internet connection, a computer with audio and video capabilities, and software, such as Skype or Facetime. Both the caller and the recipient must be using the same application or be on the same Web site. The calls can be either *voice only*, *voice-to-video*, or *video-to-video*.

Making a call over the Internet has become a very popular way to connect with friends and family, not only because the calls are usually free but also because of the video capabilities. Seeing someone while talking is better (in most cases) than just hearing a voice. Parents delight in seeing their children and grandkids, couples separated by distance feel closer, and employers save travel costs by interviewing candidates over an Internet call.

The use of Skype for international calls is growing at a faster rate than the use of a traditional telephone company (telco). In 2013, Skype captured an additional 36% of the international calling market and users spent a total of 214 billion minutes on international Skype-to-Skype calls. By March 2020, Skype was being used by 100 million people on a monthly basis and 40 million people were using it daily. Skype saw a 70% spike in spring 2020 due to the coronavirus pandemic. The jump in Skype calls is an important illustration of how technology brings family and friends closer during stressful times.

ZOOM IN 7.5

MOST POPULAR MOBILE PHONE BRANDS THROUGH TIME (1993–2019)

Check out this YouTube video that takes you through a moving timeline of the most popular mobile phones from 1993 to 2019: www.youtube.com/watch?v=IdDEVIfbGEA

A Wireless World of Entertainment

Wireless capability has already changed the way people access information. Data transfer speeds are increasing, and connecting is easy. Wi-Fi-enabled laptops, smartphones, and tablets connect to the Internet without the annoyance of cords. Mobile device users download the latest news while taking a walk, checking their email while ordering lunch, and conducting research for homework while working out in the campus gym.

In late 2014, Apple launched its Apple Pay system that lets users pay for many goods and services just by waving their cell phone in front of an electronic reader. Apple Pay digitizes and replaces the credit or debit magnetic stripe card transaction at credit card terminals. The system has seen steady increase in users and support from banking partners and retail stores. Currently, Apple Pay is available in 18 countries including all the countries in the European Economic Area.

Audio/Satellite Radio

Two companies, XM and Sirius, were formed in the early 2000s on the chance that over-the-radio listeners would pay to listen to commercial-free, or at least commercial-low, radio delivered by satellite. No matter where subscribers are located, they can listen to whatever type of music they like, and they do not have to worry about drifting out of a station's broadcast range; the signal stays with them. A jazz fan driving through rural Nebraska might have trouble picking up an over-the-air local station that plays that type of music, but SiriusXM's jazz channel will tag along for the ride across the country.

It took a few years for the concept to catch on, and both companies suffered financially, but deals made with automakers to provide in-car receivers and contracts with big-name stars like Howard Stern

Figure 7.5 Downloading music onto a portable player.
Source: Photo courtesy of iStockphoto. © abalcazar, image #7238674

boosted the companies' fortunes. In 2008, the companies merged, becoming SiriusXM. The service now has 35 million subscribers, to which it offers more than 200 commercial and commercial-free channels.

SiriusXM, however, faces a serious challenger – Internet radio. Company executives fear that drivers will not want to pay for the satellite service when they can just as conveniently tune to Spotify, Pandora, Beats, or other customizable online audio service. Even though 75% of all vehicles manufactured in the U.S. come with SiriusXM installed, a connected car, with the Internet integrated into the dashboard, could prove a formidable competitor.

Online audio also has clear advantages over traditional radio. Online listeners do not have to depend on local stations to hear their favorite music, and they can connect to hometown stations, and they have a broad range of selections. Online listeners select their favorite types of music and listen to it in the order they prefer.

Television

Program providers such as the broadcast networks, cable channels, and even some independent production houses realize that using the Internet for delivery of their products is technically viable and beneficial. Also, the broadcast networks are pushing audiences to their Web sites to encourage them to view episodes of their favorite shows. Many online television sites merely serve as promotional vehicles for their broadcast and cable counterparts. However, about 90% of television programs are now available in their entirety online, either on a network site or through streaming services, such as Netflix. Given the Internet's vast storage capabilities, it is easy to go online and watch any program at any time without being tied to a television schedule. The current television era is a complex media environment that is forcing changes to the traditional broadcast models of program delivery and how people watch 'now television.'

Over-the-Top Content/Streaming Since the changeover to digital broadcasting, viewers experience the advantages of television sets with built-in wireless Internet capabilities. Viewers are beginning to wonder if cable and satellite subscriptions are necessary if they can get their favorite programs and movies directly from the Internet but watch them on the television screen. Many young adult viewers have given up their cable or satellite subscriptions and now only use an Internet connection, their computers, and a flat-screen monitor to provide them with video entertainment. And flat-screen monitors come in various sizes and clarity. The newest '8K' display is super high definition, with 8,000-pixel horizontal resolution. In other words, the picture is almost as clear as being there in person.

With the proliferation of laptops, tablets, and smartphones that display video almost as sharply and clearly as a flat-screen digital television, comes an alternative way to watch television – on the go. Watching away from home means without cable or satellite delivery and is only possible through online streaming services, program apps, or network Web sites. Watching television by bypassing cable and satellite providers is called over-the-top (OTT) viewing. In the U.S. about 64 million U.S. households (66%) are lured by hundreds of OTT streaming services, up from 35% in 2013.

Streaming services have ushered in a 'now' way to watch television. HBO fired up HBO Now, a stand-alone, untethered-from-cable, video-streaming/OTT service in April 2015. Unlike HBO Go that requires users to have an existing subscription to the HBO television channel, HBO Now is aimed at cord cutters and cord nevers who have an Internet connection but eschew cable. HBO keeps on going with its new HBO Max, branded as an elite streaming service with access to the network's top hits.

The hopes for Apple TV are that it will become the "biggest gateway to online video – the new Comcast for the Internet" (Kang, 2015a, A14). Apple TV+ is a digital media player and OTT ad-free subscription video-on-demand service that hooks to a television via an HDMI cord to deliver digital programming from OTT providers and broadcast and cable networks. Although on the surface the service sounds like it is a good alternative to cable television, it could be that viewers end up trading cable-bundled channels for Apple TV-bundled channels but in the long run do not save any money. But Apple TV has also moved into the programming business with its first original program starring Dr. Dre in 2019, and since has produce original video-streaming shows such as *Home*, *Here We Are*, and *Defending Jacob*.

Online content viewers have their eyes on Google as it works on a follow up to Chromecast using Android TV, which will include a remote control device. Android TV, a replacement for Google TV, comes as either a set-top box or as an application on some smart television sets or devices. The switch to Android TV will make it easier to integrate other Google services such as Google voice assist.

More, however, is not always better. Too many choices keep viewers from finding new programs and keeps them watching the old. It just takes too long to shuffle through all of the options. If someone has set aside 30 minutes to watch a program, it can take almost that long to sign-in to a streaming service and look through those offerings. About half of television viewers are becoming increasingly frustrated with the number of services they need to subscribe to so they can watch the shows they like.

Despite the clumsy interface and the need to subscribe to multiple services, streaming has become so popular that, for example, Netflix itself brags of 183 million subscribers worldwide. Thanks to coronavirus shelter-at-home orders, the number of new global subscribers to Netflix way outpaced expectations by double – the service added a record 15.8 million subscribers in the first three months of 2020. With the continued outbreak, Netflix expects another 7.5 million new subscriptions in the second quarter but cautions that service could take a hit when restrictions are lifted and people get back to their social and working lives and take a break from television.

As entertainment television moves online, so does news and information programming. Media sites stream live newscasts so people away from their televisions can see the latest happenings. For example, on the day Michael Jackson was buried, the Small Business Administration's Internet

Figure 7.6 Television has come a long way.
Source: Photo courtesy of iStockphoto. © fredrocko, image #6954585

connections slowed dramatically because so many employees were watching the live newscast of the funeral online. Workers without access to a television set got to see the day's events online exactly as shown on the television networks.

Place Shifting/OTT Sling TV is an OTT provider that also brings a new way of watching television called *placeshifting*. Sling TV is a streaming device that lets users remotely watch cable and satellite channels on an Internet-connected computer or mobile device. With Sling TV, viewers never miss their favorite shows while away from home and without a television. Sling TV works through an app that can be installed on multiple devices and works with streaming media players, smart televisions, and Android and Apple devices. A major downside to Sling TV is the absence of the broadcast networks; users need a separate over-the-air antenna to pick up these networks.

But the upside is that a viewer who is vacationing in Tahiti but wants to watch a college bowl game can do so on his or her laptop through Sling TV. Sling TV's deal with ESPN to stream its games live might just be what it takes to get diehard cable subscribers to drop bundled cable channels and 'appointment television' in favor of personalized, anywhere, anytime viewing.

Time Shifting/DVR Personalized television viewing is more than just having a wide selection of programs that fulfill individual needs, but also includes choosing when to watch. Online 24/7 streaming provides individual scheduling, as do digital video recorders (DVRs). The DVR has long replaced the old-fashioned videotape using VCR, but the purpose is the same – to record programs for later viewing. A DVR digitally records and stores hundreds of hours of programming that viewers can watch whenever they want, and they can fast-forward through commercials. Some DVRs record up to four shows at once, even while channel-surfing back and forth. If a viewer is hungry but engrossed in a live program, he or she can pause the show, make a snack, and then come back and pick up the program where it left off. With a DVR, viewers have ultimate control over their viewing experience. They can fast-forward, reverse, pause, and use instant replay or slow motion and never miss a scene.

DVR viewing is popular. About 48% of U.S. households have at least one DVR, and the television is on 15% longer than in homes without DVRs. As viewers spend more time streaming shows, there is less need to record a show, but newer **cloud DVRs** that function like traditional DVRs but use recorded content stored online, act almost like a personal streaming service.

Binge Watching Streaming services and DVRs facilitate a new way to watch television – binge watching. Though there are no hard and fast criteria

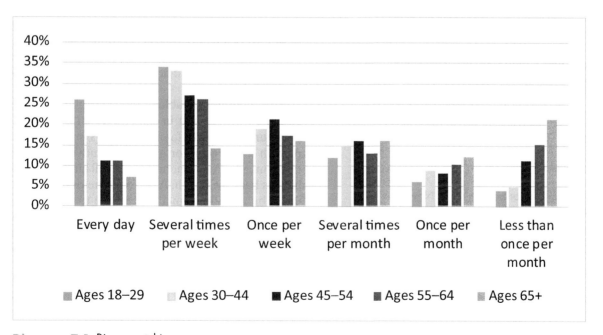

Diagram 7.1 Binge watching.
Source: Sabin, 2018

about what is considered binge watching, most experts agree that it is when a viewer watches three or more episodes of the same program for longer than two hours in one sitting. Slightly more than seven in ten Netflix viewers binge watch a title at least once a month. Binge watching can make television more like watching a movie. Binge watching is like reading a long novel instead of a book of short stories.

FYI: TOP BINGE-WATCHED SHOWS (2019)

Program	Original network (non-Netflix shows)
1 Friends	NBC
2 Grey's Anatomy	ABC
3 Brooklyn Nine-Nine	NBC
4 Game of Thrones	HBO
5 The Big Bang Theory	CBS

Program	Netflix-originated
1 Stranger Things	Netflix
2 Lucifer	Netflix
3 Orange is the New Black	Netflix
4 13 Reasons Why	Netflix
5 Marvel's the Punisher	Netflix

Sources: Epstein, 2019; Torres, 2019

FYI: AEREO

Aereo was a short-lived way to time-shift television programs. Aereo started in March of 2012 and worked by leasing a remote antenna to its subscribers. The antenna picked up broadcast television for live or recorded viewing. The antenna was what made Aereo different from Internet streaming services. In other words, it was an over-the-air service, not an Internet-based one.

Aereo filed for bankruptcy in November of 2014 after several broadcast networks sued for copyright infringement. The broadcasters asserted that Aereo needed to pay royalty fees just like cable companies. Aereo's defense was that it was obtaining the signals legally and simply making them available to the public. However, the court ruled in favor of the networks. Aereo was available in New York, Atlanta, and Boston and had about 80,000 subscribers before it was forced to shut down.

Other Personal Digital Devices

E-books

The point of having an e-reader is that hundreds of books can be downloaded onto one device, which is similar in size to a typical paperback novel. On newer e-readers, users can underline, highlight text, and write notes in the margins, and some apps work with e-readers to display photo galleries and 3D models. Amazon and the e-book maker Kindle have formed a quid pro quo partnership. Amazon provides the digitized books that are downloaded onto a Kindle. This union is a good example of the changing media environment, in which new models of monetization are being developed as new devices become available.

Even though about 52% of U.S. adults own an e-reader, most people still prefer a printed book to a digital one. In 2019, consumers in the U.S. spent nearly $26 billion on printed books and only $2 billion on e-books. The popularity of e-readers is in decline. E-books are still well-liked but a device that only serves one basic function is not suitable for today's consumer, and a printed book gives readers a break from a screen.

Webcams/Action Cameras

A Webcam is a simple, low-cost video camera that transmits live images through the Internet. Video from older Webcams was slow and jerky because cameras could only capture an image once every five or ten seconds because of bandwidth limitations. But newer Webcams operate at 30 frames per second, the same speed as a television camera. Online video looks just as sharp and flows just as easily as what is on television.

Most computers, tablets, and smartphones come equipped with a built-in camera, which is great for communicating with family members in the service or with a boyfriend or girlfriend studying abroad. Users mug for the camera and send all kinds of goofy images back and forth. A Webcam helps ease loneliness by visually connecting people to each other. The Internet abounds with live Webcams pointed at funny pets, a live colony of ants, street musicians, and just about anything or anybody imaginable. One guy even has a Webcam inside his beer refrigerator.

Webcams also have more serious uses. For example, they are used for Internet conferencing, as public relations tools (such as a city or university campus streaming live Webcam images for potential visitors), and for online courses. During the COVID-19 pandemic, Webcams connected family and friends who were socially isolated and sheltering-in-place in geographically distant locations. Some radio station Web sites entertain users by focusing cameras on their disc jockeys at work, and television news anchors and reporters can use built-in cameras to tape shows without going into the studio, as was common during the pandemic.

ZOOM IN 7.6

Check out these Webcam links:

- WebCam World: www.webcamworld.com
- EarthCam: www.earthcam.com/usa/louisiana/neworleans/bourbonstreet
- Times Square: www.earthcam.com/cams/newyork/timessquare/?cam=tsrobo1
- Animal Planet Cam: www.apl.tv/

While Webcams give others a peek into our life, action cameras take our friends along for the ride. Action cameras, such as GoPro, mount on helmets and car dashboards, strap to your forehead or chest, clip to your snorkeling mask, attach to your dog's harness or a selfie stick, and otherwise go with you anywhere, even under water. An action camera lets your friends experience what you are experiencing in real time, and recording features let you relive the wild roller coaster, white water rafting, parachuting, bungee jumping, surfing, and extreme snowboarding experiences over and over again.

Action cameras have captivated the marketplace. Though heavy clunky 'action' cameras were first used by parachute jumpers and race car drivers going back to the 1960s, the modern-day action camera was introduced to the consumer market in 2004. GoPro founder and CEO, Nicholas D. Woodman, personally hawked the first model on QVC. Brands such as GoPro caught on after they went digital starting in about 2006. By 2019, the action camera market brought in about $9 billion in revenue, selling about 20 million units.

Voice-Activated Digital Personal Assistant

The latest rage in personal gadgetry is a smart speaker voice-activated personal assistant such as Amazon Echo, Google Home, and Apple HomePod. Digital voice assistants, such as Siri (accessed through an iPhone) and Alexa (connected to through Echo), are cloud-based applications of artificial intelligence that rely on natural language and voice generation, and machine learning, to assist us in our daily lives. In addition to answering simple questions, such as sports scores, they learn to anticipate what we need by monitoring our daily habits. For example, if someone typically turns on television news in the morning but does not do so on one particular morning, his or her voice-activated assistant might voice a reminder.

When connected to a home network, digital assistants play your favorite tunes just by asking, for example, "Alexa, play 'Truth Hurts'." They can also set your thermostat, adjust lighting, phone a friend hands-free, and lock your front door, among other tasks, all just by voice command.

The Internet of Things and Wearables

The **Internet of Things (IoT)** describes **machine-to-machine (M2M) communication** that is possible because of cloud computing and networks of data-gathering sensors, which can be placed in any 'thing' – device, person, or animal. A 'thing,' in the IoT, can be a person with a heart monitor implant, a farm animal with a biochip transponder, an automobile that has built-in sensors to alert the driver when tire pressure is low, or any other natural or man-made object that can be assigned an IP address and is provided with the ability to transfer data over a network. Products built with M2M communication capabilities are often referred to as being smart.

IoT does not only refer to personal uses but commercial and industrial uses as well. For example, smart cement is equipped with sensors to monitor stresses, cracks, and warps, and will alert engineers

Figure 7.7 Action camera circa 1960s.
Source: Jim Gray/Keystone/Getty Images

to fix problems before a catastrophe arises. Smart cement also alerts drivers to road problems; the sensors in the concrete detect trouble spots and communicate the information to cars via wireless Internet. Once a car knows there is a hazard ahead, it will instruct the driver to slow down, and if the driver does not, then the car will slow down automatically.

IoT devices are all the rage as consumers and businesses discover not only the 'cool' factor but the convenient functionality. The number of IoT devices grew by more than 20% from 2013 to 2014, and as of 2020, there were about 40.9 million such devices in the U.S. alone. Worldwide, in 2019, the number of devices reached 26.66 billion. It is predicted that by 2025 there will be more than 75 billion devices connected to the Web.

FYI: FIVE TYPES OF IOT APPLICATIONS

1. Consumer IoT – light fixtures, home appliances, and voice assistance for the elderly.
2. Commercial IoT – applications of IoT in the healthcare and transport industries, such as smart pacemakers, monitoring systems, and vehicle-to-vehicle communication (V2V).
3. Industrial Internet of Things (IIoT) – includes digital control systems, statistical evaluation, smart agriculture, and industrial big data.
4. Infrastructure IoT – enables the connectivity of smart cities through the use of infrastructure sensors, management systems, and user-friendly user apps.
5. Military Things (IoMT) – application of IoT technologies in the military field, such as robots for surveillance and human-wearable biometrics for combat.

Source: Mayaan, 2020

In the last several years, '**wearables**,' any 'Internet of Things' devices that can be worn, embedded in clothing, implanted in the user's body or tattooed on the skin, instead of being carried, have grown in popularity. Smartwatches, smart clothing, and smart glasses are types of wearables. Wearables are carry-free microprocessors that can send and retrieve Internet data.

In early 2003, Microsoft introduced smart personal object technology (SPOT) software, which is driven by tiny but powerful microchips that were the basis for the first wearables in the form of a 'smart watch.' Although many brands now make smartwatches, the Apple Watch is probably the best-known. Worn on the wrist like an ordinary watch, the Apple Watch face displays the time amid app icons that pop open at a touch. The watch is paired to the wearer's iPhone and uses both Wi-Fi and Bluetooth to transfer data back and forth. Unless a wearer has super-sharp eyesight, a smart watch might be a bit awkward to use, but otherwise it is a convenient, multi-use gadget that is sure to turn heads.

Even before its release to the consumer market in spring 2015, there were already more than 4,000 Apple Watch apps to do everything from receiving news, weather, traffic reports, and recording sleep activity to sending and receiving texts, getting phone calls, listening to music, setting calendar dates and reminders, and using Siri. Apple Watch and others also monitor fitness. Like a Fitbit Flex and other such devices, smartwatches track physical activity such as number of steps taken, calories burned, and distance traveled.

Another wearable that debuted with much fanfare in 2013, was Google Glass, the high-tech eyewear. Worn like a pair of glasses, Google Glass was promoted as the wearable Internet. Put them on and have online connectivity anywhere. Google Glass functions with voice commands and by touching the glasses in different spots. Sales did not live up to expectations, and so in January of 2015, the consumer version was taken off the market. But what is old and forgotten often reappears, as has a new business-focused version of Google Glass called Google Glass Enterprise Edition 2. A long name for a pair of glasses that looks like a cross between a pair of standard black glasses and science class goggles. The updated glasses contain a faster processor and new artificial intelligence software, and promise faster charging and a longer battery life than its precursor.

Wearable smart jewelry is making a fashion statement among early adopters. Brands, such as Oura, have come out with rings that look like plain bands but are actually teeny microprocessors that track the wearer's health data, such as number of steps taken, calories burned, and heart rate. Wearable digital jewelry also comes as bracelets and bangles, and even earrings.

Smart clothing, such as microprocessor implanted t-shirts, running clothes, socks, and yoga pants, can provide more medical and lifestyle information than wearable jewelry. Swimsuits alert a wearer to put on more sunscreen. Though smart clothing and jewelry are helpful for the wearer, privacy advocates worry about wearables recording conversations or monitoring others at a meeting or gathering.

Some people are going for **implantables** – **Radio Frequency Identification (RFID)** microchips implanted under the skin. Implantables that are injected into the back of the hand can be used as substitutes for keys to the office or house, as credit cards, and even public transit cards. As with wearables, implanted chips monitor medical and health data. RFID chips are not new – they have been used for decades to track livestock and checked flight luggage, and to tag pets – but the under-the-skin human applications are new.

Critics wonder if implantables are really necessary or just a convenience. Risks could include possible death if the metal chip gets too close to an MRI machine, or false data if the chip is hacked into, or even criminals hacking off someone's hand to gain entry to a building or to get credit card information. But the biggest concern is with privacy and the future possibility of forced implantation into certain types or classes of people. No device is completely secure, and security breaches will become more frequent and disruptive as more devices are added to IoT. Common smart items such as televisions, Webcams, thermostats, sprinkler systems, home alarms, garage door openers, baby monitors, security cameras, and surgical robots have already been hacked. Security and privacy remain major concerns as IoT reaches all aspects of daily life.

FYI: IOT SECURITY: HOW TO PROTECT YOURSELF

- Install reputable Internet security software on all devices.
- Use strong and unique passwords.
- Read the privacy policy of the apps to see how they plan on using your information.
- Devices become smart because they collect personal data. Learn about what types of data are being collected, how it is stored and protected, if it is shared with third parties, and the policies or protections regarding data breaches.
- Know what data the device or app wants to access on your phone. Deny permission if you do not like the policies.
- Use a VPN to secure the data transmitted on your home or public Wi-Fi.
- Check the device manufacturer's Web site regularly for firmware updates.
- Use caution when using social sharing features, which can expose information, such as location. Cybercriminals can use this to track your movements.
- Never leave your smartphone unattended. Consider turning off Wi-Fi or Bluetooth if you do not need them.

Source: '9 Ways you can protect yourself' (Norton Security, 2020)

SEE IT LATER

More than just being fancy gadgets, smartphones, tablets, and watches serve the purpose of keeping us connected to information sources and to other people. As their functional utility increases, the personal gratifications keep people hooked. These devices become more than just tools for telling time or keeping a schedule. Rather, they keep us involved in the world. Our own personal sphere widens as we let in more and more information, which has led some people to wonder just how much we can absorb before reaching an emotional and psychological limit.

Corporate Changes

Worldwide telephone, cable television, wireless communication, and computer data networks are becoming less stand-alone systems; that is, they are converging into a powerful unified network based on the Internet protocol packet-switching system, which is versatile and can transmit any kind of information quickly and at low cost.

Rather than having people make calls to verify transactions, place orders, or move money from businesses to banks and so on, computers are doing it over high-speed broadband wires. And as the amount of information transferred continues to grow, the need for more bandwidth will become more critical as well. In early 2015, the FCC auctioned the rights to use parts of the electromagnetic spectrum to accommodate the growing use of digital media. Companies such as AT&T, Verizon, and T-Mobile bid about $44 billion for these frequencies that are

Figure 7.8 Studying on the campus green with a laptop.
Source: Photo courtesy of iStockphoto. © quavando, image #4178977

needed to further develop wireless networks, cell phones, satellite television, and other 'now' media. In 2018, the FCC started auctioning 5G services. By March of 2020, the auctions garnered over $4 trillion in sales of the upper 37 GHz, 39 GHz, and 47 GHz spectrum for 5G use. The top bidders were AT&T, T-Mobile, and Verizon.

As is the case for other delivery systems, more telecommunications consolidations are on the horizon. Mergers, such as the ones between Cingular and AT&T in 2004, and T-Mobile and Sprint (New T-Mobile) in 2020, formed huge communication companies, leading this part of the industry to an oligopoly similar to those in the cable and satellite industries. Stepping up to 5G transmission could mean more mergers, especially between wireless and cable companies. However, regulators are showing some concerns. AT&T's 2013 attempt at acquiring T-Mobile for $1.2 billion was rejected by U.S. regulators.

Lifestyle Changes

As convenient as mobile devices may be, there is the issue of feeling overwhelmed by an excess of information. Critics question the need for having such immediate and often trivial information at our fingertips and are also concerned about our overreliance on these devices. It is especially annoying to be with someone who keeps looking at his or her smartphone or, worse, using it instead of conversing. Nothing like having a friend whip out a smartphone just for the pleasure of proving you wrong about some subject. And nothing is private. Every time a smartphone is used to download an application, view a Web site, or engage in other activities, marketers are tracking movements and tailoring ads to match the user's interests and demographic profile.

Online users are bombarded with commercial messages, pelted with sound bites, and bored to death with the tedious details of friends' and family members' lives. Every beep of the phone or tablet means yet more information coming in. At some point, people begin to question how much information they really need and to wonder when keeping in touch crosses the line to being under surveillance.

Someday almost everyone in the world will own a smartphone. As it is, few people leave their homes without their smartphone in tow and some even sleep with it under their pillow. The smartphone has become our communication hub and security blanket. Smartphones were on the scene of the Boston Marathon bombing in 2013, and Internet-posted videos of the aftermath helped investigators piece together the sequence of events leading up to and after the bombs exploded. Smartphones recorded the peaceful marches and riots in Ferguson, Missouri, in 2014 after the police shooting of an African American man. In April of 2015, *The New York Times* was provided video taken with a smartphone of a police officer in South Carolina shooting an apparently unarmed man after pulling him over for a minor traffic violation. New apps are being developed to specifically monitor police activity. For example, Cop Watch is an iPhone app that automatically begins recording when the icon is tapped and automatically uploads the video to YouTube when the recording is stopped. Another app, Citizen Protect the World, gives real-time safety alerts and video of nearby incidents. Smartphones and such apps are already changing law enforcement procedures.

The new capabilities of smartphones may bring some legal and ethical problems as well. The managers of athletic clubs and other facilities with locker rooms worry about lawsuits stemming from secret smartphone photo snapping. There is concern that laws may be passed that require smartphones to include 'kill' switches that make it possible to disable the phone remotely. The kill switch could help stop cell phone theft, since the phone could be shut down if stolen, but phones could also be shut down by the government or corporate entities to stop citizen journalists from recording important events.

VoIP calling could one day render voice-only calls obsolete. The number of mobile VoIP users is projected to be three billion by 2021. The market is expected to grow 12% between 2019 and 2025, and is projected to be worth $55 billion by 2025. VoIP keeps morphing with **VoWi-Fi**, or **VoIP Wi-Fi**, which is a wireless version of VoIP. VoIP innovations provide new ways of one-to-one communication. As technological changes lead to social changes, social changes lead to behavioral changes, such as dropping a landline phone for a cell phone or eschewing telephone service providers for online services.

High-speed (broadband) Internet access is necessary to take advantage of much of the technology that affects our lives every day. It impacts regional commerce, education, health and public safety, culture, government, and provides us with many conveniences and efficiencies. Broadband brings the opportunity for direct access to education and health care for rural residents who are otherwise forced to travel long distances for

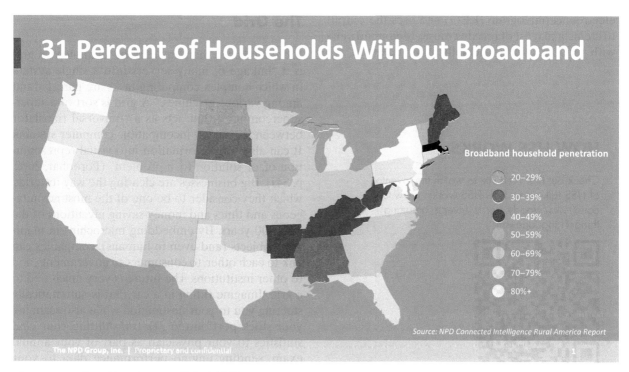

Figure 7.9 Broadband access (July 2019).

college courses and medical treatment. Rural libraries newly enhanced by high-speed Internet often experience a resurgence of community interest and participation. However, access in rural communities lags behind urban and suburban areas. Broadband penetration is lowest in North Dakota, where just 20–29% of households have access, followed by South Dakota, Mississippi, and Alabama, where it reaches only between 30–39% of homes. Access is a critical issue that needs to be addressed so that lifestyle changes will be available for all and to avoid a two-tiered system of have and have nots.

> About 30% to 40% of U.S. households lack access to broadband.
>
> Source: Moscaritolo, 2019

Television and audio remain an important part of life. Although DVR growth seems to have stalled out, large-screen televisions, HDTV, 8K screens, and Internet audio are still exciting. But it always comes down to the same question: "What's next?"

Perhaps it will be the *hypersonic sound system (HSS)*, which takes an audio signal from any source (television, stereo, CD, or computer), converts it to ultrasonic frequency, and directs it to any target up to 100 yards away. For example, one roommate could watch television while the other blasted the stereo, and neither would hear what the other is hearing. Likewise, a car full of passengers could each listen to his or her own music without hearing what others have on, or a nightclub could have several dance floors, each playing a different type of music and none interfering with the others.

The next big device could be Melomind, a headband that measures brainwaves in the same way as an

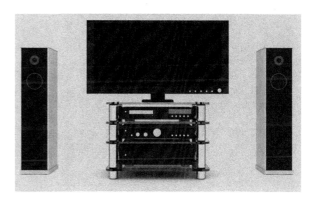

Figure 7.10 Older home entertainment systems are giving way to newer systems with fewer components. Source: Photo courtesy of iStockphoto. © 3alexd, image #5267037

electroencephalogram (EEG) and uses the output to create a playlist of relaxing music. Melomind pairs with a mobile app that actually plays the tunes.

ZOOM IN 7.7

HOW HSS WORKS

Check out this Web site to watch the inventor of HSS talk about how HSS works: www.ted.com/talks/woody_norris_invents_amazing_things?language=en

Figure 7.11 Early teen technophile.
Source: Photo courtesy of iStockphoto. © stray_cat, image #11186952

The Grid

More advanced than any current technology, a grid is a "linkage of many servers into a single system in which complex computing tasks are parceled out among various machines." A grid is sort of a super-supercomputer that acts as a "universal translator between previously incompatible computer systems. It can also turn information into visual representation of, or solution to, a problem" (Foroohar, 2002, p. 34J). Big businesses are clearing the way for grids, which they consider to be one of the most advantageous and time- and money-saving inventions of the past 200 years. By embedding microchips in inanimate objects (and even in humans), companies can link to each other, to consumers, to governments, and to other institutions. The future is very much tied to a grid. Imagine riding in a car that is automatically steering you to your destination while also scanning your stock portfolio in 3D, transmitting your vital statistics to your physician as part of your annual exam, sending engine performance data to your mechanic, and keeping an eye on your favorite television program. Also, you are wearing an outfit that was coordinated by microchips embedded in your body.

The largest grid computing network, accessing 170 computer centers in 42 countries, is used by the physicists at CERN (Conseil Européen pour la Recherche Nucléaire). It allows scientists to analyze more than 26 million gigabytes of data produced by the Large Hadron Collider (LHC) each year. Physicists access CERN's grid using mobile devices such as iPhones and iPads to submit work, monitor results, and control processes.

ZOOM IN 7.8

To visit the LHC grid, go to: http://wlcg.web.cern.ch/

The World Community Grid encourages "anyone with a computer, smartphone, or tablet to donate their unused computing power to advance cutting-edge scientific research on topics related to health, poverty, and sustainability" (World Community Grid, 2020 [www.worldcommunitygrid.org/about_us/viewAboutUs.do]). Through the contributions of more than 778,000 individuals and hundreds of organizations, the World Community Grid has supported many research projects, including searches for more effective treatments for cancer and tropical diseases, and in 2020 it launched a project with

Scripps Research to find a cure for COVID-19. Other projects include developing low-cost water filtration systems and ways to capture solar energy efficiently.

ZOOM IN 7.9

Visit the World Community Grid at: www.worldcommunitygrid.org/discover.action# undefined

There are currently several other grid computing systems in operation around the world that are being used by groups of scientists to help them with research, including the search for extraterrestrial life and identifying potential new drugs. Developers hope that as more computers join the systems and they link together, a worldwide grid will be created, giving access to resources around the world at the touch of a button.

A grid has the potential of creating a real-life version of *The Minority Report*, in which the hero, played by Tom Cruise, is flooded with personalized ads and followed at every move by retina-tracking devices. He reaches his breaking point and discovers that the only way to escape the grid is to have a black-market eyeball transplant. Privacy advocates fear such intrusion. Others claim, however, that consumers will always have the ability to control the information collected about and transmitted to them. It is predicted that the next phase of IoT is moving into mainstream use for retail, manufacturing, health care, and other industries. It will change the way consumers get real-time information, engage with each other, and interact with artificial intelligence and machine learning.

Figure 7.12 Actor Tom Cruise tries to escape the grid in the movie *The Minority Report*.
Source: Photo courtesy of Twentieth Century Fox/Photofest. © Twentieth Century Fox

Artificial Intelligence

Even more formidable than the grid is artificial intelligence (AI). Science fiction writers and scientists have long imagined a world in which computers transcend biology. AI is already being used to automate and replace some human functions. There are computers that can learn, answer questions, and solve problems. But the hopes are that computers will one day be self-aware and be of superhuman intelligence, and the thoughts and information in our brains can be downloaded or transferred to a computing environment.

AI is already taking hold in much of what we do and know, but we have barely scratched the surface of its capabilities. AI is going far beyond personal assisting by augmenting journalism and media content. AI could be used for creating more accurate content recommendations for particular users, automating stories, helping journalists sift through mounds of information, combat misinformation and disinformation, verify information and sources, and make the electronic news room more efficient and effective. In today's world of budget cuts and small staffs, if AI can help journalists and media organizations do more with less, perhaps the future of newspapers and television journalism will not be as dire as predicted.

AI is already helping to detect and combat deep fake videos by developing identifying technology that can quickly determine if a video has indeed been doctored. Moreover, once a fake video has been spotted, AI can label the video as a fake and alert the public that the video has been edited deceptively to change the context. White House press secretary Sarah Sanders shared on the Internet a doctored video showing CNN reporter Jim Acosta pushing away a White House intern, when in reality his arm did not touch her. House speaker Nancy Pelosi fell victim to a deep fake that altered her speech to make it seem like she was drunkenly slurring her words. Much work is still needed to combat deep fake video. Unfortunately, even AI software that can spot a deep fake and warning labels cannot change the minds of people who want to believe a fake is real because it harms someone or some situation they do not like.

But AI has its downsides. It is being blamed for replacing humans with robots in the workplace. For example, iHeartMedia, which owns iHeartRadio, one of the largest radio networks in the country, is laying off hundreds of station employees, such as DJs, program producers and directors, in favor of using AI to do their tasks. AI steps up to erase transition gaps between songs, layer in voice-over and sound effects, edit audio clips, and use algorithms to set a music mix that is most appealing to a station's listeners. Although AI might save iHeartMedia money in terms of salaries and benefits, it strips a local station of its 'localness,' and severs the relationship between local viewers and on-air personalities.

Further, AI is used with facial recognition technologies to capture digitized images of us. Although purportedly used for security and crime detection, facial recognition is being used to monitor where we go, what we do, where we shop, and what we buy. Our product consumption habits could then be fed to marketers to use for sending us advertisements, and to media companies to filter content that they think we prefer based on where we go – stores, sporting venues, airports, hotels, and so on.

And it is suspected that in-home digital assistants are always on and listening. Have a personal in-home conversation with a roommate about Nike shoes, and lo and behold, Nike ads will pop up the next time you connect to Amazon. Even background noise, like from a television for example, can trigger an unauthorized purchase. Although the manufacturers deny that in-home spying takes place, there have been reports of a digital assistant recording and sending private conversations to phone contacts.

ZOOM IN 7.10

DO YOU TRUST THIS COMPUTER?

Watch this video to understand the many issues we face with artificial intelligence: www.youtube.com/watch?v=aV_lZye14vs

While AI is being developed, one elderly Microsoft researcher is moving data from his brain onto a computer. Video equipment, cameras, and audio recorders are his constant companions. They capture his every move, his conversations, and his

experiences. Plus, he takes pictures of all of his receipts, event tickets, and other records. He is digitizing his life as an e-memory. It could be that someday everyone's life history will be online and searchable. Some people find this possibility very frightening, but others do not seem to mind at all.

Robots will also play an increasing role in our future, especially for medical care. Robots can move patients around a hospital so human doctors and nurses are not exposed to infectious disease, and they aid with surgical procedures. Predictions are that robots could be used as companions, especially for the elderly. Robots are ideal for combat – they are strong and have no feeling for their deadly actions. Like most technology, there will be benefits and there will be problems.

ZOOM IN 7.11

ROBOTS IN ACTION

See www.youtube.com/user/BostonDynamics to view robots in action.

ZOOM IN 7.12

AI NEWSCASTERS

Visit this Web site to see the first female AI newscaster developed by Xinhua News Agency in China: www.youtube.com/watch?v=5iZuffHPDAw

In the end, it may all come down to what the famous journalist Edward R. Murrow said in his last public speech in 1964:

The speed of communications is wondrous to behold. It is also true that speed can multiply the distribution of information that we know to be untrue. The most sophisticated satellite has no conscience. The newest computer can merely compound, at speed, the oldest problem in the relations between human beings, and in the end the communicator will be confronted with the old problem, of what to say and how to say it.

Kendrick, 1969, p. 5

SUMMARY

This chapter covered how 'now' communication tools are creating a new world of communication and changing life as we know it. The media are becoming more personalized as users determine what types of news stories they want to receive, what types of information are relevant to their lives, and when and where they want to be exposed to information. Commercial messages are tailored to personal characteristics and lifestyles.

Because of smartphones and tablets, communication among friends and strangers takes place 24/7, and users are unchained from desktop computers, landline phones, paper and pens, and even geographic locations. But along with instant communication come cultural changes. Interpersonal relations, the way information is gathered and disseminated, how situations are analyzed, problems solved, and decisions made are all influenced by technology.

Streaming, video on demand, and services such as Netflix, Hulu, and Sling TV have changed the way the audience watches television. Young viewers tend to stream television programs at times that fit their busy schedules, and they tend to binge watch multiple episodes in one sitting.

The rise of artificial intelligence and facial recognition and their use with media have the potential to isolate us further into 'filter bubbles' in which we are exposed only to news and information that we find agreeable, thus cutting us off from the other side. Mosaic developer, Marc Andreessen once said that in an Internet-connected world, there will be two types of people: Those who tell computers what to do, and those who are told by computers what to do (Roose, 2019, p. 32). As AI becomes more prominent in our everyday lives, controlling what we do, what we learn, and how we think, it could be that Andreessen's prognostication will come true.

Some of the biggest changes will happen because of the convergence of devices, accessed through wireless networks and grids. Concerns about privacy and hacking are only two of the issues that should be resolved but will probably be in constant flux.

In the modern world of personal communication devices, anything is possible. Whether through smartphones, 5G, streaming services, DVRs, e-readers, Webcams, IoT, or AI, our world is changing for the better – and for the worse. 'Now' technology, in and of itself, is neither good nor bad, it just depends on how we use it.

FYI: PARADOXES OF TECHNOLOGY

A famous French philosopher, Jacques Ellul, came up with four paradoxes of technology. Consider how your use of technology has brought both good and bad effects to your life.

- All technical progress exacts a price; it adds something and subtracts something.
- All technical progress raises more problems than it solves and tempts us to see the problems as technical in nature and to seek technical solutions to them.
- The negative effects of technological innovations are inseparable from the positive; they are not neutral.
- All technological innovations have unforeseeable effects.

BIBLIOGRAPHY

6 forms of wearable technology you must know right now. (2015, July 22). *42 Gears*. Retrieved from: www.42gears.com/blog/6-wearable-technologies-you-must-know-right-now/

9 ways you can protect yourself. (2020). *Norton Security*. Retrieved from: https://us.norton.com/internetsecurity-iot-securing-the-internet-of-things.html

15 most unique GoPro mounts for capturing your adventures. (2020). *ClickLikeThis*. Retrieved from: https://clicklikethis.com/unique-gopro-mounts/

71 key VoIP statistics: 2020 data analysis and market share. Retrieved from: https://financesonline.com/voip-statistics/

76 percent of homes have DVR, Netflix or use video on demand. (2015, January 4). Retrieved from: www.benton.org/headlines/76-percent-homes-have-dvr-netflix-or-use-video-demand

About YouTube. (2020, January). *YouTube*. Retrieved from: www.youtube.com/yt/about/

Action camera market revenue worldwide in 2018 and 2026. (2020). *Statista*. Retrieved from: www.statista.com/statistics/1059920/worldwide-sales-action-cams/

Adoption of new technology. (2008, February 18). *Visualizing Economics*. Retrieved from: http://visualizingeconomics.com/blog/2008/02/18/adoption-of-new-technology-since-1900

Albrecht, C. (2008, September 15). Study: DVRs in 27% of homes. *newteevee.com*. Retrieved from: newteevee.com/2008/09/15/study-dvrs-in-27-percent-of-tv-homes/ [January 30, 2010]

Anderson, M., & Caumont, A. (2014, September 14). How social media is reshaping news. *Pew Research Center*. Retrieved from: www.pewresearch.org/fact-tank/2014/09/24/how-social-media-is-reshaping-news/

Andersen, M., Perrin, A., Jiang, J., & Kumar, M. (2019, April 22). 10% of Americans don't use the Internet. Who are they? Retrieved from: www.pewresearch.org/fact-tank/2019/04/22/some-americans-dont-use-the-internet-who-are-they

Anderson, N. (2007, August 13). PDA sales drop by 40 percent in a single year, vendors bolt for exit. *ARS Technica*. Retrieved from: arstechnica.com/business/news/2007/08/pda-sales-drop-by-40-percent-in-a-single-year-vendors-bolt-for-exit.ars [November 1, 2009]

Anjarwalla, T. (2010, July 9). Inventor of cell phone: We knew someday everybody would have one. *CNN*. Retrieved from: www.cnn.com/2010/TECH/mobile/07/09/cooper.cell.phone.inventor/index.html

Auletta, K. (2014, February 3). Outside the box: Netflix and the future of television. *The New Yorker*. Retrieved from: www.newyorker.com/magazine/2014/02/03/outside-the-box-2

Benner, K., & Sisaro, B. (2016, February 13). Apple TV and Dr. Dre are said to be planning an original TV show. *The New York Times*, p. B6.

Berquist, L. (2002). Broadband networks. In A. E. Grant & J. H. Meadows (Eds.), *Communication technology update* (pp. 257–267). Oxford: Focal Press.

Best for last? (2009, August 3). *The New York Times*, pp. C1, C8.

Blair, I. (2019). Mobile app download and usage statistics. *BuildFire*. Retrieved from: https://buildfire.com/app-statistics/

Blumberg, S. J., & Luke, J. V. (2018). Wireless substitution: Early release of estimates from the National Health Interview Survey, January–June, 2018. Retrieved from: www.cdc.gov/nchs/data/nhis/earlyrelease/wireless201812.pdf

Blumberg, S. J., & Luke, J. V. (2014). Wireless substitution: Early release of estimates from the National Health Interview Survey, January–June 2014. Retrieved from: www.cdc.gov/nchs/data/nhis/earlyrelease/wireless201412.pdf

Bond, P. (2015, June 3). The Internet of things (you can sue about). *Forbes*. Retrieved from: www.forbes.com/sites/danielfisher/2015/06/03/the-internet-of-things-you-can-sue-about/

Brown, D. (2002). Communication technology timeline. In A. E. Grant & J. H. Meadows (Eds.), *Communication technology update* (pp. 7–46). Oxford: Focal Press.

Brown, D. (2019, May 21). Google takes another stab at Google Glass, updates the AR headsets for business customers. *USA Today*. Retrieved from: www.usatoday.com/story/tech/talkingtech/2019/05/21/google-announced-updated-google-glass/3751043002/

Brown, H. (2019, April 26). What are the most popular reasons why people use their smartphones every day? Retrieved from: www.gadget-cover.com/blog/what-are-the-most-popular-reasons-why-people-use-their-smartphones-every-day

Burger, A. (2014, April 16). Infonetics VoIP forecast $88 billion in service revenues in 2018. *Telecompetitor*. Retrieved from: www.telecompetitor.com/infonetics-voip-forecast-88-billion-in-service-revenues-in-2018/

Burrus, D. (2014). The Internet of Things is far bigger than anyone realizes. *Wired*. Retrieved from: www.wired.com/2014/11/the-internet-of-things-bigger/

By the numbers: 70 amazing Netflix statistics and facts. (2016, January 21). *DMR*. Retrieved from: http://expandedramblings.com/index.php/netflix_statistics-facts/2/

C., Andy. (2017, July 25). Top10 smart phone uses. *Mobiles*. Retrieved from: www.mobiles.co.uk/blog/top-10-smartphone-uses/

Carlson, N. (2013, December 17). The crazy stat that explains why Amazon Kindles are so cheap. Retrieved from: https://slate.com/business/2013/12/why-are-amazon-kindles-so-cheap.html

Cauley, L. (2008, May 14). Consumers ditching land-line phones. *USA Today*. Retrieved from: www.usatoday.com/money/industries/telecom/2008-05-13-landlines_N.htm [November 1, 2009]

Cell phone-only households eclipse landline-only homes. (2009, May 6). *The Tech Chronicles*. Retrieved from: www.sfgate.com/cgi-bin/blogs/techchron/detail?blogid519&entry_id539683 [January 21, 2009]

Chen, B. X. (2016, February 4). How to watch the Super Bowl when you don't have cable. *The New York Times*, p. B7.

Cisco VNI service adoption forecast 2013–2018 white paper. (2013). *Cisco*. Retrieved from: www.cisco.com/c/en/us/solutions/collateral/service-provider/vni-service-adoption-forecast/Cisco_VNI_SA_Forecast_WP.htm

Clifford, S. (2009, March 11). Advertisers get a trove of clues in smartphones. *The New York Times*, pp. A1, A14.

Cossick, S. (2019, June 19). Who's not using the internet? Ask the nearly 33 million Americans who aren't. Retrieved from: www.allconnect.com/blog/33-million-americans-dont-use-internet

The dangers of texting while driving. (n.d.). *Federal Communications Commission*. Retrieved from: www.fcc.gov/guides/texting-while-driving

DeSilver, D. (2014, July 8). CDC: Two of every 5 US households have only wireless phones. *Pew Research Center*. Retrieved from: www.pewresearch.org/fact-tank/2014/07/08/two-of-every-five-u-s-households-have-only-wireless-phones/

Dobrilova, T. (2019, May 13). 35+ must-know SMS marketing statistics in 2020. Retrieved from: https://techjury.net/stats-about-sms-marketing-statistics/#gref

Dolan, B. (2009, April 1). @CTIA: 1 trillion text messages in 2008. *MobiHealth News.com*. Retrieved from: mobihealthnews.com/1109/ctia-1-trillion-text-messages-in-2008/ [November 3, 2009]

The do's and don'ts of cell phone etiquette. (2013, July 23). Retrieved from: www.nydailynews.com/life-style/good-mobile-manners-article-1.1406873

Dugan, M., & Smith, A. (2013, September 16). Cell Internet use 2013. *Pew Research Center*. Retrieved from: www.pewinternet.org/2013/09/16/cell-internet-use-2013/

DVR state of the media report. (2010). Retrieved from: www.nielsen.com/wp-content/uploads/sites/3/2019/04/DVR-State-of-the-Media-Report.pdf

DVRs leveling off at about half of all TV households. (2013, December 6). *Leichtman Research*. Retrieved from: www.leichtmanresearch.com/press/120613release.html

Dwyer, J. (2009, August 23). Hello oven? It's phone. Now let's get cooking! *The New York Times*, p. A26.

Epstein, Z. (2019, December 28). The top 10 most binge-watched shows of 2019 on TV and Netflix. *BGR*. Retrieved from: https://bgr.com/2019/12/28/most-popular-tv-shows-2019-netflix-hbo-top-10/

Ericsson Mobility Report: 90 percent will have a mobile phone by 2020. (2014, November 18). *Ericsson*. Retrieved from: www.ericsson.com/news/1872291

Federal Motor Carrier Safety Administration. (2009). *Driver distraction in commercial vehicle operations (FMCSA-RRR-09–045)*. Washington, DC.

Fischer, S. (2015, May). 20 ways to make free Internet phone calls. *About.com*. Retrieved from: http://freebies.about.com/od/computerfreebies/tp/free-internet-phone-calls.htm

Forecast of the 5G adoption rate as share of mobile adoption rate in American from 2019 to 2025. (2020).

Statista. Retrieved from: www.statista.com/statistics/792427/5g-adoption-rate-forecast-in-the-us/

Forecast of mobile users worldwide from 2019 to 2013. (2020, February 28). *Statista*. Retrieved from: www.statista.com/statistics/218984/number-of-global-mobile-users-since-2010/

Foroohar, R. (2002, September 15). Life in the Grid. *Newsweek*. Retrieved from: www.newsweek.com/life-grid-144437

Freudenrich, C. C. (2002). How personal digital assistants (PDAs) work. *How stuff works.com*. Retrieved from: www.howstuffworks.com/pda.htm

Friedman, W. (2019, December 31). 2019–2020: TV time-shifted viewing holds steady, average viewing per show declines. Retrieved from: www.mediapost.com/publications/article/345167/2019-20-tv-time-shifted-viewing-holds-steady-ave.html

Fung, B. (2016, February 15). Google is considering a role in the future of television. *The Washington Post*, p. A13.

Gibbs, N. (2012, August 16). Your life is fully mobile. *Time*. Retrieved from: http://techland.time.com/2012/08/16/your-life-is-fully-mobile/

Gil, L. (2015, May 31). What you can do with Apple watch when your paired iPhone is out of range. *Mac Rumors*. Retrieved from: www.macrumors.com/how-to/apple-watch-with-iphone-out-of-range/

Global mobile statistics 2014. (2014). *Part A: Mobile Subscribers; Handset Market Share; Mobile Operators*. Retrieved from: https://mobiforge.com/research-analysis/global-mobile-statistics-2014-part-a-mobile-subscribers-handset-market-share-mobile-operators#subscribers

Gray, R. (2013, August 14). How CERN's Grid may place the power of world's computers in your hands. *The Telegraph*. Retrieved from: www.telegraph.co.uk/technology/news/10242837/How-CERNs-Grid-may-place-the-power-of-the-worlds-computers-in-your-hands.html

Grigonis, H. K. (2016, September 2). Camera sales are falling but studies suggest action cams are gaining ground. *Digital Trends*. Retrieved from: www.digitaltrends.com/photography/study-says-action-camera-sales-will-triple/

Hafner, K. (2009, May 26). Texting may be taking a toll. *The New York Times*. Retrieved from: www.nytimes.com/2009/05/26/health/26teen.html [May 26, 2009]

Handley, L. (2019, January 24). Nearly three quarters of the world will use just their smartphones to access the internet by 2025. *CNBC*. Retrieved from: www.cnbc.com/2019/01/24/smartphones-72percent-of-people-will-use-only-mobile-for-internet-by-2025.html

Handley, L. (2019, September 19). Physical books still outsell e-books – and here's why. *CNBC*. Retrieved from: www.cnbc.com/2019/09/19/physical-books-still-outsell-e-books-and-heres-why.html

Harwell, D. (2015, February 12). For HBO Now, 800,000 is a disappointing number. *The Washington Post*, p. A17.

Hesse, M. (2009, October 19). Worldwide ebb. *The Washington Post*, pp. C1, C8.

Hesse, M. (2020, May 6). Your dullest details, now part of history. *The Washington Post*, pp. A10, A1.

Hindo, J. (2020, April 1). World Community Grid joins in the fight against COVID-19. *IBM*. Retrieved from: www.ibm.com/blogs/corporate-social-responsibility/2020/04/world-community-grid-joins-in-the-fight-against-covid-19/

History and evolution of action cameras. (2015, November 4). *Pevly*. Retrieved from: https://pevly.com/action-camera-history/

Holson, L. (2014, July 3). Social media's vampires: They text by night. *The New York Times*. Retrieved from: www.nytimes.com/2014/07/06/fashion/vamping-teenagers-are-up-all-night-texting.html?_r=0

Holson, L. M. (2008, March 9). Text generation gap: UR2 Old (JK). *The New York Times*, pp. B1, B9.

Home landlines losing ground to cell phones. (2009, May 21). *Cell phone Shop*. Retrieved from: blog.cell phoneshop.com/2009/05/home-landlines-losing-ground-to-cell.html [January 21, 2009]

Horwitz, J. (2020, March 12). FCC's largest spectrum auction nets $4.47 billion for 5G mmWave bands. Retrieved from: https://venturebeat.com/2020/03/12/fccs-largest-spectrum-auction-nets-4-47-billion-for-5g-mmwave-bands/

Hosted PBX reseller markets seeing rapid growth. (2013, February 25). *PR Newswire*. Retrieved from: www.bizjournals.com/prnewswire/press_releases/2013/02/25/CG65326

Houston, S. M. (2014, December 31). The 10 best web series of 2014. *Paste*. Retrieved from: www.pastemagazine.com/articles/2014/12/the-10-best-web-series-of-2014.html

Hulu. (n.d.). Retrieved from: www.hulu.com

Industry Trends. (2019, April). Retrieved from: www.gminsights.com/industry-analysis/voice-over-internet-protocol-voip-market

Inside these lenses, a digital dimension. (2009, April 26). *The New York Times*, p. 4.

The Internet of Things really is things, not people. (2015). *Deloitte*. Retrieved from: www2.deloitte.com/global/en/pages/technology-media-and-telecommunications/articles/tmt-pred-the-iot-is-things-not-people.html

It's history. (2003, March 17). *Newsweek*, p. 14.

ITU releases 2014 ICT Figures. (May 5, 2014). Retrieved from: www.itu.int/net/pressoffice/press_releases/2014/23.aspx

Jarvinen, A. (2009). Game design for social networks: Interaction design for playful dispositions. *Proceedings of the ACM Siggraph Video Game Symposium*, New Orleans, pp. 95–102.

Johnston, S. J. (2002). Entering the 'W' zone. *InfoWorld*, pp. 1, 44.

Kang, C. (2014, October 17). CBS jumps on online streaming bandwidth. *The Washington Post*, p. A13.

Kang, C. (2015a, January 6). Decision to stream ESPN may be TV game-changer. *The Washington Post*, pp. A1, A14.

Kang, C. (2015b, March 19). Cord cutters look to Apple for the joy of no TV bundles. *The Washington Post*, pp. A1, A14.

Kelly, H. (2012, December 3). OMG, the text message turns 20; but has SMS peaked? *CNN*. Retrieved from: www.cnn.com/2012/12/03/tech/mobile/sms-text-message-20/

Kendall, B., & Hagey, K. (2014, June 25). Supreme Court rules Aereo violates broadcasters' copyright. *The Wall Street Journal*. Retrieved from: www.wsj.com/articles/supreme-court-rules-against-aereo-sides-with-broadcasters-in-copyright-case-1403705891

Kendrick, A. (1969). *Prime time: The life of Edward R. Murrow*. New York: Little, Brown and Company, p. 5.

Kim, E. (2015, April 15). The number of Americans paying for traditional TV peaked in 2012. *Business Insider*. Retrieved from: www.businessinsider.com/decline-of-us-tv-subscribers-2015-4

Koblin, J. (2016, February 26). All for the family. *The New York Times*, pp. B1, B6.

Koblin, J. (2020, May 2). Lockdown TV loses some of its audience after peaking in March. *The New York Times*, p. B4.

Kowlke, M. (2014, February 28). What global VoIP usage can teach us. Retrieved from: http://ip-communications.tmcnet.com/articles/371863-what-global-voip-usage-teach-us.htm

Krane, J. (2003, January 9). Microsoft's Gates touts consumer gadgets. *Excite.com*. Retrieved from: http://apnews.excite.com/article/20030109/D7OER2Q00.html

Landline phones are a dying breed. (2019, May 17). *Statista*. Retrieved from: www.statista.com/chart/2072/landline-phones-in-the-united-states/

Layton, J. (2009). How to go completely mobile. *HowStuffWorks.com*. Retrieved from: electronics.howstuffworks.com/how-to-tech/how-to-go-completely-mobile.htm [November 3, 2009]

Levy, S. (2001, December 10). Living in a wireless world. *Newsweek*, pp. 57–58.

Levy, S., & Stone, B. (2002). The Wi-Fi wave. *Newsweek*, pp. 38–40.

Liu, S. (2019, May). Tablets – Statistics & Facts. Retrieved from: www.statista.com/topics/841/tablets/

Manjoo, F. (2016, January 14). Why media titans would be wise not to overlook Netflix. *The New York Times*. Retrieved from: www.nytimes.com/2016/01/14/technology/why-media-titans-need-to-worry-about-Netflix

Markoff, J. (2009, May 24). The coming superbrain. *The New York Times*, pp. 1, 4.

Mayaan, G. (2020, January 13). The IoT rundown for 2020: Stats, risks and solutions. Retrieved from: https://securitytoday.com/Articles/2020/01/13/The-IoT-Rundown-for-2020.aspx?Page=1

McCluskey, B. (2017, March 16). An analysis of Netflix power binge watchers. *Mic*. Retrieved from: www.mic.com/articles/171147/an-analysis-of-netflix-power-binge-watchers-what-your-viewing-habits-say-about-you

McCue, T. J. (2012, December 27). Google voice stays free in 2013 but VOIP is $15 billion industry. *Forbes*. Retrieved from: www.forbes.com/sites/tjmccue/2012/12/27/google-voice-stays-free-in-2013-but-voip-is-15-billion-industry/

McKetta, I. (2019, October 1). How 5G is changing the global mobile landscape. *Speedtest*. Retrieved from: www.speedtest.net/insights/blog/5g-changing-global-mobile-landscape-2019/

McMahon, L. (2016). Keeping your eyes on the OTT prize. Retrieved from: www.multichannel.com/blog/keeping-your-eyes-on-the-ott-prize

Metz, R. (2014, November 26). Google glass is dead; long live smart glasses. *MIT Technology Review*. Retrieved from: www.technologyreview.com/featuredstory/532691/google-glass-is-dead-long-live-smart-glasses/

Meyers, D. (2011, October 11). Infonetics ups its VoIP service forecast to $76 billion by 2015. *Infonetics*. Retrieved from: www.infonetics.com/pr/2011/1H11-VoIP-and-UC-Services-Market-Highlights.asp

Miller, C. C. (2009, October 23). The cell refuseniks, an ever-shrinking club. *The New York Times*, pp. B1, B5.

Miller, C. C. (2010, May 6). In expanding reach, Twitter loses its scrappy start-up status. *The New York Times*, p. B7.

Mirkinson, J. (2014, January 28). CNN ratings plunge to near record lows. *The Huffington Post*. Retrieved from: www.huffingtonpost.com/2014/01/28/cnn-ratings-january-2014_n_4682699.html

Mitchell, A., Rosenstiel, T., & Christian, L. (2012). The state of the news media 2012. *Pew Research Center*. Retrieved from: www.stateofthemedia.org/2012/mobile-devices-and-neconsumption-some-good-signs-for-journalism/what-facebook-and-twitter-mean-for-news/

Mobile fact sheet. (2019, June). *Pew Research Center*. Retrieved from: www.pewresearch.org/internet/fact-sheet/mobile/

Mobile technology fact sheet. (2014, January). *Pew Research Center*. Retrieved from: www.pewinternet.org/fact-sheets/mobile-technology-fact-sheet/

More than 80 percent of Americans with a DVR can't live without it according to NDS survey. (2008, September 3). *NDS*. Retrieved from: www.nds.com/press_releases/NDS_DVR_Survey-US_030908.html [January 20, 2010]

Moscaritolo, A. (2015, February 2). Tablets see first-ever yearly decline. *PC Magazine*. Retrieved from: www.pcmag.com/article2/0,2817,2476199,00.asp

Moscaritolo, A. (2019, July 30). NPD: 31 percent of US households lack broadband. *PCMag*. Retrieved from: www.pcmag.com/news/npd-31-percent-of-us-households-lack-broadband

Murchu, I. O., Breslin, J. G., & Decker, S. (2004). Online social and business networking communities. *Digital Enterprise Research Institute*. Technical Report 2004-08-11.

The National Center for Health Statistics. (2019). Statista. Retrieved from: www.statista.com/chart/2072/landline-phones-in-the-united-states/

Nededog, J. (2017, March 1). Number of US homes without a TV doubled in just 6 years. Retrieved from: www.businessinsider.com/how-many-tvs-in-american-homes-number-us-department-energy-2017-3

Nielsen reports DVR playback adding to TV viewing levels. (2018, March 29). *Nielsen*. Retrieved from: www.multichannel.com/news/nielsen-reports-dvr-playback-adding-tv-viewing-levels-294196

Number of HBO Now subscribers from December 2015 to February 2019. (2019, September 17). *Statista*. Retrieved from: www.statista.com/statistics/539290/hbo-now-subscribers/

Number of Hulu's paying subscribers in the United States. (2020, February 10). *Statista*. Retrieved from: www.statista.com/statistics/258014/number-of-hulus-paying-subscribers/

Number of Netflix paying streaming subscribers in the United States. (2020, January 22). *Statista*. Retrieved from: www.statista.com/statistics/250937/quarterly-number-of-netflix-streaming-subscribers-in-the-us/

O'Dea, S. (2020, February 18). Number of mobile subscriptions worldwide 1993–2019. www.statista.com/statistics/262950/global-mobile-subscriptions-since-1993/

O'Grady, M. (2012, June 19). SMS usage remains strong in the U.S.: 6 billion SMS messages are sent each day. *Michael O'Grady's Blog*. Retrieved from: https://blogs.forrester.com/michael_ogrady/12-06-19-sms_usage_remains_strong_in_the_us_6_billion_sms_messages_are_sent_each_day

OTT streaming video playbook for advanced marketers. (2019, December 19). Retrieved from: www.iab.com/insights/ott-video-streaming-playbook-for-advanced-marketers/

OTT video penetration continues to reach new heights. (2019, July 8). Retrieved from: www.marketingcharts.com/digital/video-109123

Over the top television set to change TV consumption in the home. (2016). *BCi*. Retrieved from: www.bci.eu.com/over-the-top-tv/over-the-top-television-ott-tv/

Pachal, P. (2015, February 5). The next Google glass will probably look nothing like the first one. *Mashable*. Retrieved from: mashable.com/2015/02/05/google-glass-redesign/

Pegoraro, R. (2009, January 28). *The New York Times*, pp. A1, A18.

Percentage of adults who own an e-reader in the United States from 2009–2019. (2020). *Statista*. Retrieved from: www.statista.com/statistics/190283/penetration-rate-of-ereaders-in-the-united-states-since-2009/

Poggi, J. (2016, April 15). The CMO's guide to over-the-top TV. *Advertising Age*. Retrieved from: http://adage.com/article/media/cmo-s-guide-top-tv/298013/

Portio research mobile factbook 2013. (2013). Retrieved from: www.portio.com

Press, G. (2014, August 2014). Internet of Things by the numbers: Market estimates and forecasts. *Forbes*. Retrieved from: www.forbes.com/sites/gilpress/2014/08/22/internet-of-things-by-the-numbers-market-estimates-and-forecasts/#555e72ba2dc9

Quenqua, D. (2011, September 19). Text messaging levels off among U.S. adults. *ClickZ*. Retrieved from: www.clickz.com/clickz/news/2110274/text-messaging-levels-adults

Richtel, M. (2014, October 7). Voice activation systems distract drivers, study says. *The New York Times*, p. B4.

Rohit. (2011, October). VoIP industry to grow to $76.1 billion in 2015. *DID for Sale*. Retrieved from: www.didforsale.com/voip-industry-to-grow-to-76-1-billion-in-2015

Roose, K. (2019, November 13). Online cesspool got you down? *The New York Times Magazine*, p. 32.

Rouse, M. (2014, January). Thing (in the Internet of Things). *IoT Agenda*. Retrieved from: https://internetofthingsagenda.techtarget.com/definition/thing-in-the-Internet-of-Things

Sabin, S. (2018, November 6). Most young adults have an appetite for binge-watching shows. Retrieved from: https://morningconsult.com/2018/11/06/most-young-adults-have-an- appetite-for-binge-watching-shows/

The security and privacy issues that come with the Internet of Things. (2020, January 6). Retrieved from: www.businessinsider.com/iot-security-privacy

Seitz, P. (2015, July 8). Action camera market to see growth through 2019. *Investor's Business Daily*. Retrieved from: www.investors.com/news/technology/gopro-to-benefit-from-action-camera-market-growth-futuresource/

Seltzer, L. (2013, April 3). Cell phone inventor talks of first cell call. *Information Week*. Retrieved from: www.informationweek.com/wireless/cell-phone-inventor-talks-of-first-cell-call/d/d-id/1109376

Selyukh, A., & Nayak, M. (2015, January 29). U.S. wireless spectrum auction raises record $44.9 billion. *Reuters*. Retrieved from: www.reuters.com/article/2015/01/29/us-usa-spectrum-auction-idUSKBN0L227B20150129

Seward, Z. M. (2013, April 3). The first mobile phone call was made 40 years ago today. *The Atlantic*. Retrieved from: www.theatlantic.com/technology/archive/2013/04/the-first-mobile-phone-call-was-made-40-years-ago-today/274611

Sherman, A. (2020, February 12). Why T-Mobile's deal with spring could be the warmup to a wild decade of mergers. *CNBC*. Retrieved from: www.cnbc.com/2020/02/12/t-mobile-sprint-merger-is-a-warmup-to-more-wireless-cable-mergers.html

Sisario, B. (2016, February 21). SiriusXM fights to dominate the dashboard of the connected car. *The New York Times*, pp. B1, B6.

Sling TV review. (2015, January 26). *Consumer Reports*. Retrieved from: www.consumerreports.org/cro/news/2015/01/sling-tv-first-look/index.htm

Small crowd, young and old watch Super Bowl on Web. (2016, February 4). *The New York Times*. Retrieved from: www.nytimes.com/reuters/2016/02/04/arts/04reuters-nfl-superbowl-streaming.html

Smartphone: So many apps so much time. (2014, July 1). *Nielsen*. Retrieved from: www.nielsen.com/us/en/insights/news/2014/smartphones-so-many-apps-so-much-time.html

So many apps, so much more time for entertainment. (2015, June 6). *Nielsen*. Retrieved from: www.nielsen.com/us/en/insights/news/2015/so-many-apps-so-much-more-time-for-entertainment.html

Solsman, J. E. (2015, May 7). Sling TV may add broadcast networks but won't force you to use them. *CNET*. Retrieved from: www.cnet.com/news/sling-tv-may-add-broadcast-networks-but-wont-force-you-to-buy-them/

Sophy, J. (2014, January 19). Will most international calls be on Skype someday? *Small Business Trends*. Retrieved from: http://smallbiztrends.com/2014/01/someday-international-calls-might-skype.html

Statista Research Department. (2020, February 19). Share of adults in the United States that own a tablet for 2010 to 2019. Retrieved from: www.statista.com/statistics/756045/tablet-owners-among-us-adults/

Steel, E. (2014a, June 29). Stung by Supreme Court, Aereo suspends service. *The New York Times*, p. A21.

Steel, E. (2015, January 5). Dish network unveils Sling TV, a streaming service to rival cable (and it has ESPN). *The New York Times*. Retrieved from: www.nytimes.com/2015/01/06/business/media/dish-network-announces-web-based-pay-tv-offering.html?_r=5

Steel, E. (2016, February 11). HBO Now has 800,000 paid streaming subscribers, Time Warner says. *The New York Times*, p. B3.

Sterling, C. (2013, October). Numbers don't lie: Impressive stats on the VOIP industry. Retrieved from: https://virtualphonesystemreviews.com/numbers-dont-lie-impressive-stats-voip-industry/

Sterling, G. (2014, February 11). Nielsen: More time on Internet through smartphones than PCs. *Marketing Land*. Retrieved from: marketingland.com/nielsen-time-accessing-internet-smartphones-pcs-73683

Stevens, T. (2007, November 14). 82% of Americans own cell phones. *Switched.com*. Retrieved from: www.switched.com/2007/11/14/82-of-americans-own-cell-phones/ [November 3, 2009]

St. George, D. (2009, February 22). 6,473 texts a month, but at what cost? *The Washington Post*, p. A17.

Stone, D. (2009, August 10). Breakfast can wait. The day's first stop is online. *The New York Times*. Retrieved from: www.nytimes.com/2009/08/10/10morning.html [August 13, 2009]

Sutter, J. D. (2009, September 29). Microsoft researcher converts his brain into 'e-memory.' *CNN*. Retrieved from: www.cnn.com [September 29, 2009]

Tanaka, J. (2001, September 17). PC, phone home! *Newsweek*, pp. 71–72.

Thirty-seven smartphone addiction statistics for 2019. Retrieved from: www.slicktext.com/blog/2019/10/smartphone-addiction-statistics/ [January 4, 2021]

Thompson, C. (2008, September 7). I'm so totally close to you. *The New York Times Magazine*, pp. 42–47.

Tierney, J. (2009, May 5). Ear plugs to lasers: The science of concentration. *The New York Times*, p. D2.

Tops of 2019: Television. Retrieved from: www.nielsen.com/us/en/insights/article/2019/tops-of-2019-television-2/ [December 18, 2019]

Torres, L. (2019, December 9). The 20 most-binged TV shows on Netflix in 2019. *Insider*. Retrieved from: www.insider.com/netflix-most-binged-shows-this-year-2019-11

Total audience report. (2014, December 3). *Nielsen*. Retrieved from: www.nielsen.com/us/en/insights/reports/2014/the-total-audience-report.html

Tsukayama, H. (2014, June 29). Aereo hits 'pause' button on service, will issue refunds to customers. *The Washington Post*, p. A8.

Tsukayama, H. (2015, January 3). At tech show, expert, bigger and smarter. *The Washington Post*, p. A10.

Tyson, J. (2002). How telephony works. *How Stuff Works*. Retrieved from: www.howstuffworks.com/iptelephony.htm

Vinyals, O., Toshev, A., Bengio, S., & Erhan, D. (2014, November 17). A picture is worth a thousand (coherent)

words: Building a natural description of images. *Google Research Blog*. Retrieved from: googleresearch.blogspot.com/2014_11_01_archive.html

Watch 2014 Fall TV preview online. (2014). *Hulu*. Retrieved from: www.hulu.com/browse/picks/2014-fall-tv-preview

Watson, A. (2018, April 23). Binge watching in the U.S. Retrieved from: www.statista.com/topics/2508/binge-watching-in-the-us/

Weiss, H. (2018, September 21). Why you're probably getting a microchip implant someday. *The Atlantic*. Retrieved from: www.theatlantic.com/technology/archive/2018/09/how-i-learned-to-stop-worrying-and-love-the-microchip/570946/

Weiss, T. (2015, May 20). Tablet market continues to shrink around the world: Report. *Eweek*. Retrieved from: www.eweek.com/mobile/tablet-market-continues-to-shrink-around-the-world-report.html

Welch, C. (2020, March 10). Google's next streaming player will reportedly run Android TV and come with a remote. Retrieved from: www.theverge.com/2020/3/10/21173765/google-streaming-player-dongle-android-tv-chromecast-remote

What is telephony? (2002). *About.com*. Retrieved from: http://netconference.about.com/library/weekly/aa032100a.htm

Wortham, J. (2010, January 15). Burst of mobile giving adds millions in relief funds. *The New York Times*. Retrieved from: www.nytimes.com/2010/01/15/technology/15mobile.html?scp56&sq5%22haiti%22%20%22twitter%22&st5cse [January 17, 2010]

Zeitchik, S. (2020, April 21). Netflix adds 16 million subscribers as staying home becomes a necessity. *The Washington Post*. Retrieved from: www.washingtonpost.com/business/2020/04/21/netflix-adds-whopping-16-million-subscribers-worldwide-coronavirus-keeps-people-home/

Chapter eight

XR: Inside of Media, Inside the Mind

Charles P. ('Chip') Linscott, Ph.D., Ohio University

SEE IT THEN

Imagine being inside of your favorite movie or game, thereby becoming an active participant in your most cherished fantasy worlds. Or picture yourself visiting the Himalayas, Mars, or Jurassic-period Earth, exploring the terrain and learning the science and history firsthand without ever leaving your home. What if you could take a class from any teacher in any classroom in the world at any time? These fantastic trips are available in the short time it takes to push a button and put on a headset. These are some of the promises of **VR (virtual reality)**, where you do not just watch the adventure, you are inside of it, you *live* it.

Imagine gazing at an unfamiliar plant or animal and instantly having key information about the organism projected on top of it. Or consider learning a new skill with a virtual, professional teacher appearing next to you in the real world and in real-time. What if you could have dinner with a hologram of your best friend, who lives thousands of miles away? And think what it would be like to battle digital monsters while walking down your very own street. These are some of the promises of **AR** (augmented reality), **AV** (augmented virtuality), and **MR** (mixed reality), where you remain partially in the real world but have it magically enhanced by countless digital elements. **XR** ('extended reality' or 'cross reality') is the blanket term used to define any and all of these virtual, augmented, and mixed realities; it also refers to any combination of these realities that will not fit neatly into a single immersive media category.

ENTERING THE VIRTUAL WORLD

It is possible to understand the history of art and media as a series of attempts to get closer and closer to the object itself – to go *inside* of media until we become part of it ourselves as observers from within the fictional world, if not active participants in the story. Amphitheaters, interactive drama, video games, headphones and earbuds, wearable smart devices, surround sound, 3D films and games, widescreen and IMAX movies, multiscreen gallery installations, and amusement parks are all ways of becoming *immersed* in media. Such immersion leads to a state where there is less and less 'outside' of the medium, progressing until we are eventually surrounded by the story-world and engaging with it as a direct part of the action. XR may therefore be seen as the most recent example of the immersive drive in human beings. In this sense, XR is the contemporary culmination of this historical immersive process and is not 'new,' even though its direct roots may be clearly traced to the middle of the 20th century, if not farther back.

Take the film *Akira Kurosawa's Dreams* (Kurosawa, 1990), for instance. The anthology film features a series of vignettes inspired by the legendary Japanese filmmaker's nighttime dreams. In the entry entitled 'Crows,' the Kurosawa character magically moves from an external, staid appreciation of Van Gogh paintings in a quiet art gallery to become a part of the world of the paintings themselves. Kurosawa travels from a central vantage outside the frame – staring at a still but transcendentally beautiful oil-painted image – into Van

Figure 8.1 Akira Kurosawa enters Van Gogh's *The Langlois Bridge at Arles with Women Washing* (1888).
Source: Screenshot, *Akira Kurosawa's Dreams* (Kurosawa, 1990)

Gogh's renowned painting, *The Langlois Bridge at Arles with Women Washing* (1888) (see Figure 8.1). The director imagines the painting as a living work of art that he can enter into and in which he can act and interact, conversing with characters, exploring Van Gogh's most famous works, and eventually meeting the painter himself (playfully brought to life by director Martin Scorsese). This 'dream' is in fact the goal of VR. In VR, the user enters a mediated world without a frame, a world without monitors or traditional screens. Once in that world, VR takes the human sensorium from a grounding within the everyday world and places it inside a computer-generated environment that, as completely as possible, blocks out the 'real' world and allows the user to interact with, and to some degree affect, the virtual world. Related technologies like AR, AV, and MR allow for some of the real world to persist but place realistic digital elements within that world. Again, the sum total of all these realities, and anything in between, is often called XR, which is the subject of this chapter.

EARLY HISTORY

Writing a history of virtual media is a complex task, as the forms draw an assortment of elements from their predecessors in visual arts, literature, theater/drama, film, music, television, video games, performance, and so on. Of course, many of these fields are historically and aesthetically related to each other as well. A quick sketch of the key influences from various precursors, with reader awareness of the other chapters in this textbook, is the best approach to writing a short history of the topic at hand. The signal contribution that is largely missing from previous electronic media forms – up until video games, that is – is interactivity. While there are varying levels of interactivity possible in prior media, as hard as one tries to change a film or radio broadcast or television program, those 'worlds' remain fundamentally the same. This is not the case with games and XR, where user actions may alter the story and the mediated environment in deep and lasting ways. In fact, interactivity is one of the principal aims of such newer media, wherein users/players get to make choices, and their choices can profoundly affect the experience. Setting interactivity aside for the moment, some pivotal moments in technological history have helped pave the way for contemporary virtual media.

Stereoscopy

Stereoscopic 3D illusions – first popular as novelties in the 19th century – form the basis of most optical technology used in **HMDs** (or head-mounted displays, also known as 'VR headsets') today. The basic principle behind stereoscopy is that most humans have two eyes in two different places on their heads. This means that each eye sees the real world from a slightly different vantage point, and the human brain merges the two images into a single perception. Thus, two marginally different images may be offered, one for each eye, in order to generate the illusion of depth, or 3D. This is precisely the process that most contemporary HMDs use to generate the illusion of depth in their optics (i.e. lenses or screens). This is called stereoscopic 3D, where slightly dissimilar digital images are projected for each eye by the optical technology of headsets.

Stereo Sound

Stereophonic sound is important for modern XR technologies. First explored in Europe in the late 1800s, stereo sound provides two diverse sets of sounds, one for each ear. Monophonic sound – where the same audio is presented to both ears – held sway for most of the 20th century; it fails to generate a natural sense of space or depth, in contrast to stereo sound. 'Two-channel' stereo audio became standard in the 1970s and works to approximate natural human hearing, where two ears on two different sides of the human head receive diverse sounds at various times. Further evolution in audio technology would lead to surround sound and various types of immersive and spatial audio, the latter of which is important to XR and is covered later in this chapter. Most modern audio is presented in at least stereo format, and essentially all VR audio is heard in stereo or spatial sound on headphones. (The invention of speakers, or audio transducers, is therefore also important to XR. Audio innovation is further testament to the deeply hybrid nature of the medium.)

Animation, Cinema, Television, Games

As a profound mixture of numerous earlier forms of media, XR owes abundant historical debts. But certain precursors loom larger than others. Audiovisual and moving image media – animation, cinema, television, and video games – all inspire feelings of a sort of illusory world through manipulation of the human senses. As earlier chapters in this book demonstrate, these media emerged from numerous sources themselves, but their impacts on XR are undeniable. Stereoscopic illusions and devices such as the zoetrope and praxinoscope existed alongside the beginnings of animation and cinema in the late 1800s. Eadward Muybridge did important early work in the U.S. and England by making still photographs appear to move through their rapid successive projection. Charles Émile Reynaud was publicly exhibiting cartoon-like, hand-drawn moving images in France in the early 1890s. Reynaud's exhibitions occurred around the same time that the Lumière brothers (in France) along with Thomas Edison and W. K. L. Dickson (in the U.S.) were pioneering cinema. Of course, motion pictures would eventually pave the way for television, video game arcades, and PC/console gaming. Again, games added the crucial element of interactivity to audiovisual media, but there is at least one close antecedent to XR that persists today and is older than computer games.

Flight Simulators

Flight simulators are almost as old as airplanes themselves and were first introduced in the early 1900s. These devices are important in that they save lives and equipment by providing a realistic, *virtual* training experience before pilots ever leave the ground. The earliest flight simulators used instruments, motion, and controls to approximate the experience of flying. Edwin Link is credited with pioneering the devices. Simulators were first used by militaries around the world in the 20th century and are ubiquitous in modern flight schools. Today, advanced flight simulators are available for the home and can be purchased for just a few hundred dollars. The idea behind simulators is key to the development of VR: Media can be used to stimulate various sensory experiences, and thus users may experience a realistic facsimile of a real-world scenario that their bodies and brains will accept – and learn from – as though it were the real thing. Flight simulators *immerse* the prospective pilot in a sensory world that mimics flight. Various forms of XR – and particularly VR – work much the same way.

Morton Heilig's *Sensorama*

Morton Heilig's *Sensorama* is regularly mentioned in histories of VR technology, and with good reason. The device was an attempt by Heilig, a filmmaker, to expand interactivity to as many senses as possible, beyond simply image and sound. Devised in the 1950s and patented in 1962, Heilig's *Sensorama* was not a commercial success but should nevertheless be seen as a serious inspiration for VR. Using a world-mounted display (WMD) similar to an arcade game console, the device incorporated motion, smell, vibration, movement, wind, 3D images, and stereo sound. Thus, it was far more immersive than any cinema could hope to be and represents a very clear early attempt to place users inside of art and media. The *Sensorama* was a simulator and featured the experiences of riding various vehicles like motorcycles and helicopters. Heilig hoped to have the system widely used in places like arcades but was far ahead of his time. He also invented a head-mounted display around the same time and, all told, is clearly one of the chief influences on modern XR technology.

Ivan E. Sutherland: *Sword of Damocles* and the Ultimate Display

In 1968, Ivan E. Sutherland publicly demonstrated his work on what is generally understood to be the first successful HMD technology, the *Sword of Damocles*. Sutherland worked at institutions such as the University of Utah and Harvard University (where he was assisted by graduate student Bob Sproull, who went on to become a major figure in the history of computing). The *Sword of Damocles* was so named (after the Greek myth) due to its gigantic and unwieldy design that loomed over the user's head. Despite its cumbersome nature, the device was unequivocally groundbreaking for AR and VR. While it was essentially an AR device in that it projected digital images onto the physical world, similar head-mounted technology – with motion tracking and stereoscopic 3D optics – forms the backbone of VR headsets today.

But Sutherland is not just a key pioneer of VR/AR displays. He is also a towering figure in the history of computing – known as the 'father of computer graphics' – and he is credited as one of the inventors of the graphical user interface (GUI), which makes much of modern human–computer interaction possible. Even more importantly for VR, Sutherland published a brief theoretical paper, 'The Ultimate Display,' in 1965 that lays out the primary (if currently far-fetched) ends of VR development. In the paper, Sutherland posits a computer that can control matter itself. Therefore, a user in such an environment would be completely unable to distinguish the digital world from the physical world. In fact, the computer-generated environment would *be* a real environment, with the only true difference resting in the fact that the display-world was created by a computer rather than through processes of natural evolution and geological formation. As Sutherland argues, a bullet in such a reality would be fatal, and virtual handcuffs would be truly confining. While many contemporary audiences might find the complete computer-generated control of reality terrifying, just as many people might find it exhilarating – an opportunity to enter a 'fantasy world' where anything is possible. There is also little doubt that tech companies hope to eventually realize the fantastic aspirations detailed in Sutherland's essay.

Figure 8.2 Helig's *Sensorama*.

NASA

In the 1980s, NASA (National Aeronautics and Space Administration) conducted extensive research into VR. This U.S. government agency was interested in how VR might be used in space exploration and astronaut training. NASA's work produced important software and HMD systems, such as the Virtual Visual Environment Display, and the agency continues to be a leader in VR research, even offering free VR applications for contemporary systems such as Oculus.

Jaron Lanier

Jaron Lanier is yet another key figure in the evolution of VR technology. After leaving Atari in the mid-1980s, Lanier started a company, VPL Research, with his colleague Thomas G. Zimmerman. Lanier is recognized as having popularized the term 'virtual reality.' Although he clearly did not invent VR technology itself, Lanier made numerous pioneering contributions to the field. He was involved in NASA's 1980s' VR work and is also a crucial figure in the development of haptic (touch/force/motion-based) technology and gesture control of computer and virtual environments. His association with NASA in the 1980s led to a device that would eventually become the Nintendo Power Glove, a short-lived haptic controller for video games.

In addition to these other achievements, Lanier conceived of the vital concept of 'homuncular flexibility' and demonstrated its existence through VR experiments. Homuncular flexibility means, rather simply, that human beings can quickly adapt to the control of radically different, and radically non-human, bodies in VR. The human brain is able to rapidly learn to manipulate markedly different appendages and bodily forms when immersed in VR. Lanier's classic example is that of human control of a lobster's body, with all of its many legs and claws. Numerous experiments since Lanier's discovery of this process have proven the concept, and homuncular flexibility continues to be a key area of research in XR today.

CAVE

The 1990s saw a massive boom in commercial and consumer VR technology, but this era was also a time of continued academic research. One important breakthrough was the University of Illinois' CAVE

ZOOM IN 8.1

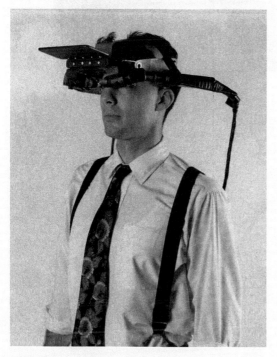

Figure 8.3 *Sword of Damocles.*

Watch this short video of the *Sword of Damocles* in action: www.youtube.com/watch?v=eVUgfUvP4uk

ZOOM IN 8.2

Learn more about 'The Ultimate Display' at: www.wired.com/2009/09/augmented-reality-the-ultimate-display-by-ivan-sutherland-1965/

system (circa 1992–1993). The Cave Automatic Virtual Environment (CAVE) was created by Carolina Cruz-Neira, Daniel Sandin, and Thomas DeFanti. It is significant in part because it employed motion tracking to allow for multiple users to traverse an open, 3D VR environment at the same time (only one user would be in control). But the device also references Plato's Allegory of the Cave, a philosophical parable that was an important inspiration for the CAVE VR system and remains a key touchstone for theories of media from ancient Greece to the present.

VR Arcades and Consumer Systems

The first real commercial VR explosion occurred in the 1990s but was essentially over before the turn of the new century. The decade saw home VR systems, stand-alone devices in normal video arcades, and destination VR facilities in malls and amusement parks. Numerous companies emerged touting the latest VR systems, including a corporation called the Virtuality Group that placed VR systems in malls around the world. Companies like Sega, Disney, and Nintendo were heavily invested, and most 1990s VR systems included innovations from previous decades, such as haptic controllers, stereoscopic 3D, stereo sound, motion tracking, and multi-user environments.

Jaron Lanier's association with Nintendo's Power Glove is far from the only connection between Nintendo and VR. In 1995, Nintendo created an early consumer VR gaming device called the Nintendo Virtual Boy. The device featured an unusual HMD that was mounted on a stand requiring placement on a table. While the graphics were 3D and the headset blocked all other images, the only possible colors were black and red. Further, the player could not move freely and was required to use a standard Nintendo game controller. Reports of motion sickness were common – this is still a problem for contemporary VR – and the device was expensive for the time. While groundbreaking for VR and gaming, the Virtual Boy was not widely adopted and was quickly discontinued, marking it as one of Nintendo's rare but highly visible commercial failures.

Academic Research and the VR Winter

Clearly, Nintendo was in good company, since every consumer VR technology from the era effectively failed or was discontinued. Companies like Virtuality went out of business, and the public lost

Figure 8.4 Nintendo's Virtual Boy: The device featured an odd design that combined an HMD with a tabletop stand and was limited to red-and-black graphics.

interest, but XR persisted. In addition to its use in flight simulators, XR research and development continued in academic environments during the commercial 'VR winter' following the economic boom of the 1990s. Several key institutions and researchers continued to work on XR throughout the 2000s, and innovation did not end with the death of VR arcades and the Virtual Boy. The University of Southern California (Albert 'Skip' Rizzo), Stanford University (Jeremy Bailenson), and the University of Washington (Hunter Hoffman) helped keep XR alive through work in scientific research, psychology, health care, military applications, and entertainment. Eventually, a new XR boom would occur, but it would take more than a decade.

SEE IT NOW

Living in Virtual Reality

There was very little commercial development in XR between 2000 and 2012. In 2012, VR began to change rapidly in large part due to the efforts of a young man named Palmer Luckey, who created the Oculus system that was eventually purchased by Facebook, Inc. for over $2 billion in February 2014. The series of events surrounding Luckey is really where the 'now' of XR begins, but we should back up a little in order to understand how all of these 'realities' – from the *Sensorama* to the hottest new XR tech – operate.

VR is a sort of digital dream that the brain treats as real. This objective existence inside a virtual environment is called *immersion*, and it is a quality that can be cultivated and enhanced by technical and narrative elements to ultimately generate a sense of *presence*, which is a partially subjective sense of existing in, and interacting with, the virtual world in which the user is immersed. Presence is somewhat subjective because each individual user is different, having different experiences, thoughts, fears, desires, and crucially, a differently functioning brain and differently constituted body than other users. For these reasons, the history of VR must be understood as the history of an *experience* and not simply the chronology of technology since, more so than in other media, the user is at the center of the action.

When we attempt to define VR, a number of key terms must be considered, including interactivity, immersion, presence, computer-generated, simulation, and the human senses. Scholars and developers of VR largely agree that it uses computer technology to create a digital simulation that plays on the various human senses (including lesser-known senses such as proprioception) to induce user feelings of being present in another world with which they can interact.

Crucially, VR systems block out as much of the outside world as possible so that users can interact with and experience the virtual world in highly dynamic ways. Just as video games are, by definition, interactive in ways that film and television are not – allowing players to change elements of the game world and have varying experiences with virtual entities – VR lets users similarly alter their digital environments. One major difference between traditional video games and VR is that VR users are generally *inside* the game or simulation – there is no traditional screen or frame, so nothing is *outside* the frame. In fact, one way to understand full VR is as the *obliteration of the frame*, a complete envelopment in the mediated world, as in Kurosawa's dream. The frame is, of course, a hallmark of most modern digital media, including television, computer/console/arcade games, film, computer monitors, smartphones, tablets, smart watches, and the like. So, while VR is not 'new,' the loss of the frame really is radically new.

FYI: DEFINITIONS

- **Interactivity.** This is the ability of users to directly participate in and alter the mediated world. It is a key feature of video games and XR. Interactivity can have both depth and breadth in varying degrees, but it is rarely found in other audiovisual media such as film and television.
- **Immersion.** Along with presence and interactivity, this is a fundamental component of XR. Immersion is an objective, measurable set of factors that is enhanced by narrative and technical considerations (image and sound quality, for instance) to make users feel an increased sense of presence. Realities like AR are not fully immersive because they maintain a significant portion of the user's real world.
- **Presence.** Presence includes immersion but adds the individual subjectivity of the user. This is a feeling that can be amplified by increasing immersion and interactivity. Presence in its simplest form means that the user feels truly and realistically inside of the virtual world. Presence will vary among users of the same applications.
- **Computer-generated.** This means that a computer device of some sort (PC, tablet, stand-alone headset, smartphone) is using a combination of user input, software/applications, content, hardware, and output to create a virtual or augmented world. The digital (computer-generated) world is in part coded by developers, artists, and software engineers but it is also affected by user interaction.
- **Simulation.** This is a facsimile, in this case created by a computer. Simulations can feel deeply real to participants but somehow lack the full reality of objective existence.
- **Human senses.** These are sight, hearing, touch, taste, smell – but also proprioception and other lesser-known senses. At present, some senses – sight, hearing, touch – are more important to XR simulations than others.
- **Proprioception.** Proprioception is a lesser-known sense that deals with human perceptions of force, bodily location, and motion. Proprioception is more easily replicated in XR than other senses such as taste and smell. It is therefore part of the digital stimuli included in many XR applications.

It is important to reiterate that VR does not have a monopoly on immersive media. Immersion occurs within non-digital (or mixed digital/live) events such as theme parks, historical reenactments, haunted houses, avant-garde theater, performance art and gallery installations, and more. Likewise, VR is surrounded by closely related media such as AR, MR, AV, CineVR (or 360 video), spatial audio, and so on. Any of these media can be placed into the category of XR, along with emerging forms that do not fit neatly into any one category. XR is thus variously defined as exponential reality, extended reality, expanded reality, or unknown reality (wherein the x functions as an unknown, like in algebra). While there is some slippage among the terms (this is one reason for the catch-all term 'XR') solid definitions are helpful. Diagram 8.1 is widely used to assist in conceptualizing the various 'realities' encompassed by XR.

In 1994, the academic researchers Milgram and Kishino published a scholarly paper detailing what they called 'the virtuality continuum' (see Diagram 8.1). This continuum provides a clear set of definitions for the various 'realities' presented by virtual and augmented technologies. Despite some room for debate, the continuum is generally accepted by scholars and developers. The simplified version of this continuum concept places the 'real' world to the far left of the line and thus outside of digitally created (or computer-generated) reality. On the far right of the scale is VR, an entirely computer-generated reality that is immersive, interactive, and occludes the real world as much as possible. In between the real world and the fully virtual world is 'mixed reality' (MR), which includes both 'augmented reality' (AR) and 'augmented virtuality' (AV). AR, put simply, allows the user to remain in touch with most of the real world but places interactive digital objects on top of physical reality. AV finds the user interacting mainly with the virtual world but with that world enhanced by the real world. On the continuum, the closer to the left a virtual or augmented technology is found, the more the real world will be available. Conversely, the closer to the right one moves, the more the real world fades away and is replaced by a computer-generated reality. XR is a term that appeared long after Milgram and Kishino published their work. XR – stretching from AR on the left to VR on the far right – includes everything *except* the real environment and is also applied when the virtual form does not clearly fall into one precise spot on the continuum.

The hardware and software requirements of each reality can be exceedingly variable, but all forms of virtuality require some sort of display, with the HMD of VR being the most familiar. Outside of VR, AR is perhaps the best known of the other forms of XR. AR also appears to be rapidly increasing in popularity and is probably the most readily accessible virtual medium, due in part to the fact that newer smart phones and tablets are AR-capable. Before moving on to how these technologies are used today, some history will be helpful, beginning again with Palmer Luckey.

Oculus Headset and Palmer Luckey's Kickstarter

Following the 'VR winter' of 2000–2012, VR technology moved forward dramatically with the invention of the Oculus headset. Its creator, Palmer Luckey, was a precocious California teenager with a love of video games. He was fond of tinkering with old gaming consoles and began to dismantle VR systems and explore their functionality around 2010. Luckey longed for contemporary VR systems to be both better (i.e. more immersive) and less expensive, and he was able to achieve both with his new HMD, which he built at home with few resources (and after much experimentation) in 2012. The Oculus was a stunning accomplishment, and the young man was just getting started. Luckey posted about his prototype HMD in online forums, eventually drawing the attention of John Carmack, who was already fabled in the gaming industry due to his work at id

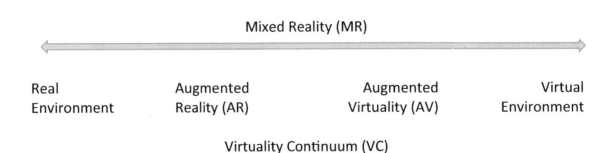

Diagram 8.1 The virtuality continuum.

Software. Carmack's name carried a lot of weight, and he pushed Luckey's invention at conferences and to associates. This outreach aroused interest from fellow enthusiasts and computer 'geeks' around the globe. In August, 2012, Luckey's Kickstarter campaign for mass-produced Oculus headsets hit the Internet with a goal of raising $250,000. The online fundraising was terminated in less than a week after having raised nearly $2.5 million, and, shortly thereafter, the nascent company received $16 million dollars in venture capital. Oculus VR proceeded to get to work.

FYI: PALMER LUCKEY: VIRTUAL REALITY PIONEER

WHY VR? WHAT DREW YOU TO REVOLUTIONIZING THIS PARTICULAR ASPECT OF MEDIA TECHNOLOGY?

Growing up, I was obsessed with virtual reality. My favorite games, movies, and books all involved VR in one way or another, and I was sure it was just around the corner. Who wouldn't be fascinated by the idea of escaping reality to worlds with totally different rules where anything is possible? When I eventually realized how stagnant the field was in the real world, I started working on designing HMDs as a hobby, learning everything I could about how the human perceptual system worked and how to stimulate it. I never expected to revolutionize anything, but rapidly improving display technology combined with a few clever ideas around real-time optical pre-distortion soon vaulted my prototypes far beyond any other headset you could buy at the time. Once that happened, I decided the fastest way to bring about my dream of mass-market VR would be starting a company to market the technology to the broader world. Terrible idea with terrible odds, considering that every other VR company up to that point had more or less failed, but it worked!

Figure 8.5 Palmer Luckey.

DOES VR HAVE TO BE CENTERED ON GAMING IN ORDER TO BE COMMERCIALLY SUCCESSFUL?

No. Gaming was an early focus for the Oculus Rift because gamers (particularly PC gamers) are more willing than the average population to adopt early technology. The games industry was also the only industry with a large pool of talent working on real-time 3D graphics, something you obviously need for VR simulation. In the long run, virtual reality will augment and even supplant almost every aspect of our daily existence, from collaborative work to travel to romance. I will always be a gamer, but VR is bigger than that.

WHAT OTHER AREAS ARE PARTICULARLY APPROPRIATE FOR VR TECHNOLOGY? OR WHERE WOULD YOU LIKE TO SEE VR GO NEXT?

Did you know that more U.S. troops die during training than in combat? I think the United States military and our allies will save countless lives by adopting VR as a training tool, which is one of the reasons I decided to work in the defense industry after being fired from Oculus by Facebook. As for where things go next, I want to see virtual reality simulations that are indistinguishable from the real world become accessible to everyone, allowing anyone to experience anything. With so many smart people now working on the problems involved, that future is inevitable.

Google Glass

Around the same time, Internet giant Google was developing XR in a different direction – through AR rather than VR. Google Glass, like the Nintendo Virtual Boy, was a famous consumer failure that nevertheless serves as an important milestone in the history of XR technology. Google Glass was released in 2013 on a very limited basis to specially selected users, whom Google called 'Explorers.' The AR device resembled ordinary, clear eyeglasses with a slightly futuristic design. It had a small camera capable of audio and video and played stereo sound. The Glass could be controlled by voice command, smartphone app, or by gestures on a small touchpad located on the Glass's arm. The device connected to the Web via Wi-Fi and was capable of displaying text and images from apps and the Internet directly on the Glass lenses without occluding the user's view of the real world. Such functions mark the device as one of the first fully functional AR headset devices, and the Glass's features are widely mimicked on contemporary MR devices such as the Microsoft HoloLens and Magic Leap 1.

Unfortunately, the ability to surreptitiously record video and take pictures proved highly controversial (and perhaps even illegal), and the sociocultural controversy led Google to discontinue the device before it reached the general consumer market. (Some users were even assaulted for wearing the glasses into public places like bars.) The device lives on in subsequent iterations as an enterprise and business technology, and Google still manufactures new versions of Glass for such purposes. Despite its very public flop, the Google Glass helped pave the way both in terms of its rich set of features and in its aim: To provide a small form factor Internet-connected AR device with significant processing power.

Smartphone VR

Luckey was not the only person dreaming of less expensive and more powerful XR systems, and tech industry giants like Google and Samsung had been working on ways to bring VR and AR technology to the masses. Google Cardboard, released in 2014, arguably remains the least expensive and most accessible VR headset.

Of course, this rather rudimentary HMD is actually made chiefly of cardboard and requires a smartphone to operate. Therefore, users without proper phone technology would still be locked out of VR. Nevertheless, many people already

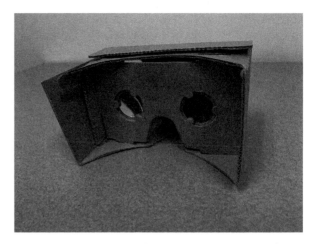

Figure 8.6A The Google Cardboard was an early inexpensive device that was used by *The New York Times* to get more readership and get people interested in VR.

Figure 8.6B The Samsung Gear was sold by Oculus and utilized a Samsung phone plugged into the headset to get a VR image.

owned an appropriate device, even in 2014, and the cheap, foldable Cardboard costs just a few dollars. It utilizes stereoscopic lenses and relies on the phone for images, motion tracking, processing power, and audio – but it truly works – and it is so affordable that *The New York Times* essentially gave the devices away to promote the paper's 360° video content. Samsung took a similar but more sophisticated approach with their Gear VR. The company had in fact been developing VR plans and patents for years but accelerated their headset program in the wake of Oculus. (The first release of the Gear VR, in 2015, was actually a partnership with Oculus, though Samsung's R&D predates Luckey's invention.) Like the Cardboard, the Gear relies on a powerful smartphone to operate, but the Samsung HMD features sturdy construction (plastic and metal rather than cardboard), more advanced optics and controls, and higher quality content. It also comes with a much steeper price – closer to $100 – but, unlike the original Oculus HMD, it does not require a PC or laptop to operate. Since the release of these devices, hundreds of inexpensive, off-brand cellphone VR headsets have flooded the market. Anyone with $20 and even a semi-recent smartphone of any brand can now access basic VR.

Facebook and Oculus

Like any hyped new company, Oculus VR experienced some growing pains. An assorted crew of venture capitalists, software developers, engineers, and entrepreneurs had come to be associated with the new corporation, but significant delays in production emerged despite their expertise. No one disputed the strength of the product, but getting it to market was a different story. Enter Mark Zuckerberg, himself once a young computer genius not unlike Luckey. At the same time, Zuckerberg was (and is) one of the wealthiest people in the world due to his position as the creator and head of Facebook. Like many computer enthusiasts, the youthful CEO of Facebook, Inc. was interested in VR. After a few demonstrations of the technology, he quickly decided to purchase Oculus VR. There was some intense haggling, but the final sale price was over $2 billion, and Luckey, Carmack, and others would come on as executives. The first commercial version of Luckey's HMD, the Oculus Rift, was released in 2016. Luckey had very quickly become incredibly wealthy, and Zuckerberg had acquired a company that might portend the future of computing. Unfortunately, the partnership was not an easy one, and Palmer Luckey left the company he created after only two years. Facebook continues to pour its vast resources into VR and appears to be the leading company pushing the medium, but there was significant competition in the field from the start.

PlayStation VR

In 2016, Sony began pursuing a similar ethos to that of the smartphone VR offered by Google and Samsung, except that they relied on their PlayStation 4 (PS4) console gaming device for processing. The PlayStation VR (PS VR) system banked on the fact that many people already owned PlayStation 4 devices and would thus be willing to pay roughly $400 for the HMD, controllers, a few games, and a motion-tracking camera. While these systems were still far cheaper than buying an Oculus Rift and a powerful computer to run it, the price was steep for many users, and the quality was lesser. The PS VR system nevertheless sold well and had the advantage of lots of content – many PS4 games were 'ported' into VR versions – and Sony appears poised to continue their engagement with VR when the PS5 releases near the end of 2020.

Race to Develop Hardware and Content

The excitement generated by the Oculus Rift spurred a new race to develop and release VR hardware, software, and content among a variety of tech companies. In the months and years following the Oculus Kickstarter in 2012, a virtual explosion of VR systems occurred. Among Oculus's chief competitors was a very similar device, the HTC Vive, which was also released in 2016 through a partnership between HTC (a Taiwanese tech giant) and Valve (an American video game titan). Both the Oculus Rift and HTC Vive used 'outside-in' (or 'room-scale') motion tracking to register user movements. This sort of motion tracking employs external sensors that beam infrared light at the user's headset and controllers. Room-scale tracking allows broad freedom of movement that is nevertheless limited by the 'tether' or cable attached to the powerful PC used to run the devices (Diagram 8.2A).

This has changed with the release of self-contained VR systems like the Oculus Go (2018) and Oculus Quest (2019), which are untethered and use 'inside-out' optical (or camera-based) room-scale tracking coming from the headset itself, allowing users to move freely but limiting processing power (Diagram 8.2B).

In 2017, Microsoft officially entered the VR market with its Windows Mixed Reality system. VR capabilities were automatically included with the Windows 10 operating system, and users would simply need to own a powerful PC and purchase an HMD and controllers. (Windows MR uses inside-out tracking, so no external trackers are necessary.) The Microsoft Mixed Reality system led to the release of a dizzying number of third-party HMDs from companies like Samsung, Lenovo, Acer, Dell, HP, Asus, and more. Many of these devices have entered their second generations, indicating some measure of success,

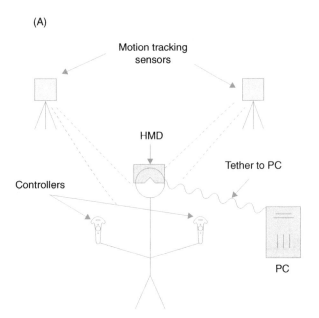

Diagram 8.2A A typical VR system: External infrared trackers sense motion of the HMD/controllers and relay that data to the PC; the headset is tethered to the PC for advanced processing power and storage.

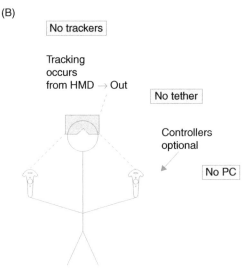

Diagram 8.2B A self-contained VR system: Tracking often occurs from the headset out; controllers are available but not required; no PC or tether is necessary, but processing power and storage are diminished as a result.

but it is notable that Microsoft has so far resisted incorporating VR into its highly successful Xbox line of game consoles, which are direct competitors with the PlayStation system.

Competition among content distribution systems emerged as well, with Valve's Steam, the Sony PlayStation Store (and brick-and-mortar game stores selling PS VR games), the Rift Store, and the Microsoft Store jockeying for position and platform dominance. (In the case of Valve's Steam digital distribution platform, the reach is so broad that most major content – Oculus, Vive, Windows MR – is offered.) Further, Google has a hand in the market as well, with YouTube being the chief distribution platform for 360° video content that plays on all devices. Facebook, in turn, has worked to host VR and 360° video content on their platform, but the company has the disadvantage of requiring accounts for most users, whereas YouTube does not.

In addition to working to surpass one another in terms of market dominance and HMD innovation, many of these companies have developed novel approaches to controller design, with '6DoF' (six degrees of freedom) controllers becoming commonplace. This sort of controller allows for more naturalistic movements along various axes: Left/right, forward/back, up/down, yaw, roll, and pitch. The Oculus Quest uses camera tracking to allow for the option of having no controllers whatsoever – directly rendering a user's hands in virtual space. Other recent movement innovations have focused on walkable VR/MR (particularly with the Oculus Quest, Microsoft HoloLens, and Magic Leap 1), omnidirectional treadmills, trackable shoes, and so on. Ultimately, the more naturally a user can move around in XR, the more believable, immersive, and realistic an experience becomes.

Some companies are noticeably absent from the competition for VR dominance. Given their spectacular failure with the Virtual Boy decades ago, Nintendo's reticence is understandable, but the Japanese company has made small nods to the VR format with its Labo toy set. In 2019, Nintendo incorporated Google Cardboard-style 3D into its existing interactive Labo toy/gaming platform and Nintendo Switch console. The company touts the systems as something of a beginner VR experience for children. Apple has thus far similarly resisted VR but has made huge investments into AR.

LIVING IN AUGMENTED AND MIXED REALITY

Apple AR

Apple's deep venture into AR appears to hinge on a simple premise: Many people already carry a sophisticated AR interface and processor with them all day, every day. The Apple iPhone is an extraordinarily powerful computer capable of astounding and diverse computational tasks. Its gyroscopes, accelerometers, motion and depth sensors, location tracking, and sophisticated cameras make it ideally suited (along with similarly advanced smartphones) to incorporating convincing digital objects into the real world. Of course, such experiences still rely on the 'framed' image of a rectangular smartphone, so immersion is limited (and always is in AR). Yet a rapidly increasing number of applications use Apple's AR features, and the latest iOS software comes standard with an augmented reality tape measure, AR emojis, and more. Copious third-party apps allow for the use of AR on iPhones and similar devices. Amazon's app allows users to place AR simulations of furniture and art in their real homes while online shopping. The Ikea app works similarly. The popular mobile game *Pokémon Go* has AR features that make Pokémon characters appear in the real world, and countless other games incorporate similar functionality. Hundreds of millions of smartphone users already employ basic AR filters that are intrinsic to social media applications such as Snapchat and Instagram. So, AR is becoming an increasingly common part of many people's lives. Rumors have swirled for years of Apple's impending release of a Google Glass-style headset that removes the frame and projects AR images directly into a user's field of view while moving about the world. We must wait and see.

MR: Microsoft HoloLens and Magic Leap 1

However, devices like the hypothetical Apple glasses already exist in some form, though they do not rely on a phone for processing. Microsoft released its self-contained MR/AR head-mounted display, called the HoloLens, in 2016 (see Figure 8.7). The HoloLens allows users to view and move freely about the real world while interacting with virtual images and sounds. Everything from Wi-Fi to processors to speakers and batteries is contained in the headset itself, and users can access the Internet, play games, watch videos, type, and much more while maintaining contact with the real world. Further, the use of the term 'mixed reality' indicates that the device is aware of both physical and virtual objects simultaneously, so virtual objects can interact seamlessly with the physical world, and the

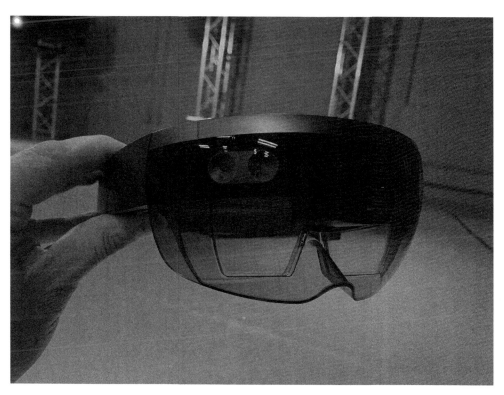

Figure 8.7 Microsoft entered the MR/AR headset industry with the HoloLens.

device can be used anywhere since it digitally scans the user's surroundings. The HoloLens was intended for research and commercial purposes rather than consumer use, and it came with a corresponding price tag: $3,000–$5,000, depending on the level of Microsoft software support.

The Magic Leap 1, released in 2018, is a similar device that does use some outside processing via an external computer small enough to fit in a pocket. Magic Leap, Inc. received over $2 billion in venture capital but faced deeply underwhelming sales upon the release of its device. On the other hand (and despite the cost), the HoloLens was a success, and the HoloLens 2 was released in 2019. Like the original HoloLens, the new iteration is not aimed at average consumers; it again runs about $3,500 plus additional monthly subscription costs for developer and enterprise support. The HoloLens 2 includes improved processing, optics, and other specifications. It also strengthens multi-user networking features, real-time mapping, and advanced interactivity while allowing the user to remain in contact with the real world. In these ways, the HoloLens is more an advanced successor to Google Glass than a pure gaming or entertainment device, though it is capable of many things. (Again, it is not capable of full VR.) Increasingly adopted by high-end manufacturing and industry, health care, and education, the device is poised to change such fields through sophisticated interactivity. For example, the U.S. military recently purchased roughly half a billion dollars' worth of modified HoloLens 2 devices for use on the battlefield.

ZOOM IN 8.3

Check out a military application for the Microsoft HoloLens system at: www.cnbc.com/2019/04/06/microsoft-hololens-2-army-plans-to-customize-as-ivas.html

MR Multiplayer Entertainment

As noted, MR (or mixed reality) can be something of a marketing term for companies like Microsoft and Magic Leap, but the term also denotes high-level interactivity (sometimes called 'scene awareness') between the real world and the virtual world. For example, the HoloLens features games where digital adversaries can hide behind a user's own furniture. This idea has been adapted by entertainment enterprises. Currently, two major companies are working to make free-roaming, destination MR into the 'next big thing.' The Void game space uses MR technology to place multiple players simultaneously inside fantasy stories involving superheroes, monsters, and aliens. These players are able to move freely about large indoor spaces while wearing lightweight computer backpacks running powerful HMDs, headphones, and controllers. The Void couples advanced VR hardware and software with real-world elements like ramps, stairs, doors, and walls. These features allow players to interact with physical objects while moving naturally – and all together – through a computer-generated world, increasing both immersion and presence. The Void has access to The Walt Disney Company's intellectual property and has used it to create popular interactive MR simulations for *Star Wars* and Marvel's *Avengers*. Similarly, Dreamscape Immersive owns licenses to DreamWorks intellectual property like *How to Train Your Dragon* and has set up destination VR facilities in movie theaters run by AMC, one of its major backers. Both of these companies operate on common assumptions: that MR needs to be a highly immersive, social experience enjoyed with others; that people will pay for a short 'ride' (approximately 20 minutes); and that recognizable intellectual property will draw users into the experience. This use of MR

ZOOM IN 8.4

Be a virtual part of your favorite film or travel to an alien zoo at this VR site

DREAMSCAPE: https://dreamscapeimmersive.com

is not unlike going bowling or paying for an 'escape room' adventure. In these cases, users do not own the expensive MR hardware themselves, but instead go out to have a destination experience in the same way that one might visit a movie theater, arcade, or water park. This kind of XR experience allows people to do something they could never do at home.

PROMINENT USES AND OTHER FUNCTIONALITY

Uses of XR

While VR gaming has clearly motivated innovators like Palmer Luckey and companies like Sony/PlayStation, entertainment is far from the only use of XR technologies. We have seen that academia, scientists, and the military have keen interests in XR. Real estate companies are increasingly using virtual and augmented realities to conduct remote tours of properties for rent or sale. The potential for education is boundless; universities are progressively incorporating XR programs into their curricula, and Google has a whole XR platform, called 'Expeditions,' dedicated to primary and secondary education. Global industries and manufacturing have made major investments in devices such as Google Glass and the HoloLens, and XR training – particularly in dangerous fields or with irreplaceable materials – is becoming conventional. Walmart even conducts employee training and evaluations using VR headsets! Quite prominently, XR is used by healthcare companies, particularly for therapy and rehab.

Figure 8.8 Henry Weber and Veena Somareddy.

VEENA SOMAREDDY AND HENRY WEBER, NEURO REHAB VR

What is your job? What do you do?

Veena: I'm the CEO and CTO of the company. I work on the day-to-day operations, product development, and set overall goals and milestones for the company, including present and future strategic vision.

Henry: I work on business development, marketing, and sales for Neuro Rehab VR, the leading provider in virtual reality physical and occupational therapy. I implement marketing strategies through various traction channels to introduce new technology and build brand recognition in the healthcare and rehabilitation industry. I'll manage the inbound and outbound conversations with the ultimate goal of generating company revenue through the sale of our XR Therapy System medical device to qualified hospitals and outpatient rehabilitation facilities.

How long have you been doing this job?

Veena: I have been working on the company for three years. In the first year, I was pursuing my Ph.D. full time along with working on the company. In 2018, I switched to working on the start-up full time.

Henry: I've been working with Neuro Rehab VR for almost two years now.

What led you to the job?

Veena: I have a background in virtual reality research and development and was working on training and simulation in VR for the medical field when Bruce Conti, co-founder, contacted me to develop VR for the patients in his neurological rehabilitation clinic. I thought it was an excellent opportunity to put my skills and knowledge to work.

Henry: I met Neuro Rehab VR co-founder Veena Somareddy at SXSW after talking about the various projects we were working on at Ohio University with VR in health care. As a graduating student from Ohio University, I was on the lookout for jobs that fell within this niche intersection of both virtual reality and health care.

What advice would you have for students who might want a job like yours?

Veena: Every entrepreneur needs to have some level of interdisciplinary skill. Take classes outside of your major or degree program, read books on the industry you want to start a business in, et cetera. Never stop learning, even after you have completed school. Take online courses and read niche blogs so you are up to date on the latest skills and trends in the industry of your choice. Intern at companies that are working on the technology or industry that you are interested in so you can get hands-on experience.

Henry: Look to work for a start-up. I enjoy the responsibility and work experience of wearing a lot of different hats to accomplish our overall goals. I enjoy being able to utilize a lot of different skill sets and talents to help make our company better off and more successful than it was yesterday.

Why use VR in therapy and in health care in particular?

Both: We have seen firsthand the limitless potential and amazing benefits for both physicians and patients when utilizing virtual reality technology in health care. Some of these use cases include educational simulations, physician training, PTSD exposure therapy, pain and anxiety alleviating experiences, concussion diagnosis, physical therapy, neurological rehabilitation, cognitive training, and more.

There is a lot of momentum going toward telehealth solutions where patients can be assessed, treated, and monitored remotely. We believe virtual reality has the potential to be a top tool to help conduct a lot of telehealth services. It's in our company's vision to eventually bring therapy to the patients at home through virtual reality technology.

What makes it special or uniquely useful?

Both: In rehabilitation, which can sometimes be tedious and mundane for patients, we provide a new, fun, and engaging therapy experience through virtual reality that incorporates activities of daily living and other functional tasks that help expedite recovery. Complex virtual therapy exercises are created with precise control over the stimulus and the cognitive load that the user experiences. This form of therapy helps patients to utilize more of their brain power to recover and rebuild neurological pathways. This is proven to be a lot more valuable than two-dimensional (2D) virtual therapy practices with no functional interactivity at all. Our customized virtual therapy exercises also record physiological and kinematic responses, therefore quantifying the progress of the patient with scores and metrics over time.

What is the state of XR in the healthcare industry, from your perspective?

Both: The healthcare industry is gradually adopting new technologies such as XR and AI. In 2017, there was still a lot of effort to educate our potential customers about virtual reality. Now, most people are aware of what these technologies are and know their potential before they even contact us. Eventually we hope to get to a point where every hospital or clinic has a VR headset with various applications to utilize.

What career options are there for XR in health care?

Both: There has been a 1400% increase in job posts for VR/AR Software Engineers in the past year, indicating a significant increase in the number of start-ups and Fortune 500 companies thinking about their VR/AR strategies. For a start-up in the healthcare field to be successful would require VR software developers, full stack engineers, marketing, sales, business development, regulatory compliance, and project management personnel. Therefore, the jobs are versatile, but prior knowledge and enthusiasm for the VR/AR tech and gaming will always increase the chances of successfully landing jobs in the industry.

How would someone break into the healthcare and XR field?

Both: For any start-up with a good healthcare application idea, clinical expertise or access to someone with the knowledge and know-how of the healthcare industry would be helpful. The healthcare field has stringent regulatory compliance, along with payment policies that vary for different verticals you might target. Understanding the payer and provider dynamic and the pain points that your customers are dealing with is critical to breaking into the healthcare XR field.

One of the key human faculties that makes XR therapy so compelling is called neuroplasticity. This sounds complex, but the term essentially means that the human brain grows and changes as people have new experiences or go through trauma. Neuroplasticity is far from unique to VR; for example, a serious accident or emotional upheaval can cause the human brain to grow and change just as easily as traveling to a new country or learning a musical instrument. But, if XR developers can create virtual experiences that the brain treats as real, then simulations can be used to induce neuroplastic responses. When in a properly functioning simulation, your brain and body largely 'forget' the physical world and inhabit the digitally created world. What occurs in the virtual world can thus induce cognitive responses that help grow new synapses in the brain. The ability of XR to use experience to change the brain is a basic principle behind VRT (virtual reality therapy), a promising area of medical and psychological rehabilitation and counseling that uses VR experiences to treat conditions such as PTSD, phobias, and neurological damage from strokes. Institutions such as the University of Southern California (USC) and Stanford University have been experimenting with neuroplasticity and VR for many years (in the case of Stanford, some of this work intersects with research on VR and homuncular flexibility). The results have been promising. Essentially, VR therapy can place the patient or client into a repeatable series of simulations that feel real to the patient's brain. These virtual scenarios can be altered in numerous ways so that the brain is forced to contend with new experiences and can thus potentially grow and change. The patient is also provided with a 'safe' environment wherein the simulation can be ended at any time. This latter capacity is especially important for those suffering from PTSD, and USC has successfully employed VRET (Virtual Reality Exposure Therapy) to assist traumatized military veterans. Beyond these uses, though, we can imagine that XR could potentially be applied to teach users *anything* so long as the simulation feels real enough.

AV

Certainly, VR and AR/MR are familiar to many users today, but AV is more obscure. Despite its rarity, augmented virtuality holds great potential. The essence of AV lies in its ability to achieve *nearly* total immersion into a virtual world. AV stops just short of full VR. It allows for users to experience a primarily virtual environment while still maintaining significant interactions in the

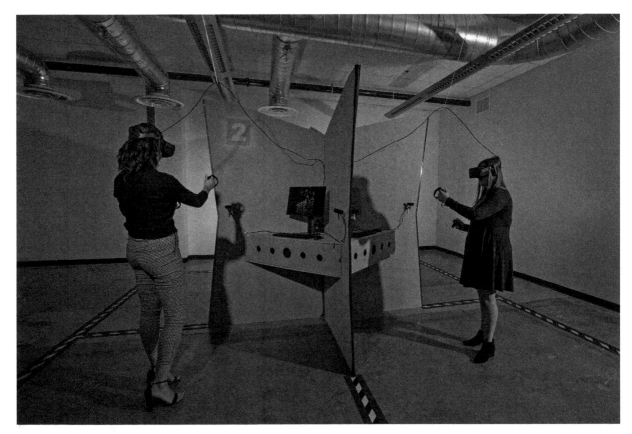

Figure 8.9 VR has also been used in education. In this picture students are using a learning arcade that allows four students to experience VR in an Organic Chemistry class. Using gamification, students learn how to build molecules from atoms and bonds in a VR setting.
Source: Courtesy Advanced Media Laboratory, Northern Arizona University

physical world. This virtual environment becomes useful for applications like remote, guided surgery or intricate repair of complex machinery and electronics. For example, imagine that a competent jet engine mechanic is on a flight far over the Pacific Ocean. Her flight is many miles from land, and there is a mechanical problem that requires immediate attention. AV would allow a virtual overlay of the engine's components – and could even provide remote virtual guidance from a more experienced mechanic via streaming video – while she repairs the engine in the real world. AV means that the virtual and real worlds fuse, and therefore it lies just next to full VR on the virtuality continuum (see Diagram 8.1).

360° Video/CineVR

CineVR is short for 'cinematic virtual reality' and is another term for 360° video, which also goes by the name of VR video, immersive film, and so on. CineVR is easily accessible in that it can be viewed without expensive computers. It functions within YouTube, Vimeo, and Facebook and can be experienced without headsets on smartphones, tablets, and browsers. The medium is both partially immersive and moderately interactive when viewed in an HMD, but 360° video is not full VR. Its interactivity is limited to allowing users control over what they see and hear when spinning around or looking up or down, so the user's vantage of the video world changes based on input, but viewers are generally unable to move into the virtual world or interact with the objects and entities within it. CineVR can be live action or animated, and 3D formats have been attempted. To be sure, the 360° format is interactive in ways that film and television are not, and it provides audiences with choices that are largely impossible in traditional audiovisual media. This level of interactivity provides great opportunities but also creates difficult situations for filmmakers. If there is a key instance of narrative importance in a CineVR film – say, a gun is dropped in a bush – it is quite likely that viewers will be looking somewhere else entirely. The audience may be drawn to a cute puppy or a striking cloud formation, completely overlooking the discarded weapon. Since the 360° format has no frame, everything is available, but nothing is specified. CineVR filmmakers have therefore had to devise new methods of forcing audience attention and perspective, including specially timed edits and the highly intentional use of sound. Because this medium requires no computer programming skills (unlike VR production) and can be undertaken with cameras that cost less than a cheap laptop computer, 360° video is highly accessible. It is used extensively in music videos, training, journalism, short documentaries, and introductory XR production, but feature-length films have so far largely eschewed the format.

Immersive Audio and Spatial Audio

Media anywhere along the virtuality continuum (Diagram 8.1) are likely to employ some form of audio. The importance of stereo sound was covered in the prior section, but audio for XR goes well beyond the innovation of merely having separate audio channels for the left and right ears. Immersive audio is produced so as to make the listener feel *inside* of sound. This category includes stereophonic sound and surround sound of various types but generally not mono sound. XR often employs binaural audio that imparts a 3D sensation by recording with special microphones and playing back on headphones in order to mimic human hearing. Of particular importance to XR is a type of immersive audio called spatial audio. These terms can be somewhat tricky, but spatial audio is essentially a three-dimensional sound space (think of being at the center of a sphere of sound) that reproduces as closely as possible a typical human listening experience. Spatial audio goes further than binaural audio by placing listeners at the center of the action and allowing sounds to change as users or their digital avatars move. A production technique called Ambisonics helps produce spatial audio, and motion tracking is essential for proper reproduction during playback. This type of sound is increasingly used in video games, virtual recording programs, movies, and especially in VR/AR and 360° videos. Spatial audio is supported by platforms like YouTube and Facebook, and Facebook provides one of the leading software programs for spatial audio production.

SEE IT LATER

Like many forms of emerging media, it is difficult to precisely predict what the future of XR will be. After all, few people could have foretold the explosive rise of VR following the Kickstarter project of an unknown teen like Palmer Luckey. Nevertheless, there are certain technologies that seem likely to make an impact on various facets of the XR field, among them 5G networks, improved haptics and nerve-signal tracking, advances in locomotion technology, increasing computational and graphics power, wearable tech (especially some sort of AR/VR glasses), next-generation computing, fuller sensory immersion, and AI/machine learning.

Challenges

In the near term, though, XR faces some unique challenges. Some prominent issues for XR include cost/accessibility, motion sickness, and – especially for full VR – the uncanny valley. The uncanny valley is an experience wherein realistic animated or robotic depictions of humanoids can feel creepy or off-putting. This phenomenon often causes discomfort. In order to avoid 'uncanniness,' computer animation will need to significantly advance in the coming years.

ZOOM IN 8.5

UNCANNY

Find out about the uncanny phenomenon and human reactions to robots at the Uncanny Valley site: https://spectrum.ieee.org/automaton/robotics/humanoids/what-is-the-uncanny-valley

Subjective difference means that individuals will experience VR in different ways. What feels real to some may feel nauseating (or uncanny) to others, and VR developers must work to create experiences that are broadly palatable, but individual variation is one of the stumbling blocks of generating presence. VR sickness (variously called cybersickness or VR-induced motion sickness) is still common, but it is less widespread than during the 1990s VR boom, when many users were disaffected by their experiences. Dizziness and nausea often result from a conflict between sensory experiences, such as sensing movement while standing still.

Similarly, unnatural movements like flying can cause VR-induced motion sickness. These sickening feelings, while not unusual, have been largely eliminated by newer VR technology. Again, what is required for effective VR simulations is, in large part, that the human brain 'buys in' to the simulation, which is what allows the virtual environment to feel real for each user.

As we have seen, XR can be quite expensive. Full VR in particular will never become a widely accepted technology if no one can afford it. One future development that is likely to impact XR is the continually decreasing cost of high-quality systems. The price has already begun to change somewhat with Oculus's self-contained headsets such as the Quest and the Go, which cost less than $500 and $200, respectively. According to Moore's law, computer technology improves exponentially every few years. This rapid increase in computing power is generally accompanied by a decrease in the cost of technology featuring older specifications. The principle applies to XR gear as well, of course, and it certainly does not hurt to have Facebook, one of the wealthiest and largest corporations in the world, pumping huge sums of money into XR research and development.

ZOOM IN 8.6

Moore's law predicts that computer power will increase and relative cost decrease. Read more about it at: www.intel.com/content/www/us/en/silicon-innovations/moores-law-technology.html

XR Companies of the Future

New companies will no doubt emerge to face the challenges of cost and motion sickness, but many of the old players will surely persist. Facebook/Oculus is leading the way with walkable, entirely self-contained headsets like the Quest and the Go, and more companies will probably follow them. (The Oculus Quest has been so popular, in fact, that the company released a super-charged new version, the Quest 2, in October 2020 while promptly discontinuing both the Quest 1 and Go HMD systems.) The ability to move about freely in an HMD and have motion tracked from the headset addresses a number of problems with tethering and limited locomotion. Users might not need omnidirectional treadmills if they can use VR outdoors and in public spaces. In wide open areas, walkable VR can mimic real-life locomotion because it will not be constrained by the generally small spaces of interior rooms. (While outdoor or public VR use may seem somewhat distasteful at present, many people use laptops and tablets in these spaces already. VR may simply be the next step.) Facebook has also developed novel approaches to haptics and control. One clear goal of XR developers is the elimination of peripherals such as mice, trackpads, game controllers, keyboards, and even screens. Many companies see XR as the next generation of computer technology – a continuation of the evolution from desktop to laptop to smartphone to wearable, wherein each successive computer is smaller, more powerful, closer to the body, and more seamlessly integrated with natural human activity. In the interest of such evolution, Facebook purchased a company called CTRL + Labs in Fall 2019. CTRL + Labs works primarily in 'nerve tracking' technology whereby a sensing device worn on the wrist or forearm reads nerve impulses sent from the brain, down the arm, and to the hands and fingers. A computer system could thus 'know' what a user is intending to do with her hands and would not have to rely on optical tracking (i.e. images) of hand movement. This technology would potentially allow for precise virtual typing without keyboards, spatial computing, seamless gestural control, and tracking of hand movements in pitch black or occluded conditions. The purchase of CTRL + Labs therefore appears to move Facebook and Oculus one step closer to the future of XR and computing, creating a world where the screen is worn on the face and the system is controlled by movement, voice, and even thought. Finally, Facebook has taken a novel approach to the geotagged photographic data that their users upload to their social media sites. Such an approach involves recreating the real world based on this datamined photographic and geopositioning information. Using what Facebook calls 'live maps,' the corporation has announced plans to virtually recreate the world for users of their impending AR smart glasses. Users could look at a real-world object, person, or location and have instant access to virtual information gleaned from the Internet. AR, Facebook seems to think, will one day be everywhere.

Apple has yet to really enter the XR market beyond the inclusion of sophisticated AR capabilities on their more advanced iPhone and iPad devices. Yet rumors have persisted for many years of the massive company's plans to create wearable XR glasses that utilize phone technology for processing. With Apple's dominance of the smart watch market, their entry into further wearables like smart glasses is but a small leap. And since Facebook has formally announced its own plans for such a device, it is difficult to imagine Apple ceding the market to a competitor. Further, Apple's code for their smart glasses appears to be buried in their most contemporary iOS operating systems for phones and tablets. Apple will likely continue its development of AR and extend its reach into new headset technology that incorporates its flagship iPhones.

Microsoft is already falling behind in its support of Windows Mixed Reality, and it does not seem inclined to keep up with Oculus's Quest and Go devices. But the HoloLens has absolutely revolutionized what AR/MR headsets can do, and it has been successful as an enterprise device in spite of its high cost. It is a safe bet that the company continues to develop the HoloLens and similar devices. With the failure to incorporate VR into the next-generation Xbox Series X console (released in November 2020), Microsoft seems willing to let Sony's PlayStation 5 (also released in November 2020) continue to be the leading VR home gaming console. Not much is known about Sony's plans to update its VR hardware – which is rudimentary compared to more powerful systems like the Oculus Quest or Valve Index – but Sony has indicated it will continue to support VR and release new titles for its next console system, which is expected to see great consumer demand.

Myriad other companies will undoubtedly continue to produce XR systems, content, and distribution platforms. Google often experiments with various hardware configurations, but its most visible investment is in supporting spatial audio and CineVR on YouTube, which is likely to endure and advance. HTC will probably keep producing headsets like the Vive Pro, which stands alongside the best headsets made by Oculus and Valve. For Valve, its new Index headsets and controllers indicate a fresh investment in VR, and the company released an award-winning and best-selling VR game, *Half Life: Alyx*, in 2020. Always enigmatic, Valve is difficult to pin down, but further AAA VR games from the company are likely, and their Steam platform is probably the most popular content distribution system for VR. This popularity shows no signs of waning.

Before the COVID-19 pandemic, the idea of more multiplayer and destination VR entertainment facilities, like The Void and Dreamscape, seemed a safe bet. That is all likely on hold for a few years, at best. What seems a near surety is the rise of AR. Since a large portion of the world has devices (smartphones and tablets) that support augmented reality, the AR/MR interface has achieved a prominence that VR headsets will take decades to match. If anything, some sort of VR/AR convergence may occur in the ensuing years, wherein one set of glasses can do both AR and VR, but powerful technology will have to get much smaller for this to be fashionable in daily life. Very few people would walk around town with the HoloLens on their faces. AR has the further advantage of maintaining sociality since the real world is visible along with the digital. For those who check their smartphones constantly, connected AR glasses might actually increase face-to-face social interaction because the phone will be *on their faces*.

One development that has the potential to change so very many things is fifth-generation wireless technology, or '5G.' 5G exists today but is not widely distributed. The subject of much hype and speculation, 5G promises to rapidly speed up all data transferred via cellular networks and will likely achieve speeds at least 100 times faster than 4G. Regardless of when this next-generation technology becomes broadly employed, there is little doubt that 5G will one day become the global norm. Many technologists argue that 5G will make 'everything' better and faster, and it is true that smartphones are now used for myriad different tasks. But the adoption of next-gen cellular tech matters for XR because most processing in XR happens locally – i.e. on a computer, inside a self-contained headset, or on a smartphone. Streaming XR requires immense amounts of bandwidth and is usually accompanied by degraded image and sound quality and deeply frustrating latency (lag between visuals and audio or delay between input and output). The global adoption of 5G will mean that many XR-related processing functions could move to the cloud, but it will also allow for instantaneous, low latency XR streaming and online gaming. Professional sports organizations and concert promoters have long dreamed of selling live-streaming tickets to popular, limited-seating events. A headset coupled with 5G tech would mean that live VR events could attain truly global, simultaneous reach. The realism offered by XR paired with the speed and power of 5G would let geographically distant users experience a live event as though they were actually there. Massively multiplayer online VR gaming will become a reality with 5G, and AR data pulled from the Web would be delivered instantaneously. A user with 5G AR glasses could walk down the street and look at a restaurant, and their glasses would instantly pull up a menu, wait times, reviews, and the like. To be clear, 5G would be a huge step toward making XR the next-generation computing platform that companies seek.

Aspirations

One definite aspiration of contemporary VR companies is increased immersion, with the ultimate goal being full immersion such that it generates the feeling of being completely present within the virtual world. Full immersion may seem simple at first, but consider everything that must be digitally replicated in order to generate the feeling that a virtual world is deeply 'real.' Science fictional depictions of fully immersive VR – such as that shown in the novel (2011) and film (2018) *Ready Player One* – are not remotely possible today. Still, Luckey's innovations with Oculus proved that rapid leaps in VR technology are possible after many years of relative idleness. Consider just two of the many things that will be needed for such a world to be created: Full sensory immersion and realistic interaction with physical objects.

One major challenge for both contemporary and future VR development rests in the ability to interact with virtual objects as though they were physically real. Imagine grasping a disposable paper cup in real life: You may feel the temperature of the liquid in the cup, notice the texture of the waxy paper material, touch condensation on the outside of the cup, and variously squeeze the cup lightly or crush it to pieces. VR lacks the ability to fully model such multifaceted sensory experiences, but it has an especially hard time convincing your brain that you are holding and manipulating something physical. Objects like guns, which gamers are generally used to controlling, can be simulated fairly well. Textures, temperatures, and gradations of pressure are much more difficult to model, and something like clasping another warm human hand, with knuckles and callouses and skin texture, is unthinkable. At present, most VR systems use sophisticated extensions of video game controllers for haptics, but much more will be needed for full immersion in the future.

Similarly, in the future, companies may move to increase incorporation of as many sensory experiences as possible into VR – a very old aspiration that can be seen in Heilig's *Sensorama*, among other places. While contemporary VR hardware and applications lean heavily on images, sounds, proprioceptive cues, and haptic feedback, the addition of smell, taste, temperature, more sophisticated haptic information, and the like will be necessary. It is true that we can enjoy and believe VR without taste or smell, but we will enjoy and believe it *more* with those sensory experiences integrated. Again, this is an area of research today. One company, Project Nourished, is developing virtual dining experiences wherein users have all their senses activated in VR, including smell and taste. Whether this proves to be the path of VR in the future or not is obviously unclear.

ZOOM IN 8.7

FULL SENSORY IMMERSION: PROJECT NOURISHED

Project Nourished, a VR start-up company, is working to incorporate every one of the human senses into its virtual reality dining applications: https://intelligence.wundermanthompson.com/2018/02/jinsoo-an-founder-and-ceo-project-nourished/

If Facebook and Apple's dreams come true, and AR is eventually everywhere, it is likely that media and advertisements will be everywhere too, so long as users are wearing their headsets (glasses). Just as smart devices featuring Amazon's Alexa assistant, Google's assistant, or Apple's Siri are always listening (a feature that may be turned off but usually is not), AR glasses would be continually scanning the real world for data. This scanning would likely include opportunities for companies to geolocate, do facial recognition, track user preferences and interests, post ads and other content, provide notifications or useful information, and so on. It is not far-fetched to imagine a world where AR is ubiquitous and the real world has a perpetual skin of digital objects overlaid on top of it. Consider a popular augmented reality game like *Pokémon Go* (2016): Users wearing AR glasses could have Pokémon jump into their field of view at any time and would catch the creatures with their hand gestures. This 'AR-everywhere' approach would be both a blessing and a curse, providing the digital advantages of constant access to games and Internet data but also creating non-stop distraction. There are obvious privacy concerns regarding perpetual facial recognition, location tracking, and data mining. (In fact, one AR glasses company, Rokid, recently incorporated heat-sensing capabilities into its devices so that users can monitor other people's temperatures in public to avoid contracting coronavirus!) Dependency would also be a concern; many people in the 21st century are, literally and figuratively, lost without their smartphones. This seems a likely outcome for AR glasses as well.

FYI:

The artist Keiichi Matsuda has created a short film called *Hyper-reality* that imagines a world where AR is everywhere and the physical and virtual worlds are difficult to separate: http://km.cx/projects/hyper-reality

Future Complications

Along with technological evolution and the introduction of new gear and content for XR platforms, there will inevitably be social, economic, ethical, behavioral, and legal developments related to XR. One of the chief concerns for the future of XR will be attempts to reckon with the ethics and legality of these powerful new technologies and the ways in which they are used.

Take 'deep fakes,' for instance. Machine learning algorithms and sophisticated programmers have already been able to create realistic audio and video facsimiles of real people, and now this technology has been extended to mimic the handwriting and even writing style of essentially anyone.

ZOOM IN 8.8

Advanced computer technology and artificial intelligence (AI) make digital falsification ('deep fakes') nearly indistinguishable from the real thing. What will happen when deep fakes are incorporated into XR?
Deep Fakes: www.theguardian.com/technology/2020/jan/13/what-are-deepfakes-and-how-can-you-spot-them
   See if you can tell what is real and what is a deep fake here: www.youtube.com/watch?v=-QvlX3cY4lc

All of these facsimiles can be achieved without the subject's permission. There are serious ethical pitfalls here, clearly, but what happens when such techniques are extended to VR? Science fiction television programs such as *Black Mirror* have already explored some of the far-reaching implications of such technologies. Imagine what 'sentient code' (software programs that can think and feel) might be like when paired with a deep fake VR representation of a celebrity or a deceased loved one. What laws and ethical frameworks might apply to a realistic VR simulation of a real person who has not consented to his or her own simulation? (This is already a pointed question for purveyors of virtual performer programs of deceased celebrities like Tupac, Ronnie James Dio, and Michael Jackson.) In fact, VR is already being used for these purposes today. In 2020, a South Korean company simulated a 'living' virtual version of the deceased child of a grieving mother. And, while the VR child could not think or feel, her simulation modeled her voice, appearance, demeanor, and movement so realistically that the mother could 'touch' and talk to her child in VR. The grieving mother was moved to tears. We can only expect more of this, and our existing ethical and legal frameworks have so far done very little to incorporate the advanced technologies we have now, much less what might occur as XR and AI rapidly advance. Similarly, we can imagine privacy concerns, ceaseless advertisements, virtual crime, virtual bullying, and virtual torture to appear with advancing and widely adopted XR technologies, just as all of these maladies have accompanied earlier innovative and ubiquitous technologies such as the Internet.

FYI: AFTERLIFE IN VR

Virtual Afterlife: To aid in the grieving process, a South Korean mother was 'reunited' with her deceased daughter using VR technology. This practice poses ethical dilemmas but may become increasingly common in the future: www.reuters.com/article/us-southkorea-virtualreality-reunion/south-korean-mother-given-tearful-vr-reunion-with-deceased-daughter-idUSKBN2081D6

Any professor will tell you that laptops and smart phones are major problems in the classroom. While these devices do much to enhance learning, they also provide instant access to innumerable distractions: Social media, email, shopping, videos, games, and so on. What happens when all of these same distractions are placed in wearable headsets that are virtually indistinguishable from contemporary eyeglasses? Further, if much of the world's population spends too much time on our phones – looking at them at dinner, when crossing streets, when other people expect our attention – what will happen when XR technology is potentially always on our faces? A related future concern is addiction. Gaming addiction is a real but limited problem today, and it is easy to imagine that advanced, highly immersive VR would be preferable to the real world for many users. In fact, this is a major tension in the popular book and movie *Ready Player One*: The world's problems have gotten so severe, and the virtual world has become so compellingly realistic, that most people have retreated to their virtual lives and let the outside world fall into dystopian decay. Clearly VR is neither properly advanced nor widespread enough for this to happen today, but look at what has happened since Facebook bought Oculus just a few short years ago.

There is a very long history of suspicion, fear, and anxiety regarding the impact of art and media on both individuals and culture more broadly. Plato was adamant that art should be stringently censored in order to avoid what he saw as its toxic influence on society. Many readers will recall the ongoing debates about violence and sex in film, television, music, and video games. Sometimes, these anxieties are about the content of the medium, and at other times, people fear the medium itself (of course, it can also be both). The introduction of photography in the 19th century and radio in the 20th century was accompanied by spiritualist claims that these new technologies allowed connections between the realms of the living and the dead. And who can forget that the evil spirits in Tobe Hooper's film *Poltergeist* (1982) snatched a little girl through the family television set? Clearly, these historical fears regarding new media are often represented through narratives in contemporary art and cultural productions. VR technology is, in this way, no different. The 1990s' VR boom was accompanied by a multitude of pre-millennial films expressing deep trepidation about the mind-altering and reality-warping capabilities of VR. The most famous of these films is no doubt *The Matrix* (The Wachowskis, 1999), but many other films traveled the same ground: *Lawnmower Man* (Leonard, 1992), *Virtuosity* (Leonard, 1995), *Dark City* (Proyas, 1998), *The 13th Floor* (Rusnak, 1999), *Existenz* (Cronenberg, 1999), and many more. Each film has a slightly different take on how VR technology works, but fear of the medium is common to all, and the message of each film is acute: Immersive, computer-generated realities are powerful, potentially deceptive, and have the capacity to control or destroy humans by erasing the lines between the real world and the digital world. Like most science fiction, the capabilities of the technologies are exaggerated, but the worries are very real. It is no shock to find similar themes cropping up to accompany the current XR boom. Novels like Ernest Cline's *Ready Player One* (2011) and television programs such as *Westworld*, *The I-Land*, *Kiss Me First*, *Upload*, and *Black Mirror* are again exploring the 'dangers' posed by sophisticated XR technologies, and many of these intensify the earlier historical anxieties by adding fears about AI and robots. Again, while Sutherland's dream of the 'Ultimate Display' seems far-fetched at present, few people predicted the current XR revolution spawned by Palmer Luckey's device. When will the next revolutionary development appear, and how much do human beings really have to fear from intelligent machines and mind-altering media?

SUMMARY

This chapter began with the recognition that human beings have long sought to have art and media become seamless parts of their lives. Living in an alternate digital reality, where our surroundings and interactions are fantastically imagined and creatively constructed in part through digital means, provides nearly limitless possibilities. These are the promises of VR. Similarly, many people also enjoy remaining grounded in their everyday realities but having their experiences enhanced with interactive digital content. These are the promises of AR and MR.

This chapter has covered the history of how earlier media technologies have contributed to the deeply hybrid medium of XR and how closely related forms – such as cinema, television, and games – have approached the unique aspects of XR but stopped short. Immersion, presence, and interactivity combine in 'extended reality' to create something that is new, or at least it feels that way. Then, it thoroughly defined what all of these 'realities' mean and how they work. It traced a history of innovation and commercialization of the technologies, emphasizing the explosion that occurred after the Oculus Rift emerged in 2012. Finally, this chapter detailed prominent uses for XR and the problems and possibilities it presents.

XR is not a new technology, to be sure, but many of the advancements of the last decade are in fact new. The rudimentary XR technologies of the 1950s–1970s and the exciting but stunted innovations of the 1980s–1990s paved the way for the rapid expansions kicked off by Luckey's invention of the Oculus in 2012. While XR technologies are not yet ubiquitous parts of most people's daily lives, that time will soon come. Much of the world is old enough to remember a time when carrying a phone in a pocket or bag all day, every day seemed completely ludicrous. Now, for better or worse, most people cannot live without their smartphones. How long will it be until humans cannot imagine a world without XR?

BIBLIOGRAPHY

Bailenson, J. (2018). *Experience on demand: What virtual reality is, how it works, and what it can do*. New York: W.W. Norton and Company.

Cline, E. (2011). *Ready Player One*. New York: Broadway.

Ewalt, D. M. (2018). *Defying reality: The inside story of the virtual reality revolution*. New York: Blue Rider Press.

Fink, C. (2018). *Charlie Fink's metaverse: An AR-enabled guide to VR & AR*. Washington, DC: Cool Blue Media.

Garner, T. (2018). *Echoes of other worlds: Sound in virtual reality*. Cham, Switzerland: Palgrave Macmillan.

Ghaffary, S., & Molla, R. (2020, February 11). *How tech companies are trying to make augmented and virtual reality a thing again*. Vox. Retrieved from: www.vox.com/recode/2020/2/11/21121275/augmented-virtual-reality-hiring-software-engineers-hired

Harris, B. J. (2019) *The history of the future: Facebook, Oculus, and the revolution that swept virtual reality*. New York: Dey Street.

Inoue, M., Kurosawa, H. [Producers], & Kurosawa, A. [Director]. (1990). *Akira Kurosawa's Dreams* [Motion Picture]. Japan: Warner Brothers.

Jerald, J. (2015) *The VR book: Human-centered design for virtual reality*. London: Morgan and Claypool.

Lanier, J. (2017). *Dawn of the new everything: Encounters with reality and virtual reality*. New York: Henry Holt and Company.

Levy, S. (2020) *Facebook: The inside story*. New York: Blue Rider Press.

Milgram, P., & Kishino, F. (1994). A taxonomy of mixed reality visual displays. *IEICE TRANSACTIONS on Information and Systems*, 77(22), 1321–1329.

Parisi, D. (2018) *Archaeologies of touch: Interfacing with haptics from electricity to computing*. Minneapolis: University of Minnesota Press.

Plato., Bloom, A., & Kirsch, A. (2016). *The Republic of Plato*. New York: Basic Books.

Rubin, P. (2018). *Future presence: How virtual reality is changing human connection, intimacy, and the limits of ordinary life*. New York: HarperOne.

Scoble, R., & Israel, S. (2017). *The fourth transformation: How augmented reality and artificial intelligence change everything*. USA: Patrick Brewster Press.

Chapter nine

Advertising: From Clay Tablets to Digital Tablets

Advertising plays an important role in the U.S. economy and fulfills many consumer needs. Most people tend to think of advertising as a modern-day phenomenon when actually it has been around for thousands of years. Today, advertising is a very complex business that employs principles of psychology, sociology, marketing, economics, and other sciences and fields of study for the end purpose of selling goods and services.

In today's world, advertising is not limited to the typical radio or television commercials but also includes awareness-creating endeavors such as T-shirts, pens, and coffee mugs imprinted with a brand, store, or Web site logo, free products, sponsorships of 5k runs, and tweeting customers a product-related message. Advertising is no longer 'non-personal' and is not always 'mass mediated.' New online and personal technologies have transformed advertising into an "art of engagement" (Othmer, 2009).

Getting products and services to customers is a multistep process that starts with **marketing**, which is "the process of planning and executing the conception, pricing, promotion, and distribution of ideas, goods, and services to create exchanges that satisfy individual and organizational objectives" (Vanden Bergh & Katz, 1999, p. 155). Products and services are then advertised to potential customers. Thus, advertising is a function of marketing.

Traditional industry models define *advertising* as "non-personal communication for products, services, or ideas that is paid for by an identified sponsor for the purpose of influencing an audience" (Vanden Bergh & Katz, 1999, p. 158), and similarly, advertising is any "form of non-personal presentation and promotion of ideas, goods, and services usually paid for by an identified sponsor" (Dominick, Sherman, & Copeland, 1999, p. 397) or "paid, mass mediated attempt to persuade" (O'Guinn, Allen, & Semenik, 2000, p. 6).

This chapter begins with an overview of the origins of advertising and moves to the contemporary world of advertising on radio, television, cable, and the Internet. The chapter describes the purpose and function of advertising agencies and addresses advertising regulation, and wraps up with a look at the criticisms aimed at contemporary advertising.

FYI: FOUR Ps OF MARKETING

1 Product
2 Price
3 Place
4 Promotion
    4a. Advertising

SEE IT THEN

Advertising: 3000 BCE TO 1990

The earliest form of advertising has been traced back to about 3000 BCE, where tradesmen inscribed their names on clay tablets in the city of Babylon in ancient Mesopotamia. Ancient Egyptians advertised on papyrus (a form of paper) that was much more portable than clay tablets, and the ancient Greeks used town criers to advertise the arrival of ships carrying various goods. In ancient Rome and medieval England, tavern and shop owners distinguished their establishments with creative names and signs. Even back then, merchants recognized the need to get the word out about their products or services.

It was Johannes Gutenberg's invention of the printing press around 1450 that gave people the idea of distributing *printed* advertisements. Toward the end of the 1400s, church officials were printing handbills and tacking them up around town. The first printed advertisement is thought to have appeared in Germany around 1525; it promoted a book about some sort of miracle medicine.

The growing popularity of newspapers in the late 1600s and early 1700s led to the further development of print advertising. In 1704, the *Boston Newsletter* printed what is thought to have been the first newspaper ad, a promise to pay a reward for capturing runaway slaves. Benjamin Franklin, one of the first publishers of colonial newspapers increased the visibility of ads placed in his papers by using larger type and more white space around the ads. Until the mid-1800s, most newspaper ads were in the general form of what is known today as classified ads, or simply lines of text.

Advertising took on a new role in the 1800s with the Industrial Revolution. People moved away from their rural farms and communities into the cities to work in factories. The swelling population of cities, and the mass production of goods gave rise to mass consumption and a mass audience. Advertising provided the link between manufacturers and consumers. City dwellers found out about new products from newspaper and magazine ads rather than from their friends and family.

Manufacturers soon realized the positive effect advertising had on product sales. They also realized they needed help designing ads, writing copy, and buying media space in which to place the ads. Volney B. Palmer filled the void by opening the first advertising agency in the U.S. in Philadelphia, PA in the early 1840s. Palmer primarily contracted with newspapers to sell advertising space to manufacturers. Several decades later, another promotions pioneer, Francis Ayer, opened the first full-service agency in Philadelphia. Ayer's agency produced and placed print ads in newspapers and magazines. As new agencies opened and advertising as a business took off, there came new insights into consumerism and eventually new advertising strategies and techniques. For example, by 1860, full-color magazine advertisements flourished thanks to new cameras and linotype machines.

Early Radio Advertising

A new form of advertising was born with the advent of radio. At first, radio was slow to catch on with the public, because there were not very many programs on the air. One of the first radio broadcasts was of the 1921 heavyweight boxing championship between Jack Dempsey and Georges Carpentier. As each punch was thrown in Hoboken, New Jersey, on-site telegraph operators tapped out the action to station KDKA in Pittsburgh, Pennsylvania, where telegraph operators translated the signals and vocally reported what was happening over the airwaves. Such live broadcasts piqued interest in radio, and as more people purchased radio sets, the demand for programming increased, and broadcasters were left grappling with how they were going to finance the endeavor.

In the early 1920s, station and ham (amateur) radio operators were airing programs but not generating any revenue for their efforts. Given the costs of operating a station, there was a collective call among broadcasters to figure out a way to generate income. In 1924, *Radio Broadcast* magazine held a contest with a $500 prize for the person who could come up with the best answer to the question, "Who is going to pay for broadcasting and how?"

Although the idea of commercial radio was undergoing serious discussion, radio advertising was largely considered in poor taste and an invasion of privacy. Additionally, radio had to comply with the Communications Act of 1934, which stated that there must be a recognizable difference between commercials and program content, such that a station is required to disclose the source of any content it broadcasts for which it receives any type of payment. Many listeners were resistant to over-the-air commercialism and even then-Secretary of Commerce Herbert Hoover, who later became president of the United States, claimed that radio programming should not be interrupted with senseless advertising.

But then in 1922, AT&T-owned radio station WEAF came up with the idea of **toll broadcasting** – payment for using airtime. On August 28, a Long Island, New York, real estate firm paid $50 for 10 minutes of time to persuade people to buy property in the New York area. Although toll advertising may seem like a modern-day infomercial, AT&T did not consider these toll messages 'advertising' but simply courtesy announcements, because the prices of the products and services were never mentioned.

Radio advertising helped boost product sales. The Washburn Crosby Company (now known as General Mills) saw the sales of Wheaties cereal soar after introducing the first singing commercial on network radio in December 1926. In geographic areas in which the Wheaties jingle was aired, the cereal became one of the most popular brands, but in the areas in which

Figure 9.1 Ad from *Radio Broadcast* magazine, May 1924.

WHO IS TO PAY FOR BROADCASTING AND HOW?
A Contest Opened by RADIO BROADCAST in which a prize of $500 is offered

What We Want

A workable plan which shall take into account the problems in present radio broadcasting and propose a practical solution. How, for example, are the restrictions now imposed by the music copyright law to be adjusted to the peculiar conditions of broadcasting? How is the complex radio patent situation to be unsnarled so that broadcasting may develop? Should broadcasting stations be allowed to advertise?

These are some of the questions involved and subjects which must receive careful attention in an intelligent answer to the problem which is the title of this contest.

How It Is To Be Done

The plan must not be more than 1500 words long. It must be double-spaced and typewritten, and must be prefaced with a concise summary. The plan must be in the mails not later than July 20, 1924, and must be addressed, RADIO BROADCAST Who Is to Pay Contest, care American Radio Association, 50 Union Square, New York City.

The contest is open absolutely to every one, except employees of RADIO BROADCAST and officials of the American Radio Association. A contestant may submit more than one plan. If the winning plan is received from two different sources, the judges will award the prize to the contestant whose plan was mailed first.

Judges

Will be shortly announced and will be men well-known in radio and public affairs.

What Information You Need

There are several sources from which the contestant can secure information, in case he does not already know certain of the facts. Among these are the National Association of Broadcasters, 1265 Broadway, New York City; the American Radio Association, 50 Union Square, New York, the Radio Broadcaster's Society of America, care George Schubel, secretary, 154 Nassau Street, New York, the American Society of Composers and Authors, the Westinghouse Electric and Manufacturing Company, the Radio Corporation of America, the General Electric Company, and the various manufacturers, and broadcasting stations.

Prize

The independent committee of judges will award the prize of $500 to the plan which in their judgment is most workable and practical, and which follows the rules given above. No other prizes will be given.

No questions regarding the contest can be answered by RADIO BROADCAST by mail.

---

the commercial did not air, sales were stagnant. By the late 1920s, the initial resistance to advertising gave way to the increasing cost of operating a radio station, and eventually broadcasters and the public endorsed the idea of advertising-supported radio, though largely in the form of sponsorships.

Sponsored Radio

In 1923, the Browning King clothing company bought weekly time on station WEAF to sponsor the Browning King Orchestra. Whenever the name of the orchestra was announced, so was the company's name. But in keeping with WEAF's anti-commercial sentiment, the announcers were careful not to mention that Browning King sold clothing. Other companies and then later advertising agencies took the lead from Browning King and began producing radio programs in turn for being recognized program sponsors. The **sponsor identification rule (Section 317)** of the Communications Act of 1934 protects listeners from commercial messages coming from unidentified sponsors by requiring broadcasters to reveal sponsors' identities.

At first, sponsored radio seemed like a good idea for the audience and the broadcasters. Programs, usually 15 minutes in length, were produced by an ad agency in partnership with the sponsoring company. The ad agency benefitted by being paid for its creative work, the company benefitted by gaining brand recognition and hopefully sales, and the audience benefitted by being made aware of a product but without being subjected to blatant promotional messages.

ZOOM IN 9.1

- Learn more about WEAF's first commercial and listen to a short clip of the spot at: www.npr.org/2012/08/29/160265990/first-radio-commercial-hit-airwaves-90-years-ago

- Listen to old-time radio commercials within the *Old Time Radio* site: www.old-time.com/commercials/index.html

ZOOM IN 9.2

Listen to the Wheaties singing commercial. The spot starts with an announcer talking and then ends with the first singing jingle: www.youtube.com/watch?v=GLy5tANvXhY

Rather than running a commercial per se, a sponsor's name would either be part of a short narrative read by an announcer at the beginning, middle, and end of a program, or it would be mentioned in the script. For example, in *Oxodol's Own Ma Perkins* (1933–1960), the sponsor's laundry products were woven into the storyline when characters did the clothes washing. The American Tobacco Company sponsored *The Lucky Strike Hit Parade* (1935–1953), a program that played the best-selling records of the day. Every time the announcer said the name of the program, the audience would hear the name of the brand Lucky Strike. By naming the program after Lucky Strike, brand recognition increased, and smokers who listened to the program were presumably likely to purchase Lucky Strike cigarettes.

Radio stations began to resent that ad agencies and advertisers were fully responsible for program content, the times that programs aired, and even had complete control over performers. The ultraconservative Rexall Drugs would not let its spokesman, entertainer Jimmy Durante, appear on a **campaign** program that was soliciting votes for Democratic President Franklin D. Roosevelt.

By the mid-1940s, station executives were tired of ceding program control to ad agencies and advertisers, they wanted more control over program content and scheduling so they could draw the largest audience possible. CBS radio network owner William S. Paley devised a plan for programming and advertising. He set up a programming department that was charged with developing and producing new shows. In turn, the network would recoup its expenses by selling time within the programs to advertisers. Although CBS liked the plan, the ad agencies vigorously opposed it, as they wanted to keep control over programs and advertisers. Eventually, programming became the network's responsibility, but the agencies that bought time within programs still controlled casting and scheduling. Then, as radio shows moved to television, radio program sponsorships and agency control over programs diminished, and advertising was increasingly sold not as sponsorships but as commercials within and between programs. The number of advertising spots increased substantially between 1965 and 1995, as did advertising revenue.

Early Television Advertising

The public got its first glimpse of television at the 1939 New York World's Fair, but television did not immediately catch on. World War II interrupted television set manufacturing and program creation and transmission, stalling consumers' adoption of television until about three years after the war. In 1948, the economy was booming, American consumers were very enthusiastic about television, and the industry exploded with new stations, new programs, new sets, and new viewers, realizing growth of more than 4,000% in that year. Television quickly became a mass medium and promised to become as popular as radio, which left industry executives struggling with how to make television a financially lucrative medium.

ZOOM IN 9.3

To listen to old radio sponsorships for programs such as *Little Orphan Annie* (sponsored by Ovaltine) and *Your Hit Parade* (sponsored by Lucky Strike), go to: www.old-time.com/weekly

Sponsorship

At first, television advertising was based largely on radio's sponsorship model. Television programs were much more expensive to produce than radio programs, and agencies and advertisers were spending more than they were generating in increased product sales. Advertisers and their agencies produced sponsored programs such as *Kraft Television Theater* (1947–1958), *Texaco Star Theater* (1948–1953), *The Dinah Shore-Chevy Show* (1951–1957), and *The Colgate-Palmolive Comedy Hour* (1950–1955). The return on investment (ROI) did not always satisfy the sponsors and they were not always happy with how their products looked on television.

Many television programs and commercials were produced live, but unrefined production techniques left viewers watching coarse black-and-white images. Soap sponsors touting the whitening power of their products discovered that on black-and-white television, viewers could not tell the difference between 'white' and 'whiter.' Commercial actors had to hold up a white shirt next to a blue one for contrast and pretend that it had been washed with a competing soap.

Live product demonstrations did not always go off as planned, either, embarrassing the advertiser and the product spokesperson. Aunt Jenny, a character on the *Question Bee*, dripped beads of perspiration from the hot studio lights onto the chocolate cake that was freshly baked using Spry vegetable shortening. To make matters worse, she licked some cake off the knife blade and then cut more slices with the same knife. Spry was not happy, to say the least, neither was *Variety* magazine, which called the whole business of television 'unsanitary.'

In another bungled demo, a hand model demonstrated live Gillette's new automatic safety razor, except it was not so automatic. The new disposable blade, which was supposed to twist open for easy and safe replacement, became stuck, and the television audience watched as the hands desperately struggled to unstick it, to no avail. That was the last live product demonstration for Gillette. From then on, it prerecorded its commercials. Refrigerator doors sticking shut, can openers that would not open cans, and spokespeople holding up one brand but trumpeting another are just some of the other debacles of live television commercials. Despite these bloopers, advertisers continued to promote their products on the airwaves, and television executives cheered as they watched advertising revenues rise rapidly throughout the 1950s.

The high cost of television sponsorship kept all but a few of the largest and most financially stable companies off the air. It soon became apparent that the model for television advertising had to change, especially after the quiz show scandal that rocked the television industry in the late 1950s. At the time, television quiz shows were the most popular programs on the air. The competition for viewers was fierce, as there were many quiz shows on the air, and they were often on at the same time on different channels. To keep popular contestants on the air (and thus to keep viewers watching the shows), sponsors and producers started to secretly give well-liked contestants the answers to the questions before the game. One contestant, who had lost to a competitor who had been given the right answers, came forward and exposed the quiz shows as fraudulent.

When the scandal made the headlines, the public was outraged at the deception and blamed the networks, even though it was the sponsors who had coerced the networks to cheat under threat of losing sponsorship. The networks figured that if they were

Figure 9.2 Westinghouse commercial with Betty Furness, 1956.
Source: Photo courtesy of Photofest

going to be held responsible for televised content, then they should take over programming from the advertising agencies. Sponsors and agencies gradually gave up control over production, scripts, and stars, and instead of one advertiser sponsoring an entire program, various advertisers bought commercial time throughout the show.

Spot Advertising

NBC television executive Pat Weaver (father of *Alien* and *Imaginary Heroes* star Sigourney Weaver) extended William Paley's idea of selling ad time within radio programs to television and came up with what is known as the **magazine style** of television advertising. This approach later became known as **spot advertising**. Weaver had figured out that television could make more money by selling one-minute spots within and between programs to several sponsors than by relying on one sponsor to carry the entire cost. His idea was similar to the placement of magazine advertisements between articles. Weaver's plan also promoted the production of programs by network and independent producers, keeping advertisers out of the business of programming.

With Weaver's plan, advertisers found it much less expensive to purchase one minute of time rather than the entire 15 or 30 minutes of a program, and they did not have to concern themselves with program content. Affordable airtime brought many more advertisers (especially smaller, lesser-known companies) to television for the first time, much to the chagrin of larger, wealthier sponsors and ad agencies, which were concerned about losing their broadcast dominance.

Figure 9.3 Andy Griffith (Sheriff Taylor) and Clint Howard take a break from filming the *Andy Griffith Show* to sell Jell-O brand pudding.
Source: Photo courtesy of CBS Photo Archive/Getty Images

The Bulova watch company was the first advertiser to venture to television and also the first company to purchase spot radio time (as opposed to a program sponsorship). Starting in 1926, the United States ran on Bulova time, with its well-known radio commercial announcements: "At the tone, it's 8:00 p.m. B-U-L-O-V-A. Bulova watch time." Bulova later adapted its radio spot to television. On July 1, 1941, Bulova paid $9 to a New York City television station for a 20-second ad that aired during a Dodgers versus Phillies baseball game. The Bulova commercial showed a watch face with the current time but without the audio announcement. Bulova later kept the close-up of the watch face but added a voice announcing the time.

Throughout the 1960s, most television commercials were available in 60-second units (**spots**). Thus, many different commercials featuring different brands and products were advertised during the course of one program, initiating a more competitive marketplace. For many advertisers, this was the first time they had faced strong competition for the audience's attention. They had to come up with creative ways to make their product or brand stand out from the others. Slogans, jingles, and catchy phrases started to make their way into commercials.

ZOOM IN 9.4

Watch classic television commercials at these sites:

- Classic TV ads: www.youtube.com/watch?v=zokomJZT40

- Saturday morning commercials (1960–1970): www.tvparty.com/vaultcomsat.html

A New Look

Television advertising took on a slightly new look in 1971 after the federal government banned commercials featuring tobacco products. Cigarette companies had been among television's biggest advertisers until they were forced to transfer their advertising dollars from the airwaves to print, leaving broadcasters scrambling to fill unsold time. Television networks quickly discovered that many other companies simply could not afford to buy commercial time in 60-second blocks, but they could buy 30-second units. This arrangement proved profitable for the networks, as they could sell two 30-second spots for more money than they could one 60-second spot. The effect on an hour of television programming was that programs became infiltrated with shorter spots from more advertisers.

In 1965, about 70% of all commercials were 60 seconds in length, but by four years after the 1971 ban on tobacco advertising, only 11% of commercials were 60 seconds in length and almost 80% were 30-second spots. By 1985, almost 90% of all commercials were 30 seconds in length and only

Figure 9.4 Cindy Crawford drinks a Pepsi in a commercial, 1991.
Source: Photo courtesy of Getty Images Entertainment

2% ran for a full minute. Throughout the 1980s, the length of commercials began to vary and included 10-, 15-, 20-, and 45-second spots. The 15-second spots especially caught favor, and by 1990 they accounted for about one-third of all commercials.

Although they are rare, sponsorships do still occur. For example, Ford Motor Company was the sole sponsor of the movie *Schindler's List* when it made its television debut in 1997, and in the early 2000s Kleenex aptly sponsored several 'tearjerker' movies, such as *An Officer and a Gentleman* and *Steel Magnolias*. *The Biggest Loser* has been sponsored by Wrigley's Extra sugarless gum, and BMW's early sponsorship of AMC's *Mad Men* limited commercials to brand mentions at the beginning and end of an episode plus a few breaks, which contained only BMW spots and network promos.

Creative Strategies

With more commercial spots on the air, advertisers needed to be creative to remain competitive. Faced with new creative challenges, many commercials took on a more narrative approach. Rather than just showing a product and reciting its features, commercials became more like 30-second mini-movies that tell a story with characters and plots that build a product or **brand image**. A brand image is a consumer's perception about a product or service. A brand image is different for each product and makes the product stand out from its competitors. This approach, called *image advertising*, goes beyond simply promoting a product; instead, it attempts to set a product or brand perception in the consumer's mind. For example, Whole Foods has cultivated a brand image committed to organic food and sustainability.

Figure 9.5 Brooke Shields.
Source: Advertising Archives

Target Advertising on Broadcast and Cable Television

Throughout the 1970s and into the early 1980s, the three broadcast television networks (ABC, CBS, and NBC) shared about 90% of the viewing audience and television advertising dollars. As more homes hooked up to cable television in the 1980s, broadcast television found a new rival.

Cable offered advertisers many new and specialized channels that attracted niche audiences. Rather than pay a large amount of money to reach a mass audience through broadcast television, advertisers could easily target their consumers on specialty cable networks. For example, it is more efficient and effective for Ping golf clubs to target golfers by running commercials on the Golf Channel or ESPN instead of on a broadcast network where only a small percentage of viewers may be interested in golf equipment.

As more cable channels dedicated to particular interests became available, the television viewing audience became more fragmented, and each cable and broadcast network found itself competing for a diminishing share of the overall audience. Fragmented audiences and more channel choices have contributed to a decline in advertising effectiveness. With the touch of a button, a viewer can easily switch to another channel during a commercial break, which hampers a network's ability to deliver a steady audience to its advertisers.

SEE IT NOW

Advertising: 1990 to Present

Advertising is more than just creating a commercial, it is all about creating a commercial that will resonate with the buying public and generate revenue. Advertising is an art of creativity and also a science of consumer buying behavior and product sales. It takes a team of artists and advertising experts to build a successful campaign.

From a Business Perspective

Clearly, no business today could survive without using some form of advertising. Some companies and professional services might rely entirely on word-of-mouth referrals, while others might use a combination of referrals, the Yellow Pages, and ballpoint pens inscribed with the company's name. Other businesses might choose to create full-blown media advertising campaigns. Regardless, most businesses rely on some type of advertising.

Advertising plays several crucial roles in an entity's marketing efforts – it tells the audience about the benefits and value of a product or service, and it generates revenue by persuading people to purchase products and services. But persuading people to buy a product is psychologically complex; the process includes building a positive brand image and relies on generating **brand loyalty**, the repeated purchase of a product.

New lessons in advertising were learned during the coronavirus pandemic of 2020. As the manufacturing, retail, and service sectors were shut down for months and people were largely confined to their homes, the demand for 'unnecessary' products and services (e.g. hair and nail salons, clothing stores, concerts) plummeted, as did advertising spending. Television ad revenue is expected to have dropped about 25%, representing tens of billions of dollars, in the first half of 2020 alone, mostly due to the cancellation of the March Madness tournament and all other live sporting events. Newspapers, which were already suffering from decreasing ad revenue, will probably come out of the pandemic devastated, with at least a 50% decline in ad sales.

Even though people were watching more television and reading and watching more news (cable and broadcast network news viewership soared) while sheltering-in-place, advertisers were not willing to risk keeping their name out in front of consumers in a time of low demand for products and services, a strategy that could backfire in the long run. During times of economic downturn (e.g. after 9/11, the Great Recession), brands that maintained their advertising and marketing strategies and kept themselves highly visible to consumers, fared better in economic recovery than those that cut back on ad spending.

It is not just the traditional media that are hurting from loss of ad revenue, 98% of online publishers expect a sharp decrease from 2019. Social media are hurting as well. Twitter experienced a loss in ad sales despite increased use as followers eagerly turned to tweets to find out about the virus. From April 2019 to April 2020, Twitter saw a 24% jump in users but a 27% decrease in ad revenue. As businesses cut back, the bidding price of online ads slumped, lowering revenue for online publishers.

FYI: U.S. ADVERTISING REVENUE BY MEDIUM – WITH DIGITAL SHARE (2018) (IN $ BILLIONS)

Internet	$107.5
Television (broadcast and cable)	$71.0
Television digital share	$5.5
Radio	$17.7
Radio digital share	$2.2
Consumer magazine	$16.4
Consumer magazine digital share	$7.1
Newspaper	$15.8
Newspaper digital share	$5.5
Out-of-home	$10.0
Out-of-home digital share	$5.5
Trade magazine	$4.2
Trade magazine digital share	$2.8
Video games	$1.5
Cinema	$.9
Podcast	$.4
Total	$237.7 billion

Sources: IAB, 2019; Marketing Charts, 2018

From a Consumer Perspective

Although advertisements may be annoying, they are also beneficial. Advertisements serve social and economic purposes. Advertising is also educational – it is the way consumers learn about new products and services, sales, and specials. Advertising benefits the economy by promoting free enterprise and competition. The results of these forces are product improvements, increased product choices, and lower prices.

Advertisements also serve a social function in that they reflect popular culture and social values and give people a sense of belonging. Spots often include the hottest celebrities, the latest trends, and the most popular music. Products are advertised within the cultural environment. For instance, after the 9/11 terrorist attacks on Washington, DC and New York City, many commercials contained shots of the American flag and other symbols of national unity.

FYI: TOP U.S. ADVERTISERS BY AD SPENDING: 2018 (IN $ BILLIONS)

1	Comcast Corp.	$5.75
2	Proctor & Gamble	$4.39
3	AT&T	$3.52
4	Amazon	$3.38
5	General Motors	$3.24
6	Verizon Communications, Inc.	$2.64
7	Ford Motor Company	$2.45
8	Charter Communications	$2.42
9	Alphabet, Inc.	$2.41
10	Samsung Electronics	$2.41

Source: Patel, 2018

Buzz Marketing

Buzz marketing, "the transfer of information from someone who is in the know to someone who isn't" (cf. Gladwell, 2003), is a contrived version of word-of-mouth endorsement. True word-of-mouth advertising is traditionally a highly trusted and very effective type of communication. 'Word of mouth' implies that someone who has used a product or service is giving their honest opinion about it of their own volition. With buzz marketing (also known as *viral marketing*), a company pays people to pass themselves off as ordinary consumers using a product and to promote its features and benefits regardless of whether they believe in it. For example, Vespa hired good-looking, hip young men and women to cruise around Southern California hot spots on its scooters. As admirers asked about the scooters, the paid endorsers touted the product and even handed out the address and phone number of the nearest Vespa dealer.

Advertising Agencies

Advertising agencies have long been the hub of the advertising industry. Agencies are hotbeds of creativity – the places where new ideas are born and new products come to life. Agencies are where many minds come together – account representatives, copywriters, graphic artists, video producers, media planners, researchers, and others who collaborate to devise the best campaigns possible.

Advertising agencies are not all alike. There are large agencies that serve national advertisers, and there are small agencies that serve locally owned businesses. Some agencies have offices around the world, some have offices around the nation, and others have an office or two in some city or town. Some agencies have staffs of thousands, and others are one- or two-person shops. Although agencies exist to serve advertisers, how they do that is not the same from agency to agency.

Full-Service Agencies

Full-service agencies basically provide all of the advertising functions needed to create an advertising campaign. Agency employees plan, research, create, produce, and place commercials and advertisements in various media. They often provide other marketing services as well, such as promotions, newsletters, and corporate videos. Some full-service agencies are very structured; each group or department focuses on its strengths, and projects are basically moved down the line in each step of the process. Other agencies are more collaborative, often grouping employees with different fields of expertise together on a project. For instance, to create a campaign, an account representative may team with copywriters, graphic artists, researchers, planners, and media buyers.

Creative Boutiques

Some advertisers have an in-house staff that plans, researches, and buys media time and space but needs help with the actual creation of a campaign. These advertisers contract with a creative boutique rather than with a full-service agency. Creative boutiques focus specifically on the actual creation of ads and campaigns and are therefore staffed with copywriters, graphic artists, and producers. Advertisers benefit by hiring a group of people with expertise in creative work.

Media-Buying Services

Some advertisers have in-house creative departments that write and produce their commercials and advertisements, but they might not know the best media in which to place their spots and print ads. These advertisers depend on a *media-buying service*. Once the in-house agency finishes producing the ads, they rely on media buyers who are experts in media placement to maximize exposure and sales. Media buyers are market experts who know what media will help their clients achieve their advertising and sales goals.

ZOOM IN 9.5

For a clearer understanding of the differences among creative boutiques, media-buying services, and full-service agencies, visit each of these agency Web sites and read about the types of services it offers:

- Creative boutique – Jugular: http://jugularnyc.com/
- Media-buying service – Capitol Media Solutions: https://capitolmediasolutions.com/
- Full-service agency – Tombras Group: http://tombras.com/

Interactive/Cyber Agencies

Many advertisers today see the need to expand their advertising to the Internet, DVDs, smartphones, tablets, and other interactive and mobile platforms, so they are turning to media/interactive agencies with expertise in Web design and interactive technology. These *interactive agencies* (or *cyber agencies*) create and maintain client Web sites, create and place banner ads, and produce and distribute other interactive advertising materials. Interactive shops tout expertise in many areas that full-service and other types of agencies might not provide.

Advertising Campaigns

Airing one or two commercials here and there is not, in most cases, an effective way to sell a product or service. That is why advertising agencies specialize in planning and implementing advertising **campaigns**, a series of commercials and print and online ads that follow the same basic theme. Successful recent campaigns include Burger King's, 'the Whopper detour' and Microsoft's 'We all win.'

Strategizing a campaign involves setting advertising and marketing objectives, analyzing a product's uses as well as its strengths and weaknesses, determining the target audience, evaluating the competitive marketplace, and understanding the media market. The creative staff of copywriters, graphic artists, and video producers confer, toss around ideas, argue, and change their minds until finally they come up with a campaign that meets their client's marketing and advertising objectives.

Producing a campaign is often a long and stressful process. Difficult-to-please clients sometimes think they know more than the advertising experts. Deadlines come up much too quickly, and the marketplace changes in a flash. Moreover, the advertising business is very competitive, which requires agencies to frequently pitch new advertisers for their business. Just one failed campaign can cost an ad agency a multimillion-dollar account. On the positive side, being part of a successful ad campaign is very satisfying. Agencies and their creative staffs often build client relationships that last for years. Creative staffs are recognized for their excellent work by national and international associations and by receiving **Clio Awards**, which are the advertising equivalents of Emmy Awards.

ZOOM IN 9.6

For a historical look at the development of Coca-Cola campaigns, go to: www.cocacolacompany.com/content/dam/journey/us/en/our-company/history/coca-cola-a-short-hisotry-125-years-booklet.pdf

ZOOM IN 9.7

Check out the Clio Award winners and learn more about this prestigious award at: www.clioawards.com

Using the Media

Each medium – radio, television, and the Internet – has strengths and weaknesses as a marketing tool. Smart media buyers know which products do best on which medium and in which market. They also know which creative strategies and appeals work best for the different media audiences. It is not possible to say that one medium is *always* better than the others.

RADIO ADVERTISING

Radio commercials are generally 15, 30, or 60 seconds in length. Radio commercials are classified as **local, national, and network spots**. One of radio's strengths is the ability to reach a local audience. About 80 cents of every dollar of time is sold to local advertisers who wish to have their message reach the local community. Local restaurants, car dealerships, and stores know that by advertising on the radio, they are reaching the local audience that is most likely to visit their establishments.

Advertisers also buy national spots in which they place their ads on individual stations in certain markets. For example, McDonald's might buy commercial time on selected radio stations in many different geographic regions. If it has a special promotion going on in the South, it will run its commercials on stations located in southern markets.

Radio stations affiliate with a radio network that provides them with programming. When advertisers buy time on a network, they are buying spots within the network's shows that are aired on its affiliated stations. The advertiser benefits from this one-stop media purchase of many stations that reach its target audience. Some of the larger radio networks are iHeartMedia, Fox Sports Radio and Fox News Radio, Radio Disney, and Cumulus Media Networks, which merged with Westwood One. Advertisers who want to reach the largest audience, say for automobiles or fast food, will place their commercials on network programs.

Advantages of Radio Advertising

- *Local* – Radio spots reach a local audience, the most likely purchasers of local products and services.
- *Flexible* – A radio spot can be sold, produced, and aired within a few days. Copy can quickly be changed and updated.
- *Targets an audience* – The various program formats make it easy for advertisers to reach their target markets.
- *Low advertising cost* – Radio is less expensive in terms of the number of listeners reached, than television or the Internet.
- *High exposure* – Advertisers can afford to buy many spots, and through repeated exposure, listeners learn the words to jingles, memorize phone numbers, and remember special deals and other commercial content.
- *Low production costs* – Radio commercials are generally inexpensive to produce.
- *High reach* – About 92% of U.S. adults listen at least once a week.
- *Portable and ubiquitous* – With small, lightweight radios, stations can be picked up from most anywhere.
- *Commercials blend with content* – Commercials with background music and jingles often sound similar to songs, and commercials with dialogue sound similar to talk show conversations.

Disadvantages of Radio Advertising

- *Audio only* – Radio involves only the sense of hearing. Listeners are easily distracted by what else they may be doing or seeing, and it is difficult for listeners to visualize a product.
- *Background medium* – Radio is often listened to while people are engaged in other activities, so they do not always hear or pay attention to commercials.
- *Short message life* – Radio ads are typically 30 seconds in length, which is not much time to grab attention, especially if a listener is involved in

another activity. Also, unlike print, where people can go back to an ad and write down the information, once a radio spot has aired, the information is gone.
- *Fragmented audience* – Most markets are flooded with radio stations, all competing for a piece of the audience. Fragmentation forces many advertisers to expand their reach by purchasing time on several stations in one market.

Television Advertising

Television is considered the most persuasive advertising medium. The combination of audio and visual components captures viewers' attention more so than other media do. Plus, almost everyone watches television. Advertisers reap the benefits of an audience of millions. Despite its strong points, broadcast television is not the best advertising outlet for all advertisers. Not everyone needs to reach a large mass audience, and not everyone has the budget to produce television spots. For many advertisers, cable television offers more audience for less money.

Television Commercials (Network, National, Local) It often seems that commercials take up more television time than programs themselves. But in 2017, commercials on broadcast networks accounted for 24.0% of all programming, and commercials on cable networks took up 28.0% of time. Although some viewers like commercials, others think they are intrusive and click to other channels when they break into a program or appear between shows.

Television commercials are typically 30 seconds in length, but they can run to 15, 45, or 60 seconds. Shorter spots began taking hold in the 1990s and by 2017 about four in ten spots were 15 seconds in length, and Fox experimented with 6-second spots for NFL games. Most commercials air in **clusters**, or **pods**, of several commercials between and within programs. Advertisers may be guaranteed that their commercials will not air in the same pod as a competitor's commercial. In other words, a Ford commercial may be guaranteed not to air in the same pod as a Chevrolet commercial. Commercials aired within a particular program are known as **spots**, or **participations**, and commercials that air before and after programs are known as **adjacencies**.

Like radio commercial time, television time is categorized as **network, national, and local** buys. Network spots are very expensive to purchase, so

FYI: COST OF A 30-SECOND SPOT ON PRIME-TIME BROADCAST NETWORK PROGRAMS

2016–2017 Season		
Sunday Night Football	NBC	$650,000
Thursday Night Football	CBS	$524,047
The Walking Dead	AMC	$470,410
This is Us	NBC	$394,314
Empire	FOX	$294,141
2015–2016 Season		
Sunday Night Football	NBC	$603,000
Empire	FOX	$497,360
Thursday Night Football	CBS	$464,625
Big Bang Theory	CBS	$348,300
How to Get Away with Murder	ABC	$252,934
2014–2015 Season		
Sunday Night Football	NBC	$637,330
Empire	FOX	$521,794
Thursday Night Football	CBS	$462,622
The OT	FOX	$303,200
Big Bang Theory	CBS	$289,621
2013–2014 Season		
Sunday Night Football	NBC	$593,700
American Idol	FOX	$355,946
Big Bang Theory	CBS	$316,912
The Voice	NBC	$294.038
American Idol Results	FOX	$289,942
2008–2009 Season		
Sunday Night Football	NBC	$339,700
Grey's Anatomy	ABC	$240,462
Desperate Housewives	ABC	$228,851
Two and a Half Men	CBS	$226,535
Family Guy	FOX	$214,750
2005–2006 Season		
American Idol (Wed.)	FOX	$705,000
American Idol (Tues.)	FOX	$660,000
Desperate Housewives	ABC	$560,000
CSI	CBS	$465,000
Grey's Anatomy	ABC	$440,000
2000–2001 Season		
ER	NBC	$620,000
Friends	NBC	$540,000
Will & Grace	NBC	$480,000
Just Shoot Me	NBC	$465,000
Everybody Loves Raymond	CBS	$460,000

Sources: Steinberg, 2009, 2015, 2017

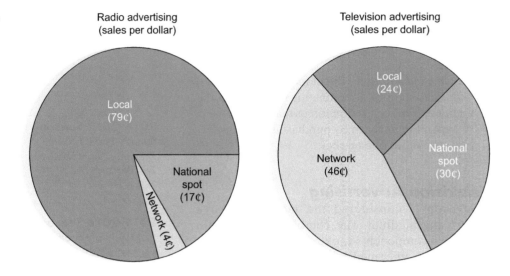

Diagram 9.1 Radio versus television dollars.

network time is usually reserved for a limited number of big-time advertisers, national buys are used when an advertiser wants to reach select markets in the country, and local ads are for connecting advertisers to local consumers.

Buying and Selling Television Broadcast networks typically sell time in what are called the **upfront market**, the **scatter market**, and the **opportunistic market**. The 'markets' are actually time frames throughout the year. With the upfront market, the networks begin selling time each spring for programs that will air during the fall season. Advertising costs are based on ratings projections – the higher the estimates, the higher the price. Because advertisers are taking a risk by buying ahead, the networks guarantee a minimum audience size. If a program falls short, the network reimburses the advertiser through additional spots. Networks strive to sell about 70–80% of their inventory up front.

With the scatter market, unsold time is offered to advertisers four times a year for the upcoming quarter. But because advertisers wait until they have a better idea of how a show is doing, their risk is lower. Advertisers who buy in the scatter market typically pay a higher rate and are not guaranteed a minimum number of viewers. Advertisers also buy opportunistic time that comes up at the last minute. For example, maybe time slots open up in a popular program or a news event prompts a special program that is sure to draw a large audience.

Product Placement Sometimes, rather than buying commercial time, companies resort to **product placement**, paying to have their product used in or visible within a program scene. Product placement, also known as 'product integration' and 'stealth advertising,' is a subtle but effective way of exposing viewers to products, often without their conscious knowledge or realization. After falling out of favor for a number of years, product placement is making a comeback (especially on cable networks and in movies) as a creative way to promote a product apart from the typical 30- and 60-second spots and to thwart digital video recorder (DVR) users from escaping product exposure.

Tens of thousands of products are embedded in television shows at a cost to an advertiser that can range from several thousands of dollars to over a million. In 2019, revenue from product placement hit about $11 billion as just over 600 brands were integrated into television programs.

Going beyond merely putting a box of cereal in some character's kitchen, streaming services are experimenting with **virtual product placement** in which different viewers are shown different integrated products based on their purchasing and viewing habits. Streaming services collect personal data every time a subscriber logs-in to watch a show. By using subscriber data, the service can show to some viewers, for example, a character walking in front of a billboard advertising Coca-Cola, while Pepsi lovers will see the same scene but the billboard will instead show a can of Pepsi. Walmart-owned streaming service, Vudu, and the Roku Channel are going so far as to let viewers click on a product or commercial to find out more information or to purchase directly from their smart television or computer.

Stranger Things has revived sales of Eggo waffles, the favorite sustenance of the show's character, Eleven. Contestants on the *Biggest Loser* gabbed about Ziploc bags, *Top Chef* countertops were stacked with Glad products, Raj and Barry demonstrated Siri on *The Big Bang Theory*, and Luke Dunphy on *Modern Family* ate an Oreo cookie. Some of the highest recalled product placements in older shows include Alan driving a bright-red Porsche on *Two and a Half Men*, and David Lee offering Veronica M&Ms on the *Good Wife*.

The potential drawback of product placement is that advertisers give up control over how their product will be presented, risking it being shown in an unfavorable light. However, it is in the producers' best financial interest to convey product benefits in a positive manner. Further, scriptwriters might not like being compelled to weave products into storylines, thus subliminally persuading viewers. And many viewers object to this practice of covert marketing.

The Federal Communications Commission (FCC) requires that programmers disclose placement sponsors in a show's credits. But the credits are typically in very small print, and they roll by so quickly that they are almost impossible to read. The FCC is considering new regulations for informing viewers about product placement. The FCC wants to extend the disclosure of product placement to make it more prominent and ban all placements in programs targeting children under the age of 12. The FCC believes that viewers have the right to know when they are being sold a product.

Bonus Programming Viewers are becoming weary of the 50-some-year-old advertising format of 30-second spots airing during commercial pods. Moreover, the proliferation of on-demand services is tugging viewers to advertising-free options. Viewers wonder why they should pay for commercial-laden cable or satellite programs when they can pay for ad-free shows or at least those with minimal commercial content on services such as Hulu.

Television executives are scrambling to keep viewers tuned to the broadcast and cable networks, but they know that, to do so, they must reduce the number of commercials. 'Bonus programming,' such as behind-the-scenes interviews and extended segments, are shown in lieu of the typical spots. Instead of going to commercials at the end of a program, viewers are treated to what they think is 'extra' programming but is in fact 'sponsored' time. Strategists see this new formula as a win situation for all; viewers enjoy the extra programming and prefer the subtle promotional messaging over blatant commercial spots, advertisers have a new and effective way to reach viewers, and television retains viewers who might otherwise have abandoned commercial-dependent programming for ad-free services.

Public Service Announcements Public service announcements (PSA) promote non-profit organizations, such as the American Lung Association and the United Way, as well as social causes, such as 'Friends Don't Let Friends Drive Drunk' and 'Click It or Ticket' seatbelt advocacy. Most stations air PSAs at no charge and whenever they can fit them into the schedule.

Radio and television stations also promote their own programs on their own stations. Frequently, a promo airing in the morning will promote an afternoon drive-time show, or an afternoon television promo will alert viewers about a special program coming up later that evening.

Advantages of Advertising on Television

- *Visual and audio* – Television's greatest advantage is its ability to bring life to products and services. The combination of sight and sound grabs viewers, commands their attention, and increases their commercial recall.
- *Mass appeal* – Television commercials reach a broad, diverse audience.
- *High exposure* – Although the high cost of a television spot generally limits the number of times it can be aired, one exposure will reach many people simultaneously.
- *High reach* – About 99% of U.S. households have at least one television set, and 82% have two or more sets. There are almost as many television sets in the United States as there are people – 120.6 million television homes, each averaging 2.6 sets, comes to 313.5 million sets for 330 million people. In an average U.S. household, the television is on for 7 hours and 40 minutes per day.
- *Ubiquitous* – Television is everywhere. It is rare to go someplace where there is not a television set.
- *Commercials blend with content* – Commercials are cleverly inserted within or between programs to try to make them less obvious and blatant, and brands are blended into a story line with product placement.
- *Variety* – With so many program types, advertisers have many options for commercial placement to reach their customers.
- *Entertaining* – Television is highly entertaining, which spills over into commercials. With some programs, such as the Super Bowl and the Emmys, viewers actually find the commercials more exciting than the programs themselves.

- *Persuasive* – Television is the most persuasive commercial medium. Catchy audio, spectacular visual effects, and interesting product demonstrations often get the most skeptical of viewers to try new products.
- *Emotional* – Television engages our emotions, and we witness the intrinsic rewards that come from purchasing a product.
- *Prestige* – Many viewers believe that if a company is wealthy enough to buy commercial time and a product is good enough to be advertised on television, then it is good enough to buy.

Disadvantages of Advertising on Television

- *Channel surfing, zipping, and zapping* – These are three terms that advertisers hate to hear. Viewers channel surf, zip, and zap when they move all around the television dial and change the channel when a commercial comes on. Advertisers end up paying for an audience that does not even see their commercials.
- *Digital video recorders (DVRs)* – These devices are commercial-avoidance culprits. About 90% of DVR owners fast forward through commercials.
- *Fragmented audience* – The average household receives about 206 channels, of which viewers only watch about 20 on a regular basis.
- *Difficult to target* – Advertisers often end up paying for wasted coverage – paying for a large audience when they really only wanted to reach a smaller subset of viewers.
- *High cost* – Dollar for dollar, broadcast television is the most expensive medium, especially when considering both production costs and airtime.
- *Clutter* – Broadcasting and cable non-programming time (commercials, station ID, station promotion) takes up an average of 15 minutes in a typical prime-time hour. Early-morning television airs 18 minutes per hour of non-programming materials, and daytime television has nearly 21 minutes devoted to non-program fare. When so many commercials are cluttered together, viewers tend to pay little attention to any of them, thus hampering message recall.

Cable Advertising

Cable television has been around for many years, but only recently has it challenged broadcast television for audience share. For one week in the summer of 1997, basic cable channels for the first time edged out ABC, NBC, and CBS with 40% of the prime-time audience, compared to the networks' 39% share. During the 2001–2002 season, the advertising-supported cable networks drew for the first time a larger prime-time audience than the seven broadcast networks (ABC, CBS, NBC, Fox, UPN, WB, and PAX) combined. Now it is common for the cable networks to outpace the broadcast networks.

Cable offers viewers select program options on which advertisers target niche audiences. Advertisers have the option of buying commercial time on

FYI: THE AVERAGE COST OF A 30-SECOND SUPER BOWL COMMERCIAL

It often seems that the best part about watching the Super Bowl is seeing the commercials. The Super Bowl delivers a huge audience, so advertisers pay big money to promote their products and services. Below are the average costs of a 30-second spot through the years:

Year	Average cost
2020	$5,600,000
2019	$5,200,000
2018	$5,000,000
2017	$5,000,000
2016	$5,000,000
2015	$4,500,000
2014	$4,000,000
2013	$3,700,000
2012	$3,500,000
2011	$3,000,000
2010	$2,650,000
2009	$3,000,000
2008	$2,700,000
2007	$2,400,000
2006	$2,500,000
2005	$2,400,000
2004	$2,300,000
2003	$2,200,000
2002	$2,200,000
2001	$2,200,000
2000	$2,100,000
1997	$1,200,000
1992	$850,000
1987	$600,000
1977	$125,000
1967	$37,500

Sources: Baumer, 2011; BNC Sports, 2020; Statista, 2019)

ZOOM IN 9.8

MOST MEMORABLE SUPER BOWL COMMERCIALS

View some of the most iconic commercials – ones that will be remembered years after they have aired. See some of the best at: www.vogue.com/article/best-super-bowl-ads-of-all-time

Squarespace, 2018, Keanu Reeves
Old Spice, 2010
Hyundai Smart Park, 2020, John Krasinski, Chris Evans, and Rachel Dratch
Pepsi, 1992, Cindy Crawford
Pepsi, 2002, Britney Spears
McDonalds, 1993, Larry Bird vs. Michael Jordan
Budweiser, 2002, Bud Frogs
Budweiser, 2014, Puppy Love
Budweiser, 2015, Lost Dog
Mountain Dew, 2016, Puppy Monkey Baby
Apple *1984*, 1984
E Trade Buby, 2008–2013
Snickers, 2010, Betty White
Volkswagen, 2011, The Force
Clash of Clans, 2015, Liam Neeson

Source: Vogue, 2020

---

specialty cable networks such as Golf Channel, HGTV (Home & Garden Television), Nick at Nite, MTV, and hundreds of others. These channels are perfect advertising venues for marketers who want to target specific audiences. For example, there is probably not a better place to advertise kitchen appliances or cooking products than on Food Network or gardening supplies on HGTV. Even though advertising-supported cable networks tend to have small audiences, advertisers are attracted to these specialized and often loyal markets.

Most cable buys take place at the network level, wherein advertisers buy time on a cable channel, such as ESPN, and the spots are shown in selected locations or throughout the country. The remaining program time is then sold locally by the cable service. Local cable reps from one service (e.g. Comcast) also team with other cable service reps (e.g. Cox) to sell **interconnects**. Large cities such as New York may have several cable providers, each sending out cable programming to a specific part of the city. Interconnects allow advertisers to purchase local cable time with several providers with one cable buy. In doing so, a local advertiser could simultaneously run a commercial on all of New York's cable systems for a larger audience reach.

Advantages of Advertising on Cable Television

- *Visual and audio* – Like broadcast television, cable television's primary strength is its ability to attract viewers through sight and sound.
- *Select audience* – Cable delivers specific consumers to its advertisers.
- *Upscale* – The cable television audience tends to be made up of young, upscale, educated viewers with money to spend, making cable an ideal venue for specialized and luxury items.
- *Variety* – With hundreds of cable networks to choose among, it is easy for an advertiser to match its product with its target audience.
- *Low cost* – With so many cable networks competing for advertisers, they rarely sell all of their available commercial time, which keeps the cost per spot low.
- *Seasonal advantage* – Cable networks have learned to take advantage of the broadcast networks' summer 'vacation,' when broadcast television is airing stale reruns, cable is counterprogramming with shows that attract larger-than-normal audiences.
- *Local advantage* – National spot buyers and local businesses take advantage of cable's ability to reach specialty audiences within certain geographic areas.
- *Media mix* – Cable's low-cost and niche programming makes it an ideal supplement in the media mix.

Disadvantages of Advertising on Cable Television

- *Zipping, zapping, and channel surfing* – Zipping, zapping, and channel surfing are also the enemies of cable television.
- *Fragmented audience/low ratings* – Cable audiences are fragmented and spread across many cable networks.
- *Lack of penetration* – Cable's household penetration is ebbing as more viewers subscribe to satellite services and watch programs on the Internet.
- *Churn* – Cable audience size is affected by *churn*, which is the ratio of new subscribers to the number who disconnect their cable service.

Internet Advertising

There is some debate as to what constitutes **Internet advertising**, but it is generally considered to be such when a company pays or makes some sort of financial or trade arrangement to post its logo, product information, or service with the intent of generating sales or brand recognition on someone else's Internet space. For example, when Macy's pays to place its banner on *The Washington Post* Web site, this is considered Internet advertising. However, when Macy's sells clothing and other products on its own Web site, it is considered *marketing*. But the lines are becoming blurred. Marketers tweet, create Facebook pages, host blogs, use social networks, and send images and text through mobile phones. There is a question as to whether these are marketing or advertising strategies.

No other medium is as interactive as the Internet. Clicking on ads, watching and listening to in-banner videos, reading product recommendations, playing games that are disguised ads, or tweeting about recent purchases are ways that consumers interact with brands. Advertisers are up on the advantages of reaching consumers online – so much so that they spend about $135 billion a year doing so. More importantly, online advertising expenditures surpassed broadcast television for the first time in 2013. Television advertising is forecast to increase by only 1.3% per year, while online advertising is expected to grow at a rate of 7.7% per year. But this trend does not mean that television will disappear; rather, the Internet and television will settle as complementary media, with online recognized as the better way to target consumers, and television the better way to reach a mass audience.

FYI: FIRST ONLINE BANNER AD

Online advertising was born on October 27, 1994, when HotWired (www.hotwired.com), the online version of *Wired* magazine, posted the first-ever banner ad, which was sponsored by AT&T. The banner was something new and intriguing, and thus it was clicked on by 44% of users who saw it. With no idea of how much traffic HotWired would attract or of its audience's demographic profile, HotWired set a price of $30,000 for a 12-week commitment and was thrilled when other advertisers – MCI, Volvo, and Sprint – signed up.

Figure 9.6 First banner ad.

Source: Edwards, 2013

FYI: PUSH/PULL STRATEGIES

Internet advertising includes *push* and *pull* strategies. A *push strategy* means that an ad is pushed to consumers, whereas a *pull* refers to consumers pulling, or seeking, the message. Web marketers are taking advantage of Internet technology by pushing products and other promotional messages through banner ads, email, and mobile ads. Consumers pull product information when they access Web sites and sign-up for subscription services and notifications.

Types of Banner Display Ads There are various types of online ads. Some of the most common types of online ads are described in the paragraphs below. Most online ads can be adapted across platforms for Internet, smartphone, and tablet delivery. Online ads are measured in terms of **pixels**, tiny dots of color that collectively form an image. On a computer screen, about **72 to 96 pixels equals one inch**.

Traditional banner ads are static or non-animated, with little more than an advertiser's logo with some embellishment. To increase consumer interest and to make purchasing easier, a banner usually contains some sort of animation or moving image and a link to the advertiser's landing page that relates to the ad or to its homepage.

Rich-Media Ads. The static and animated banner ad has given way to a more exciting visual presence, the rich-media ad. It is not a type of ad per se but describes how an ad is designed. Different from static or animated ads, rich-media banners contain audio and video. Advertisers are concerned that with so many banner ads dancing on Web sites, users might get annoyed at the distraction and ignore them all.

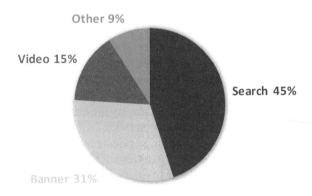

Diagram 9.2 Internet advertising format share of revenue.
Source: IAB, 2019

Interstitial/Superstitial. To make sure that their messages are seen, many online advertisers opt for rich-media **interstitials** and **superstitials**. The word 'interstitial' means 'in between.' Thus, both interstitials and superstitials appear in a separate browser window in between Web pages or on top of Web page content. Interstitials and superstitials are both rich-media ads, but interstitial is a general term for any type of ad that runs between pages, while the term superstitial is trademarked by Unicast as a descriptor of its rich-media technology. Once referred to as 'polite' ads, because they only played when fully downloaded and initiated by the user, some video superstitials now rudely self-start while a page is loading up, and it is often impossible to turn off the ad until after the page is fully loaded. An interstitial/superstitial that appears on top of Web page content is called a **pop-up**. If it lurks behind the content browser window, it is a **pop-under**, also called the 'evil cousin of a pop-up,' because it surprises users when they close the browser window. Both dazzle the eyes with animation, graphics, interactive transactional engines, and near-television-quality video.

Expanding Rich Media. **Pencil pushdown/sliding billboard ads** are skinny, full-width ads that expand lengthwise when clicked on. Pencil pushdowns are highly visible, and when clicked and opened, they push down Web site content rather than covering it up. An **expanding banner** looks like a square banner but doubles in size horizontally and covers up adjacent content when the cursor swipes over it.

Billboards. These high-impact display banner ads offer prominent positioning and custom performance. Billboards are large in size (970 × 250 pixels), and are generally placed on home, section front, and/or subsection front Web pages. Billboards are not collapsible, in other words they display only as large banners, and thus are mainly used for display on desktop/laptop platforms.

Gravity Ads. As full-screen animated display with video-capable overlay, Gravity ads take up the entire desktop space. Users scroll through and off the ad, and are pulled into the Web site's content, thus the name 'Gravity.' Gravity ads virtually offer an immersive experience, and thus are best for desktop/laptop placement.

Paramount Ads. Similar to Gravity ads but smaller in size, these high-impact display ads offer cross-platform versatility. Paramount ads often contain video, and are customizable for such features as store locaters and email captures.

Leaderboard, Mobile Leaderboard and Interscroller. At 728 × 90 pixels, the **leaderboard** ad spans the width of a Web page and is typically placed at the top of a page ('above the fold') between the masthead (the Web page title) and content. When users have to scroll to see an ad, it means it is placed 'below the fold.' The terms come from newspaper advertising in which printed display ads are literally placed above or below the paper's horizontal fold. While the leaderboard is best for laptop and desktop viewing, the **mobile leaderboard** (320 × 50 pixels) delivers a brand's message exclusively on mobile devices. Like the mobile leaderboard, the **Interscroller** also serves exclusively on mobile platforms/mobile Web, and offers display and video. The interscroller moves vertically on and off the display screen, similar to how a Gravity ad scrolls away.

In-Banner Audio/Video. An **in-banner audio** ad looks like a banner but contains audio that either plays automatically or must be initiated by the user. An **in-banner video** ad is a display ad but has the look and feel of a television commercial, usually 15 seconds in length. Pre-roll in-banners are video spots that play before non-commercial video content starts up, like on many YouTube videos.

Corner Peel. A **corner peel** ad is a double-image ad in which the top layer peels down from a corner when a user runs the cursor over a corner tab. Additional product information is displayed when the corner rolls down and a clickable link sends the users to the product's Web site.

Hover Ads. Hovers are full-screen-width ads that slide up about one inch from the bottom of a Web page when it is opened. They cover up content and must be clicked on to 'collapse' them.

Floating Ads. This type of ad features an image, perhaps a butterfly, flower, plane, or bird flittering around the screen for a few seconds, and then it 'lands' on the Web page and morphs into a small, very short animated ad that covers up the content and then, 'poof,' disappears on its own. At their most annoying, floaters challenge users to try to nab them with the mouse as they dance inside a browser window and even turn cursors into ads, making it almost impossible to use the page. Frustrated users find themselves playing cat and mouse while desperately trying to sink the floater.

Big Box, Skyscraper/Tower, Extramercial. These display banners take several forms. They can be non-animated (static), contain an animated image or Flash animation, or contain rich-media elements. The **big box** is the most popular type of banner. Measuring 300 × 250 pixels, big-box ads are highly visible and flexible in terms of placement on the Web page. The aptly named **skyscraper/tower ad**

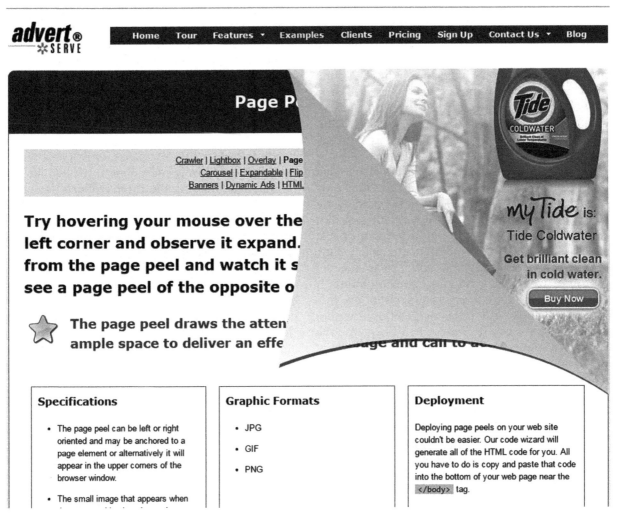

Figure 9.7 Example of corner peel ad.
Source: Photo courtesy of Shutterstock

extends vertically (160 × 600 pixels) along one side of the browser window. The right-hand placement makes it visible and high impact but subtle and nonintrusive. An **extramercial**, which is placed in the 3-inch space to the right of the screen that is usually not visible unless the user scrolls sideways or has his or her monitor sized at 1,024 × 768 pixels or higher. Extramercials have given way to newer formats and are not as commonly used as they once were.

Wallpaper and Homepage Takeover. These types of ads occupy Web page architecture by actually becoming part of the page. **Wallpapers** are high visibility ads that frame Web content. When users click on the frame, they are sometimes taken to the product Web site, but not all wallpaper ads are clickable links. Wallpaper ads do not block content as do **homepage takeover ads**. Users are taken aback when they click on a Web site, and instead of seeing the familiar landing page, they get a full-page advertisement that completely covers the content. For example, a few years ago Yahoo! users were awestruck when a few seconds after opening the homepage, a Ford Mustang came careening across the screen, obliterating the site's content. As the Mustang swerved around the screen, it changed colors and morphed into different models to urge visitors to customize their own vehicle on the Ford Web site. A few seconds after the ad finished, the landing page reappeared. A homepage takeover ad can be thought of as a 'road blocking' strategy, in which one advertiser buys all of the ad space on a page.

Banner Displays Go Mobile Mobile is the newest way to reach consumers on the go by pushing ads directly to a smartphone or tablet. The sheer number

Figure 9.8 Example of floating ad.
Source: The Oklahoman Media Company

of mobile devices in the world advantages mobile advertising. In the U.S. alone, there are about 177 million tablet users and about 272 million smartphone users. Worldwide, the number of tablet users had soared 1.4 billion and smartphone users to 3 billion by 2019.

The various types of mobile ads include banners placed within apps, top- or bottom-of-page banners, ads placed within mobile games and videos, full-screen interstitials that come up before content appears, and even audio-only jingles. But mobile ads do not have the same visual impact as ads on larger screens, nor are they very interactive, so they are easier to overlook than desktop ads.

Other Types of Online Advertising In addition to banner display ads, marketers promote their goods and services in other ways, which are sometimes so subtle that consumers cannot distinguish an ad from other content. Search engine marketing, native ads, product placement, buzz marketing, advertorials, word of mouth, social media marketing, and mobile ads, are other forms of online advertising and promotion.

Search Engine Marketing (SEM). SEM is the most common form of online advertising strategy, accounting for about 45% of all online revenue. SEM optimizes ad targeting with paid placement on search engine results pages. For example, companies like Nike or New Balance pay to have their ads appear as part of the search results whenever a user enters the term 'athletic shoes.' Directories, such as Zillow.com, which lists homes for sale and rent, are another type of SEM.

Native Ads (Branded Content) Advertorial. **A native ad**, also known as **branded content**, or

ZOOM IN 9.9

Check out this Web page for examples of standard Internet ads: www.masternewmedia.org/online_standard_ad_formats_official_advertising_formats/

Check out this Web page for examples of interactive ads: https://marketingland.com/state-interactive-advertising-new-formats-infusing-digital-ads-creativity-gets-results-228655

FYI: TOP FIVE COMPANIES, RANKED BY U.S. DIGITAL AD REVENUE SHARE, 2019

1	Google	38.2%
2	Facebook	28.2%
3	Amazon	6.8%
4	Microsoft	4.1%
5	Verizon	3.4%

Source: Wagner, 2019

advertorial is an established blend of commercial messages and editorial content, but the ad mimics a Web site by using the same headline style, type font, layout, and so on. Branded content helps sponsors/advertisers stand out in a crowded marketplace by telling the sponsor's story so it resonates with consumers. Native/branded content ads draw in even the most experienced online users who think they are reading real news stories, but instead are subjected to highly persuasive native ads. For example, BuzzFeed once embedded the headline, '14 People Making the Best of Bad Situations' among other news stories, except the link went to a Volkswagen-created 'story' that tied into its theme, 'Get in. Get happy,' without overt indication that it was really an advertisement.

Celebrities and companies often create **Webisodes** (short narrative videos) as part of their digital marketing campaigns. For example, Feed Us is a company that develops Web-based content-management applications. Instead of a hard, direct pitch about the features and benefits of its products, Feed Us produced humorous Webisodes about two of its technicians who answer customer problems over the phone.

Advertisers like to use native ads because they are not subject to '**banner blindness**,' which is when users become so accustomed to seeing banners that they ignore or simply do not see them. Going native also hinders **ad blockers**, which are software products that prevent advertisements from appearing in a browser window. Native/branded content 'ads' are supposed to be marked as sponsored content, so readers know they are seeing an advertiser-sponsored content summary/link to more information and not editorial content. Because a native product endorsement is written into the content, there is no ad to block as such, and so viewers are exposed to the advertising message. Savvy Web site designers know how to integrate native ads into their content seamlessly.

Product Placement/Cooperative Advertising. The strategy of placing a product within content has become more common in movies and television over the last several decades, and now products are being placed within Web site content.

Cooperative advertising is a strategy in which a marketer promotes a product on a Web site, usually in the form of a review or endorsement, and shares sales profits with the hosting site. For example, Amazon and other online booksellers charge book publishers to promote their titles on their sites. The publisher writes a promotional review of its own book, and Amazon places the review next to an image of the book either for a set fee or for a percentage of sales made from the site. This type of co-op advertising, also known as **product placement**, has been sharply criticized for failing to make it clear to customers that the reviews are promotional pieces, not unbiased editorials written by Amazon's staff. Amazon now posts a page that explains its publisher-supported placement policy and lets users know which ads are part of the co-op program.

Buzz Marketing Online. Buzz marketing is when a company's employees pretend they are everyday people who just happen to go online to talk up a well-liked product. These hired promoters present themselves as ordinary consumers to infiltrate chat rooms, blogs, and social media to push a particular brand. Product information may be a sales pitch disguised as a helpful tip or a 'review' that is really a paid-for promotion rather than an unbiased viewpoint. Unsuspecting Internet users might be duped into buying a product or service based on buzz marketing rather than honest word-of-mouth promotion. To curb the obfuscation of paid online endorsements, the Federal Trade Commission is now requiring that bloggers disclose whether they have received any kind of payment or free samples of products they review or endorse.

Word of Mouth. Word-of-mouth promotion is very powerful. People buy products that are endorsed by people they trust. Music promotion has exploded online. Musicians promote their music, and music aficionados tout their favorite artists and songs and blast the ones they do not like on such sites as MP3.com, OurWave.com, and last.fm. One of the first examples of successful online word-of-mouth promotion was with the fringe-rock band Weezer. The band's popularity soared even when it was on a recording hiatus – about one-quarter of those who bought the band's 1996 album in 2002 did so because of online word-of-mouth recommendations.

Social Media Marketing and Advertising. Using social media is another way to generate Web site attention, traffic, and sales. Social media sites, such as Facebook, are advantageous for marketers to connect with and establish a 'social' relationship with consumers. Social media ads are often 'wanted' in the sense they are sent to consumers who have expressed interest in a product or brand by going to the Web site, signing up for a newsletter or an alert, or by 'friending' a company on Facebook. Consumers seem to have accepted the thinking that if they have to be exposed to ads they might as well be for products they like or are thinking of purchasing. A young woman will probably be happier getting an ad from the shoe company Zappos than one from Just For Men, which sells hair coloring to cover the gray.

Marketers strive for engagement – the longer a user stays on a Web site or social media page, the more likely he or she is to make a purchase. Advertisers are not only designing attractive and persuasive ads but ads that are compelling and interesting enough to generate social media cachet and get consumers to share them among their contacts. Marketers depend on consumers to spread the word to their own social groups through 'likes,' reposted comments, and tweets. Conversations about products and services instill brand loyalty and leave customers believing that the company cares about them and values their feedback, which in turn leads to increased and repeat sales. Marketers know that online users trust ads and recommendations that come from people they know more than any other source. Social media is the modern version of word of mouth, and the larger the number of social media groups marketers connect with, the larger the number of followers to spread the word to others.

Further, new software that monitors the conversations and analyzes their general tone cues marketers about brand image and buying intention and helps them identify and target particular consumers who are most likely to purchase or replace an old item.

FYI: U.S. AD SPENDING ON SOCIAL MEDIA, 2006–2019

2019	$23.6 billion
2018	$21.3 billion
2017	$18.2 billion
2016	$14.7 billion
2015	$11.2 billion
2014	$8.5 billion
2013	$6.1 billion
2012	$4.7 billion
2011	$4.2 billion
2010	$3.3 billion
2009	$1.8 billion
2008	$1.4 billion
2007	$920 million
2006	$350 million

Sources: emarketer, 2019; Hoelzel, 2014; Infographic: Rise of social media, 2010; Stambor, 2013.

Display Banner Targeting: Primary Strategies As presented so far in this chapter, advertisers have an array of advertising formats to best promote their products and services. They might focus on several formats or use many in different combinations for different brands and to reach different audiences. Their choices, however, are made with an overall marketing plan or outcome in mind. And the ads are not just put out on any old Web site and left to

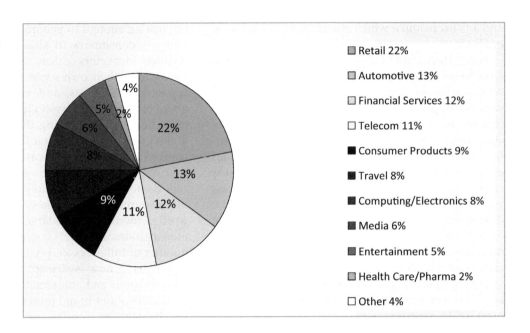

Diagram 9.3 Internet ad revenue by major industry categories, 2019. Source: Which industries spend, 2019

- Retail 22%
- Automotive 13%
- Financial Services 12%
- Telecom 11%
- Consumer Products 9%
- Travel 8%
- Computing/Electronics 8%
- Media 6%
- Entertainment 5%
- Health Care/Pharma 2%
- Other 4%

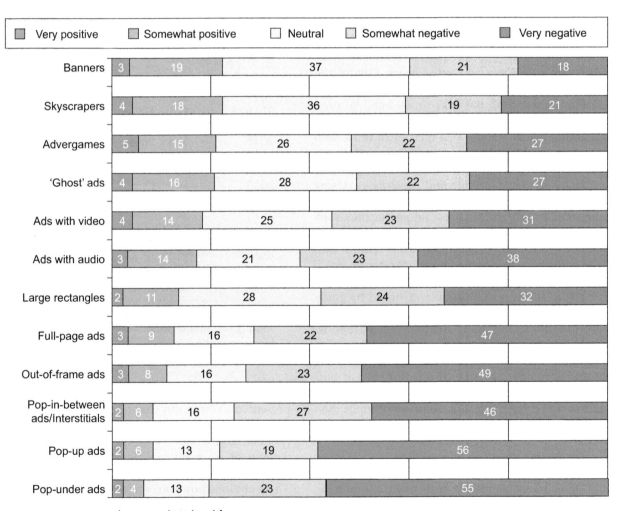

Diagram 9.4 Attitudes toward Web ad formats.

chance that someone will see them and make a purchase, rather they are placed and designed according to a plan that will yield the most buyers for the least amount of ad spending.

There are 12 basic targeting strategies to ensure that ads go to online users who are most likely to be interested in the product: run of network, keyword contextual, category contextual, geofencing, addressable geofencing, over-the-top (OTT)/Connected TV (CTV), Web site retargeting, dynamic inventory retargeting, list retargeting, search retargeting, lookalikes, and blacklisting/whitelisting. Each strategy is explained below:

1 *Run of Network.* This strategy is used to target consumers within a specific geographic location, regardless of demographics, shopping habits, or the device they use most often to access the Internet. For example, if two persons, one living in Los Angeles, CA and the other in Knoxville, TN, are both on the same news site, the person from L.A. might see an ad for a local restaurant while the person in Knoxville might see an ad for a nearby clothing store. Run of network targeting is possible because most publishers work with a large network of Web sites and apps that connects advertisers to their local community.

2 *Keyword Contextual Targeting.* Keyword contextual targeting is used to target consumers who have entered specific keywords into search engines. The thinking is that users who searched by keyword must be interested in the topic and/or related product. Keyword contextual targeting matches the Web page content of what a user is reading, then places relevant ads on that Web page.

3 *Category Contextual Targeting.* Category contextual targeting is used to reach consumers who have recently visited Web sites or apps that are relevant to an advertiser's business. For example, if a user visits a travel site, a competing cruise line ad might appear on the next site the user opens. The psychology behind category contextual targeting is that recent online behavior suggests a possible interest in a brand and product.

4 *Geofencing and Geotargeting.* **Geofencing** works by targeting consumers based on their mobile device's physical location whose boundaries the advertiser has predefined. When a consumer carrying a mobile device with the phone location setting turned on, enters the advertiser's predefined area, the advertiser's ad pops onto their mobile browser or ad-enabled apps. **Geotargeting**, on the other hand, delivers ads to people who have entered the specified geofencing area **and** meet specific criteria (e.g. age, income, hobbies).

For example, a Mercedes dealer that wants to target affluent adults might set a geofenced area, which includes high-end restaurants, jewelry, and clothing stores, and send its ad to the smartphones of everyone who enters the geofenced area. The Mercedes dealer, could

WHAT IS KEYWORD CONTEXTUAL TARGETING?

Target users viewing content with keywords that have been designated by the advertiser.

Figure 9.9 Example of keyword contextual targeting.
Source: Wordstream (www.wordstream.com/contextual-advertising)

Figure 9.10 Geofencing vs geotargeting.
Source: Malone, 2020

also geotarget people within the geofence, by sending its ad only to those whose income is above a certain level. The Mercedes dealer could also add a layer of tracking by measuring a 'conversion zone,' – counting the number of people who visited the Mercedes lot after receiving the ad when they were in the geofenced area.

Geofencing and geotargeting advantage the advertiser by zeroing in on consumers based on the physical places they have visited, which shows a potential interest in the advertiser's brand. For example, a veterinarian can target users who are romping with their pet at a nearby dog park.

5 *Addressable Geofencing.* This subtype of geofencing uses property line data to reach consumers at their home address. Think of it like direct mail, only it is digital and trackable. Rather than getting a printed flyer in a home mailbox, an online ad is sent to residents in a particular neighborhood.

Addressable geofencing is best used to drive foot traffic into the advertiser's business from nearby neighborhoods. Better yet, an advertiser can target multiple consumers within a single household across their devices, including Internet-connected television, and for up to 30 days after they have left the address, keeping the advertiser's brand at the top of the consumer's mind.

Addressable geofencing makes good use of a business's advertising dollars, because the ads are targeting consumers who are most likely to shop near where they live. More important to

Figure 9.11 Addressable geofencing.
Source: Adzscape (https://adscapz.com/solutions/geo-fencing/addressable-geo-fencing.html)

the advertiser is the ability to track 'conversion zones.'

6 *Over-the-Top (OTT)/Connected TV (CTV).* OTT/CTV is a video ad delivery method that targets users who are exposed to streaming video ads when watching television via an Internet connection.

OTT stands for 'over-the-top' content and CTV stands for 'connected TV,' which is any device that plays OTT content. CTV is a smart television set or one that connects to the

Internet through Apple TV, a Roku Box, or Sling TV. iPads, smartphones, and other devices that can stream online content are also considered CTV.

By combining OTT and CTV to stream programs on an Internet-connected player, advertisers can send their messages to viewers who are likely to buy their products. For example, an advertiser of running shoes could target a sports fan who is using her Roku connected television set (CTV) to stream a football game from the SEC (Southeastern Conference) site (OTT).

Advertisers can even use OTT/CTV to reach cable subscribers who also stream content. Many network television sites have agreements with cable companies to allow subscribers to stream full episodes. By entering their cable subscription user name and password, subscribers can watch a program on their laptop or any Internet-connected device. Advertisers can then target these cable subscribers while they are streaming a show, and viewers cannot fast forward through the streamed spots. OTT/CTV is an excellent way to reach a targeted group of video consumers that advertisers cannot target with traditional television commercials.

7 *Web Site Retargeting.* This strategy works by targeting consumers online who have recently visited an advertiser's Web site. A site's unique identifier stays with users as they browse other ad-serving Web sites, the original site then knows what other content and sites their users have visited. Site retargeting is an effective 'always on' strategy that keeps a business's message in front of users when they leave a Web site and move on to the next site. Site retargeting helps keep top-of-mind awareness throughout a web surfer's buying journey.

8 *Dynamic Inventory Retargeting.* Using this strategy, consumers are targeted when they visit a certain page within an advertiser's Web site. For example, a user is targeted when he or she visits a car dealer's Web site and clicks on the link for the Toyota Prius page, or searches within the advertiser's online inventory or product feed list (such as a search for 'Prius'). If a user links to another site, he or she is targeted with ads for a Toyota Prius. Dynamic Inventory also works with search engines. For example, if a user who is looking for a new car enters 'new Toyota Prius Clinton, TN' in the Google search bar, dynamic inventory pulls new car availability directly from the dealership's inventory feed,

Figure 9.12 Over-the-top (OTT)/Connected TV (CTV).
Source: MartechToday (https://martechtoday.com/nielsen-catalina-now-targets-ads-using-purchase-data-on-connected-tv-and-ott-226495)

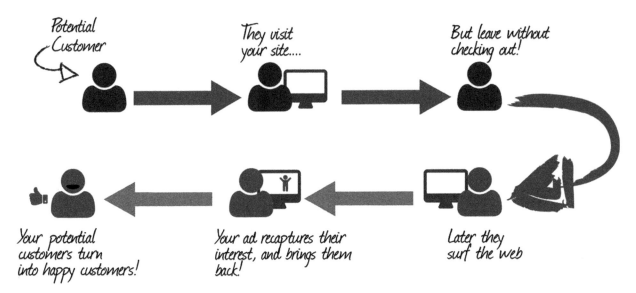

Figure 9.13 Retargeting.
Source: ReTargeter (https://retargeter.com/what-is-retargeting-and-how-does-it-work/)

and generates an ad showing how many new Priuses, or as some would say, Prii, are on the lot. Dynamic inventory retargeting benefits both the marketer and the consumer. The advertiser is reaching only those users who are interested in a particular product, and consumers are targeted with exactly the product(s) they have recently viewed and expressed interest in.

9 *List Retargeting.* With list retargeting, only online users who are already in the advertiser's database of prospects or customers are targeted with display ads. Online users who have purchased a product from an advertiser or have, for some other reason, entered their email address on the advertiser's site are ripe for targeting. Once a business has a consumer's email address it can send the person promotional emails, though reputable companies give consumers the opportunity to 'opt in' or 'opt out' when they no longer wish to receive promotional materials.

List retargeting is highly effective because advertisers are reaching product-friendly customers who have either expressed interest in what they have to offer or have already purchased an item. This strategy works best for businesses that sell products or services bought on a regular basis, and when consumers buy the same brand each time (e.g. toothpaste, laundry soap).

10 *Search Retargeting.* This strategy targets users who have searched for keywords on a third-party Web site (i.e. local news site, WebMD, or About.com) relevant to an advertiser's brand in the last 30 days. If the search matches one of an advertiser's keywords, that person will later be served one of the advertiser's ads when the viewer visits another Web site or app. For example, when someone looks for airline information on travel.com and then clicks on a news site, an ad for Delta airlines might pop up. Search retargeting is a good way to reach potential customers who are interested in an advertiser's types of products and services but may not yet be familiar with that particular brand.

11 *Lookalikes.* This strategy is used by advertisers to target consumers whose Internet use and purchases resemble the advertiser's existing customers. By matching a person's search history, Web behavior, and geographic and demographic data to existing customers, a marketer is betting that buyers with similar qualities are interested in similar products. Lookalike marketing is best used to introduce a new brand to a new audience.

12 *Whitelist/Blacklist Targeting.* This type of targeting both identifies sites that potential customers might visit (whitelist) and rules out sites that are not worth advertising on (blacklist). This strategy gives advertisers control of placing their ads on the best sites to reach their type of consumer. Knowing which sites are appropriate for the product and which to avoid, saves money and ensures that the message is delivered to users who find it relevant. For example, it might not make good sense for a

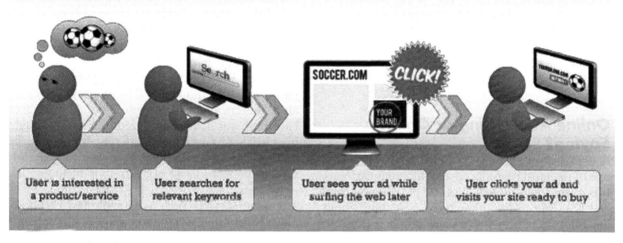

Figure 9.14 Search retargeting.
Source: TargetOnStar (https://targetonstar.com/search-retargeting/)

conservative politician to run an election ad on a medical marijuana Web site. On the other hand, *who* sees the ad might be more important than *where* they see it. It could be that some conservative voters are pro medical marijuana and thus would be happy to see which candidates are also supportive.

Advantages of Advertising on the Internet

- *Worldwide marketplace* – The Internet serves as a worldwide marketplace, delivering a vast, diverse and physically distant audience to advertisers.
- *Targeting consumers* – The Internet's ability to carry messages to targeted groups is one of its most effective marketing tools.
- *Exposure and run time* – Internet ads have longer exposure and run times than ads in traditional media.
- *Low production costs* – Web advertisements are generally less expensive to produce than ads in traditional media.
- *Updating and changing ad copy* – Updating and changing the copy and graphics of online ads can be accomplished fairly quickly and they can be posted within a relatively short period of time.
- *Prestige* – The prestige of online advertising casts a positive image on advertisers and their products.
- *Competition* – The generally lower cost of online advertising puts companies with small advertising budgets in a competitive position with companies with more advertising resources. Online, small businesses are not so small.
- *Quick links to purchases* – Online purchases are often made by simply clicking on a banner ad and following the trail of links to an online order form, bypassing the product Web site.
- *Mobility* – Users can check their mobile devices for coupons and specials, for prices, and for products and services from anywhere.

Disadvantages of Advertising on the Internet

- *Banner blindness* – Consumers see so many ads that they have learned how to screen them out. About 30–40% of ads are never even noticed, and only about 1% are clicked on.
- *Creative restrictions* – Online ads are still somewhat restricted in creative terms. Many banners are nothing more than the equivalent of a roadside billboard.

- *Fragmentation* – With hundreds of thousands of Web sites and thousands of pages within each site, it is difficult to determine ideal ad placement.
- *Unreliable audience measurement* – Unreliable and unstandardized measurement techniques limit an advertiser's knowledge of how many users are exposed to a message, thus hindering effective advertising buys.
- *Questionable content* – The online audience is already tired of deceptive content and the over-commercialism of the Web, and so they resent advertising popping up all over their screens, which makes them not very receptive to what they do see.

Online Advertising or Online Spying?

A 1993 issue of *The New Yorker* printed what is now one of the most reproduced and infamous cartoons. It is of two dogs looking at a computer screen, and one is saying to the other, "The best thing about the Internet is they don't know you're a dog." In the Internet chapter (Chapter five) of this book, a 2000 version of the cartoon adds to the earlier one with a second frame in which the computer screen flashes the dog's exact demographic profile. The two cartoons show how quickly online privacy diminished. Throughout most of the 1990s, online users were fairly confident of online privacy; now nothing is private. If a user does not want anyone to know something, he or she should not post it online, and that includes on email, blogs, Facebook, or another social medium.

Marketing companies, data services, retailers, and manufacturers figured out how to best use online technologies to monitor who uses their sites and what products they purchase. Within a few milliseconds of opening a Web site, marketers are scanning the user's activity and sending ads based on past online travels. Google alone handles about 3.5 billion search queries per day or about 1.2 trillion searches per year. The typical U.S. user conducts 110 to 130 searches per month, and each time companies like Yahoo!, Google, Microsoft, Bing, and others are there to process, monitor, track, and collect that information.

Cookies Most online users do not know how much and what kinds of data are being collected. The use of cookies is one of the most common methods of collecting online audience data. A **cookie** is a type of software that surreptitiously installs itself on a user's hard drive and allows a Web site operator to track and store the user's movements throughout the Web site. The cookie creates a personal file that the company then uses to customize online information to target that individual. Cookies leave a bad taste because they collect personal data that users do not always intend to give. Data could include where someone lives, past online purchases, sites visited most often, hobbies, music and entertainment preferences, and – with social network sites – even friends' names and their online travels. For the most part, information collection goes on without a user's knowledge or permission. Despite the fact that 85% of users believe that sites should not be allowed to track their online movements, fewer than 15% opt to turn off cookies. Bowing to public pressure, Facebook discontinued its Beacon program, which alerted users about their friends' online purchases. Facebook was heavily criticized after a guy bought his girlfriend an engagement ring, but before he had the chance to propose, Facebook had already spread the word to all of his friends, including the girlfriend.

Online services such as Google and Facebook make it easy to find and share photos and personal information. For example, Google happily exhibits 'shared endorsements' on its advertising network of over two million sites. If a user follows a shoe store or a particular restaurant or rates a new album, that user's name, personal information, and photo could show up in ads for that shoe store, restaurant, or album without his or her explicit permission or knowledge. Although Google and Facebook claim they provide ways to opt out of such endorsements, most people are not aware of this option and do not even know how to go about doing so. Online companies say that they are helping consumers by personalizing online ads, but they are butting heads with privacy advocates who claim they have gone too far in breaching personal security.

All this tracking and personalization, coined '**the long click**,' is under fire from advocacy groups, the Federal Trade Commission (FTC), legislators, and Internet users who are concerned that personal information is being collected without direct consent or knowledge. These groups are calling for sites to notify users when information is collected and to offer a way to block the process.

California is the first state to enact legal restrictions on data collection. The California Consumer Privacy Act (CCPA), which came into effect on January 1, 2020, gives residents of the state the power to find out what personal data is being collected and by what company, instruct the companies to stop selling their personal data, and request companies to delete

FYI: ADS THAT WATCH YOU

A poster for domestic abuse at a Berlin bus stop proclaims, "It happens when nobody is looking," and means what it says. When someone is looking at the poster, it shows a happy couple with a man's arm lovingly encircling the woman, but when no one is looking, the image switches to the man raising his fist to strike the woman, who is leaning away and protecting her face with her hands.

A camera with face-tracking software is recording whether anyone in the bus shelter is looking at the poster, and it measures attention and gender of the onlooker.

The technology is also being used at a German rental car company. When a man sees the poster, he sees an image of a limousine; when a woman sees the same poster, the image changes to a Cabriolet.

Source. Advertisement that watches you, 2009

---

the data. Although the law covers state residents, major companies such as Netflix, Microsoft, and Starbucks are extending those same rights to all Americans. Although this is a big step in protecting privacy it is not easy to make it happen. To exercise CCPA rights, a consumer must fill out request forms for each online company.

Consumers who are tired of being pelted with online ads, are fending off the onslaught of digital promotions with **ad-blocker software**, which uses an algorithm to identify ads, block them from appearing on a Web site, and place a blank space where the ad would have appeared without the ad blocker. Makers of ad-blocker software claim they are as intent on protecting consumers' privacy as marketers are on monitoring them. Users benefit by the absence of all those pesky ads, but Web sites lose out on advertising revenue. About 25% of online users take advantage of ad-blocker software, putting hundreds of millions of advertising dollars at stake. Online businesses and marketers disdain ad blockers and often provide users to their sites with instructions on how to selectively unblock them.

Spam Commercial messages for Viagra, weight loss, hair loss, body-part enhancement, get-rich-quick schemes, medical cures, and a host of other products and services clog millions of email boxes every day. Email may have started as a promising method of delivering commercial messages, but it has captured the wrath of users, who are up in arms at receiving these unsolicited sales pitches, commonly known as **spam**. According to Internet folklore, there are two origins of the word 'spam.' Some say the term comes from the popular *Monty Python* line, "Spam, spam, spam," pronounced with an English accent, of course, which refers to the canned sandwich filler: A whole lot of junk but no real meat.

Regardless, spam is any unsolicited message or any content that requires the user to opt out. Advertisers often justify sending spam by tricking customers into signing up for the information. Sometimes when users are making a purchase, filling out an online poll, or just cruising through a Web site, they inadvertently click on or run their mouse over a link or icon that signals permission to send email messages. Most legitimate businesses offer users a way to opt out of unwanted messages; others make it almost impossible to block out spam.

From a marketing standpoint, sending out promotional material via email is much more efficient than waiting for potential customers to stumble upon a product's Web page or view one of the product's banner ads. Besides, it takes only a few seconds for users to recognize and delete unwanted promotional messages, so marketers figure that recipients are spared any real harm. This kind of thinking can backfire on the advertisers, because customers who are spammed might harbor negative feelings and even boycott these companies' products and services. Despite spam's bad reputation, marketers are spamming full force. About 150 billion spam messages are sent globally per day, or about 50% of the almost 300 billion emails.

But what is old is new. Marketers are getting tricky, they are now using new online data collection methods to stuff physical home mailboxes with circulars, coupons, and printed ads, tailored especially for the resident. Marketers have fine-tuned the retro approach of mass mailing with digital targeting. By using data collected by smartphones and computers, marketers know favorite stores and products bought, or even products looked at but not purchased, videos viewed, and other online activities. But instead of sending digital ads, marketers are sending printed mail. For example, smartphone data could pick up location information indicating that a person is stopping in at realtor open houses or touring houses that are for sale. Coupled with Zillow searches, online algorithms could figure what type of house and the price range a person is considering. A real estate company would then mail printed ads of houses for sales that match the price, location, and style of the potential buyer.

FYI: HOW SPAM WORKS

1. Spammers first obtain email addresses either from low-cost 'spambots' (software that automatically combs the Web), social media, and other online resources or from businesses that sell their customers' personal information.
2. By changing Internet accounts to avoid detection, spammers send out millions of pieces of spam from one or several computers.
3. The spam messages are then sent to stealth servers, which strip away the clues that could identify their origin and add fake return addresses, and ordinary subject lines like 'From Me', 'How are you?', and 'Returned message.'
4. Spam then gets sent to unregulated blind-relay servers which redirect the spam internationally, making it more difficult to trace.
5. The spam finally travels back to the United States. The circuitous route fools ISPs and spam blockers into thinking spam is legitimate email.

Source: Stone & Lin, 2002

FYI: STOPPING SPAM

ISPs scramble for new ways to outfox spammers' underhanded means of circumventing spam blockers. Here are some examples:

ISP's strategy: Block messages from known spammers.
Spammer's counter-response: Set up new email addresses. *Example:* When Mary@offer4U.com is blocked, the spammer simply changes the address to Mary@goodoffer4U.com
ISP's strategy: Use a spam blocker to cross-check the address and verify the sender.
Spammer's counter-response: Mask its identity in the email header, making it seem as though the message is coming from someone else.
ISP's strategy: Use anti-spam software to block messages containing marketing terms.
Spammer's counter-response: Alter the spellings of words or add invisible HTML tags to confuse the spam blockers. V*I*A*G*R*A, V1AGR@, VI,B.,/B.AGRA
ISP's strategy: Use anti-spam software to check messages with altered text and to block mail sent to multiple addresses. Require the sender of suspected spam to access a Web site and enter a code before the message can be sent. If the message is spam, there will not be anyone to respond so it does not get sent.
Spammer's strategy: Send spam out from many computers through stealth servers, so origin and content cannot be detected.

Source: Stone & Weil, 2003

The **Federal Trade Commission** (FTC) receives tens of thousands of complaints and hundreds of thousands of examples of Internet spam each week. In accordance with the 1938 **Wheeler Lea Act**, the FTC is responsible for the regulation of advertising and has the power to find and stop deceptive advertising in any communication medium. The FTC, however, does not have regulatory power over email, but it is politically influential. Although legislation has been introduced to combat this persistent nuisance, federal legislators have been unable to totally stop these cyber intruders, but headway has been made. Currently, most states have passed some type of spam-related law, ranging from Delaware's outright banning of spam with false return addresses to requiring 'remove me' links and demanding subject alerts on sex-related spam. Some states that do not have specific anti-spamming laws have sued spammers, using laws pertaining to deceptive advertising.

On the federal level, Congress passed the **Controlling the Assault of Non-Solicited Pornography and Marketing Act**, or *CAN-SPAM*, in December 2003. CAN-SPAM abolishes the most offensive tactics used by spammers, including forging email headers and sending pornographic materials. Email marketers are now required to have a functioning return address or a link to a Web site that can accept a request to be deleted from the emailing list. The Act was updated in 2008 and again with the **M-Spam Act of 2009**, which prohibits spamming on mobile devices. Other help comes from Internet service providers that offer **spam blockers**, special software that filters out spam, and from organizations, such as the Coalition Against Unsolicited Commercial Email (CAUCE, www.cauce.org), that are fighting for legislation to protect the online community from unwanted email.

Criticisms of Advertising

Advertising is highly criticized, not so much for its very nature but because of its content, its negative influences on society, and the types of products it promotes. Advertising is blamed for inducing people to buy products they do not need, at prices

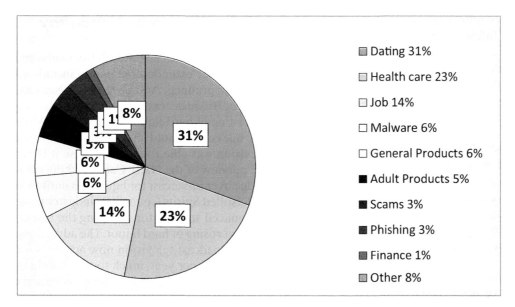

Diagram 9.5 What do spammers sell?
Source: Spam Statistics, 2019

they cannot afford. It is also blamed for creating a society in which someone's self-worth depends on how many 'things' they own and where they shop. But like it or not, our economic system depends on advertising, which fuels purchasing, and thus economic growth.

Advertising Is Misleading and Deceptive

Critics assert that commercials are often exaggerated and misleading; for instance, basketball players wearing Nike shoes jumping higher than their rivals wearing other brands. Nike does not directly claim its shoes can make people jump higher. Even so, critics claim that people may associate these products with these unlikely outcomes and that such hyperbole is unethical.

Advertising defenders, on the other hand, assert that such depictions are nothing more than harmless puffery: exaggerated claims that a reasonable person knows are not true. Defenders claim that most people know that Nike shoes do not make people jump higher, no matter what an ad might suggest; thus, puffery is harmless to consumers.

The FTC regulates deceptive advertising, and even though many ads do not break the law, they do breach ethical standards. The FTC promotes and supports a free marketplace by ensuring that advertising is not untruthful or deceptive using the following three criteria:

1. There must be a representation, omission, or practice that is likely to mislead the consumer.
2. The act or practice must be examined from the perspective of a consumer acting reasonably in the circumstances.
3. The representation, omission, or practice must be a material one ('FTC Policy Statement,' 1983).

The FTC issues sanctions against advertisers found guilty of deceptive advertising. Complaints are most commonly settled by requiring the advertiser to sign a consent decree, agreeing to stop misleading advertising practices. An advertiser who breaches the decree may be fined up to $10,000 a day until the deceptive advertising stops. If an advertiser refuses to sign a consent decree, the FTC has the authority to take the matter further by issuing a cease-and-desist order, demanding an end to the deceptive advertising. In this situation, the case would be brought before an administrative law judge, who could impose penalties or overturn it.

Infomercials Look Too Much Like Television Programs

Prior to the 1980s, commercials longer than two minutes in length were considered infomercials and thus were barred from radio and television. The FTC dropped the provision for radio ads in 1981 and for television spots in 1984. Modern-day infomercials usually air on early-morning, late-night, and weekend television. Infomercials promote everything from exercise equipment to get-rich-quick real estate schemes. Viewers often think they are watching a program when they are actually watching an infomercial filled with product promotions.

Advertising Encourages Avariciousness and Materialism

Many people complain that advertising encourages them to buy items they do not need just for the sake of amassing goods. 'Whoever dies with the most toys wins' was a popular bumper-sticker slogan of the 1980s and reflects the type of thinking that encourages greed and competition among friends and neighbors based on the amount of material goods they collect. All too often, people are persuaded to spend money on goods that they cannot afford and do not need, because product promotions have convinced them that their self-worth depends on these purchases. No one is immune to the persuasive power of advertising. In 2018, the average U.S. credit card holder was carrying about $6,000 in credit card debt. Many critics insist that runaway debt is the direct result of advertising promoting materialism.

Advertising Reinforces Stereotypes

Commercials are under fire for their unrealistic and often demeaning portrayals of women, minorities, and other individuals. Stereotypical images of women mopping floors, men being bosses, smart people being nerds, blondes being dumb, and old people being fools pervade many advertisements. Unfortunately, people's beliefs are shaped by what they see on television, and when they are repeatedly exposed to stereotypes, they come to believe what they see. Recent pressure on advertisers has brought about some changes in how groups are depicted in commercials and other types of promotional materials.

Advertising Exploits Children

Parents and advocacy groups are concerned about the negative effects of advertising on children. Most children see about 30,000 commercials each year, and marketers spend about $4 billion each year pushing products to children under the age of 12. Critics claim that children cannot interpret the purpose of a commercial, judge the credibility of its claims, nor differentiate between program content and a sales message.

Advertising Promotes Unhealthy Behaviors

In 1971, Congress banned cigarette ads from radio and television and later extended the ban to include all tobacco-related products. Additionally, the National Association of Broadcasters Code imposed a ban on the broadcasting of hard-liquor commercials. While broadcast television could not accept liquor ads, cable television was able to do so because it is not under the purview of the NAB. Cable had the competitive edge over broadcast for liquor ads until 1996 when the Distilled Spirits Council, an industry trade group, announced support for reversing the ban on broadcast advertising of hard liquor. The advertising of liquor on broadcast television now accounts for about $306 million per year, much to the chagrin of critics, who claim that such advertising encourages drinking, especially among minors.

Unhealthy eating is also heavily advertised. In 2017, food companies spent about $11 billion on television spots alone, about 80% of those dollars went for advertising soda, fast food, candy, and other sugary and salty snacks. Research studies show that ads for unhealthy foods contribute to obesity, diabetes, high blood pressures and other ailments.

Even though parents are the strongest influence on children's eating habits, commercials help set food preferences. With the number of overweight children (6–11 years of age) doubling and adolescents (12–19 years of age) quadrupling from the early 1980s through the first two decades of the 2000s, there is no denying that the United States is witnessing an epidemic of obese children. Today, about 18% of children and 21% of adolescents are considered severely overweight.

Blame is often placed on the food industry for making sugar- and fat-laden foods and on the advertising industry for making such foods attractive to children. Food commercials make up 50% of all advertising on children's shows, and most ads are for unhealthy fare. In comparison, only a small fraction, about 3%, of ad dollars goes to promoting healthy food.

During Michelle Obama's time as First Lady, her campaign against childhood obesity helped change the food industry's attitudes toward advertising junk food to children. Subway has set aside over $40 million to promote healthy eating for children, and Unilever (owner of such brands as Ben & Jerry's and Klondike) announced that it will no longer direct food-related social media ads at children, nor air food and beverage ads aimed at children under the age of 12 in markets where they represent more than 25% of the television audience.

Online Product Ratings Systems

Online product ratings systems are often more effective in pushing a product than an advertisement, and in some ways, act as an ad. The number of stars or the quality of a review on Yelp can sway someone to buy a product. Even an increase of just one star on Amazon boosts a product's sales by 26%, and moves it up to the top of search engine listings. Good reviews are not always warranted and bad reviews not always deserved.

The deception runs deep as unscrupulous Web site operators pay online users to write good reviews in exchange for free products or monetary reimbursement. Online reviewers game the system by buying inexpensive goods then writing a 5-star review and receiving remuneration that is more than the cost of the products. Unsuspecting buyers are led to believe by a 'verified purchaser' label that the reviewer wrote an honest rave, when in fact it is phony.

SEE IT LATER

Although advertising is in many ways beneficial, it also contributes to *information overload*. Consumers are bombarded with ads practically everywhere they look. Television, radio, and the Internet are all packed with ads; the nation's roadways are cluttered with billboards that urge people to pull over and eat, to buy gas, or to listen to a particular radio station, and it is almost impossible to buy a product that does not prominently display its brand or logo. Consumers are persuaded to spend their money almost everywhere they look.

Online Product Rating Restrictions

Consumers, honest retailers, and the FTC itself are calling for stricter reviews of deceptive advertising disguised as objective online product rating systems. Some companies, such as Amazon, remove phony reviews but it is like playing the game 'Whack-a-mole'; when one phony review is taken down, another one goes up. With the help of companies such as Fakespot that detects fake or paid reviews by looking for language patterns, account creation dates, and the types of items reviewed, perhaps deceptive reviews will become an anomaly, but until then online shopping is a 'buyer beware' world.

The lack of a monetary consequence for faking the reviews signals to companies that it is worthwhile to pay for positive fake reviews in exchange for the uptick in sales, and then cry "we're sorry" to the FTC and the online world. Until there are financial penalties, there is not much stopping unethical purveyors of online goods from deceiving the public about the quality of their products.

Mobile on the Rise

Mobile advertising is on the brink of perhaps becoming the ultimate method of consumer targeting. With GPS and the tracking capabilities of apps and browser cookies, users' every move, likes and dislikes, attitudes, lifestyles, and product preferences are collected, sorted, categorized, and analyzed by algorithms that spit out personal consumer profiles, which are then sold to marketers who partner with creative teams to design ads

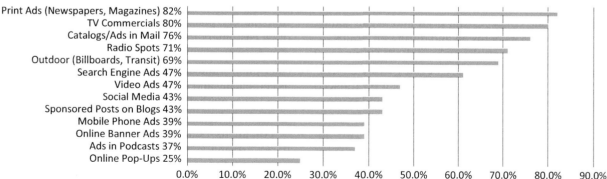

Ad Medium Most Trusted for Making Purchasing Decisions.
(Percent of Respondents)

- Print Ads (Newspapers, Magazines) 82%
- TV Commercials 80%
- Catalogs/Ads in Mail 76%
- Radio Spots 71%
- Outdoor (Billboards, Transit) 69%
- Search Engine Ads 47%
- Video Ads 47%
- Social Media 43%
- Sponsored Posts on Blogs 43%
- Mobile Phone Ads 39%
- Online Banner Ads 39%
- Ads in Podcasts 37%
- Online Pop-Ups 25%

Diagram 9.6 Trust in advertising.
Source: Marketing Chart, 2017

that resonate with users and compel them to buy products.

Mobile advertising surpassed desktop for the first time in 2016, and it now accounts for almost two-thirds of online revenue ($69.9 billion mobile, $37.6 billion desktop). There are still several hurdles facing mobile ads – they are too passive, non-interactive and small to really grab users' attention. Location-based app technology and other types of mobile spying, and payment security issues make users uneasy about mobile advertising and reluctant to make purchases through mobile devices. Despite lingering limitations, mobile advertising is on the rise, and its power to persuade should not be underestimated. Even newer variations of wearables such as smart watches, clothing, and eyeglasses could soon be transmitting ads directly onto us and into our retinas.

SUMMARY

Advertising is not a modern-day phenomenon but has been around since the days of clay tablets. Although it has changed in many ways since its origins, its purposes have remained the same: to get the word out about a product or service and persuade people to buy.

Radio was never meant to be an advertising medium, but high operating costs drove entrepreneurs to come up with a way to raise money for their over-the-air ventures. Stations experimented with toll advertising and later with sponsored programs. When television emerged as a new medium, there was no doubt that advertising was going to fund its existence. The only question was how to do so effectively. Early television experimented with radio's style of sponsored programs, which was fairly successful until the quiz show scandal damaged the reputation of several sponsors. The television industry then implemented the magazine concept from print and began selling commercial spots between and during programs.

Most commercials throughout the 1960s were 60 seconds in length, but they began getting shorter toward the end of the decade. The ban on tobacco advertising in 1971 left many networks and stations with unsold commercial time. Stations and networks were happy to find that they could sell two 30-second commercials to two different advertisers for more money than one 60-second spot to one advertiser. Throughout the 1970s, 30-second commercials began to dominate the airwaves, as they still do today.

Advertising benefits both marketers and consumers. Marketers are able to promote their products and services, and consumers get to learn about them. Advertising on radio and television and their 'now' online counterparts, the Internet, including mobile devices, has both advantages and disadvantages. Some products and services are best served by advertising on network television; some will reach their audiences in a more cost-efficient way on radio or cable television, and for others banner ads, interstitials, superstitials, and extramercials, among other types, work best. But even with the personalization of online advertising, and the advantages of radio and print ads, television advertising is still the most persuasive and effective advertising medium. As a research analyst once summed up, "Television is unique, in all forms of advertising, in being able to deliver that emotional power." And as another analyst pointed out, "After all, when was the last time a banner ad made you cry? If you're a greeting card company, it sure better" (Harwell, 2015, p. A16).

Placing a commercial here and there is usually not as effective an advertising strategy as constructing a campaign. Most campaigns are created by advertising agencies. There are basically four types of advertising agencies: Full-service agencies, creative boutiques, media-buying services, and interactive or cyber agencies.

Effective advertising begins with an understanding of the market. Creative teams understand consumer psychology and what motivates purchasing decisions. Knowing which medium best provides access to the target audience is the responsibility of media market researchers. Measuring a commercial's effectiveness is a crucial part of the advertising process, as it provides the feedback needed to reassess and update a campaign.

Advertising on 'now media' has taken a whole new meaning with display banner targeting. These strategies are newly developed just for targeting the audience of 'now media' users. Advertisers have discovered that merely placing ads online is not enough to draw customers to their products and services, they now have to reach out and grab them through their smartphones and other devices by using strategies such as geofencing and addressable geofencing. Zeroing in on a target market is an efficient and effective way to maximize advertising on 'now media.'

For all its benefits, there are some downsides to advertising as well. Critics claim that too much commercialism makes us greedy and materialistic. Advertising is blasted for exaggerating product benefits and for misleading consumers into buying items that they are later disappointed in or simply do not need. Regardless of how consumers may feel about advertising, without it, commercial media would not exist, and the economy would be dramatically different.

Given all the new technological developments, critics of advertising scream for relief from the over-commercialized world that it creates. And although some people try to shield themselves from advertising, others have come to accept that it is an everyday part of life. In the future, it is likely that advertising will creep into areas that were once regarded as being above such peddling. Even though public protest may stem the tide, commerce will probably prevail. Advertising will probably become increasingly ubiquitous, increasingly influential, and increasingly controversial.

BIBLIOGRAPHY

63 fascinating Google search statistics. (2018). *Bluelist*. Retrieved from: https://bluelist.co/blog/google-stats-and-facts/

Advertisement that watches you. (2009, December 13). *The New York Times Magazine*, p. 27.

Advertiser expenditure in the largest advertising categories in the United States. (2016). *Statista*. Retrieved from: www.statista.com/statistics/275506/top-advertising-categories-in-the-us/

Advertising spending of the distilled spirits industry in the United States. (2018). *Statista*. Retrieved from: www.statista.com/statistics/259642/advertising-spending-of-the-distilled-spirit-industry-in-the-us-by-medium/

Ah, the good-old days: The best Super Bowl ads. (2010, February 10). *MSNBC*. Retrieved from: www.msnbc.msn.com/id/35174030 [February 10, 2010]

American ad spending on online social networks. (2009, November 10). *TechCrunchies.com*. Retrieved from: techcrunchies.com/american-ad-spending-on-online-social-networks/ [February 3, 2010]

American advertisers increasingly take aim at children. (2009, August 31). *Washington Profile*. Retrieved from: www.washprofile.org/en/node/3396 [September 12, 2009]

Arango, T. (2009, October 6). Soon, bloggers must give full disclosure. *The New York Times*, p. B3.

Average cost of a 30-second TV commercial in selected formats in the United States in 2016. (2016). *Statista*. Retrieved from: www.statista.com/statistics/599147/tv-commercial-ad-price-usa/

Balderston, M. (2018, February 23). Report: Average number of TVs in U.S. homes declining. *TVTechnology*. Retrieved from: www.tvtechnology.com/news/report-average-number-of-tvs-in-us-homes-declining

Barclay, E. (2016, January 29). Scientists are building a case for how food ads make us overeat. *NPR*. Retrieved from: www.npr.org/sections/thesalt/2016/01/29/462838153/food-ads-make-us-eat-more-and-should-be-regulated

Barnes, B. (2009, July 27). Lab watches Web surfers to see which ads work. *The New York Times*, pp. B1, B6.

Baumer, K. (2011, February 3). Here's a look at the cost of Super Bowl ads through the years. *Business Insider*. Retrieved from: www.businessinsider.com/cost-super-bowl-ads-through-the-years-2011-2

Berthon, P., Pitt, L. F., & Watson, R. T. (1996). The World Wide Web as an advertising medium: Toward an understanding of conversion efficiency. *Journal of Advertising Research*, *36*(*1*), 43–54.

Billboard. (2020). *Digital Song. Year-end Sales 2019.* Retrieved from: www.billboard.com/charts/year-end/digital-songs

Bulova story. (2003). *Bulova Watch Company*. Retrieved from: www.asksales.com/Bulova/Bulova.htm

Can Spam Act: A compliance guide for business. (n.d.). *Federal Trade Commission*. Retrieved from: www.ftc.gov/tips-advice/business-center/guidance/can-spam-act-compliance-guide-business

Chen, A. (2002, August 19). How to slam spam. *E-week*, pp. 34–35.

Chen, B. X. (2016, February 4). How to watch the Super Bowl when you don't have cable. *The New York Times*, p. B7.

Costera Meijer, I., & Groot Kormelink, T. (2015). Checking, sharing, clicking and linking: Changing patterns of news use between 2004 and 2014. *Digital Journalism*, *3*(*5*), 664–679.

Credit card debt statistics for 2019. (2019, October 4). *The Motley Fool*. Retrieved from: www.fool.com/the-ascent/research/credit-card-debt-statistics/

Davis, W. (2009, September 20). Facebook to wind down Beacon to resolve privacy lawsuit. *MediaPost News*. Retrieved from: www.mediapost.com/publications/?fa5Articles.showArticle&art_aid5113848 [September 25, 2009]

Deal, D. (2018, January 27). How we listen to music in our cars. *Medium.com*. Retrieved from: https://medium.com/@davidjdeal/how-we-listen-to-music-in-our-cars-fb659ecec47

Digital ad business withers under COVID-19. (2020, April 17). *PYMNTS.com*. Retrieved from: www.pymnts.com/news/retail/2020/digital-ad-business-withers-under-covid-19-crisis/

Domain numbers of the top ten largest domains. (2019). *Statista*. Retrieved from: www.statista.com/statistics/262947/domain-numbers-of-the-ten-largest-top-level-domains/

Dominick, J. R., Sherman, B. L., & Copeland, G. A. (1996). *Broadcasting cable and beyond* (3rd ed.). New York: McGraw Hill.

Edwards, J. (2013, February 13). Behold: The first banner ad ever – from 1994. *Business Insider*, www.businessinsider.com/behold-the-first-banner-ad-ever--from-1994-2013-2

Elkin, T. (2003, September 22). Spam: Annoying but effective. *Advertising Age*, p. 40.

eMarketer. (2019). U.S. social media ad spending. Retrieved from: www.emarketer.com/content/us-digital-ad-spending-2019

Ember, S. (2015, September 2). With technology, avoiding both ads and the blockers. *The New York Times*, p. B3.

Ember, S. (2016, February 27). TV networks are recasting the role of commercials. *The New York Times*, p. B3.

Emergence of advertising in America. (2000). *John W. Hartman Center for Sales, Advertising and Marketing History, Duke University*. Retrieved from: www.scriptorium.lib.duke.edu/eaa/timeline.html [June 4, 2001]

Fact book: A handy guide to the advertising business. (2002, September 9). *Advertising Age* [Supplement].

Fahri, P. (2013, February 2). Blurring the line between news and advertising. *The Washington Post*, pp. A1, A9.

Freeman, L. (1999, September 27). E-mail industry battle cry: Ban the spam. *Advertising Age*, p. 70.

Forecast of the 5G adoption rate as share of mobile adoption rate in American from 2019 to 2025. *Statista*. Retrieved from: www.statista.com/statistics/792427/5g-adoption-rate-forecast-in-the-us/ [February 3, 2021]

Fowler, G. (2020, February 9). Stop selling my data! Now you have some say. *The Washington Post*, G1, G5.

Freier, A. (2019, April 26). Facebook mobile ad revenue jumps 30% to 13.9 billion. *Business of Apps*. Retrieved from: www.businessofapps.com/news/facebook-mobile-ad-revenue-jumps-30-to-13-9-billion/

Friedman, W. (2017, October 4). Shorter-duration TV commercials on the rise. *MediaPost*. Retrieved from: www.mediapost.com/publications/article/308248/shorter-duration-tv-commercials-on-the-rise.html

FTC policy statement on deception. (1983). *Federal Trade Commission*. Retrieved from: www.ftc.gov/bcp/policystmt/ad-decept.htm [August 12, 2004].

Giles, M. (1998, October 18). Fighting spammers frustrating for now. *Atlanta Journal-Constitution*, pp. H1, H4.

Golden Super Bowl now grown up from modest beginning. (2016, January, 30). *The New York Times*. Retrieved from: www.nytimes.com/reuters/2016/01/30/sports/football/30reuters-nfl-superbowl-history.html

Google Search Statistics. (2020, February 10). *InternetLive*. Retrieved from: www.internetlivestats.com/google-search-statistics/

Gorman, B. (2010, June 4). Basic cable's primetime audience share remains 59%, to broadcast's 39% for 2009–2010. *TV by The Numbers*. Retrieved from: http://tvbythenumbers.zap2it.com/2010/06/04/basic-cables-primetime-audience-share-remains-59-to-broadcasts-39-for-2009–10/53203/

Grant, A. E., & Meadows, J. H. (2012). *Communication technology update and fundamentals* (13th ed.). New York: Focal Press.

Guttman, A. (2019, August 7). Kids advertising spending worldwide. *Statista*. Retrieved from: www.statista.com/statistics/750865/kids-advertising-spending-worldwide/

Harwell, D. (2015, November 17). Holiday ads moving from TV to digital. *The Washington Post*, p. A16.

Harwell, D. (2019, June 1 12). Top AI researchers race to detect 'deepfake' videos: 'We are outgunned.' *The Washington Post*. Retrieved from: www.washingtonpost.com/technology/2019/06/12/top-ai-researchers-race-detect-deepfake-videos-we-are-outgunned/

Helft, M. (2009, March 11). Google to offer ads based on Interests, with privacy rights. *The New York Times*, p. B3.

Herlihy, G. (1999, September 26). Deliver me from spam. *Atlanta Journal-Constitution*, p. P1.

Hoelzel, M. (2014, December 2). The social media advertising report. *Business Insider*. Retrieved from: www.businessinsider.com/social-media-advertising-spending-growth-2014–9

Hsu, T. (2019, December 20). You see Pepsi, I see Coke: New tricks for product placement. *The New York Times*. Retrieved from: www.nytimes.com/2019/12/20/business/media/streaming-product-placement.html

IAB Internet advertising revenue report. (2019, May). *IAB*. Retrieved from: Full-Year-2018-IAB-Internet-Advertising-Revenue-Report.pdf

Industry data. (2015). *National Cable & Telecommunications Association*. Retrieved from: www.ncta.com/industry-data

Infographic: Rise of social media. (2010, September 8). *Digital Buzz*. Retrieved from: www.digitalbuzzblog.com/infographic-rise-of-social-media-ad-spending/

James, M. (2020, April 21). Coronavirus could wipe out $12 billion in TV ad spending. *MSN.com*. Retrieved from: www.msn.com/en-us/sports/nba/coronavirus-could-wipe-out-2412-billion-in-tv-ad-spending/ar-BB12WpKH

Johnson, S. (2014, September 29). New research sheds light on daily ad exposures. *SJ Insights*. Retrieved from: https://sjinsights.net/2014/09/29/new-research-sheds-light-on-daily-ad-exposures/

Kang, C. (2015, November 10). Online advertising predicted to top TV by 2017 as viewing habits evolve. *The Washington Post*, p. A16.

Kapner, S. (2001, August 31). Kleenex to sponsor movies to cry by on TV. *The New York Times*. Retrieved from: www.nytimes.com/2001/08/31/business/the-media-business-advertising-addenda-kleenex-to-sponsor-movies-to-cry-by-on-tv.html [September 1, 2009]

Kaye, B. K., & Medoff, N. J. (2001a). *Just a click away*. Boston, MA: Allyn & Bacon.

Kaye, B. K., & Medoff, N. J. (2001b). *The World Wide Web: A mass communication perspective*. Mountain View, CA: Mayfield.

Kelly, H. (2020, February 9). Targeted ads are now showing up in your mailbox. *The Washington Post*, G3.

Kelly, H. (2020, April 27). Advertisers change messages – fast. *The Washington Post*, A17.

Khermouch, G., & Green, J. (2001, July 30). Buzz marketing. *Business Week*. Retrieved from: www.businessweek.com/print/magazine/content/01_31/b3743001.htm

Lynch, J. (2015, June 9). Why TV is still the most effective advertising medium. *AdWeek*. Retrieved from: www.adweek.com/tv-video/why-tv-still-most-effective-advertising-medium-165247/

Maheshwari, S. (2017, August 30). Six-second commercials are coming to N.F.L. games on Fox. *The New York Times*. Retrieved from: www.nytimes.com/2017/08/30/business/media/nfl-six-second-commercials.html

Maheshwari, S. (2019, November 29). The online star rating system is flawed ... and you never know if you can trust what you read. *The New York Times*, pp. B1, B4.

Maller, B. (2004, May 5). Baseball sells out to Spiderman. *Ben Maller.com*. Retrieved from: http://benmaller.com/archives/2004/may/05-baseball_sells_out_to_spiderman.html

Malone, M. (2020). Geo-fencing and geo-targeting: What's the difference? *Vici Media*. Retrieved from: www.vicimediainc.com/geo-fencing-and-geo-targeting-whats-the-difference/

Marketing Charts. (2017, January 17). Which advertising channels consumers trust most and least when making purchases. *MarketingSherpa*. Retrieved from: www.marketingsherpa.com/article/chart/channels-customers-trust-most-when-purchasing

Marketing Charts. (2018, June 25). U.S. online and traditional media advertising outlook, 2018–2022. Retrieved from: www.marketingcharts.com/featured-104785

McCambley, J. (2013, December 12). The first banner ad. Why did it work so well? *The Guardian*. Retrieved from: www.theguardian.com/media-network/media-network-blog/2013/dec/12/first-ever-banner-ad-advertising

Miller, C. C., & Goel, V. (2013, October 12). Google to sell users' endorsements. *The New York Times*, pp. B1, B2.

Miller, C. C., & Sengupta, S. (2013, October 12). Selling secrets of phone users to advertisers. *The New York Times*, pp. A1, A4.

Mobile banners continue to boast high click rates. (2012, August 27). *eMarketer*. Retrieved from: www.emarketer.com/Article/Mobile-Banners-Continue-Boast-High-Click-Rates/1009299

Murphy, K. (2016, February 21). The ad blocking wars. *The New York Times*, p. SR 7.

NBC Sports. (2020, February 2). Super Bowl commercials 2020: How much do ads cost. Retrieved from: https://sports.nbcsports.com/2020/02/02/super-bowl-commercials-2020/

Nielsen. (2019, August 27). Nielsen estimates 120.6 million TV homes in the U.S. for the 2019–2020 season. Retrieved from: www.nielsen.com/us/en/insights/article/2019/nielsen-estimates-120-6-million-tv-homes-in-the-u-s-for-the-2019-202-tv-season/

Nielsen reports television tuning remains at record levels. (2001, October 17). *Nielsen*. Retrieved from: en-us.nielsen.com/main/news/news_releases/2007/october/Nielsen_Reports_Television_Tuning_Remains_at_Record_Levels [September 1, 2009]

Nielsen tops of 2012: Advertising. (2012, December 17). *Nielsen*. Retrieved from: www.nielsen.com/us/en/insights/news/2012/nielsen-tops-of-2012-advertising.html

Number of smartphone users in the United States from 2010 to 2023. (2020). *Statista*. Retrieved from: www.statista.com/statistics/201182/forecast-of-smartphone-users-in-the-us/

Number of smartphone users worldwide 2016 to 2020. (2020). *Statista*. Retrieved from: www.statista.com/statistics/330695/number-of-smartphone-users-worldwide/

Number of tablet users in the United States from 2014 to 2020. (2020). *Statista*. Retrieved from: www.statista.com/statistics/208690/us-tablet-penetration-forecast/

O'Guinn, T. C., Allen, C. T., & Semenik, R. J. (2000). *Advertising*. Toronto, Canada: South-Western.

Orlik, P. B. (1998). *Broadcast/Cable copywriting*. Boston, MA: Allyn & Bacon.

Othmer, J. P. (2009, August 16). Skip past the ads, but you're still being sold something. *The Washington Post*. Retrieved from: www.washingtonpost.com/wp-dyn/content/article/2009/08/14/AR2009081401629.html

Patel, Neil (2018). Which U.S. brands are spending the most on advertising? *NeilPatel.com*. Retrieved from: https://neilpatel.com/blog/top-ad-spenders/

Perez, S. (2019, September 4). YouTube creators may also be liable for COPPA violations, following FTC settlement. *TechCrunch*. Retrieved from: https://techcrunch.com/2019/09/04/youtube-creators-may-also-be-held-liable-for-coppa-violations-following-ftc-settlement/

Poggi, J. (2013, October 9). TV ad prices: Football is still king. *Advertising Age*. Retrieved from: http://adage.com/article/media/tv-ad-prices-football-king/244832/

Pogue, D. (2002, June 27). Puncturing Web ads before they pop up. *The New York Times*. Retrieved from: www.nytimes.com/2002/06/27/technology/circuits

Purkayastha, A. (2020, February, 12). Digital marketing for corporate. *Digital Marketing University*. Retrieved from: www.digitalmarketinguniversity.com/category/digital-marketing/

Radio facts and figures. (2019). *News Generation*. Retrieved from: https://newsgeneration.com/broadcast-resources/radio-facts-and-figures/

Ritchie, M. (1994). *Please stand by*. Woodstock, NY: Overlook Press.

Romm, T., Dwoskin, E., & Timberg, C. (2019, June 20). YouTube under investigation over child privacy complaints. *The Washington Post*, pp. A1, A21.

Rosen, J. (2012, December 2). Who do they think they are? *The New York Times Magazine*, pp. 40–45.

Rovell, D. (2004, May 7). Baseball scales back movie promotion. *ESPN*. Retrieved from: www.sports.espn.go.com/espn/sportsbusiness/news/story?id51796765

Russell, J. T., & Lane, W. R. (1999). *Kleppner's advertising procedure*. Upper Saddle River, NJ: Prentice Hall.

Schumacher, L. (2020, April 23). What is OTT, CTV, vMVPD and all the rest of those new digital video terms? *Vici Media*. Retrieved from: www.vicimediainc.com/what-is-ott-ctv-vmvpd-and-all-the-rest-of-those-new-digital-video-terms/

Scola, N. (2014, August, 23). Tune in to satellite TV? The political ads may be tuned in to you. *The Washington Post*, p. A10.

Scott, N. (2015, January 28). 10 best Super Bowl commercials of all time. *USA Today*. Retrieved from: http://ftw.usatoday.com/2015/01/10-best-super-bowl-commercials-of-all-time

Semuels, A. (2009, February 24). Television viewing at all-time high. *Los Angeles Times*. Retrieved from: articles.latimes.com/2009/feb/24/business/fi-tvwatching24 [September 1, 2009]

Sherman, E. (2017, October 25). What Eggo Waffles mean to Eleven from 'Stranger Things.' *Food & Wine*. Retrieved from: www.foodandwine.com/news/stranger-things-eggo-waffles

Sivulka, J. (1998). *Soap, sex and cigarettes*. Belmont, CA: Wadsworth.

Smartphone, tablet uptake still climbing in the U.S. (2013, October 14). *eMarketer*. Retrieved from: www.emarketer.com/Article/Smartphone-Tablet-Uptake-Still-Climbing-US/1010297

Spam statistics: Spam email traffic share 2019. (2019, December 4). *Statista*. Retrieved from: www.statista.com/statistics/420391/spam-email-traffic-share/

Stambor, Z. (2013, April 12). Social media ad spending will reach $11 billion by 2017. *Internet Retailer*. Retrieved from: www.internetretailer.com/2013/04/12/social-media-ad-spending-will-reach-11-billion-2017

Steinberg, B. (2009, October 26). 'Sunday Night Football' remains costliest TV show. *Advertising Age*. Retrieved from: http://adage.com/article/ad-age-graphics/tv-advertising-sunday-night-football-costliest-show/139923/ [February 2, 2010]

Steinberg, B. (2015, September 29). TV ad prices: Football, 'Empire,' 'Walking Dead,' 'Big Bang Theory,' top the list. *Variety*. Retrieved from: http://variety.com/2015/tv/news/tv-advertising-prices-football-empire-walking-dead-big-bang-theory-1201603800/

Steinberg, B. (2017, November 2). TV ad prices: Football hikes, 'This is Us' soars, 'Walking Dead' stumbles, 'Empire' falls. *Variety*. Retrieved from: https://variety.com/2017/tv/news/tv-ad-prices-football-walking-dead-empire-1202602792/

Stelter, B. (2009, November 11). TV News without the TV. *The New York Times*, pp. B1, B10.

State laws relating to unsolicited commercial or bulk e-mail (SPAM). (2015, January 9). *National Conference of State Legislatures*. Retrieved from: www.ncsl.org/research/telecommunications-and-information-technology/state-spam-laws.aspx

Stibich, M. (2020, February 4). Top reasons to turn off your TV. *VeryWellMind*. Retrieved from: www.verywellmind.com/top-reasons-to-turn-off-your-tv-2223895

Stone, B. (2002, October 24). Those annoying ads that won't go away. *Newsweek*, pp. 38J, 38L.

Stone, B., & Lin, J. (2002). Spamming the world. *Newsweek*, pp. 42–44.

Stone, B., & Weil, D. (2003, December 8). Soaking in spam. *Newsweek*. Retrieved from: www.newsweek.com

Story, L. (2008, March 10). To aim ads, Web is keeping closer eye on what you click. *The New York Times*, pp. A1, A14.

Super Bowl average cost of a 30-second TV advertisement from 2002 to 2019. (2020). *Statista*. Retrieved from: www.statista.com/statistics/217134/total-advertisement-revenue-of-super-bowls/

Super Bowl TV ratings. (2009, January 18). *TV by the Numbers*. Retrieved from: Tvbythenumbers.com/2009/01/18/historical-super-bowl-tv-ratings/11044

Taylor, C. (2019, March 18). Addressable TV is creating new advertising capabilities … and they are not going away. *Forbes*. Retrieved from: www.forbes.com/sites/charlesrtaylor/2019/03/18/addressable-tv-is-creating-new-advertising-capabilities-and-they-are-not-going-away/#556ac5ee15ce

Tedeschi, B. (1998, December 8). Marketing by e-mail: Sales tool or spam? *The New York Times*. Retrieved from: www.nytimes.com [February 19, 1999]

Television watching statistics. (2013, December 7). *StatisticsBrain.com*. Retrieved from: www.statisticbrain.com/television-watching-statistics/

Terms for the trade. (2014). *Nielsen*. Retrieved from: www.arbitron.com/downloads/terms_brochure.pdf

Thompson, D. (2013, March). The incredible shrinking ad. *The Atlantic*, pp. 24, 26, 28.

Tsukayama, H. (2014, July 24). Facebook profits surge, largely from mobile ads. *The Washington Post*, p. A14.

TV costs and CPM trends – Network TV primetime. (2015). *Television Advertising Bureau*. Retrieved from: www.tvb.org/trends/4718/4709

Twitter usage up, but advertising sales slow. (2020, May 1). *The Washington Post*, p. A 14.

Usage statistics of top level domains for websites. (2019). *W3Tech*. Retrieved from: https://w3techs.com/technologies/overview/top_level_domain

Valentino-DeVries, J., Singer, N., Keller, M. H., & Krolik, A. (2018, December 18). Your apps know where you've been, and can't keep a secret. *The New York Times*, pp. A1, A20–21.

Vanden Bergh, B. G., & Katz, H. (1999). *Advertising principles*. Lincolnwood, IL: NTC Business Books.

Verna, P. (2018, August 23). Connected TV advertising. *eMarketer*. Retrieved from: www.emarketer.com/content/connected-tv-advertising

Vogue. (2020, February 1). The 16 most iconic Super Bowl ads of all time. Retrieved from: https://www.vogue.com/article/best-super-bowl-ads-of-all-time

Wagner, K. (2019, February 20). Digital advertising in the U.S. is finally bigger than print and television. *Vox*. Retrieved from: www.vox.com/2019/2/20/18232433/digital-advertising-facebook-google-growth-tv-print-emarketer-2019

Warner, C. (2009). *Media sales*. West Sussex, UK: Wiley-Blackwell.

Weiss, A. S. (2013). Exploring news apps and location-based services on the smartphone. *Journalism & Mass Communication Quarterly*, *90*(3), 435–456.

Which industries spend the most on digital advertising? (2018, July 18). *Marketing Charts*. Retrieved from: www.marketingcharts.com/advertising-trends/spending-and-spenders-105020

White, T. H. (2001). United States early radio history. Retrieved from: www.ipass.net/~whitetho/part2.htm [June 4, 2001]

Who's watching how many channels? (2016, October 3). *Marketing Charts*. Retrieved from: www.marketingcharts.com/television-71258

Wolf, C., & Schnauber, A. (2014). News consumption in the mobile era: The role of mobile devices and traditional journalism's content with the user's information repertoire. *Digital Journalism*, *3*(5), 759–776.

Yarow, J. (2014, May 8). This is the thing that people hate about paying for TV more than anything. *Business Insider*. Retrieved from: www.businessinsider.com/average-number-of-channels-watched-per-household-2014-5

Chapter ten

Audience Measurement: Who's Listening, Who's Watching, Who's Surfing?

Let's pretend you have started an online news site. You spent all your savings to purchase computers and other digital equipment, but operating the site involves many other expenses, including your employees' salaries. How are you going to come up with the money to cover these costs? By selling digital ad space, or selling your site users to your advertisers. And to do so, you must know something about your users: How many there are, who they are, what news do they want to know, and so on. The need to know about audience members is what prompted the development of audience measurement techniques.

Knowledge of audience characteristics, including their use of media, is at the core of selling advertising, whether for radio, television or online sites. This knowledge translates into various figures, such as **ratings** and **shares**, and cost per click, which are then used to determine the price of commercial time and space. Generally, the program, station, or Web site that draws the largest or most desirable audience can charge the most for its airtime or space.

This chapter begins with an overview of the early methods of obtaining audience feedback and then examines contemporary ways of monitoring radio, television, and Internet audiences. The explanations of these measuring techniques include the mathematical formulas that show how ratings, shares, and other measures are calculated and reported. The chapter then ties ratings to advertising sales by demonstrating how stations and Web sites use audience data to price commercial time and space.

SEE IT THEN

Early Ratings Systems

In the late 1920s, when radio advertising was just beginning to catch on, stations were stumped as to how much to charge for commercial time. Radio stations were setting prices for time, but advertisers were hesitant to buy unless they had information about the listeners. The void was filled in 1929 when pollster Archibald M. Crossley called on advertisers to sponsor a new way of measuring radio listenership: The **telephone recall system**. With this method, a random **sample** of people was called and asked what radio stations or programs they had listened to in the past 24 hours. Memory failure was the biggest drawback to Crossley's system, as listeners could not accurately remember what stations and programs they had tuned in to.

Crossley's fiercest competitor was C. E. Hooper, who used the **telephone coincidental method** of measuring radio listenership. Using this method, Hooper's staff telephoned a random sample of people and asked, "Are you listening to the radio just now?" The next question was something like, "To what program are you listening?" followed by, "Over what station is the program coming?" (Beville, 1988, p. 11). Many thought Hooper's method was superior to Crossley's because it did not rely on listeners' memories but instead surveyed exactly what they were listening to at the time of the call. Of course, the drawback to Hooper's method was that someone might have listened to a station for hours that day but just happened to have the radio turned off or tuned to another station at the time of the telephone call. Both the Crossley and Hooper methods had their flaws, but at the time, they were the best ways to measure listenership.

In the 1930s, A. C. Nielsen was also working on ways to measure the radio audience. Nielsen took a slightly different approach by using an electronic metering device, the *audimeter*, which attached to a radio and monitored the stations that were tuned to and for how long. The audimeter was the precursor to today's audience-metering devices.

In the late 1940s, Hooper supplemented telephone coincidental calling by asking listeners to keep **diaries** of their radio use. The American Research Bureau (ARB), which changed its name to Arbitron in 1973, also championed the diary method for measuring radio listenership.

The Crossley, Hooper, and Nielsen companies were the leaders in radio ratings into the late 1940s. Crossley left the ratings business in 1946, and in 1950, Nielsen bought out Hooper and thus eliminated its biggest competitor. Hooper himself, however, met a tragic death four years later when, while on a duck hunting trip, he stumbled into a rotating airplane propeller.

As television came to life in the late 1940s, the ratings services – mainly Nielsen and ARB/Arbitron – adapted their radio research methods for the new medium. Although many other ratings services emerged, none was able to overcome the market dominance of these two companies.

SEE IT NOW

Gathering Radio and Television Audience Numbers

Nielsen

Media outlets need to know who is listening to their stations and who is watching their programs. Radio station managers must stay tapped into their number of listeners and who listens to what music when. These types of data reflect how consumers use the media and help managers set the price for commercial time.

Arbitron once measured both local radio and television audiences but dropped its less profitable television services in 1994, leaving television audience measurement in the hands of Nielsen. For almost 20 years, Nielsen measured television audiences while Arbitron monitored radio listeners.

In September 2013, Nielsen acquired Arbitron and renamed it Nielsen Audio. Nielsen is interested primarily in how the public uses the media. If a particular television program has consistently low ratings, it might be moved to another time slot or perhaps dropped. If a radio station is highly rated in the market, management knows that it has a winning format that can command top dollar for commercial time. Nielsen is now the premiere audience measurement company for television (broadcast, cable, satellite, DVR) and radio (over-the-air, online, podcast).

ZOOM IN 10.1

For information about Nielsen, visit www.nielsen.com/us

Nielsen measures radio and television audiences by geographic regions called media markets. It geographically segments radio listeners and television viewers to gain a clearer understanding of audience differences and similarities across the nation.

Nielsen divides the United States into 210 **designated market areas** (**DMAs**) based on geographic location and television and radio stations. The largest DMA has long been New York, New York, with 6,824,120 households, and the smallest is Glendive, Montana, with 3,630 households. Market areas are sometimes redrawn if overall viewing behavior shifts because of changes in cable penetration or satellite television subscriptions. When Arbitron was a separate company from Nielsen, it had developed its own market designation scheme, known as the **area of dominant influence**, but it then switched to Nielsen's DMAs.

Nielsen Audio also breaks out radio markets by **metro survey areas** (**MSAs**) and **total survey areas** (**TSAs**). There are 263 distinct geographic **MSAs**, often referred to as *metros*. An MSA is generally composed of a major city or several cities and their surrounding county or counties. In 2019, New York, New York, was the largest MSA (16,110,500 population, ages 12+) and Grand Forks, North Dakota, the smallest (78,600 population, ages 12+).

ZOOM IN 10.2

To find your city's DMA ranking and to see whether you live in a metered market, visit https://mediatracks.com/resources/nielsen-dma-rankings-2020/

Nielsen Audio also measures listenership in 215 designated geographic regions known as **total survey areas (TSAs)**. A TSA is made up of a major city and a larger surrounding area than a metro area. For example, the MSA population for Knoxville, Tennessee, is 739,500, but Knoxville's TSA population is 1,356,700. Smaller metros help low-powered stations compete in the ratings game against high-powered stations by providing smaller geographic areas of measurement.

FYI: TSA/MSA

Station 1 is a 3,000-watt station whose signal is limited to its metro survey area (MSA). Station 1's ratings are low compared to those of Station 2, a 50,000-watt station whose signal reaches the entire total survey area (TSA).

Station 1 may be very popular throughout the smaller MSA; thus, measuring Station 1 within the MSA more accurately reflects its popularity and boosts its ratings. Conversely, measuring Station 1 throughout the TSA will lower its ratings, because its signal does not reach that population.

- TSA = 50,000 population
- MSA = 25,000 population
- Station1 = 2,500 listeners
- TSA rating = 2,500/50,000 = 0.05 or 5% of listeners
- MSA rating = 2,500/25,000 = 0.10 or 10% of listeners

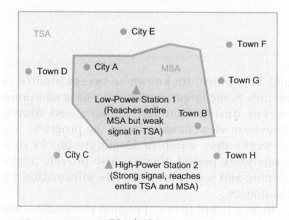

Diagram 10.1 TSA/MSA.

Counting Radio Listeners and Advertising Exposure

Radio listening is measured using listener diaries and portable people meters. Each method has its advantages and disadvantages but, in the end, each offers, at best, estimates of the number of listeners, as well as demographic data. But the estimates are crucial as radio stations set the price of their commercials, and brand their market position and station image based on who is listening.

Radio Diary Nielsen Audio collects radio-listening data by sending diaries to a sample of listeners within the top 50 TSAs. Nielsen surveys most radio markets at least once a year, though larger market areas may be surveyed up to four times per year or even on a continuous basis. Nielsen mails each randomly selected listener a seven-day diary, with each day divided into 15-minute time blocks. In exchange for a small monetary reward (usually less than $5), each participant records when he or she starts and stops listening and the station (identified by call letters and/or dial position) that he or she listens to during each 15-minute time block. Additionally, each participant is asked demographic information, typically age, gender, income, education, and ethnicity.

The Portable People Meter Hailed as a revolutionary audience-metering device, the **portable people meter (PPM)** is capable of capturing radio-listening data through wireless signals. The PPM is used to measure radio listenership in 48 markets. The PPM is a small, pager-sized device that picks up and decodes inaudible signals that radio broadcasters embed in their programs. The PPM clips on to a survey participant's clothing or purse, and the PPM decodes the unique signal emitted by each radio station as the wearer gets into signal range, whether at home, in a shopping mall, in a car, or anywhere else. The PPM is the first electronic metering device that monitors in- and out-of-the-home exposure to over-the-air radio. The PPM stores each station's signals, and at the end of each day, survey participants place the device in an at-home base station that sends the codes to Nielsen for tabulation. PPM data suggest that radio-listening habits may be different from what has been reported through diaries. PPM numbers indicate that many more people listen to the radio but spend less time listening and they change stations more frequently than previously known.

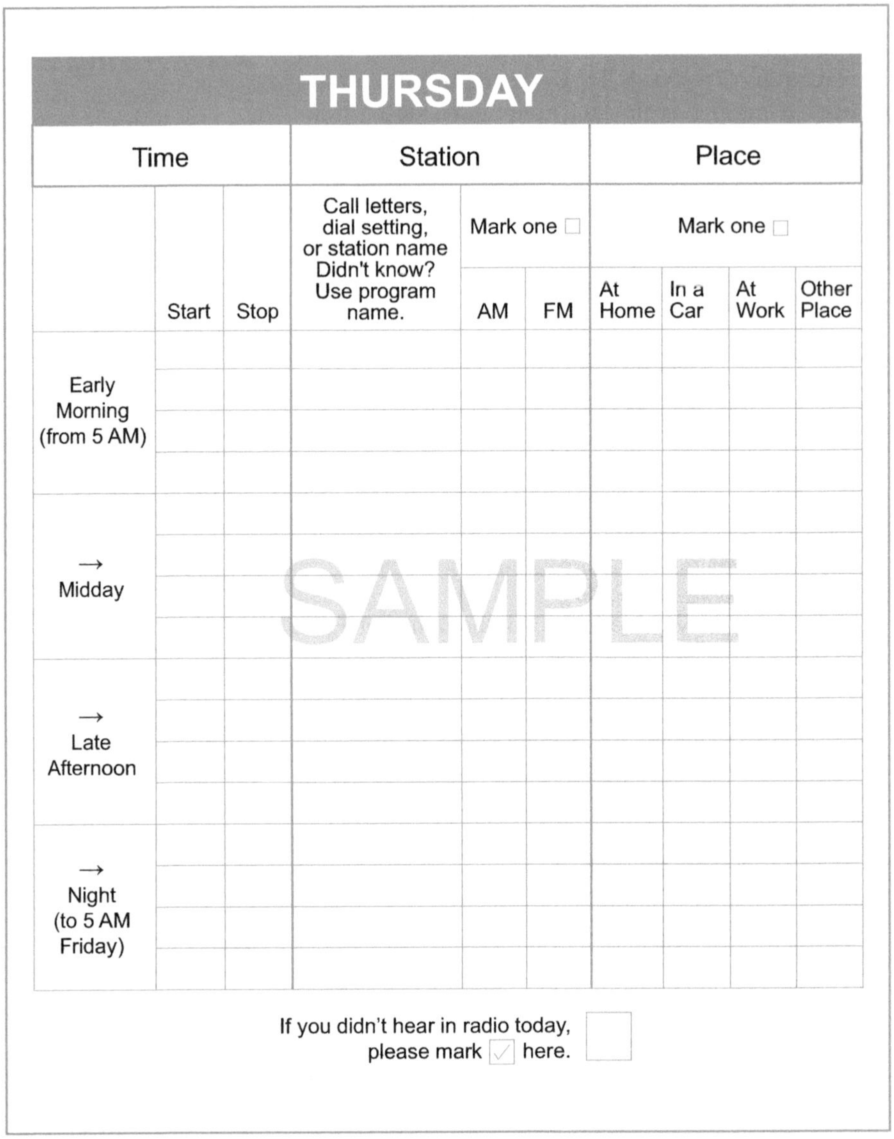

Diagram 10.2 A page from a Nielsen radio diary. Source: Photo courtesy of Nielsen

THURSDAY

Time			Station			Place			
	Start	Stop	Call letters, dial setting, or station name Didn't know? Use program name.	Mark one ☐		Mark one ☐			
				AM	FM	At Home	In a Car	At Work	Other Place
Early Morning (from 5 AM)									
→ Midday									
→ Late Afternoon									
→ Night (to 5 AM Friday)									

If you didn't hear in radio today, please mark ✓ here. ☐

Counting Television Viewers and Advertising Exposure

Television viewing is also measured using diaries and meters. As with radio, each method has its advantages and disadvantages, and each also delivers estimates of the number of viewers, as well as demographic data. These estimates are just as important to television as to radio. Television stations use the numbers to set the price of their commercials, and to brand their market position and station image based on who is watching. Television networks rely on audience numbers to set the price of national commercials, which are inserted within a program, and to know which shows to keep on the air and which to cancel.

Television Diary Nielsen diaries are sent out to sample homes each November, February, May, and July, which are known as **sweeps** months or periods. Some larger markets may have additional sweeps during October, January, and March. Television viewers write down the programs and networks they watch in 15-minute blocks over a one-week period. Viewers also provide demographic and sometimes lifestyle information for the diaries.

In an attempt to raise program viewership – and hence ratings – television and radio station networks use sweeps periods to present special programs, season finales, and much-awaited episodes that resolve a mystery. When these special programs and episodes are aired during sweeps, the ratings may be inflated, because viewers might watch only these heavily hyped episodes and not tune in during the regular season.

Figure 10.1 A Nielsen portable people meter.
Source: Photo courtesy of Nielsen

Figure 10.2 A Nielsen people meter.
Source: Photo courtesy of Nielsen

Set-top and People Meter In addition to diaries, Nielsen also uses metering devices to record viewing habits in randomly selected television households in the United States. The **set meter**, first used in 2,000 homes in 1987, attaches to a television set and records when it is turned on and off and to what broadcast, cable, and satellite programs and channels the set is tuned. The set meter requires an accompanying diary to track what programs household members are watching.

The **people meter** is a newer, more sophisticated version of a set meter. Each member of the metered household is assigned an identification button on the accompanying remote control unit. Household members are supposed to push their identification number when they start and finish watching television. The people meter records both what programs are being watched and who is watching them. Diaries are used alongside the meter but only to match viewers with their demographic characteristics.

FYI: TO METER OR NOT TO METER

In October 2002 when Knoxville, Tennessee, was the 59th-largest market in the United States with 552,000 television households, it converted from diary reporting to the Nielsen meter system. Area television station executives expressed mixed feelings about the meters, especially as the $500,000 price tag for each station was about three times what they had paid for diaries. Beyond cost, some were concerned that the meters would not make any difference in the stations' market positions and thus were not worth the expense. Yet others argued that the meters would give more accurate audience data than the diaries. Knoxville, home of the University of Tennessee, has a younger-than-average population that is less likely to fill out diaries, which could hurt the Fox affiliated station that tends to draw younger audiences. Meters, which have a high participation rate among young viewers, could more accurately reflect larger audiences for Fox.

The big advantage of being a metered market is that such a market is more likely to draw national advertisers. The financial outlay for the metered ratings could be made up in increased revenue. One station's general manager said that meters "are the best thing for the market," but another sighed, "They are coming, and we are going to deal with it" (Flannagan, 2002; Local television market universe estimates, 2016; Morrow, 2002).

Challenges of Gathering Radio and Television Audience Numbers

Ratings are based on *estimates* of numbers of viewers and listeners and are thus not absolutes. Each method of measurement is limited in some way. Although ratings companies do their best to ensure that the most accurate data possible are gathered,

Figure 10.3 The Nielsen people meter is attached to the viewer's television set.
Source: Photo courtesy of Nielsen

there are still many concerns and criticisms over how audience data are collected.

Samples One of the most serious concerns about ratings data is the sample of viewers and listeners whose media habits are monitored. It is not possible to poll the entire U.S. population, so a subset, or *sample*, of listeners and viewers is selected. For a long time, Nielsen sampled 5,000 U.S. households, but many critics contended that the sample size was too small to produce accurate results. Others claim that the sample size was large enough as long as it represented all of the demographic groups within a population. Nielsen has since increased its sample to about 47,000 homes. It has also changed some of its sampling methods to ensure demographic representation. In other words, in a DMA survey area in which 30% of the people are elderly, diaries should be sent to a group of participants that includes that same percentage of elderly people to be sure that the data accurately reflect media use. Ratings companies make every attempt to ensure that samples accurately reflect the survey population and often mathematically weight samples to correct over- or underrepresentation of particular groups. Even so, ethnic and other specific demographic groups are underrepresented in most samples, yielding an inaccurate picture of television viewing.

Location Television and radio are not used only in the home but also in bars and restaurants, in hospitals, at work, in hotels, and so on. But out-of-home television and radio use is not recorded. Because most people listen to the radio at work or in cars, not recording such use is a huge detriment to radio stations, and the ratings may not accurately reflect the actual number of listeners. In comparison, television ratings are minimally affected, because people do most of their watching at home. The PPM should minimize this concern because it provides out-of-home radio-listening estimates.

Accuracy Ratings companies depend on participants to fill out diaries and use people meters correctly. Yet too many listeners and viewers fail to follow diary instructions, forget to record the stations they have listened to, and write down the wrong station or channel number. All of these mistakes result in inaccurate data and thus create an incomplete picture of media use and users.

Television-metering devices are also problematic. Although set meters record what is being watched, Nielsen researchers depend on participants pushing their ID buttons before and after each viewing session, but many forget to do so.

Data Collection Method How data are collected also influences responses. People often over-report and under-estimate their media use. For example, telephone survey participants may feel a bit shy, embarrassed, or hesitant to tell a researcher how much television they watch or how long they spend surfing the Internet, and so they tend to report they use the media much less than in reality. For example, participants reported, via telephone survey, that they watched television an average of 121 minutes per day, but direct observation showed that they actually watched for 319 minutes per day, a reporting difference of 198 minutes. Participants also overestimated by diary that they devoted 26 minutes per day reading a newspaper, when indeed direct observations showed they only spent 17 minutes (34.6% fewer minutes). Because of these types of self-reported errors, researchers often conduct several studies using various protocols to research the same issue.

Calculating Radio and Television Audience Numbers

Once ratings companies have gathered audience numbers and know how many people are watching television or listening to the radio, they conduct more meaningful analyses. For example, just knowing that 10,000 people listen to Station A does not have much meaning unless it is considered in the context of the total audience size. If 10,000 people listen to Station A, does that mean many people listen or just a few? The answer depends on the market size. If in a market of 100,000 people, 10,000 are listening to Station A, then that station has captured 10% of the population. However, if Station A is in a market with a population of 200,000, it has drawn only 5% of the population. Translating the number of viewers or listeners into a percentage provides comparison across markets. These percentages are commonly referred to as ratings and shares.

Ratings and Shares Radio and television both rely on ratings and shares to assess their market positions. For both radio and television, the mathematical computations are identical. **Ratings** and **shares** are simply percentages. A *rating* is an estimate of the number of people who are listening to a radio station or watching a television program divided by the number of people in a population who have a radio or a television. Another way to think of a rating is that 1 rating point equals 1% of the households in the market area that have a television set or radio. A radio *share* is an estimate of the number of people who are listening to a specific radio station or radio program divided by the number of people who are listening to radio. A television *share* is an estimate of the number of people who are watching a specific television program or station divided by the number of people who are watching television. Thus, the only difference between a rating and a share is the group that is being measured. Ratings measure everyone with a radio or television, whereas shares only consider those people who are actually listening to the radio or watching television at a given time.

The examples show that ratings and shares for both television and radio are calculated in the same way but that radio measures individual listeners **(People Using Radio, or PUR)** and television measures households with televisions **(Households Using Television, or HUT)**. Because the average household consists of 2.6 persons and contains 2.6 televisions, and because household members are increasingly watching television separately from others in the home, the HUT measure is giving way to Persons Using Television (PUT), a measure that may be more valuable to the media industry.

Notice that when ratings are calculated, all members of the population who have radios or televisions are included in the calculation, even if they're *not* listening or watching. When shares are figured, however, only members of the population who *are* listening or watching are included in the

FYI: HOW TO CALCULATE RADIO AND TELEVISION RATINGS AND SHARES

Radio ratings = Number of people listening to a station/Number of people in a population with radios

Radio shares = Number of people listening to a station/Number of people in a population listening to radio

TV ratings = Number of households watching a program/Number of households in a population with televisions

TV shares = Number of households watching a program/Number of households in a population watching television

Example: In City Z, **5,000 people have radios but only 2,500 are listening**. Compare Station A's rating and share to Station B's rating and share.

Radio Ratings

500 listeners tuned to Station A/**5,000** people with radios = 0.1 or 10%, expressed as '10 rating.'

400 listeners tuned to Station B/**5,000** people with radios = 0.08 or 8%, expressed as '8 rating.'

Radio Shares

500 listeners tuned to Station A/**2,500** people listening to radio = 0.20 or 20.0%, expressed as '20 rating.'

400 listeners tuned to Station B/**2,500** people listening to radio = 0.16 or 16.0%, expressed as '16 rating.'

Notice that when calculating ratings, everyone in the population with radio (5,000) is included in the denominator. But when calculating shares, only those who are listening to radio (2,500) are included in the denominator.

calculation. Shares, therefore, are always larger than ratings, because they include a smaller segment of listeners or viewers rather than the entire population. In the unlikely event that every person in the market watched or listened to the same show, then the rating and share would be equal.

One way to visualize the relationship of ratings to shares is to think of a chocolate birthday cake. If there are ten people at a party, the cake must be cut into ten slices (ratings). But if two partygoers are on a diet, another one is allergic to chocolate, and another one is too full to eat a piece of cake, then the cake will be 'shared' among six people, and thus each slice will be larger than if the cake was cut into ten smaller slices.

> **FYI: A POINT TO REMEMBER**
>
> One rating point represents 1% of the population being measured. For example, one rating point represents 1,206,000 households (1.20 million) or 1% of the nation's estimated 120,600,000 (120.6 million) television homes.
>
> Source: Nielsen, 2013

Average Quarter Hour In almost all markets, there are more radio stations than television stations; therefore, radio station ratings are generally lower than television station ratings, because there are more radio stations competing for the same number of listeners. To compensate for lower rating points, radio measures audience in terms of average quarter hour (AQH).

There are several AQH measures: Average quarter hour *persons*, average quarter hour *rating*, and average quarter hour *share*. Nielsen defines AQH persons as "**the average number of persons listening to a particular station for at least 5 minutes during a 15-minute period**" (Terms for the Trade, 2014). AQH persons are **duplicated listeners** who can be counted up to four times an hour, one time for each quarter hour. For example, if Jim listens to Station A for at least five minutes during the 1:00 p.m. to 1:15 p.m. quarter hour, he is counted as one listener, and if Jim listens during the 1:15 p.m. to 1:30 p.m. quarter hour, he is counted again and thus is considered two listeners.

AQH rating and share are calculated as follows:

- AQH rating:
  (AQH persons/Survey area population) × 100 = AQH rating (%)

- AQH share:
  (AQH persons/Listeners in survey area population) × 100 = AQH share (%)

Cumulative Persons Whereas AQH persons represent *duplicated* listeners, **cumulative persons** (or *cume*) represent *unduplicated* listeners. Nielsen defines cume as "**the total number of different persons who tune to a radio station for at least 5 minutes during a 15-minute period**" (Terms for the Trade, 2014). No matter how long the listening occurred, each person is counted only once. For example, if Sharon listens to Station A for at least five minutes during the 1:00 p.m. to 1:15 p.m. quarter hour, she is counted as one listener, but if she also listens during the 1:15 p.m. to 1:30 p.m. quarter hour, she is not counted again and is considered one listener.

Cume rating is calculated as follows:

- Cume rating:
  (Cume persons/Survey area population) × 100 = cume rating (%)

Although cumes are more commonly used to assess radio listenership, household television meters also produce cumes to measure program viewership. Because many prime-time television shows air once a week, four-week cumes are sometimes reported to assess how many people watch a program over a month-long period.

> **ZOOM IN 10.3**
>
> Practice your media math with the online media math calculator at: www.srds.com/frontMatter/sup_serv/calculator/cpm/index.html
>
>

Reporting Radio and Television Audience Numbers

Nielsen data are reported in standardized ratings reports or are tailored to meet the specific needs of a particular client. The standardized radio market report provides radio audience estimates for each ratings period as well as market information (station and population profiles) and audience information. Rating, share, cume, and AQH numbers are broken out by age, gender, and time of day. Eyeing these figures gives station managers a good idea of how their station compares with other stations in the market. Whereas one station may be highly rated among men, another may draw more women. Or a station may have strong ratings during one part of the day but be weak in another. One station might draw the 25- to 54-year-old crowd, while another might dominate the teen listening group.

Television: Nielsen Services Nielsen provides several standard television reports and customizes others for its clients. Below is a sampling of Nielsen's reports:

- **Nielsen Television Index (NTI)** provides metered audience estimates for all national broadcast television programs and national syndicated programs. Nielsen breaks out station and program rating and share figures by age and gender. These program ratings are used to compare and rank shows against one another. If a program is consistently rated poorly, there is a good chance it will not be on the air for very long.

Figure 10.4 A page from an NSI ratings book.

- **Nielsen Station Index (NSI)** provides local market television-viewing data through diaries and continuous metered overnight measurement. Metered information is processed overnight, and a report is sent to the station the next morning. Television executives rely on these overnights for immediate feedback on how their programs and stations rate compared to their rivals. Overnights are often used to make quick programming decisions to beat out the competition. NSI provides rating and share figures for programs that air on local affiliates, such as newscasts, as well as for network shows.

Broadcast and Cable Television Program Ratings The large number of cable networks has diluted the viewing audience, and, when coupled with spotty coverage (as not all cable systems carry the same cable networks), cable network programs typically garner lower ratings than broadcast programs. For example, the highest-rated cable program for the week of February 10, 2020, was TNT's 2020 NBA All Star game with a 4.1 rating (7.3 million viewers), whereas CBS's *NCIS* led the pack of broadcast shows with a 7.3. rating (11.7 million viewers). The highest-rated syndicated show was *Judge Judy* on CBS with a 6.5 rating (nine million viewers).

Of the 100 top-rated programs during the 2018–2019 television season, 87 aired on the broadcast networks and 12 on cable nets, with *Thursday Night Football* airing on both Fox and cable's NFL Channel. In contrast, in the 2008–2009 season, the 80th-ranked National Football League's regular-season games shown on ESPN was the only cable program in the top 100. But by the 2012–2013 season, three cable shows had made it into the top 25 – *The Walking Dead* (AMC, 10th), *Monday Night Football* (ESPN, 16th), and *Duck Dynasty* (A&E, 21st).

Primetime has always drawn more viewers than any other television time slot. But between streaming and DVR recording, prime-time viewership is declining. In 2019, the average total viewership of an individual prime-time series had dropped 14% to 5.6 million from 6.5 million a year earlier. Fewer prime-time shows are topping the 10 million mark in viewers.

DVR playback, at about 14 hours per month, can throw off ratings by a large margin. For instance, sporting events, such as Monday Night Football, are usually viewed live and so immediate ratings will place the game in first place. Entertainment programs, however, are often recorded for later viewing. When those numbers come in days later, they often catapult the program over the

FYI: TOP 10 PRIME-TIME REGULARLY SCHEDULED TELEVISION PROGRAMS OF 2018

Program	Average season audience
Roseanne (ABC)	20 million
NBC Sunday Night Football (NBC)	19.6 million
The Big Bang Theory (CBS)	18.3 million
NCIS (CBS)	16.7 million
This Is Us (NBC)	16.6 million
Young Sheldon (CBS)	15.7 million
Manifest (NBC)	14.6 million
The Good Doctor (ABC)	14.5 million
America's Got Talent (NBC)	14.3 million
Bull (CBS)	14.3 million

Source: Levin, 2018

sporting event. Nielsen adds the live viewing audience to the number of DVR viewers within three days, designated as either Live +3 (L3) or C3 (Commercials watched plus 3 days) and/or seven days (Live +7 or L7 or C7) after the original broadcast. DVR viewing can boost a program's viewership and thus ratings often increase substantially. For example, for one evening's broadcast in 2015, *Empire* (Fox) had 16.2 million viewers, but DVR L3 viewers bumped that total to 20.8 million (28.4% increase), and *Quantico* (ABC) saw its viewership increase from 7.1 million live viewers to 11.2 million when DVR viewers were counted in the total.

People cannot always watch a program from home, yet viewing in such places as hotels, airports, and bars are not counted, which reduces the viewership numbers and ratings. In an effort to boost ratings, or perhaps to get a more accurate count, Nielsen began measuring **out-of-home** viewing in 2020. Programs such as the Super Bowl, the World Series, the Oscars and other special events are often viewed in public spaces or with friends at private parties. A program airing in a restaurant or bar with five television sets could be watched by 100 patrons but viewership will only be counted as five persons, but with new ways to measure out-of-home viewing, a more accurate count of viewers could lift sports viewership numbers by about 10% and news viewing by about 7%.

Ratings and Product Placement. Remember, Nielsen is counting audience members for the purpose of informing advertisers of the number of people who are exposed to their commercials. And it is not just the within and between program spots that are of interest but also branded products that are used as part of a scene. The growing business of product placement has induced Nielsen to develop a special measurement system to track if viewers notice and react to products that are embedded into program content. Initial results show that 49% of viewers took some kind of purchasing action after

FYI: SERIES FINALES

Series finales often draw large audiences. Below are examples of some of the most-watched last episodes:

Final episode	Year	Number of viewers	Percent	Share of households of all TV viewers
M*A*S*H	1983	121.6 million	60.2	77
Cheers	1993	93.1 million	45.5	64
Seinfeld	1998	76.3 million	41.3	58
Friends	2004	52.5 million	29.8	43
Magnum, P.I.	1988	50.7 million	32.0	48
The Tonight Show (Johnny Carson)	1992	50.0 million	n/a	n/a
The Cosby Show	1992	44.4 million	28.0	45
All in the Family	1979	40.2 million	26.6	43
Family Ties	1989	36.3 million	20.8	35
Home Improvement	1999	35.5 million	21.6	34

Source: Most watched series finales, 2010

seeing a product placement, and if a commercial for the same brand airs during the program, product recall increases by 18%. In a test of brand recognition, 46.6% of the participants recognized a brand when exposed to only the commercial, but 57.5% recalled the brand when exposed to a commercial and product placement. It may seem circular, but as ratings systems are better able to track exposure to placed brands, the more likely advertisers are to seek placement opportunities, and the more opportunities, the more crucial it becomes to monitor exposure.

Broadcast and Cable Network Ratings Prior to the proliferation of cable networks and thus cable subscribers, ABC, NBC, and CBS vied for the largest share of the audience. Now hundreds of cable networks compete against each other and against the big three broadcast networks for the audience's attention. In addition, DVD viewing, satellite, and streaming delivery systems are stealing viewers away from both broadcast and cable networks.

The broadcast and cable networks have been duking it out for years for ratings both on the program level as well as on the overall network level. Although broadcast network ratings and shares are still far ahead of those for cable, cable is making serious inroads. For example, of the 15 most-watched networks in 2018 the big four broadcast networks ranked the highest: (1) NBC, (2) CBS, (3) ABC,

FYI: DVRS BOOST VIEWERSHIP. 2019. LIVE +7. RANKED BY PERCENT CHANGE IN VIEWERS

Rank	Program	Originator	Percent increase in viewers from time shifted viewing
1	Yellowstone	Paramount	365.6%
2	American Horror Story	FX	340.3%
3	Real Housewives of Beverly Hills	Bravo	232.6%
4	Vikings	History	214.3%
5	The Orville	Fox	182.8%
6	The Rookie	ABC	176.1%
7	Blacklist	NBC	169.8%
8	Project Blue Book	History	167.5%
9	Walking Dead	AMC	167.2%
10	Emergence	ABC	157.4%

Source: Nielsen, 2019a

FYI: DVRS BOOST VIEWERSHIP. 2019. LIVE +7. RANKED BY CHANGE IN NUMBER OF VIEWERS

Rank	Program	Originator	Absolute difference in viewership (000)
1	Game of Thrones	HBO PRIME	9,221
2	The Big Bang Theory	CBS	8,485
3	This Is Us	NBC	7,172
4	The Good Doctor	ABC	7,024
5	Manifest	NBC	6,672
6	NCIS	CBS	6,457
7	New Amsterdam	NBC	5,969
8	911	FOX	5,915
9	Blue Bloods	CBS	5,804
10	Young Sheldon	CBS	5,577

Source: Nielsen, 2019a

(4) Fox, followed by the cable networks, (5) Fox News Channel, (6) MSNBC, (7) ESPN, (8) USA, (9) HGTV. Broadcast networks Univision came in 10th place, followed by The CW, then back to cable TBS, broadcast Ion, cable TNT, and 15th place cable History. Cable networks collectively threatened the broadcast networks for many years and eventually took over. By 2010, cable nets were hooking about 59% share compared to broadcast's 39%. In general, both broadcast and cable networks are experiencing a decline in audience members as more viewers, especially Millennials, cut the cord or shift their viewing to streaming services.

FYI: MICHAEL JACKSON'S FUNERAL

Michael Jackson was always good television, and even in death in 2009 he still had the power to draw a large audience. An estimated 31 million (20.6 rating) U.S. viewers tuned into his funeral to say goodbye to one of the most popular singers of all time.

What does 31 million viewers mean? It is more than the number of viewers who tuned into the funerals of former presidents Ronald Reagan (20.8 million, 15.7 rating, 2004) and Gerald Ford (15 million, 11.3 rating, 2007), and Pope John Paul (8.8 million, 7.2 rating, 2005).

Although an audience of 31 million is quite impressive, Michael Jackson's funeral drew about as many viewers as the first 2009 *American Idol* episode (30 million, 18 rating) and lagged behind Princess Diana's funeral (33 million, 26.6 rating, 1997), and the O. J. Simpson verdict (53.9 million, 42.9 rating, 1995).

Separating Cable and Satellite Ratings

Cable networks such as Nick at Nite and ESPN uplink their shows to satellite and then they are either downlinked by a cable service and then delivered to subscribing homes or downlinked directly to homes with satellite dishes. Thus, when ratings are calculated, they include both the number of viewers who watched programs via cable and those who also watched programs via satellite. This double counting increases the cable networks' ratings by making it seem as though they are drawing larger audiences. Moreover, combining the data does not give a clear picture of the differences between cable and satellite viewers' characteristics or viewing patterns. To remedy this problem, Nielsen signed on in 2016 to use DISH set-top box data to strengthen and refine television-viewing measurement methods. ComScore, originally an online measurement company, which has entered the television audience measurement business, is also counting satellite viewing through a proprietary company.

When Audience Numbers Are Not Enough

Numbers cannot reveal personal characteristics that determine media choice. For the radio, television, and online industries to really know their audiences, they need to look at them in ways that numbers cannot provide. For example, ratings and shares might reveal which programs people watch or listen to most frequently, but they do not tell why viewers prefer one program over another, whether people prefer certain characters on a program, or why and when they like to listen to certain types of music.

Nielsen and other companies study the personal characteristics of media consumers. **Psychographics** categorize listeners based on their overall lifestyle, including hobbies, travel, interests, attitudes, and values. For example, a company advertising snowboards probably wants to reach young, athletic, daring individuals, and diaper manufacturers aim for parents. Another segmentation scheme is based on **technographics** – the use of and attitudes toward technology. For example, the number of television sets someone owns or how often someone watches a video on a smartphone is a technographic measure. From all this information, ratings companies obtain a clearer profile of the types of people who watch or listen to particular programs. Because advertisers depend on the media to reach their target consumers, it is crucial to match up with a medium, station, or program that is used by their customers. Psychographic and technographic information guides advertisers to programs that draw viewers who are most likely to use their products.

The media industry also uses specialized types of research such as music preference, pilot and episode testing, television quotient data, and online consumer profiling to further understand its users and their media preferences.

Music Preference Research

Used by radio stations, music preference research assesses radio listeners' musical likes and dislikes to determine station playlists. Music testing is an expensive venture but often well worth the effort. Music research is conducted either over the telephone, over the Internet, or in an auditorium-like setting. Researchers either call a sample of listeners or bring them together (usually hundreds) in one location to play song hooks (5 to 20 seconds each of various songs). Participants are asked to evaluate how much and why they like or dislike each song. Sometimes, smaller focus groups of listeners, usually fewer than 20, are convened to listen and discuss their preferences in depth.

Television Program Testing

Television programs are subject to rigorous concept testing before they make it to the air – if they ever do make it that far. Even after programs are on the air, the testing is not over.

New programs and episodes of current programs are tested to gauge audience reactions to the plot, characters, humor, and other program elements and to the overall program itself. The test audience watches perhaps the pilot episode of a new program or the season finale of an existing show. Audience members indicate their levels of like or dislike on the test meter at any time during the show. Executives and writers rely on these screening data for content decisions. If, for example, most viewers move their dials down after hearing a punchline, the program's writers might delete the joke.

On a broader scale, test meter information is matched up with demographic characteristics – such as age, gender, education, and income – for information about what types of viewers prefer what types of programs. For instance, the data might show that males over the age of 40 who have high incomes and are sports minded prefer action-oriented programs, that women between the ages of 18 and 34 prefer situation comedies over police dramas, or that children prefer shows with teen characters. Given this information, the producers might add a character or a storyline that appeals to a certain type of viewer.

Television Quotient Data. Television quotient data (Q Scores) focus on viewers' perceptions of programs and stars and are often used to supplement Nielsen ratings and other data. Q Scores are collected through mail and telephone surveys, diaries, and panel discussions that examine the familiarity and popularity of programs and actors. A low Q Score could mean the death of a character, whereas a high Q Score could shift a secondary character to a more prominent role.

FYI: TELEVISION CITY

Perhaps one of the most well-known testing centers opened in 2001 in Las Vegas. CBS's Television City Research Center is housed within the MGM Grand Hotel and attracts tens of thousands of program-rating volunteers a year from all over the United States. CBS and Viacom partnered with A. C. Nielsen Entertainment to provide the testing equipment and audience feedback reports. Basically, groups of about 20 to 25 volunteers are ushered into one of six studios to watch and rate commercials and new program pilots as well as episodes of existing shows from CBS, MTV, Nickelodeon, and other Viacom networks. Each screening lasts for about one hour. Learn how to participate in program testing at www.vegas.com/attractions/on_the_strip/televisioncity.html

Physiological Testing. Advanced laboratories are commissioned to measure human response to advertising and programs. Participants are wired to instruments that measure heart rate, eye movement, skin temperature, facial muscle movement, and other physiological changes that are induced by exposure to television and online content. For example, researchers measure whether participants attend to or ignore advertising, how long their eyes rest on an ad, and whether a scene evokes an emotional response.

Researchers also assess viewers' opinions of television celebrities. For example, researchers may want to know whether Anderson Cooper, Rachel Maddow, and Stephen Colbert are 'cool,' 'friendly,' 'fun,' 'intelligent,' 'trustworthy,' 'witty,' and so on. Presumably, viewers are drawn to celebrities they like. Researchers may also ask viewers how much they like/love a particular program and about their 'emotional connection' to a show. Many users are happy to fill out surveys or participate in focus groups, especially if they get a discounted price or some other reward for doing so.

Translating Audience Information to Advertising Sales

Media outlets develop audience profiles based on audience numbers (such as ratings and shares), audience information (such as age, ethnicity, gender, and other personal characteristics), and lifestyle preferences (such as program and music likes and dislikes). Once an audience profile has been established and the media outlet knows who listens to its station or watches its programs, how it rates in the market, and what its most popular programs are, it can establish commercial rates for its airtime. Again, airtime is nothing but empty seconds unless someone is listening to or watching the spot; thus, it is really the media *audience* that is being sold. Viewers and listeners are not the consumers but the product being sold to advertisers.

Not only are advertisers interested in buying media that reach their target consumers, but they also want to know how they can reach the largest number of target consumers for the lowest cost. Although the price of commercial time is largely based on program and station rating points, advertisers often need more information when planning a media buy. Advertisers compare the costs of reaching their target audience using various stations and types of media. For example, perhaps teens primarily listen to three radio stations in their town and also watch a television affiliate's Saturday-morning line-up. Advertisers need to have some way to compare the price of airtime on each of the three radio stations and the television program. To do so, they often rely on figures such as cost per thousand, gross ratings points, and cost per point.

Cost per Thousand (CPM) Cost per thousand is one of the most widely used means of comparing the costs of advertising across different media, such as television and radio. Many people often wonder why *cost per thousand* is abbreviated *CPM* instead of *CPT*. It is because *thousand* is derived from the Latin word *mil*, which means 'one thousandth of an inch.'

Formula for cost per thousand:

(Cost of a spot or schedule/population × 1000)
$10,000 cost for ads/200,000 viewers × 1000 = $50 per thousand.

Gross Rating Points (GRP) Gross rating points are the total of all ratings achieved by a commercial schedule. Put more simply, it is the *reach* (average rating) multiplied by the *frequency* (the number of spots).

Gross rating points:

Reach (ratings) × Frequency (number of spots)

Cost per Point (CPP) Cost per point is the cost of reaching 1% (1 rating point) of a specified market. The CPP gives advertisers a way to compare the costs of rating points in various markets. The CPP can be determined using this formula:

CPP = Cost of schedule/Gross rating points

Advertisers are interested in knowing the cost per thousand viewers or listeners and with cost per point, the cost of reaching 1% of the market population. The advertisers' goal is to pay the least amount of money for the largest possible audience. They want to know how many viewers or listeners will see or hear their ads (number of viewers) and how many times on average they are exposed to the ads (gross ratings points). With that knowledge, they calculate the CPP and CPM to compare the cost of various outlets.

FYI: COST PER THOUSAND – PRIME-TIME BROADCAST NETWORK TELEVISION

Year	Cost per 30 seconds	Cost per 1,000
1965	$19,700	$1.98
1970	$24,000	$2.10
1975	$32,300	$2.39
1980	$57,900	$3.79
1985	$94,700	$6.52
1990	$122,000	$9.74
1995	$95,500	$8.79
2000	$82,300	$13.42
2005	$129,000	$21.45
2010	$103,600	$19.74
2014	$112,100	$24.76
2018	$115,000	$31.97

Sources: Friedman, 2018; TV costs and CPM trends, 2015

Radio and Television Ad Sales

Radio and television stations set base prices for their commercial spots. Some stations, especially television, do not give out their base rates to their advertisers but rather use them as a starting point for negotiation. With the availability of new computer software, many television and radio stations have maximized their revenues through **yield management**, in which commercial prices change each week or even each day, depending on the availability of airtime. If there is high demand to advertise on a particular program, the last advertiser to commit may have to pay a higher price than the first advertiser. The station may also sell a high-demand time slot to whichever advertiser is willing to pay the highest price.

Dayparts Radio is usually sold by **dayparts**, which are designated parts of a programming day. Radio breaks out the 24-hour clock into these five segments (Eastern Standard Time):

Morning drive:	6:00 a.m. to 10:00 a.m.
Midday:	10:00 a.m. to 3:00 p.m.
Afternoon drive:	3:00 p.m. to 7:00 p.m.
Evening:	7:00 p.m. to midnight
Overnight:	Midnight to 6:00 a.m.

Television is also sold by dayparts but more commonly by specific program. Dayparts and program ratings determine commercial costs. For example, a commercial spot aired during the daytime daypart will cost less than the same spot aired during primetime. Further, the cost of commercial time during a daypart may vary with the popularity of the program in which it is inserted. The cost of running a 30-second spot during a top-rated prime-time program will be higher than for airing the same spot during a lower-rated prime-time show.

Here are common television dayparts (Eastern Standard Time), though there is no universal agreement among stations.

Early morning (Morning News)	6:00 a.m. to 9:00 a.m.
Morning:	9:00 a.m. to noon
Afternoon (Daytime)	Noon to 4:00 p.m.
Early Fringe:	4:00 p.m. to 6:00 p.m.
Early evening (Evening News)	6:00 p.m. to 7:00 p.m.
Prime Access:	7:00 p.m. to 8:00 p.m.
Primetime:	8:00 p.m. to 11:00 p.m.
Late Fringe (Late News)	11:00 p.m. to 11:30 p.m.
Late night:	11:30 p.m. to 2:00 a.m.
Overnight:	2:00 a.m. to 6:00 a.m.

Source: SmartPlus Dayparts, n.d.

Run of Schedule Radio stations further classify dayparts by advertiser demand and the number of listeners. The most desired and thus most expensive time is categorized as Class AAA, followed by Class AA, Class A, Class B, and down to Class C, the lowest priced. Because most advertisers do not want to buy what they perceive as second-rate time, some stations do not even designate a Class B or Class C time; some stations even prefer to classify their rates by daypart instead of by class.

One of the most common discounted buys is called **run of schedule** (ROS). In this type of buy, the advertiser and the salesperson agree on the number of times a commercial is going to air, but the station decides when to broadcast the spot, depending on available time. For example, on radio, instead of purchasing five spots in Class AAA time, ten in Class AA time, and ten in Class B time, the advertiser is offered a discounted rate for the package of 25 spots, and the station will air them during whatever time classes are available. In many cases, ROS benefits the advertiser, which may end up getting more spots on the air during prime dayparts than it actually paid for or otherwise could afford. Television ROS works similarly with radio, except that spots are rotated by daypart rather than by time classification.

Fixed Buys Radio and television stations charge higher rates for **fixed buys**, in which advertisers specify what time they want their commercials to run. Perhaps a fast-food restaurant wants to dominate the 10:30 a.m. to 1:00 p.m. period and thus wants its commercials broadcast every 15 minutes. Most stations, especially radio stations, can accommodate these specific needs, but they charge a premium price for doing so.

Careful planning and buying do not guarantee that commercials will run during agreed-upon times or that the audience will be as large as promised. Suppose a radio station inadvertently runs a spot at the wrong time or a television station overestimates the number of viewers it anticipated would watch a particular program, but the station has already charged the advertiser for the commercial time. In these and other situations in which the terms of advertising agreements are not met, advertisers are offered **make-goods** as compensation, which usually consist of free commercial time, future discounts, or return payment.

Frequency Discounts Radio and television stations are willing to offer frequency discounts when advertisers agree to air their commercials many times during a given period, such as six months. Advertisers who buy a large number of spots reduce their cost per commercial, and the more commercial time they buy, the bigger the discount per spot. For example, suppose an advertiser could buy ten 30-second spots at $200 each for a total cost of $2,000. But if the advertiser agrees to run the spot 20 times instead of ten times, the station may lower the cost per spot from $200 to $125 for a total of $2,500. So, for an extra $500, the advertiser has doubled the number of commercial spots. Other frequency discounts might include six-month or yearly deals or other longer-term advertising agreements.

Local Discounts Radio and television stations often offer discounts to local businesses. National chains, for example, benefit from reaching a station's entire market area. McDonald's may have 20 or more restaurants in a market, any of which might be visited by the station's listeners. In contrast, a local business, such as a produce market with one store on the east side of town, will probably draw only shoppers who live in close proximity, and thus the business will not want to pay to broadcast to the west side of town. Because a large part of the station's audience may live too far from the produce market, the station may offer it discounted airtime.

Bartering Some radio and television commercial time is **bartered** rather than sold. Also known as **trade** or **trade-out**, bartering basically involves trading airtime in exchange for goods or services. For example, instead of charging a promoter to advertise an upcoming concert, a station might exchange airtime for tickets of equal value. Or a station in need of office furniture might air a local office furniture store's spots in exchange for desks and chairs.

Cooperative Advertising Radio and television cooperative (co-op) advertising is a special arrangement between a product manufacturer and a retailer; namely, a manufacturer will reimburse a retailer for a portion of the cost of advertising the manufacturer's product. For example, Sony might reimburse a local retailer for promoting Sony televisions during the retailer's commercials. Co-op agreements vary, depending on the manufacturer, time of year, new-product rollouts, previous-year sales, and other factors. For example, Sony might have a 3% of sales co-op agreement with a retailer. If the local retailer sells $100,000 worth of Sony television sets, then Sony will reimburse the retailer $3,000 of its advertising expenditure that was spent promoting Sony televisions.

Co-op agreements are often very specific and require close attention to detail to get sufficiently reimbursed. Many stations hire a co-op coordinator, whose primary responsibility is to help the station's advertisers track co-op opportunities. The more reimbursement advertisers receive from manufacturers, the more commercial time they can afford to purchase.

Cable Television and Interconnects Selling commercial time on cable television follows a slightly different process from selling network or affiliate time. Cable network programming is transmitted through a local cable company, such as Comcast.

There are hundreds of cable program networks, and most cable companies provide subscribers with hundreds of cable networks plus the major broadcast networks. Because there are so many more cable networks than broadcast networks, the audience for each cable network is much smaller; thus, the price for commercial time is lower on cable than on broadcast networks.

Just over two-thirds of cable buys take place at the network level such that advertisers buy time on a cable network, such as ESPN, and the spot is shown in selected locations or throughout the country. The remaining time is then sold locally by the cable service. For example, Comcast sales representatives at the local company will sell time on cable channels to local advertisers. Local cable companies sell advertising spots within and between cable network programs, similar to the way broadcast affiliates sell time. Where broadcast and cable differ, however, is with multiple cable system buys. Large markets are often served by several different cable companies, each assigned to a specific geographic region. For example, one cable company might serve the east side of the New York market, another the west side, another the north side, and yet a fourth the south side. If a local advertiser has stores located all over the market, then it will want to reach the entire market. The advertiser can do so through an **interconnect buy**, in which it buys time on all the area cable companies through a one-buy pricing agreement. In some cases, arrangements are made for commercials to appear at the same time across the city.

Gathering Internet Audience Numbers and Ad Views

Some of the data collection methods used by the television and radio industries are also used to measure Internet audiences, and new methods have been developed specifically for the medium as well. As with any medium, online advertisers are concerned with attracting the most eyeballs for their money. However, they face unique challenges in buying online space, calculating the number of users who saw their ads or clicked on a banner, and reaching their target audience. Web site operators struggle to get a clear picture of their consumers whose characteristics are critical for convincing advertisers to buy space on their sites.

A number of **third-party monitors** are in the business of supplying Web sites and advertisers with online ratings and customer profiles. Third-party

CAREER TRACKS: TREY FABACHER, DIRECTOR OF SALES, SINCLAIR BROADCASTING, SAN ANTONIO, TX

WHAT ARE YOUR PRIMARY RESPONSIBILITIES?

In the role of Director of Sales, I am responsible for all revenue aspects of the television station operation. These responsibilities include managing a team of sales managers and sellers who work with clients and develop revenue – television commercial revenue as well as digital revenue, a vital growth part of our business.

WHAT WAS YOUR FIRST JOB IN ELECTRONIC MEDIA?

In my 36 years since graduation from the University of Tennessee, I have taken a progressive sales track from account executive to national sales manager, local sales manager, general sales manager, station manager, VP/general manager, and now Director of Sales for the largest revenue-generating market within Sinclair Broadcasting. We are always focusing on servicing our clients and the overall sales performance. My first job in broadcasting was as an overnight radio DJ in Lafayette, LA, so I can really say that I started my career in the graveyard. I was able to create spec spots for salespeople, which gave me my first taste of sales. From that point, I knew I wanted to sell media, and the Department of Broadcasting (now School of Journalism & Electronic Media) at the University of Tennessee exposed me to the connections to make that happen. Through various scholarships and intern programs, I was able to make the proper connections to get a foot in the door and demonstrate my abilities.

Figure 10.5 Trey Fabacher.

WHAT LED YOU TO YOUR PRESENT JOB?

By the time I had graduated in 1989, I had worked two years in radio as a DJ, one year in a sales research internship at the local ABC station, and one summer at NBC New York as a sales pricing analyst for the NBC-owned and -operated stations (O&Os). That aggressive experience allowed me an opportunity to enter a sales training program with Blair Television and get my first sales job six months later in Minneapolis. While there, I moved to CBS, where I managed three TV stations in three different markets. I left CBS to join Meredith to manage the CBS affiliate in Atlanta. I moved from that role after two years, and we decided as a family to prioritize our location, with our kids attending college in Georgia, which led me to Columbus, Georgia, and Raycom Media. Once a move was available, we chose San Antonio and the Sinclair TV stations that include NBC, FOX, CW, and CompulseDigital.

WHAT ADVICE WOULD YOU HAVE FOR STUDENTS WHO MIGHT WANT A JOB LIKE YOURS?

This business is awesome, and real-world experience is critical to getting started. But more important is a dedicated work ethic to learning and growing. Seek out a mentor or two that can help guide you and replicate the work effort they demonstrate in their jobs each and every day. You have to be focused on learning and growing.

monitors are companies that excel in tracking online usage with proprietary software. For example, a site that sells women's shoes might not have the technical ability to set up a system that tracks its shoppers or how many times they click on an ad posted on the site. Such a site will contract with a third-party monitor, such as Nielsen, ComScore and Webcast Metrics, to provide it with whatever tracking data it needs.

Third-party monitors employ various auditing techniques and audience-measuring methods, such as monitoring the number of times a banner ad is clicked on; providing site traffic reports by day, week, month; monitoring Web site activities; and developing customer profiles. Web site operators use this information to sell space to advertisers, and advertisers use this information to find the best sites on which to place their ads. Nielsen's Total Internet Audience panel of 500,000 online users across 30,000 sites measures digital media use including computers, smartphones, and tablets.

Counting Internet Users and Ad Exposure

Hits In the Web's early years, the number of Web site visitors was largely determined by the number of hits received on a page. A **hit** is typically defined as the number of times a page is accessed. A hit is a very poor measure of the number of site visitors, however. Fifteen hits on one site could mean 15 visitors, one user visiting 15 times, or a small number of users visiting several times. It is virtually impossible to say that the number of hits equals the number of visitors.

Caching presents another challenge in determining the number of visitors to a site. Computers often store a copy of a visited Web page as a **cache file** (or temporary storage area) on the hard drive rather than on the server, so when a user returns to a Web page he or she previously visited during the same Internet session, the browser retrieves the information from the cache file rather than from the server. Consequently, the site does not record the second hit, even though it is a repeat visit or second exposure.

Adding to the confusion are newer **push technologies**, which rely on automated searches (also called **robots** and **spiders**) that comb the Web looking for information. Each time an automated agent scans a Web page, it is counted as a page view, although a human never saw any of the content. Even if automated searches account for only a small percentage of all Web site traffic, a small overestimation of visitors could cause an advertiser to pay millions of dollars more than warranted for a banner ad. Conversely, deflating the number of human page views could cost Web site publishers millions of dollars in lower ad rates.

Cost per Click Rather than count people (or hits) who visit a site and might not even notice a particular banner ad, many sites and advertisers rely on *click-through* counts, which identify the percentage of visitors who actually click through a banner ad.

Click-through rates (CTRs), also called **cost per click (CPC)** or **pay per click (PPC)**, are based on the percentage of Web site visitors who click through a banner ad. Advertisers are charged only for the people who actually expressed an interest in the ad, not for all the others who landed on the site but did not click on the banner.

Advertisers hoping to reach the click-happy consumer are often disappointed when numbers show very few click-throughs. In 1994 when the Internet was still new and a curiosity, about 10% of users clicked on banners. By 2009, only about 3% of Web users clicked on a banner ad, and by 2019, that percentage dropped to about 0.86%.

Measuring the effectiveness of mobile ads is also a bit elusive. The CTR for mobile banners in the U.S. hovers around 0.9%, and in Europe, the Middle East, and Africa it is up to 1.4%, but the rate only tells part of the story. It is difficult to know how many people might see an ad on their smartphone but not take purchasing action until later when they get to their computer.

Cost per click is calculated by multiplying the number of people who click on an ad by an agreed-on rate. For example, if a Web site charges $0.10 per click and 1,000 users click on the ad, the site would charge the advertiser $100 ($0.10 × 1,000 users). For example, YouTube, charges about $0.10 to $0.30 per click.

Cost per Transaction An online advertiser who is charged by cost per transaction (CPT) pays only for the number of users who respond to an ad. Web sites and advertisers negotiate a fee or a percentage of the advertiser's net or gross sales based on the number of inquiries that can be directly attributed to a banner ad. Online advertisers are assessed a minimal charge or, in some cases, no charge at all for ad placement, but they are charged for the number of people who ask for more information or buy the product as a result of seeing the banner. CPT is a good pricing system for online merchants who are skeptical of investing advertising dollars that probably will not lead directly to sales, and for high-traffic Web sites that are eager to earn a commission on top of a small fee for delivering an audience.

ZOOM IN 10.4

MILLION DOLLAR HOMEPAGE

A 21-year-old British student created what is probably the cleverest use of click-through banners. He came up with the idea to sell click-through logos on one page for $100 each. For their money, each advertiser got about a 10 × 10 pixel space, which is very small considering that about 72 to 96 pixels equals one inch. It took the student only about five months to sell a million dollars' worth of logos. Check out the colorful page at www.milliondollarhomepage.com

Cost per Impression (CPI)/Cost per Thousand Impressions (CPM) Web publishers also use CPM to sell online space – about one-third of online revenues are priced on a cost-per-thousand basis. Online CPM is slightly different from CPM for non-online media. With non-online media, CPM reflects the cost of reaching 1,000 consumers, but for online ads the measurement is called CPI (cost per impression) – the cost of reaching an individual consumer.

Formula for cost per impression:

(Cost of a spot or schedule/number of impressions)
$10,000 cost for ads/200,000 impressions = $.05 per impression

Dollar for dollar, Web advertising can be more expensive than television advertising, though recently online CPMs seem to be trending downward. But in fact, lower CPMs and CPIs might actually translate into higher costs when advertisers pay for online site visitors who are not within their target audience.

Time Spent Online Online users who spend long periods of time reading a page and traveling within a site are more valuable to advertisers than those who just land on a page for a few seconds and then move on to another site. The longer a user spends on a page, the more likely he or she is to click on a banner; thus, some online pricing schemes are based on how long the ad stays on the screen and how long the average user stays on the page.

Size-Based Pricing Borrowing from the conventional newspaper advertising pricing structure, the cost of a banner ad is sometimes based on the amount of screen space it occupies. The charge for a newspaper ad is assessed according to a specific dollar amount per standard column inch area, whereas the rate for a banner ad is measured by pixel area. **Pixels** are tiny dots of color that form images on the screen. Size-based pricing is calculated by multiplying an ad's width by its height in pixels. The price is determined per pixel or by the total number of pixels.

Challenges of Gathering Online Audience Numbers

The biggest challenge faced by Web sites and online advertisers is the lack of standardized data collection methods. Different third-party monitors use different techniques and data collection methods to arrive at their figures, which often lead to contradictory and confusing reports. For example, they use different metering techniques, select their samples differently, and have different ways of counting Web site visitors. Some companies might count the number of **duplicated** visitors, for example, one person who accesses the same site five times is counted as five visitors. Yet other third-party monitors might count only the number of **unduplicated** visitors, for example, one person who accesses the same site five times is counted as one visitor. The unit of analysis may also be defined differently from one measurement service to another. For example, an *Internet user* could be defined as 'someone who has used the Internet at least ten times in the last week' or 'someone who has been online at least once in the past week.' These variations can yield very different results. For example, back in 2009, Nielsen reported 8.9 million users for Hulu, yet ComScore gave the site 42 million users.

The Internet's struggles with audience measurement have resulted in imprecise ratings and inconsistent pricing schemes. It is often difficult to measure how many people see an online ad and thus how much it should cost. When setting prices, most Web sites and auditors rely on the number of hits, number of click-throughs, and other means of counting their visitors. The process is often based on estimates, sales are often made through middlemen, and pricing is inconsistent across Web sites, thus online publishers are concerned that they are not getting the prices they should and online marketers are worried that they are paying too much. Several companies have put out the call for pricing transparency and efficiency through initiatives for sharing data and rules for verifying content and audience.

Several associations and organizations are taking the lead in establishing Web auditing and measurement standards. The Coalition for Innovative Media Measurement (CIMM) was formed in 2009 for developing new methods of monitoring media audiences. The council is a collaborative effort between media research companies and networks to come up with innovative ways to measure how consumers are using new media delivery systems. Further, the **Interactive Advertising Bureau (IAB)** and the **Advertising Research Foundation (ARF)**, along with several third-party monitors, are at the forefront of developing new and effective means of gathering audience data and standardizing definitions of terms. The IAB, along with prominent Web publishers and several advertising technology firms, has issued voluntary guidelines for online advertising measurement and many other aspects of Internet advertising.

Internet Ad Sales

Online sites use their number of Web users/impression to set the prices of advertising on their sites. But the numbers are a starting point for negotiation. Online ad buying and selling is a complicated transaction, which sometimes requires going directly through the site, while at other times uses ad buyers and sellers. As with radio and television, high-demand sites cost more money than less popular ones. The advantage of advertising on the Internet is that there are millions of Web sites whose audience can match most any product. By selecting the right kind of site for a product means that an advertiser is paying for potential customers who are most likely to buy its product rather than just sending out a promotional message to a mass audience who might or might not be interested, such as with network television spots.

ZOOM IN 10.5

Browse through *USA Today* media kits to get an idea of how much it costs to run an ad in print and online. Look for CPMs, guaranteed impressions, maximum run times, and other information on audience demographics, Web site traffic, production specifications, and sales contracts.

- *USA Today*'s Online Media Kit: http://static.usatoday.com/en/advertising/

Ad Exchanges/Ad Auctions/Discounts Online advertising can be a risky venture, especially for small businesses with thin wallets that are hesitant to spend advertising dollars on a new and unproven medium. But there are several online buying strategies that advertisers can employ without risking their finances. Some Web sites offer discounted rates if online ads come **online ready**, or fully coded in HTML and complete with all graphic, audio, and video files. In addition, sliding fees are formulated based on how much formatting the Web site manager has to do to an ad before posting it online. Frequency discounts are also available on many Web sites.

Although many small businesses have created successful Web sites, they typically cannot afford aggressive online campaigns. One way to get the word out is through an **advertising exchange** – advertisers place banners on each other's Web sites free of charge. For example, a company selling beauty products could place its banner on a site that sells women's shoes, and in turn, the shoe company could put a banner on the beauty product site. Neither company charges the other; they simply exchange ad space. Advertising exchanges are gaining in popularity, especially among marketers who do not have much money and who do not have a large sales team. By trading space, advertisers find new outlets that reach their target audiences that they would not otherwise be able to afford.

Ad space is available on millions of Web pages. Web sites with unsold banner spaces put them up for auction at reduced rates through specialized third-party sites. The thinking behind this arrangement is that getting discounted rates for unsold space is better than not getting any money at all. Plus, advertisers are more willing to take chances on less popular sites if they are paying a discounted price for them. Just like the name implies, ad auctioneers put the space up for bid, and the advertiser with the highest bid buys the space, sometimes at less than half price.

Sophisticated target marketing technologies have moved the auction model beyond selling discounted space to matching advertisers with targeted purchasers. **Ad auctions**, such as those conducted by Google and Yahoo! Ad Exchange, help advertisers identify Web sites that attract their target market. The auctions are set up so that advertisers determine how much they are willing to pay for targeted clients.

The key players involved are as follows:

Audience: Web site users who will see the ads
Sellers: Web site publishers with space to sell
Bidders: Advertisers who bid on the ad space
Ad Exchange: Operates the auction by accepting and processing bids.

Typically, the seller auctions space based on impressions – the expected number of viewers who will see the ad. Advertisers bid on the space/impressions. How much money they bid depends on how likely they think the audience is to buy their products. **Ad auctions** are high-stakes, high-energy bidding used by Google, Facebook, Microsoft, and others where prices fluctuate in an instant and buys are made at the moment visitors land on a page. In real-time bidding, advertisers adjust the price-per-click they are willing to pay based on the competition and the value of impressions. A consumer is no longer just a consumer; some are more valuable than others. A wealthy, 40-year-old risk taker who likes luxury cars is worth more to BMW than an unemployed 25-year-old who drives an older-model sedan. At one point, after learning that Mac users spend more on hotel rooms than PC users, the travel site Orbitz started sending Mac users ads for hotels that were 11% more expensive than the ads it sent to PC users.

Google Ads (known as Google AdWords until July 2018) offers the following example of how its ad auctions work. An ad is given a quality score based on its past click-through rates, how well the ad matches the auctioning site's audience, the quality of the advertiser's landing page (the page a user is connected to after clicking on the ad) and ad formats, and enhancements to search ads that prominently display information such as a phone number or product descriptions. A business whose ad garners a high-quality score might get it placed on a Web site even though it bid less than a business whose ad has a low-quality score but is willing to pay more per click.

Example: Three advertisers with the same quality score

Advertiser	Price-per-click bid	Quality score
Company A	$5	10
Company B	$3	10
Company C	$1	10

In this case, Company A wins the auction to place its ad on the Web site. But because the actual amount of money Company A needed to bid to beat the competitors was $3.01, that is the amount Company A will actually pay (About the ad auction, 2015).

ZOOM IN 10.6

To learn more about GoogleAds and Facebook ad auctions watch the three videos listed here: www.youtube.com/watch?v=SZV_J92fY_I

www.youtube.com/watch?v=JLf2LwA2oek

www.youtube.com/watch?v=-XtaHBfVd14

CAREER TRACKS: JOHN C. MONTUORI, FOUNDER AND PRESIDENT, SALES ACADEMY OF LEADERSHIP AND EMPATHY

WHAT LED TO YOUR PRESENT JOB?

At 40 years old this year, my career has taken its most climactic and profound turn yet.

The progression of my media sales career was relatively conventional. After discovering a knack for selling in my media sales class at the University of Tennessee in 2003, I slid right into my first job at a radio broadcasting group immediately following graduation. Quickly switching to the TV side, I worked as an account executive for seven years before moving into my first leadership role as a National Sales Manager for the CBS affiliate in Knoxville, TN.

In 2013, I answered a call from an E. W. Scripps recruiter who asked me to join a brand-new corporate team selling customized digital strategies. It was the best decision I ever made. The position came with a base salary and a rewarding commission plan, but zero accounts. The role was 100% acquisition based, and for me, it meant completely starting over. I knew that for me to stay relevant in the media field, I would need to take this leap of faith, learn a whole new industry, and build a new book of business from scratch. The digital world is unlike anything I had previously known. The language was foreign, and the accountability to data was

Figure 10.6 John Montuori.

overwhelming. I spent days and nights learning the technology. Through hard work, dedication, and probably some luck, I became one of the top-producing reps in the company.

By my one-year anniversary, I was promoted to digital sales manager. I oversaw our corporate remote digital team as well as 30 account executives and managers on the newspaper side of our company. I was responsible for 100% of the digital revenue associated with our local property. I coached our teams to create, articulate, and execute effective digital strategies for our clients and partner agencies. We conducted thorough customer needs analysis, identified and analyzed their profit centers, revenue streams, and ideal prospects and customers, and offered customized executions to help clients reach their goals.

In 2016, I ascended from a local property digital sales manager role to a division-wide sales manager role, overseeing revenues across eight markets throughout the country. In 2018, I leveled-up again during a corporate acquisition, and took charge as the Vice President of Digital Sales and Market Development. I would travel to our markets training sellers and managers in the digital space, as well as conducting workshops and seminars for local business owners.

In late 2019 though, I was in for an abrupt career disruption. In a surprising turn of events, my company changed direction and decided against investing in corporate digital oversight. For the first time in 17 years, I found myself unemployed.

As I licked my wounds and dove deep into some soul searching, the puzzle pieces began to come together. The common denominator of my success in sales and management had always been education and my ability to relate to others. With the support of colleagues and loved ones, I decided to create a business out of the very traits that defined me: leadership and empathy.

As president of **The Sales Academy of Leadership and Empathy**, I focus on six services: Sales Process Development, Product Training, New Business Development, Leadership Development, Field Coaching, and Cultural Evaluation and Strategy. I build bridges between salespeople and their leadership teams, sales teams and operations teams, and sales people with their clients. The ability to open up communication enhances productivity in a measurable way.

WHAT ADVICE WOULD YOU HAVE FOR STUDENTS WHO MIGHT WANT A JOB LIKE YOURS?

Learning doesn't cease after the conclusion of your formal education. Throughout your career you will continue to sharpen the skills of your trade. My advice for any young professional though is not to neglect your personal development. It took me years to figure out that my passion wasn't in 'sales.' My passion was in mentoring, coaching, and education. Through those vehicles, I was able to achieve success in sales. Try to figure out what you're passionate about as soon as possible so you can parlay that into a successful career.

Mobile Marketing/Advertising Pricing Similar to traditional media and the Internet, the price for mobile advertising is based on audience numbers. Mobile advertising is most commonly priced by cost per impression (the cost of reaching one mobile consumer) and by cost per click (advertisers pay per consumer who clicks on their ad). Advertisers might also be charged on conversion (price per number of consumers who buy the product) or other interactive measure. Because consumers pay less attention to mobile ads and the conversion rate from exposure to purchase is about one-half that of desktop conversion, online companies must price them at a lower rate to attract advertisers. The average price per mobile viewer is about one-half that per desktop view. Mobile revenue, however, surpassed desktop revenue for the first time in 2016, with Google and Facebook alone generating about 57% of all U.S. mobile ad revenue.

Ads placed on social media sites are charged according to a site's model. For example, LinkedIn and Facebook commonly charge by cost per click or cost per impression, but Facebook also has special prices for embedded sponsor stories, wall photos, or for placing an app on a page, and Twitter commonly charges per click. But like other Web sites, social media sites might also charge by cost per transaction, size, or other method. But audience numbers are not enough to gauge a successful mobile or social media campaign. More important is the level of audience activity.

SEE IT LATER

The Never-Ending Quest for Ratings

The radio, television, and Internet industries are all on a quest for accurate and reliable ways to measure their audiences. It is critical for media to know as much about their audiences as possible so they can get a true measure of exposure to advertising so they can match their audience to advertisers' products and services, and charge for ad space accordingly. New technologies and new ways of reporting data hold promise for more accurate information.

Moving Target Audience

One of the biggest challenges to marketers is how to reach a moving target audience. Think of a target audience – those an advertiser wants to reach – but one that resides in a mobile world. **Global positioning system (GPS)** technology is being developed to track consumers' exposure to billboard and other out-of-home advertising, such as a commercial shown on a gas pump video screen or a print ad posted in a bathroom stall. For GPS ad tracking to work, motorists and pedestrians must wear a small battery-operated meter that tracks their movements every so many seconds. The GPS data, along with travel diaries, are then matched up to a map of billboard and other outdoor ads to determine the 'opportunity to see' an outdoor advertisement. GPS systems hold promise to deliver "reliable reach, frequency, and site-specific ratings data on real people, passing real sites, in real time" (Wi-Fi & GPS Combined, 2008).

Years of research has honed methods of reaching consumers through television and radio and other media, but catching the attention of those who flit among mobile phones, tablets, and laptops is a formidable task. The old ways of targeting must be redesigned for a new, increasingly mobile world. Smart marketers are tapping into methods of measuring mobile users who see their products and brand names. For example, establishing a smartphone app to locate restaurants or even recipes, in a sense exposes users to advertising. A Dockers campaign encouraged consumers to shake their iPhone to get a model wearing a pair of khakis to dance. Exposure to the ad was measured by tracking how long users kept shaking their phones.

The proliferation of smart televisions that hook into the Internet has initiated major changes in the way Nielsen counts television viewers and commercial exposure. Knowing that the number of cord cutters, viewers who have canceled their cable service, and those who have never subscribed is growing, industry executives can begin strategizing about new ways to reach viewers. If, for example, the same commercials shown on an original broadcast of an episode of *The Simpsons* are also shown when the episode is streamed on the Fox Web site or through its program app or on Hulu, Nielsen would capture ad exposure to both the broadcast and online views and count them equally. But the technology for counting the streaming audience has not yet caught up to streaming technology, and thus billions of hours of viewing on Netflix and Hulu might not be captured accurately. Nielsen is perhaps best positioned to come up with an effective way to measure audience viewing across television and digital platforms, and it is improving its monitoring systems, but the pressure is on its executives to perfect a technique before an upstart measurement company beats them to it.

Talking About Television

Smart television and online streaming have redefined the meaning of 'television viewer' and 'television program,' especially considering that a television set is a screen and that computers and tablets are also screens that one day could be considered 'television sets.'

Television ratings and program popularity are no longer about just the number of viewers but also about what they are talking about. Research shows that the more viewers talk about a program, the more engaged and involved they are, and thus the more they watch it; and more viewers translates to higher ratings. Talking about television, whether on Twitter, Facebook, or other social media is known as **Social TV**. Nielsen Social tracks television-related conversation, and there is a lot of it. For example, hundreds of millions of tweets about television shows go out each year. Nielsen estimates that the size of the audience that reads the tweets is 50 times the number of those who write the tweets. And the ratio of tweets of live programming vs non-live is about 2:1.

Nielsen continues to research the connection between Social TV and the audience, and it is refining its way of identifying users who use social media to 'talk television.' Although studies are beginning to show that Social TV has the potential to boost ratings, the degree of influence is still unclear. It will take a few more years of monitoring to get a sense of who is most likely to engage in social media chatter about television and how it affects ratings.

FYI: SOCIAL TV – MOST SOCIAL PRIME-TIME TELEVISION SERIES (2019)

Series	Average total interactions per episode
Game of Thrones (HBO)	5.3 million
WWE Monday Night Raw (USA)	1.9 million
WWE Friday Night SmackDown (FOX)	1.6 million
America's Got Talent (NBC)	1.4 million
American Idol (ABC)	1.3 million
WWE SmackDown (USA)	1.1 million
The Bachelorette (ABC)	937 million
The Bachelor (ABC)	768 million
Grey's Anatomy (ABC)	729 million
Dancing with the Stars (ABC)	715 million

Source: Nielsen, 2019a

Tuning In to Online Music and Podcasts

The music industry is also finding ways to monitor listenership. In early 2013, *Billboard* teamed with Nielsen Music to include YouTube video streams with authorized audio in its calculations of most popular songs. Nielsen's methodology calculated digital download sales, on-demand audio streaming, and online radio streaming. Online sales data helps music producers get a better reading about a song's popularity. For example, the hip-hop single 'The Harlem Shake' was one of the first songs to benefit from the Nielsen/*Billboard* partnership. On sales totals alone, it would have been placed within the top 15 of the Hot 100 list, but with downloads, it ranked number one on both the Hot 100 and Streaming Songs charts. Monitoring music listenership took a new spin when, in 2019, Valence acquired Nielsen Music with the goal of developing a more comprehensive way of collecting listenership data for both online and physical sales.

Online music services know who is listening by counting their number of subscribers. For 2018, the three most popular music services in the U.S. were Apple Music (49.5 million monthly users), Spotify (47.7 million), and Pandora (36.8 million). Thanks to music services, listenership had jumped from 23.5 hours in 2015 to 32 hours per week in 2017, but then leveled off to 27 hours in 2019. The number of subscribers, however, is only one measure of the audience. Music services hope to gain a competitive edge for advertisers for their ad-supported services by monitoring time spent listening, a measure that is also used by radio stations. The more time users spend listening, the more ads they will encounter. For example, listeners spend an average of 25 hours per month on Spotify, a figure the service could use to sell against its competitors on which listeners spend less time.

It is also important to monitor the number of podcast listeners. Many podcasts generate revenue through ad sales. Podtrac is the leading podcast measurement and advertising company, providing advertisers with audience numbers and monthly rankings of podcasts. For instance, based on the average number of downloads per episode, the top three new podcasts of 2019 were (1) The Shrink Next Door – Wondery, (2) The Ron Burgundy Podcast – iHeartRadio, and (3) Over My Dead Body – Wondery. Podtrac developed and is refining its proprietary analytics to provide the most accurate picture of podcast users.

SUMMARY

Researchers started measuring radio audiences in 1929 to give stations a base on which to price time for commercial spots and to give advertisers a way to compare stations by number of listeners and demographic characteristics. Many of the techniques used for measuring radio audiences were later used to measure audience numbers for broadcast stations and cable and broadcast networks. Today, audience ratings figure prominently in the media world and often determine the success of a program or a station. New methods are being developed for measuring Internet usage, online ad exposure, time spent listening to music services, and programs watched on streaming services, to get a clearer picture of changing media habits.

Nielsen is the most prominent media audience measurement company. To get a sense of the number of listeners and viewers, Nielsen relies on media use diaries and metering devices. Nielsen has updated its methodology and measuring techniques to capture uses of digital media. The Internet poses new challenges to traditional audience measurement techniques. Metering computer use is more complex than metering television or radio use, and the definition of an *Internet user* has not been standardized. Researchers are grappling with how users should be counted, such as by number of hits or number of click-throughs. Although several companies monitor traffic at Web sites, there are many discrepancies regarding how the audience is measured.

Audience numbers are not the only way to evaluate a media audience. Online and traditional media and research companies often use methods that delve into users' likes and dislikes as well as their motivations for using a particular medium or program. Twitter conversation reveals what viewers really think about television shows and might even forecast a program's success or failure. Online marketers use customer profiling to deliver banner ads to promote products similar to those on other sites users have visited.

It is essential for media outlets to understand their audiences because, after all, the media are in the business of selling audiences to their advertisers. Television and radio stations and networks and online sources depend on audience numbers and profiles to establish how much they should charge for advertising. Calculations such as cost per thousand, cost per point, gross rating points, average-quarter hour, and cume figures are used to set ad costs. Advertisers use these figures when making media buys to compare costs among various media.

Radio and television offer a variety of pricing structures, depending on availability, daypart, program, and other factors. Most stations offer fixed buys, run-of-schedule, frequency discounts, and barter arrangements. Online advertising pricing structures also include CPM and discount buys. Unique to the medium, banners are charged by click-through rates, size-based pricing, cost per transaction, and hybrid deals.

Audience measurement tools yield, at best, gross estimates of audience characteristics and media use, but they are only as accurate as the technology allows. Technological innovations may one day produce audience numbers that are far more accurate and reliable than what is currently available.

'Now' media technologies have forced marketers to rethink tried-and-true methods of reaching an audience. With so many new ways of communicating product information and for persuading an audience to make a purchase, marketers need to expand their media choices. But it is difficult to measure a mobile media audience and to know how to capture its attention yet not annoy customers with ads that are pesky and intrusive.

BIBLIOGRAPHY

About the ad auction. (2015). *Google.com*. Retrieved from: https://support.google.com/adsense/answer/160525?hl=en

AC Nielsen Entertainment partners with CBS for real-time audience research [press release]. (2001, April 18). *Nielsen*. Retrieved from: www.acnielsen.com/news/corp/2001/20010418.htm

Amazon Alexa Echo recorded conversation and then sent to contact. (2018, May 24). *YouTube*. Retrieved from: www.youtube.com/watch?v=S7ta4CfXu1Y

Atwell, L. (2013). Social media advertising: The total cost breakdown of a 'like'. *Adpearance*. Retrieved from: http://adpearance.com/blog/social-media-advertising-the-total-cost-breakdown-of-a-like

Bachman, K. (2010, January 22). Nielsen preps TV and PC report. *Adweek*. Retrieved from: www.adweek.com/news/television/nielsen-preps-tvandpc-report-101308

Barnes, B. (2009, July 27). Lab watches Web surfers to see which ads work. *The New York Times*, pp. B1, B6.

Bauder, D. (2015, October 6). DVR usage changes Nielsen ratings picture. *Daily Herald*. Retrieved from: www.dailyherald.com/article/20151006/business/310069772/

Beville, H. M., Jr. (1988). *Audience ratings*. Hillsdale, NJ: Erlbaum.

Boyce, R. (1998, February 2). Exploding the Web CPM myth. *Online Media Strategies for Advertising* [Supplement to *Advertising Age*], p. A16.

C3 TV Ratings show impact of DVR ad viewing. (2009, October 15). *Nielsen*. Retrieved from: www.nielsen.com/us/en/insights/article/2009/c3-tv-ratings-show-impact-of-dvr-ad-viewing/

Chaffey, D. (2020, February 12). E-commerce conversion rates – how do yours compare? *Smart Insights*. Retrieved from: www.smartinsights.com/ecommerce/ecommerce-analytics/ecommerce-conversion-rates/

Clifford, S. (2009, March 11). Advertisers get a trove of clues in smartphones. *The New York Times*, pp. A1, A14.

DiPasquale, C. B. (2002, October 8). Nielsen to test outdoor ratings system. *Advertising Age*. Retrieved from: www.adage.com/news.cms?NewsId536254#

Dogtiev, A. (2019, January 18). Mobile app advertising rates. *Business of Apps*. Retrieved from: www.businessofapps.com/ads/research/mobile-app-advertising-cpm-rates/

Flannagan, M. (2002). Local TV will pay big bucks for Nielsens. *Knoxville News-Sentinel*, pp. A1, A13.

Friedman, W. (2018, July 31). Analyst: TV upfront revenues rise 5.2%, CPMs are 10% higher. *MediaPost*. Retrieved from: www.mediapost.com/publications/article/322976/analyst-tv-upfront-revenues-rise-52-cpms-are-1.html

Fung, B. (2015, December 22). Survey: Many cord-cutters don't have broadband. *The Washington Post*, p. A12.

Glaser, M. (2007a, July 25). The problem with web measurement, part 1. *PBS*. Retrieved from: www.pbs.org/mediashift/2007/07/the-problem-with-web-measurement-part-1206.html [February 5, 2010]

Glaser, M. (2007b, August 1). The problem with web measurement, part 2. *PBS*. Retrieved from: www.pbs.org/mediashift/2007/08/the-problem-with-web-measurement-part-2213.html [February 5, 2010]

Gorman, B. (2009, January 18). Super Bowl TV ratings. *TV by the Numbers*. Retrieved from: tvbythenumbers.com/2009/01/18/historical-super-bowl-tv-ratings/11044 [September 25, 2009]

Graham, K. (2018, November 13). OTT vs CTV (Over the top vs connected TV). *MonetizeMore*. Retrieved from: www.monetizemore.com/blog/ott-vs-ctv/

Hagey, K. (2016, April 4). Nielsen to include set-top-box data in ratings for first time. *The Wall Street Journal*. Retrieved from: www.wsj.com/articles/nielsen-to-include-set-top-box-data-in-ratings-for-first-time-1459764001

Hall, R. W. (1991). *Media math*. Lincolnwood, IL: NTC Business Books.

Hansell, S. (2001, July 23). Pop-up ads pose a measurement puzzle. *The New York Times*, pp. C1, C5.

Hemdev, A. (2019, January 31). There is a difference between OTT and CTV. *Zypmedia*. Retrieved from: http://blog.zypmedia.com/ott-vs-ctv

Hot 100 news: Billboard and Nielsen add YouTube video streaming to platforms. (2013). *Billboard*. Retrieved from: www.billboard.com/articles/news/1549399/hot-100-news-billboard-and-nielsen-add-youtube-video-streaming-to-platforms

Ibarra, S. (2009, February 2). Super Bowl TV ratings change: Last minute victory. *TV Week*. Retrieved from: www.tvweek.com/news/2009/02/super_bowl_tv_ratings_change_l.php [September 25, 2009]

Improving their swing. (1998, February 2). *Online Media Strategies for Advertising* [Supplement to *Advertising Age*], p. A45.

James, M. (2012, August 31). Nielsen 'people meter' changed the ratings game 25 years ago. *Los Angeles Times*. Retrieved from: http://articles.latimes.com/2012/aug/31/entertainment/la-et-ct-nielsen-people-meter-anniversary-20120830

Kang, C. (2015, November 10). Online advertising predicted to top TV by 2017 as viewing habits evolve. *The Washington Post*, p. A16.

Kaye, B. K., & Medoff, N. J. (2001). *The World Wide Web: A mass communication perspective*. Mountain View, CA: Mayfield.

Koblin, J. (2019, May 2). Nielsen now measures a lucrative audience: Same sex households. *SF Gate*. Retrieved from: www.sfgate.com/business/article/Nielsen-now-measures-a-lucrative-audience-13815058.php

Lafayette, J. (2018, March 16). Nielsen launches TV product placement measurement system. *Broadcasting+Cable*. Retrieved from: www.broadcastingcable.com/news/nielsen-launches-tv-product-placement-measurement-system-171571

Levin, G. (2018, December 17). 2018 in review: The year's most popular TV shows according to Nielsen. *USA Today*. Retrieved from: www.usatoday.com/story/life/tv/2018/12/17/2018-review-nielsen-ranks-years-most-popular-tv-shows/2339279002/

Local television market universe estimates. (2016). *Nielsen*. Retrieved from: www.tvb.org/media/file/2015-2016-dma-ranks.pdf

McIntyre, H. (2017, November 9). Americans are spending more time listening to music than ever before. *Forbes*. Retrieved from: www.forbes.com/sites/hughmcintyre/2017/11/09/americans-are-spending-more-time-listening-to-music-than-ever-before/#7d696e022f7f

Mitchell, R. L. (2014, January 15). Ad blockers: A solution or a problem? *Computerworld*. Retrieved from: www.computerworld.com/article/2487367/e-commerce/ad-blockers-a-solution-or-a-problem.html

Mitovitch, M. (2019, October 7). Fall TV ratings: Which new shows are enjoying the beefiest DVR gains? Retrieved from: https://tvline.com/2019/10/07/new-fall-tv-show-ratings-biggest-dvr-playback-increases/

Morgan, B. (2018, February 5). Are digital assistants always listening? *Forbes*. Retrieved from: www.forbes.com/sites/blakemorgan/2018/02/05/are-digital-assistants-always-listening/#3550d1414eeb

Morrow, T. (2002, May 14). Knox stations replacing Nielsen system. *Knoxville News-Sentinel*, p. D1.

Most popular music streaming services in the U.S. 2018, by audience. (2019, November 20). *Statista*. Retrieved from: www.statista.com/statistics/798125/most-popular-us-music-streaming-services-ranked-by-audience/

Most watched series finales. (2010). *Wikipedia*. Retrieved from: en.wikipedia.org/wiki/List_of_most-watched_television_episodes#Most-watched_series_finales [February 10, 2010]

Nastic, G. (2013, July 1). Nielsen launches Twitter TV ratings. *CSI Magazine*. Retrieved from: www.csimagazine.com/csi/Nielsen-launches-Twitter-TV-Ratings.php

Nielsen. (2013). Terminology and definitions. Retrieved from: www.arbitron.com/downloads/terms_brochure.pdf

Nielsen. (2019a). Tops of 2019: Social TV. Retrieved from: www.nielsen.com/us/en/insights/article/2019/tops-of-2019-social-tv/

Nielsen. (2019b). Tops of 2019: Television. Retrieved from: www.nielsen.com/us/en/insights/article/2019/tops-of-2019-television-2/

Nielsen. (2019, August 27). Nielsen estimates 120.6 million TV homes in the U.S. for the 2019–2020 season. Retrieved from: www.nielsen.com/us/en/insights/article/2019/nielsen-estimates-120-6-million-tv-homes-in-the-u-s-for-the-2019-202-tv-season/

Nielsen and Twitter establish social TV rating. (2012, December 18). *Nielsen*. Retrieved from: www.nielsen.com/us/en/press-releases/2012/nielsen-and-twitter-establish-social-tv-rating/

Nielsen DMA rankings. (2020). *Media Tracks Communication*. Retrieved from: https://mediatracks.com/resources/nielsen-dma-rankings-2020/

Nielsen Media. (2020). *Online*. Retrieved from: www.nielsen.com/us/en/solutions/measurement/online/

Nielsen Media. (n.d.). Glossary of terms. Retrieved from: www.nielsenmedia.com/glossary/terms/R/R.html

Nielsen Social TV. (n.d.). Retrieved from: www.nielsen.com/us/en/solutions/measurement/social-tv.html

The portable people meter. (2009). *Arbitron*. Retrieved from: www.arbitron.com/portable_people_meters/ppm_service.htm [September 23, 2009]

Porter, R. (2019, September 21). TV long view: Five years of network ratings decline in context. *The Hollywood Reporter*. Retrieved from: www.hollywoodreporter.com/live-feed/five-years-network-ratings-declines-explained-1241524

Price of a 30 second ad in network TV. (2009). *Media Literacy Clearinghouse*. Retrieved from: www.frankwbaker.com/mediause.htm [September 9, 2009]

Rosen, J. (2012, December 2). Who do they think they are? *The New York Times Magazine*, pp. 40–45.

Schneider, M. (2018, December 27). Most-watched television networks: Ranking 2018's winners and losers. *IndieWire*. Retrieved from: www.indiewire.com/2018/12/network-ratings-top-channels-espn-cnn-fox-news-cbs-nbc-abc-1202030597/

Schneider, M. (2019, May 22). Top 100 shows of 2018–2019, adults 18–49. *Variety*. Retrieved from: https://variety.com/2019/tv/news/most-watched-tv-shows-highest-rated-2018-2019-season-game-of-thrones-1203222287/

Scott, N. (2015, January 28). 10 Best Super Bowl commercials of all time. *USA Today*. Retrieved from: http://ftw.usatoday.com/2015/01/10-best-super-bowl-commercials-of-all-time

Seals, T. (2013, July 1). TV cord-cutters, cord-nevers total 19 percent of US adults. *Cable Spotlight*. Retrieved from: http://cable.tmcnet.com/topics/cable/articles/2013/07/01/344134-tv-cord-cutters-cord-nevers-total-19-percent.htm

Singer, N. (2012, November 18). Your attention, bought in an instant. *The New York Times*, Business pp. 1, 5.

Small crowd, young and old watch Super Bowl on Web. (2016, February 4). *The New York Times*. Retrieved from: www.nytimes.com/reuters/2016/02/04/arts/04reuters-nfl-superbowl-streaming.html

SmartPlus Dayparts (n.d.). *Nielsen.* Retrieved from: http://en-us.nielsen.com/sitelets/cls/documents/smartplus/SmartPlusDayparts.pdf

So many ads, so few clicks. (2007, November 12). *BusinessWeek.* Retrieved from: www.businessweek.com/magazine/content/07_46/b4058053.htm [September 23, 2009]

The state of the news media. (2013). *The Pew Research Center's Project for Excellence in Journalism.* Retrieved from: www.stateofthemedia.org/2013/network-news-the-pace-of-change-accelerates/network-by-the-numbers/

Steel, E. (2016, February 2). Nielsen plays catch-up as streaming era wreacks havoc on TV raters. *The New York Times.* Retrieved from: www.nytimes.com/2016/02/03/business/media/nielsen-playing-catch-up-as-tv-viewing-habits-change-and-digital-rivals-spring-up.html

Steinberg, B. (2019, September 9). TV networks plan to include out-of-home audiences in national ratings in 2020. *Variety.* Retrieved from: https://variety.com/2019/tv/news/out-of-home-audiences-nielsen-tv-ratings-2020-1203328231/#!

Stelter, B. (2009, December 2). Nielsen to add online views to its ratings. *The New York Times,* p. B4.

Stelter, B. (2009, May 14). Hulu questions count of its audience. *The New York Times.* Retrieved from: www.nytimes.com/2009/05/15/business/media/15nielsen.html?_r51&ref5media [February 5, 2010]

Stelter, B. (2009, September 11). Media group to research new methods for ratings. *The New York Times.* Retrieved from: www.nytimes.com/2009/09/11/business/media/11ratings.html [September 25, 2009]

Stelter, B. (2013, February 21). Nielsen adjusts its ratings to add Web-linked TVs. *The New York Times,* pp. B1, B2.

Stross, R. (2009, February 8). Why television still shines in a world of success? *The New York Times.* Retrieved from: www.nytimes.com/2009/02/08/business/media/08digi.html?_r51&scp51&sq5%22Why1television1still1shines1in%22&st5nyt

Terms for the trade. (2014). *Nielsen.* Retrieved from: www.arbitron.com/downloads/terms_brochure.pdf

Thompson, D. (2013, March). The incredible shrinking ad. *The Atlantic,* pp. 24, 26, 28.

Top ten list. (2020, February 10). *Nielsen.* Retrieved from: www.nielsen.com/us/en/top-ten/

Top time-shifted series of 2018. (2018, December 17). *USA Today.* Retrieved from: www.usatoday.com/story/life/tv/2018/12/17/2018-review-nielsen-ranks-years-most-popular-tv-shows/2339279002/

Top 20 new podcasts of 2019. (2020). *Podtrac.* Retrieved from: https://analytics.podtrac.com/top-20-new-podcasts-of-2019

Tops of 2015: TV and Social Media. (2015, December 8). *Nielsen.* Retrieved from: www.nielsen.com/us/en/insights/news/2015/tops-of-2015-tv-and-social-media.html

TV audience of 31 million for Michael Jackson memorial. (2009, July 8). *Radio Business Report.* Retrieved from: www.rbr.com/tv-cable_ratings/15685.html [September 25, 2009]

TV costs and CPM trends – Network TV primetime. (2015). *Television Advertising Bureau.* Retrieved from: www.tvb.org/trends/4718/4709

Valance media acquires Nielsen Media and Insights product suite and creates a new MRC data division. (2019, December 18). *Nielsen.* Retrieved from: www.nielsen.com/us/en/press-releases/2019/valence-media-acquires-nielsen-music-data-and-insights-product-suite-and-creates-new-mrc-data-division/

Versaw, R. (2017, August 29). The future of mobile apps is now. *Forbes.* Retrieved from: www.forbes.com/sites/forbestechcouncil/2017/08/29/the-future-of-mobile-apps-is-now/#6798c2b91152

Voight, J. (December, 1996). Beyond the banner. *Wired,* p. 196.

Walmsley, D. (March 29, 1999). Online ad auctions offer sites more than bargains. *Advertising Age,* p. 43.

Watson, A. (2019, August 1). Time spent listening to Spotify content among active monthly users 2015–2017. *Statista.* Retrieved from: www.statista.com/statistics/813876/spotify-monthly-active-users-time-spent-listening/

Webster, J. G., Phalen, P. F., & Lichty, L. W. (2000). *Ratings analysis.* Hillsdale, NJ: Lawrence Erlbaum.

When does your audience engage? (2019). *Nielsen.* Retrieved from: www.nielsensocial.com/

Wi-Fi and GPS combined move outdoor audience measurement indoors. (2008, April 10). *Business Wire.* Retrieved from: www.businesswire.com/portal/site/google/?ndmViewId5news_view&newsId520080410006247&newsLang5en [September 23, 2009]

Chapter eleven

The Business of Entertainment and Media Ownership

The authors thank Jeffrey Wilkinson, Ph.D. (Florida A&M University) for his contributions to this chapter.

College students wishing for careers in media industries select their major because they believe the field is creative and exciting. Others are hoping to achieve fame, success, and even wealth. Almost everyone sees that our reliance on media is growing even as the applications expand. Time spent with media continues to rise along with the devices we use to consume media content.

It is important to remember that, first and foremost, all media are businesses. Producing quality content takes creativity, talent, and hard work, but all the important decisions are governed by money and the revenue that can be brought in. That is why it is called *show business* rather than *show art*.

SEE IT THEN

Finding a Business Plan That Worked

Since the early 20th century, electronic media have delivered information from one point to another. Wireless radio began with ship-to-ship, ship-to-shore, and shore-to-shore communications. There was no revenue-building strategy at first because early broadcasters used radio for non-commercial purposes. Like many technologies when they are new, many early adopters had little expectation of making money. Mostly they enjoyed using the new technology to do things in a new and exciting way. Early radio pioneers sent messages via wireless to other experimenters and anyone with a receiver.

Over time, things changed and early broadcasters grew the hobby into a business. From simply transmitting Morse code signals between ships and amateur radio enthusiasts, broadcasting evolved to the transmission of the human voice, live music, and entertainment programming. Equipment had to be purchased, studios were built, and people were hired to help operate and manage the stations. By the early 1920s, broadcasters searched for ways to make money, not only to reimburse themselves for their expenses but also to sustain their new business.

Business Models

Whenever a new electronic medium is invented, a new model emerges for monetization. Everyone involved watches to see where the money will come from. It may be from advertising, subscribers paying monthly or annual fees, government taxes, sales taxes, community support (fundraisers), or some other unforeseen area. In the 1920s, owners and operators of radio stations considered several financial models for turning a profit. In fact, other forms of electronic communication offered models for making radio pay for itself. These models are discussed in the following sections.

The Telegraph Model Telegraph companies bought their own equipment and network (telegraph lines) and also had to pay for professional telegraph operators. To recoup their expenses and generate additional revenue, the companies charged for each word in a message sent by customers. Shorter messages were less expensive than longer ones. Similar to today's tweets, telegraphed messages were often cryptic and written using abbreviations to keep the messages as short as possible. Both the sender and the receiver were charged for the cost of the message. The telegraph was overtaken by faster, cheaper technologies and went out of business in the 1930s.

The Telephone Model The original telephone system was an interconnected series of wires stretching across the country (via 'telephone poles') and managed by companies run under Bell Telephone and its inventor, Alexander Graham Bell. Telephone service was only offered by 'Ma Bell' ('Mother Bell'), as numbers were assigned to locations and messages were 'patched' by operators manually to connect one telephone to another. By the 1970s there were concerns about the monopolistic power of the company (by then renamed AT&T), and the federal government pushed the company to divest itself into eight regional 'Baby Bells' for local service, and – for the first time – long-distance service could be provided by outside competitors.

Up until the 1980s, telephones were wired into the home by one of the telephone companies and users simply 'rented' the home telephone receiver. For decades, there was one model of phone, and the phones were basic black. When a service was disconnected, the phone had to be returned. After the 1980s' breakup of the Bell system, extension telephones were sold on the consumer market, and purchasers could simply hook one up to existing wiring in their homes. Telephone companies are equipment intensive, and typically charge for various services, certain types of federal taxes, flat rate fees, and some variation in amount of data used per month.

The Per-Set Tax Model Radio observed how telegraph and telephone companies charged for services. Telegraphs and telephones were both one-to-one models of communication, limited to one sender and one receiver. Radio was different. One person could reach many people at once, and the broadcaster did not know who was actually listening. Radio companies had to find a different way to stay solvent financially. In the earliest days of radio, RCA's David Sarnoff proposed a 2% tax on each radio a consumer purchased. The money collected was to be sent to broadcasters to pay for programming. But this idea did not get the needed approval of legislators for it to become part of sales tax law. This licensing model was subsequently adopted in England, however, and for many years provided funding for the British Broadcasting Corporation (BBC).

The Voluntary Audience Contribution Model Another idea for generating revenue was to get listeners to pay for radio programming. In 1922, a station in New York tried this plan, seeking $20,000 in listener donations. But when it only collected $1,000 from listeners, the station ended up returning the money. The contribution model persisted however, and today public broadcast stations regularly rely on viewers or listeners to send in contributions. This precursor to crowdfunding helps non-commercial stations raise money by selling printed program schedules and other promotional items. Online Web sites such as Kickstarter and GoFundMe use a similar appeal for community help to fund individual projects.

The Government Subsidy or Ownership Model After World War I, the U.S. government debated whether to keep broadcasting under the control of the Navy for security reasons. The government observed the fierce competition between telegraph and telephone companies and opted to avoid that situation among broadcast companies. Despite some strong lobbying for the Navy to maintain control, the idea of government control was dropped. This approach is common in other countries, particularly authoritarian regimes where the government prefers to own and operate the media because they can manage and control the messages.

The Toll Broadcasting Model Toll broadcasting started in 1922 at WEAF, the AT&T station in New York. The station borrowed from telephone's revenue model by charging an advertiser based on minutes of airtime used. A Long Island real estate firm was the first company to pay for radio time. It bought ten minutes of airtime to promote real estate it was selling. The broadcast is often thought to be the first infomercial. The toll model also included the concept of network buying. Because WEAF was part of a 13-station group owned by AT&T, toll advertising was sold for all stations at one purchase price, which was less expensive than buying time on each station individually.

The toll model is still used today in the form of the infomercial, in which program-length time is bought from a media outlet. The station airing the program does not produce it but simply schedules and airs it in exchange for a fee.

Another version of the toll model is today's telephone data plan. Mobile phones are being used for increasingly data-rich applications such as video and gaming. Companies have found customers understand and accept the idea that the more data they use, the more they will have to pay.

The Sponsorship Model In the early 1920s, advertisers were discouraged from including direct sales pitches in their messages. Instead, they were encouraged to sponsor entire programs and performers by paying for the costs involved. Often, this practice led to naming the program (*The Kraft Music Hall*) or the performers (Astor Coffee Dance Orchestra) after the sponsor. Mixing advertising with entertainment programming was not initially acceptable to many people but, by 1928, advertising sponsorship had established itself as the primary model for providing financial support to the radio broadcasting industry. Sponsorship remains popular across media today.

The Spot Advertising Model In the late 1950s, the networks started moving away from single-advertiser sponsorship of a program. The main reason for this shift was that sponsors wanted too much control over program production, stars, and content. Program sponsors had quite a bit of power, including the ability to rig the outcomes of the television quiz shows. In short, pressure from sponsors to gain huge audiences led to the television quiz show scandal of the late 1950s.

The move away from sponsorship also came about because the advertising marketplace was experiencing maturation. Advertisers were willing to pass up 60-second spots in favor of less expensive 30-second spots. The spot advertising model gave the networks more commercial inventory to sell to sponsors. Rather than feature a single sponsor for a program, multiple sponsors, buying shorter 30-second ads, could be featured in the programming. Shorter, less expensive spots gave smaller advertisers the opportunity to reach a network audience and also gave the networks the opportunity to generate more revenue for selling the same amount of airtime.

Spot ads are typically part of a media plan which considers the message length, scheduling frequency, and time of day. Messages are often created to fit standard 30- and 60-second lengths, but shorter lengths are also commonly used. For example, 15-second ads can be effective, and online video commonly has six-second ads which run before the desired clip. A second factor is scheduling frequency. Companies associated with seasonal events (such as Halloween costumes or Super Bowl snacks) obviously want more exposure as the event date approaches. The third factor, time of day, is also important because cost is associated with audience size and composition. The audience watching a show at noon may be totally different from the audience watching at midnight.

The Subscription Model The subscription model for radio borrowed from the print media, which have used product subscriptions as a revenue stream for years. Readers pay a monthly or annual fee to subscribe to a magazine or newspaper.

The print media derive a greater portion of their income from audience subscriptions than from single-issue rack sales. Because the vast majority of newspapers and magazines also generate revenue from advertising, subscription income is part of a *dual revenue stream*, with the subscription fee usually supplementing the overall revenue picture. The subscription model has proven popular with several platforms because revenue is generated from both the audience and the advertisers.

Ownership by Broadcast Networks

The Report on Chain Broadcasting

As the early radio networks grew more profitable, the affiliates found themselves with less control over their programming. The affiliation contracts forced the local stations to broadcast network programming even if they had to cancel their own shows. After several stations complained, the FCC investigated this relationship beginning in 1938, and released their report in 1941. *The Report on Chain Broadcasting* found that NBC and CBS constituted an unfair monopoly of prime-time radio programming across the country. The report mandated several changes, including shortening affiliation contracts to three years, allowing affiliates to reject network programs, and limiting networks to one affiliate per market.

The networks challenged the FCC's new rules, and the case eventually went before the U.S. Supreme Court (*NBC v. United States*, 1943). But the Supreme Court ruled in favor of the FCC, and the 1943 decision changed the network-affiliate model forever. Broadcast giant NBC had internally divided stations into the Red network and the Blue network. This ruling forced NBC to divest itself of the smaller Blue network which it sold to LifeSavers candy mogul Edward J. Noble. The Blue network was subsequently renamed the *American Broadcasting Company (ABC)*. The three major broadcast networks, ABC, CBS, and NBC, dominated broadcasting for the next 50 years.

Network Ownership Since 1945

After World War II, all three networks added the new medium of television to their holdings. The story-themed programs such as dramas, thrillers, and situation comedies were shifted to television. The radio networks changed rapidly to accommodate television. There was substantial overlap in ownership with local radio and television stations in various markets. Affiliated television stations were contractually committed to airing the networks' programs in exchange for **network compensation**, or money paid to them for their airtime. In return, the networks gained an audience large enough to sell national advertising time. This arrangement helped the television networks and the affiliates prosper and grow.

The television networks owned and operated their own stations in many of the largest markets. Network programs drew huge audiences which the network could sell to advertisers for the highest prices. Owning the stations allowed the networks to maximize their profits. In the 1950s, one network, the short-lived Mutual Broadcasting System (MBS), did not own stations and operated as a programming cooperative.

In the early 1950s, a fourth television network briefly emerged. Allen DuMont, the owner of a television manufacturing business, started the DuMont network. But since most popular television stations were already under contracts with the other three networks, DuMont faced an uphill battle. The DuMont network affiliates were all weaker stations and its audience never ranked higher than fourth place behind ABC, NBC, and CBS. Since many markets only had three television stations, DuMont could not even distribute programs in those markets. The DuMont network ceased operating in 1955.

When NBC divested itself of the ABC stations, it left ABC with the weakest line-up of stations. ABC affiliates tended to be those leftover stations not already under contract with either NBC or CBS. Because ABC stations had lower audience share, the network struggled to stay profitable. In 1951, ABC was negotiating with potential partners and, in 1953, it merged with United Paramount Theaters. The merger brought much-needed cash to ABC, which continued to struggle financially for another 20 years. Still, under the big three networks, the television industry stabilized, grew, and became a familiar and integral part of American life.

A series of changes began in the 1980s which would have lasting effects. In 1985, group owner Capital Cities Communications took control of ABC. Capital Cities brought stability and profitability by controlling expenses. The following year, GE bought RCA, the parent company of NBC. Afterward, GE sold off NBC radio to program syndicator Westwood One.

Also in 1986, Australian media mogul Rupert Murdoch launched the Fox television network. Murdoch had become a U.S. citizen the year before in order to comply with media ownership rules. Named after Murdoch-owned Twentieth Century Fox film studio, the new network at first lost money – $80 million in 1988 alone. Fox eventually caught on and five years later its operation became similar to that of the big three networks when it began scheduling prime-time programming seven nights per week.

In 1995, CBS was acquired by Westinghouse Electric Corporation. Westinghouse sold its electronics business (appliances) and renamed itself CBS Corporation. Then in 1996, ABC was sold again as the Walt Disney Company bought Capital Cities/ABC for $18.5 billion. Disney became vertically and horizontally integrated, a full-service media company with television, film, entertainment brands, and, of course, the theme parks. In 1999, CBS was sold again, this time to Viacom, the company that owned media brands including MTV and pay channel Showtime.

Radio Networks

With the rise of television, the networks put most of their resources into producing news and entertainment programs for the visual medium. With fewer shows from the network, radio stations embraced local announcers and recorded music programming. Radio networks continued, but quickly became little more than top-of-the-hour news briefs until the 1980s. During the late 1970s and early 1980s, audiences shifted from AM to FM. By the late 1980s, AM stations were struggling to stay viable. During this time, conservative talk radio host Rush Limbaugh gathered new listeners back to the AM band and became syndicated through EFM Media Management. He grew in popularity through the 1990s, and stations installed other conservative hosts around his midday show.

The other significant radio network programming that has emerged has come from National Public Radio (NPR). Many radio stations had preproduced material, but the majority of that material was supplied by syndicators rather than by networks. Audiences did not care about a radio station's network affiliation, but they did care about a station's format. Radio station ownership included some small groups and many stations owned by small companies, often local to a market or region. 'Mom-and-pop' radio stations were common for many years.

Deregulation's Influence on Broadcast Media

After decades of increasing its control over broadcasting, the FCC stepped back in the late 1970s, allowing congressional factions to spar over the appropriate level of FCC media oversight. Liberal legislators championed the idea that broadcasting operate in the "public interest, convenience, and necessity." Conservative lawmakers argued the media marketplace was self-correcting and stations needed freedom in order to be able to compete with emerging challenges from cable television. The conservative view prevailed and, in 1981, the radio industry began deregulation. Radio stations were free to program whatever would draw the biggest audience. They did not have to consider the public interest or even be involved in the local community. License renewal became little more than filling out and returning the designated postcard.

Deregulation next took on the Fairness Doctrine, which, beginning in 1949, had required stations to air all sides of local issues. The FCC agreed to stop enforcing it in 1987. The Fairness Doctrine was different from the 'equal time' provision which is still in place. The 'equal time' provision requires stations to give political candidates equal air time. The spirit of deregulation reflects the First Amendment and the marketplace of ideas concept. Broadcasters sought freedom from intrusive government regulation and felt programming should be determined by the marketplace and the audience, not bureaucrats.

For example, the FCC had been using the code of ethics from the **National Association of Broadcasters (NAB)** as a measure for judging whether stations were operating in the public interest. In a 1983 decision, the courts ruled that the code set an 'arbitrary limit' on how many minutes of commercials could be available for advertisers. Both the broadcasters and the advertisers were unhappy with those limits.

In sum, the FCC sided with broadcasters about the marketplace concept and developed more opportunities for programming and advertising to reach audiences. The FCC looked to reduce the barriers for entry into electronic media businesses. The result was more stations and increased competition from cable, satellite, and multipoint multichannel delivery systems.

By the late 1980s, Congress felt there had been enough deregulation and it should play "a more direct role in the policy sandbox" (Sterling & Kittross, 2002, p. 560). *De*regulation had crossed into *Un*regulation; there were fewer public service programs required, no limit on the number of commercials per hour, a much easier license renewal process, and few restrictions on selling stations. Previously, an owner had to hold a broadcast station at least three years before selling it. In 1984, the FCC increased ownership limits from the **rule of sevens** (which allowed owning only seven stations in any service: AM, FM, or television) to 12 stations in either radio service and up to 25% of the total television households nationwide. In 1992, the limits were raised again, allowing up to 40 radio stations per owner and some relaxation of the duopoly rules. The Telecommunications Act of 1996 relaxed ownership rules even more.

Cable Growth and Ownership

In the 1960s and 1970s, cable ownership was mostly in small and medium-sized communities that did not have a television station or only had one or two local television broadcast stations. Big name companies like TelePrompTer (maker of teleprompters for television studios), Westinghouse, and Cox began investing in cable. The number of cable subscribers nationwide increased from one million in 1963 to 4.5 million in 1970. In the 1970s, the number of cable operators grew steadily as cities adopted franchising agreements allowing them to be wired by cable companies.

In 1975, HBO unveiled its plans to use satellite distribution to send its programming to cable systems nationally. HBO triggered an explosion of interest and investment in cable television ownership nationwide. Larger companies with deeper pockets bought small companies that could not afford the satellite equipment. By the end of the 1980s, more than 50 million households had cable subscriptions. The number of cable-only (not a broadcast station or network) channels grew from 28 in 1980 to 79 by 1989. By the end of the 1990s, seven in ten households (more than 65 million) subscribed to cable.

By the late 1990s, the bulk of cable subscribers were held by the top ten **multiple-system operators** (MSOs). Although not a true oligopoly, the industry was consolidating. A typical example of the ownership consolidation was Tele-Communications, Inc. (TCI), the leading cable company. TCI had almost 14 million subscribers, and in 1999 it was purchased by AT&T. Charter Communications then acquired some of the cable systems owned by AT&T.

The Internet Becomes a Mass Medium

The development of the Internet continues to change the media landscape, extending its reach into almost every aspect of people's lives. The worldwide computing system links hardware and allows anyone to transmit digital information using Web pages, browsers, and common operating systems. Governments, businesses, and consumers use the Internet in countless ways, accessing it through wired or wireless connections.

The Internet emerged as the network of research computers became connected over the course of the 1960s and 1970s. By the early 1980s, this networked system of computers became known as the Internet. Companies called Internet Service Providers (ISPs) enabled others to connect to the Internet and by the late 1980s, commercial businesses and consumers became interested in its potential.

In the 1990s, innovations including HTML (invented by Tim Berners-Lee) allowed the creation of Web pages, resulting in the World Wide Web. As personal computers became affordable, connecting to the Internet became popular in homes throughout the U.S. Personal computers were popular Christmas gifts and a variety of dial-up plans were sold. Home connectivity relied on the telephone wires installed across the nation by the telephone company. These wires were made of copper, which allowed basic voice transmission but limited Internet speeds well below modern broadband. Commercial and consumer demand for ever faster connectivity has been a driving force behind fiber optic cable and current efforts to adopt 5G wireless.

Broadcasters and cable were not threatened by early Internet use. Dial-up connections could not sustain audio or video, and even computer games were disk-based and low resolution. One sector of the media industry – the music labels – became concerned when it discovered massive illegal sharing of music occurred through Napster, a peer-to-peer sharing site which launched in 1999. Young people were sharing digital music files for free in violation of copyright laws, and the labels saw a significant drop in their corporate profits. The Recording Industry Association of America (RIAA) sued Napster on behalf of the American music industry and effectively shut it down in 2001.

By the early 2000s network upgrades enabled streaming video and audio over the Web to become common. YouTube was launched in 2005, allowing anyone to upload video for others to see. YouTube was purchased by Google in 2006 and had to quickly remove thousands of television show clips that had been uploaded by consumers in violation of copyright law. In 2007, Hulu was launched, ushering in the age of online video streaming services. Hulu was initially a joint venture between News Corporation and NBC Universal, Providence Equity Partners, and later the Walt Disney Company. Hulu featured recent episodes of television series from their respective television networks.

SEE IT NOW

The Broadcast 'Star' Model

Radio first used the toll broadcasting model, then changed to the sponsorship model, and finally adopted the spot advertising model. Broadcasters have relied on spot advertising since the late 1950s, and newer online media have also adopted a blend of toll, sponsorship, and spot advertising as well as subscription approaches.

No media property is an island. For example, for a local radio or television station to be successful, other segments of the industry must be accommodated. It is strategically important to see how related groups, organizations, and businesses function to make the system work. Together these elements visually form a star. In the example shown in Diagram 11.1, a network-affiliate television station in the center must work closely and engage in frequent dialogue and communication in order to remain competitive.

At the center of the star model are television stations, consisting of three different types: Network owned and operated (O&Os), network affiliates, and independent stations. These types of stations differ by their relationship to the network. A network O&O is owned and operated by a network. A network affiliate is by far the most common type of station and has a long-term agreement with a network to run network programs and commercials. An independent station does not have an agreement with a network and thus must find other sources for programs.

Diagram 11.1 shows the basic interrelationships between stations, program suppliers, regulators, advertisers, and the audience. These stakeholders must work closely to anticipate changes within the industry which affect each one of them. Audiences' changing consumption patterns and behaviors are driving programming to the Internet. Programming sources may still come from standard television,

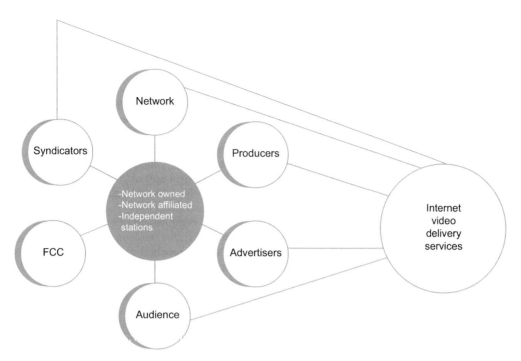

Diagram 11.1 Broadcast star model.

FYI: NETWORK, SYNDICATION, AND OFF-NET PROGRAMS

The process by which new shows are produced continues to change in the era of OTT. More programs are being proposed than ever before, and the almost limitless number of channels make it a seller's paradise. The business is experimenting with new models to monetize programs and get programs to pay for themselves. In the past, network programs were created by independent producers with some support from the networks, but the producers maintained control of the program. After a program accumulated a few years of episodes (about 100), the producer could contract with a syndicator who would license (or rent) the program directly to stations or groups of stations at whatever price the market would bear. The program would continue to air on the network until no new episodes were being made. At that point, the program would be considered **off-network**.

Networks can also own a controlling portion of the shows they air, and that has changed the life cycle of a television program. The network decides where and when the program will be shown after (or even during) its network run. A network like Fox, for example, might also agree for the show to run on a sister channel (such as Fox's FX) at a discount, thus helping the network brand while preventing a rival from potentially getting a hit show.

At the 2019 National Association of Television Program Executives (NATPE) conference in Miami, industry buyers admitted that finding ways to monetize content is much harder than it used to be. President and CEO for the CBS Global Distribution Group and Chief Content Licensing Officer for the CBS Corporation Armando Nuñez said program buyers (distributors) agreed the changes have forced everyone to be more aggressive (Berman & Hayes, 2019). In the days of traditional broadcasting, the networks had most of the control. Network rejection was the kiss of death.

Today there are far greater opportunities. If rejected, a program creator may consider going with Netflix, HBO, a streaming channel or negotiate for it to be on another network property (such as NBC's Peacock streaming channel).

Proposals for new shows today must include ideas for spinoffs and future programs. For example, in 2019 *Star Trek: Discovery* was offered as a streaming option on CBS All Access, which is now in additional seasons. The program creators announced they will launch a spin-off, still untitled, based on Michelle Yeoh's character Phillippa Georgiou. Nuñez said "*Star Trek: Discovery* was a win because it brought a slew of new subscribers to our digital extension and we did so without having to dilute our linear model."

Nuñez said there is so much original content being made the business model continues to evolve. Both the traditional networks and the newer start-ups are hoping there is enough for everyone to make a profit.

FYI: RECORDING INDUSTRY: ADJUSTING TO NEW CONSUMPTION PATTERNS

Globally, music is a huge profit-center. The umbrella organization helping musicians, producers, and companies receive the royalties they deserve is the International Federation of the Phonographic Industry (IFPI), a non-profit members' organization based in Europe with regional offices in Brussels, Hong Kong, and Miami. In the United States, the music industry is represented nationally by the RIAA.

For many years, music was recorded on discs and sold as record albums. Consumers were satisfied because it could be brought home and enjoyed either alone or with friends. With some effort, young people could make multiple home copies and the cassette tape (the infamous 'mix tape') was an integral part of youth culture. Countless poor-fidelity copies of favorite songs were gifts that were traded and shared in the name of friendship (and courtship).

When recording went digital in the 1980s, consumers began to push back, complaining about the quality and price of Compact Disks (CDs). A great tug of war ensued between consumers, the music industry, and the go-between distributors, the 'Record Stores' such as Turtles and HMV. This tension reached its apex in the 1990s with the rise of peer-to-peer networks like BitTorrent. The most famous was probably Napster, which resulted in legal action against several users. Some artists learned to circulate their music online, hoping that giving away content would lead to future dividends. Bands like the Grateful Dead and Phish encouraged their fans to record their concerts and distribute them to others as cheaply as possible.

Apple parlayed a temporary truce when it launched the iPod in 2001. Apple brought stability and a new pricing structure by declaring that any song could be purchased for $0.99. The rest of the industry soon followed. As time has passed, streaming music services and countless new artists and songs have been distributed online. The RIAA continues working with ASCAP and BMI to monitor song play and royalty distribution.

The move to digital has changed the way consumers think about music. Instead of vast personal libraries of purchased records, tapes, disks, or files, consumers have increasingly moved toward a subscription model. According to the RIAA end-of-the-year 2019 report, 80% of music revenues came from streaming music services. People are more willing to buy access to on-demand music services which allows them to listen to any artist or song at any time.

U.S. Music Industry Year-End Revenues (billions)

	2017	2018	2019
Retail	8.8	9.8	11.1
Wholesale	5.9	6.6	7.3

Source: Friedlander & Bass, 2020

Revenues from streaming grew at almost 20%, accounting for almost 80% of all music revenues. The categories include premium subscription services, ad-supported on-demand (YouTube, Vevo, ad-supported Spotify), and streaming radio services (Pandora, SiriusXM, others). Paid subscriptions to on-demand services generate by far the most revenue. Actual purchases of music continues to rapidly decline and may one day fade away:

U.S. Permanent Digital Download Purchases (millions)

Year	2015	2016	2017	2018	2019
Revenues $	2,314	1,849	1,404	1,039	856

Source: Friedlander & Bass, 2020

In sum, it seems that commercial music has changed from a commodity to a service. Music used to be something tangible, a gift that a person could buy and keep, or give as a gift to someone special. Today, music is a subscription, a service to manage your mood or keep up with new artists. Radio stations, therefore, operate as 'service stations' for music and information.

cable, and DBS, but the networks no longer simply depend on local affiliate stations to reach audiences. The Internet can occupy various positions in the star system. Internet Delivery Services such as Amazon Prime, Hulu, or Netflix can be competition or may occupy the center of the star. Also, program providers like the networks are going 'over the top' to reach audiences directly through the Internet and may leave affiliated stations with smaller audiences and less leverage with advertisers.

The star model shows the interdependence that exists across media industries. Media convergence brings together Internet companies, computer businesses, network developers, telephone, broadcast, cable, as well as advertising, marketing, and regulatory agencies. As illustrated in the model, each line connects the entities in a direct, two-way relationship. For example, the network supplies the affiliate with programs, and the station provides the network with the local audience. The Internet provides a way to reach audience members who may have moved to another city but hold roots in their hometown. Advertisers represent clients who are local businesses. These people are also invested in the community and work with the station to reach people who can use the product or service being advertised. The star model is a useful way to visually conceptualize the stakeholders in a particular media solar system.

The Telecommunications Act of 1996

For much of the 20th century, the most significant legislation guiding media was the Communication Act of 1934, passed at a time when the only electronic medium was AM radio. For 60 years, the regulations for FM radio, television, and even cable were stitched together in piecemeal fashion. By the early 1990s, with the rise of the Internet, it became obvious a major change was needed. It took Congress a full year or more of political negotiating and compromise before passing the Telecommunications Act of 1996 by an overwhelming majority vote. Signed on February 8, 1996, the Act led to dramatic changes in ownership, products and services, and how audiences use media.

The biggest immediate change brought by the Act was the removal of radio ownership restrictions. A station group can now own as many stations as it wants, as long as it has less than 50% of the stations in any single market. This provision led to a decade of frenzied buying-and-selling and, by 2002, roughly 50% of the 15,000 U.S. radio stations were controlled by six companies. The largest, Clear Channel, reported owning 1,216 stations in 190 markets. As Internet growth increasingly shifted significant advertising dollars away from traditional media like radio, television, and newspapers, radio groups had to rethink their business strategy, particularly as things worsened during the recession in 2008 and 2009. In 2014, Clear Channel rebranded itself iHeartMedia, and in 2020 still reported owning 850 stations in 150 markets. According to InsideRadio (2020), other notable groups owning at least 100 stations in 2020 are Cumulus Media (428 stations in 87 markets), Townsquare Media (321 in 67 markets), Entercom (235 stations in 48 markets), Salem Media Group (117 stations in 38 markets), and Saga Communications (113 in 27 markets).

The practical impact of group ownership was to minimize operational costs by sharing personnel and resources among stations. As such, radio stations became more corporate and formulaic, simulcasting the same music playlists and announcers across several different markets. These standardized formats reflect genres like Country, Pop, Urban, Soft Rock, and even Alternative. Much of the programming and recording tasks are automated and done on a computer. It is now standard practice to do **voice tracking**, where the announcer talks, makes announcements, and mixes the music into a show that can be aired on any or all of the group stations. Voice tracking can be accomplished from anywhere in the world with a microphone, a computer, and an Internet connection. This has impacted radio's localism, as many hosts have never been to the markets where their voices are broadcast.

The Telecommunications Act of 1996 brought changes in five major areas: Telephone service, telecommunications equipment manufacturing, cable television, radio and television broadcasting, and obscenity and violence in programming.

Telephone Service

The 1996 Act overruled state restrictions on competition in providing local and long-distance service, which allowed the regional Bell (Telephone) operating companies (also known as Baby Bells and **RBOCs**, pronounced 'ree-boks') to provide long-distance service outside their regions and required them to remove barriers to competition inside their regions. New universal service rules guaranteed that rural and low-income users would be subsidized and that equal access to long-distance carriers would be maintained.

Telecommunications Equipment Manufacturing

The RBOCs were allowed to manufacture telephone equipment. The FCC was empowered to enforce non-discrimination requirements and restrictions on joint manufacturing ventures and monitor the setting of technical standards and accessibility of equipment and service to people with disabilities.

Radio and Television Broadcasting

The 1996 Act relaxed media concentration rules for television by allowing any one company to own stations that could reach 35% of the nation's television households, an increase from the previous limit of 25%. The major broadcast television networks (ABC, CBS, NBC, and Fox) were even allowed to own cable systems as long as they did not exceed the national household limit. Media concentration rules for television changed again in 2003, when television groups were allowed to reach 39% of the national audience.

Limits on radio station ownership were also repealed, but some restrictions remained on the number of local stations a company could own in any one market. In markets with 45 or more stations, one company could own eight stations but no more than five in either AM or FM. In markets with 30 to 44 stations, a company could own seven or fewer with only four in any one service. In markets with 15 to 29 stations, one company could own only six stations (four in any or either service). In a market with less than 15 stations, only five could be owned by one company – three maximum in either AM or FM and only 50% or less of the total number of stations in that market. In addition, the U.S. Department of Justice's Antitrust Division was assigned to look closely at large transactions to prevent any group that already had 50% of the local radio advertising revenue or was about to obtain 70% of the local advertising revenue in any one market from acquiring more stations in that market.

Cable Television

The Telecommunications Act of 1996 relaxed the rules set down in the 1992 Cable Act, essentially removing rate restrictions on all cable services except basic service. In addition, telephone companies were permitted to offer cable television services and to carry video programming. Other provisions were that cable companies in small communities were to receive immediate rate deregulation, that cable systems had to scramble sexually explicit adult programming, and that cable set-top converters could be sold in retail stores (instead of being available only as rental units from the cable companies).

Obscenity and Violence

Part of the Telecommunications Act included a provision titled the Communications Decency Act. This was an effort by the United States Congress to regulate pornographic material on the Internet. The Act was a response to parents' groups concerned about protecting their children from accessing sexually explicit content on the Internet. But the CDA was immediately challenged by free speech advocates, and in 1997, the U.S. Supreme Court struck down the anti-decency provisions in the landmark case of *Reno* v. *ACLU*.

FYI: CABLE SUBSCRIPTIONS 2010–2023

2010	99.0 million households
2011	100.9
2012	100.8
2013	99.3
2014	98.0
2015	97.2
2016	97.7
2017	94.3
2018	90.3
2019	86.5
2020	82.9
2021	79.4 (est.)
2022	76.0 (est.)
2023	72.7 (est.)

The overall number of cable subscriptions has been decreasing steadily and is projected to be fewer than 83 million by 2021.

Source: Spangler, 2020

The overall number of cable subscribers has declined steadily the past few years. As video viewing migrates to the Internet and smartphones, cable companies have tried different strategies to keep people from 'cutting the cord.' But customers have a long history of dissatisfaction with the pricing and strategy of bundling unpopular channels with favorites.

FYI: TOP 10 VIDEO SUBSCRIPTION SERVICES

Netflix	60.1 million
Hulu	28.0
Amazon	26.0
Comcast	21.6
DirecTV	21.6
Charter	15.8
Dish	12.0
Verizon FiOS	5.8
Altice	4.9
Cox	4.0

Amazon, Netflix, and Hulu are Internet video subscription services; DirecTV and Dish are satellite subscription services; Altice, Comcast, Time Warner, Charter, and Cox are cable subscription services; and Verizon is a telecommunication company that also offers a multichannel subscription service.

Source: National Cable Television Association, March 2020

The Role of Government and the FCC

Government Guidance

The U.S. government is routinely asked to consider enacting media-related policies. For example, in 1998, Congress passed the **Internet Tax Freedom Act** (H.R. 4328) which placed a moratorium on Internet taxes. In 2014, Congress passed the Permanent Internet Tax Freedom Act, banning state and local taxation of Internet access and on multiple or discriminatory taxes on electronic commerce. But as e-commerce grew, state governments voiced concern about the loss of significant online sales tax revenues. In 2018, the United States Supreme Court ruled in *South Dakota* v. *Wayfair Inc.*, that states were allowed to collect sales tax from Internet-based retailers.

Each state has its own considerations. Massachusetts, for example, has drafted H.4000. H.4000 defines 'marketplace' as

> a physical or electronic forum, including a shop, a store, a booth, a television or radio broadcast, an Internet web site, *a catalogue or a dedicated sales software application, where the tangible personal property or services of a marketplace seller is offered for sale, regardless of whether, in the case of tangible personal property, such property is physically located in the commonwealth.*
>
> Cole, 2019

Another tax that Congress has been asked to consider is a proposed *Performance Tax* which is vigorously opposed by the NAB. Music companies wish to impose a tax on radio stations for airing free music. Radio stations pay licensing fees to the organizations representing composers and songwriters (e.g. BMI, ASCAP) when their songs are broadcast. Internet radio stations pay a fee to stream music. The music labels have pushed Congress to impose a tax on radio stations to compensate performers when the stations air their music. In response, the NAB argued that broadcasters already pay one licensing fee for the music, and it is wrong to levy a second fee, plus it would hurt smaller stations who are struggling financially. The counter-proposal, the Local Radio Freedom Act, gained support from Congress throughout 2019, and at year's end over 200 representatives and 25 senators had signed on to oppose the tax (Wharton, 2019).

The FCC

The Federal Communications Commission coordinates with Congress and helps address a number of legal issues regarding media companies and their audiences. The President of the United States nominates the commissioners, and the Senate must approve them. Originally there were seven commissioners until that number was reduced to five in 1982. The composition must be bipartisan and only three commissioners can come from any one political party. The FCC helps lawmakers consider broad policies about media regulation regarding issues such as Spectrum Space, implementation of 5G wireless broadband, licensing of low-power FM radio, and investigating Internet scams such as the numerous 'cures' sold online in 2020 during the coronavirus pandemic.

There are seven bureaus or divisions: The Consumer and Governmental Affairs Bureau, the Enforcement Bureau, the International Bureau, the Media Bureau, the Public Safety and Homeland Security Bureau, the Wireless Telecommunications Bureau, and the Wireline Competition Bureau.

ZOOM IN 11.1

For more information about the FCC and its bureaus, go to www.fcc.gov/about

> **FYI: LICENSE RENEWAL**
>
> When renewing a station license, the owners of the station must certify the following:
>
> - The station has been and continues to be on the air.
> - The station maintains a current public file.
> - The owners have filed the required ownership reports.
> - The station can show that it has complied with the children's programming requirement of three hours per week.
> - The owners of the station have not committed any felonies.

Media Finances Because of convergence, media continue to be highly subjective and competitive businesses. Program content is important, but advertising sales and financial management keep the station from going dark. Enough income must be generated to offset the bills: Items such as payroll, facilities (electricity and water), equipment purchase, repair, and depreciation, and even added expenses for affiliations and additional outside news or entertainment programming.

Given the vast sums of money needed to launch and sustain media, an often overlooked element is debt service. Debt service is repaying money borrowed from investors. It is a huge concern for most organizations. For example, during the economic collapse of 2008–2009 and 2020, advertising revenue dropped even while media consumption went up. Depressed consumer spending has a negative effect on advertising spending and platform or station profitability.

Typically the business department tracks all money coming in and going out. A profit and loss statement, referred to as a P&L, is regularly generated which lists all revenue (i.e. money received) and all expenses (i.e. money spent) during a given time period. The balance sheet shows the net result, either profit or loss.

LMA, Duopoly, and Cross-ownership Media properties today work in a variety of configurations and companies may turn to others in order to survive. For example, when a media company like a radio station is owned by another one in the same city, they may engage in a relationship known as a **local marketing agreement (LMA)**. An LMA lets one station take over programming and advertising sales for another station in the same market. LMAs are used by both television and radio stations, and are similar to the newspaper JOA (joint operating agreement). Operating one content stream from two channels helps to control costs.

Another type of sharing is the Joint Sales Agreement (JSA), which is similar to an LMA but limited to sales in that everything is separate except the sales representatives also sell time for another station in exchange for a percentage of the revenue. Some stations also produce newscasts for independent stations that want to air an evening newscast but cannot afford to staff a newsroom. Depending on the relationship, the station producing the newscast may buy the airtime on the independent station or reach an agreement for revenue sharing.

The term *duopoly* describes when one licensee owns two or more stations in a local market area. The Telecommunications Act of 1996 eased ownership restrictions, and although public interest groups typically argue against relaxing ownership rules limits, financial considerations often rule the marketplace.

Cross-ownership refers to an individual or company owning media which occupy different platforms, such as a newspaper and a broadcast station in the same market. In 2018, the federal government eliminated the prohibition on cross-media ownership by stating there were sufficient diversity, voices, and viewpoints due to cable and online platforms. The FCC rules allow cross-ownership of a daily newspaper and a full-power broadcast station (AM, FM, or television), underscoring that ownership issues are driven by bottom-line concerns.

Ethics and the Law

Organizations create formal ethical codes or guidelines to help people make good decisions in the work environment. Sometimes referred to as *applied ethics*, such guidelines can give group members specific information about what is considered acceptable or ethical behavior. Many industries adopt and publish their own codes of ethics as a way to guide members regarding common practices or concerns.

FCC regulations are meant to guide the operations of broadcast television and radio stations and

(A)

BROADCASTING COMPANY
Balance Sheet
As of _____

ASSETS	This Year	Last Year
Current Assets:		
Cash and cash equivalents		
Marketable securities		
Accounts receivable, less allowances		
Program rights—current		
Prepaid expenses and deferred charges		
Other current assets	_____	_____
Total Current Assets		
Property, plant, and equipment at cost		
Less: accumulated depreciation		
Net Property, Plant, and Equipment		
Program rights—long-term		
Other non-current assets		
Intangible assets, net of amortization	_____	_____
Total Assets	======	======

LIABILITIES AND STOCKHOLDERS' EQUITY

	This Year	Last Year
Current Liabilities:		
Accounts payable		
Notes payable		
Accrued expenses		
Income taxes payable		
Program rights payable—current		
Other current liabilities	_____	_____
Total Current Liabilities		
Long-term debt		
Deferred income taxes		
Program rights payable—long-term		
Other non-current liabilities		
Total Liabilities		
Stockholders' equity:		
Capital stock		
Paid-in capital		
Retained earnings		
Treasury stock	_____	_____
Total Stockholders' Equity	_____	_____
Total Liabilities and Stockholders' Equity	======	======

Diagram 11.2a The balance sheet gives a picture of the station's overall financial health at one point in time. It shows what the station owns (assets) and what the station owes (liabilities and stockholders equity).

networks. FCC rules cover technical and ownership issues, hiring practices, programming content, and ethics to ensure that stations and networks operate in the interest of the public. For example, broadcast stations may not conduct lotteries (although stations may accept advertising for government-run lotteries), operate with more power than their license permits, or refuse to sell advertising to a political candidate if his or her opponent has already purchased advertising on that station. Although most regulations are fairly straightforward, situations arise which may be legal but considered unethical. For example, is it ethical to schedule sugary cereal ads during a children's program? Or promote a heavy metal band in a show aimed at elderly senior citizens? If a radio station promotes music performed by an employee's

Diagram 11.2b A statement of operations, or profit and loss statement, tracks a station's financial performance over a set period of time. The difference between station revenue and station expenses yields the profit or income of the station for that period of time.

```
                                    (B)
                         BROADCASTING COMPANY
                           Statement of Operations
                         As of _____
                                  (Date)
MONTH: _____                              YEAR-TO-DATE

Actual    Budget    Last Year                    Actual    Budget    Last Year
                              OPERATING REVENUE
                              Local
                              National
                              Network
                              Production
                              Other Broadcast
                              Misc. Revenue
_____    _____    _____                          _____    _____    _____

                              Gross Revenue

                              Less:
                              Commissions
_____    _____    _____                          _____    _____    _____

                              Net Revenue

                              OPERATING EXPENSES
                              Technical
                              Programming
                              News
                              Sales and Traffic
                              Research
                              Advertising
                              General and Admin.
                              Depr. and Amort.
_____    _____    _____       Total Op. Expenses  _____    _____    _____

                              Op. Profit (Loss)
_____    _____    _____         Before Taxes      _____    _____    _____

                              Provision for Taxes

=====    =====    =====       Net Income          =====    =====    =====
```

girlfriend or boyfriend it can be a conflict of interest. And what are the ethics of a reporter deciding not to pursue a potentially negative story about a local official considered a friend?

These are ethical, rather than legal issues. Ethics deal with moral principles or values that guide behavior. In situations where there are no specific laws, ethics will guide our decisions.

For example, the Radio Television Digital News Association (**RTDNA**) has a code of ethics. This code addresses issues regarding newsgathering practices. The NAB has had a code of ethics for radio since 1929 and another code for television since 1952. Both codes address practices related to programming and advertising. They have been revised over time to keep up with changes in technology and current practices.

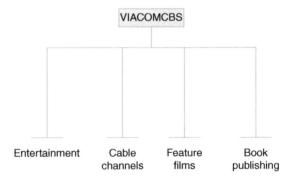

Diagram 11.3 Examples of the media-related business of VIACOMCBS. Entertainment: CBS Television Network, CW Network, CBS Television stations; Cable: Comedy Central, Showtime, MTV, BET, Nickelodeon; Feature Films: Paramount Pictures, Paramount Film Studios; Book Publishing: Simon and Schuster, Pocket Books.
Source: ViacomCBS, 2020

ZOOM IN 11.2

Go to www.rtdna.org/content/rtdna_code_of_ethics to see the RTDNA Code of Ethics.

Ethical codes must be practical to be useful. In the 1980s the NAB code set limits on minutes per hour of advertising which hampered stations struggling to make money. This advertising limitation led to complaints and an antitrust suit alleging that limiting advertising minutes per hour was actually forcing higher advertising prices. The provision was revoked in 1983.

Three common ethical perspectives are rules-based, consequence-based, and virtue, or character-based (Day, 2006). Rules-based ethics rely on standards of right and wrong, like lying versus telling the truth. Consequence-based ethics emphasize the outcome and what is best for the most people. Virtue ethics are concerned with moral character and how a person grows through good habits and behaviors over time.

Deciding which works best in a given situation will depend on the context. Situations we call *ethical dilemmas* are those where you are unsure what to do; it is a dilemma because not knowing what to do prevents you from moving forward. Sadly, chronic liars do not have ethical dilemmas about lying. It can be a dilemma when you have to confront a dishonest client or admit you did something wrong that costs money or your company's reputation. When there is widespread agreement that some action or behavior is unethical, it can become a law (such as fraud or libel).

Competition among media organizations can lead to actions that seem illogical until they suddenly

FYI: COMMON ETHICAL DILEMMAS

Media people frequently face ethical dilemmas, and some common situations are discussed below.

- **Reporting**: Reporters are often pressured to cover or not cover a certain story. For example, an advertising client is opening a new store and insists that the grand opening is a newsworthy event. They want publicity in the form of a news story. Although news organizations typically won't cover it as journalism, what happens if the client threatens to withdraw their advertising?
- **Paying for interviews**: Most reputable news operations will not pay interviewees, having learned from past practices. Paying for an exclusive interview is termed **checkbook journalism**. An increasingly common practice is for celebrities and newsmakers to write books so they can be 'invited' to be interviewed about that work. The organization gets the interview and the newsmaker gets media attention. Another related practice is called **catch and kill**. For example, former Playboy model Karen McDougall agreed to give the story of her affair with Donald Trump to the *National Enquirer* for $150,000. Chairman and CEO of American Media Inc., Publisher David Pecker, admitted to working with the Trump campaign to hide the story and avoid bad publicity during the presidential election race (Swaine, 2018).
- **Reenactments**: In November 1992, *Dateline NBC*, a prime-time news magazine produced a story investigating whether General Motors (GM) pickup trucks were susceptible to fiery explosions in side-impact collisions. The producers of the story tried to show how gas tanks on the side could ignite when hit by another vehicle. But when they could not recreate it, the producers arranged for a small incendiary device inside to explode on impact. The dramatic video was challenged by GM, and when it was discovered the scene was staged, they sued the network. NBC publicly apologized in a *Dateline NBC* program and paid almost $2 million to GM to settle the suit. The president of NBC's news division was fired, as were others associated with the story, and the network's reputation was considerably damaged.
- **Using unnamed sources:** Using unnamed or anonymous sources is another practice that is common in American journalism to protect sources who would otherwise be reprimanded or even fired. But when overused, it undermines the credibility of a news program or journalist. The problem with anonymous sources is that the audience cannot judge whether the information is true.
- **Dirty tricks:** Media are highly competitive and rely on audience ratings to set ad rates. Continuous audience measurement can show a station's popularity from the night before. Media companies may do whatever they can to increase ratings, such as calling and pretending to be a viewer asking 'innocent questions' about some anticipated changes, or even hiring people to dig through the trash of a competitor hoping to find discarded internal memos, drafts of contracts, old research, and other bits of information that might give valuable insights.

> **FYI: ETHICAL DILEMMAS: WHAT WOULD YOU DO?**
>
> - A salesperson from a radio station has sold airtime to a local restaurant, whose manager wants to see if advertising will bring in new customers for the 'two for the price of one' special. The restaurant is running spots on the station for only one day, and it has not placed ads for the special with other media. Obviously, the salesperson wants the commercial to prove successful and feels pressure to get a large crowd out to the restaurant. One option is to call his or her friends and relatives and tell them to go to the restaurant that night, mentioning that they heard the spot on the radio. Another option is for the salesperson to add bonus spots for the restaurant to that day's commercial announcements but not tell the client. Both options involve questionable ethics on the part of the salesperson.
> - The musical director of a college radio station has the responsibility for calling record companies to request songs for airplay but knows that some companies will not provide free service to small stations, especially college stations. The director is under a great deal of pressure to get new music for the station. Should he or she simply avoid mentioning to the record companies that the station is a college station to get the free records?
> - During a sweeps month, a television station has a contest to win $50 in free gas during a time when gas prices have jumped significantly. Viewers can win if they know a secret code word that will be given during a newscast. The station has its news anchors deliver numerous teases, and reporters do live shots in front of gas pumps during the newscast. In a one-hour newscast, the station devotes ten minutes to contest information and talk. Is this information worthy of being in the newscast, or is it strictly promotional?
>
> In addition to professional ethics, there are also personal ethical issues that involve electronic media, especially Internet usage. Here's another example to think about:
>
> - Since Napster gained popularity in 1999, Internet users have downloaded songs without paying for them. As a consequence, the songwriters, copyright holders, performers, and record companies that produced the songs are not paid for their work. Movies can also be downloaded from numerous sites. Should individuals continue to download music, movies, and other copyrighted material without paying for them?

make sense. For instance, in a market where a popular news anchor headed up the number-one news show, the number-two station tried to find her a position in a bigger market. The competing station even recorded her newscast and, without her knowledge, sent copies to stations in distant markets that were looking to hire a news anchor.

It is not always easy to draw the line between aggressive competitive behavior and being unethical. No two situations are alike, so an ethical foundation can help you decide what action is best at that moment.

Defamation: A Legal Concern

Defamation is a legal term referring to an attack on a person's reputation that causes humiliation, ridicule, or loss of good name, or makes him or her the target of hatred. Defamation is further classified as either **libel** or **slander**. **Libel** is the printed publication of offending words or images and **slander** involves words, sounds, or gestures that injure a subject's character or reputation. Libel is considered more serious because print is more permanent than the spoken word. Much hinges on whether the person being attacked is a public or private figure. Public figures must prove actual malice in order to win their cases; private figures only need to successfully argue the defendant acted with negligence. Media companies can alleviate the threat of legal action by providing retractions if they find they aired false information.

Online communication is treated a bit differently, as when one individual 'cybersmears' another. According to Ernst (2003), a divided Georgia Supreme Court ruled in *Mathis* v. *Cannon* (573 S.E.2d 376) that online postings have some protection even if they are defamatory. In the midst of a heated debate about garbage collection, a Crisp County man anonymously posted defamatory messages about a waste management company executive on an Internet message board. In the post, the executive was called a 'thief' who was fired from his previous job for being a 'crook,' and anyone doing business with him was also labeled a 'crook.' When the poster was identified, the executive sued for libel. The majority ruled that this type of 'cybersmearing' should be interpreted broadly to protect free speech. The court argued that the executive was a public figure in this particular situation and could not prove malice, that the smears were emotional outbursts from an angry and frustrated citizen.

The constitutional standard for defamation comes from the U.S. Supreme Court case *New York Times* v. *Sullivan* (1964). In this case, a public official (Sullivan) sued *The New York Times*, accusing it of publishing false and malicious attacks on his character. A full-page advertisement "Heed Their Rising Voices" described actions (some inaccurately) taken by the Montgomery police against Dr. Martin Luther King Jr. and other civil rights activists. Public Safety Commissioner L. B. Sullivan sued on behalf of his officers, although he himself was not named.

In a defamation case, the person who is attacked must prove that the media source was in error or negligent in publishing the story. The Supreme Court decided that the *Times* was not guilty of defamation and did not show a reckless disregard for the truth. The ruling established that there needed to be actual malice in the false reporting of a story for the newspaper to be held liable. *New York Times* v. *Sullivan* made it harder for public officials to win libel suits against media and the case is considered a landmark in protecting a free press. The case affirmed that public officials cannot sue the media for publishing negative stories about them.

Although the First Amendment and *New York Times* v. *Sullivan* protects journalistic practices, there are also protections for people who have been wronged by the media.

There are three criteria for determining libel or slander: Was the material published? Does the material defame a person (or persons)? Was the person identified clearly? If the answer to all three is 'yes' that person may have a legal case to pursue. But other important conditions must be considered, such as whether the statement was true; whether there is a 'statute of limitations' (typically two years); if the comments resulted from coverage of statements made by a reliable speaker such as a government official; or whether the comments are stated as an opinion rather than a fact.

Media Group Ownership

As of 2019, there are more than 120.6 million television households in the U.S. (Kurtz, 2019). Consolidation has led to a market that is increasingly **oligopolistic**. Some very large companies, including all four major networks, own several television stations. The big four typically own stations in the largest U.S. television markets. The five largest markets are New York, Los Angeles, Chicago, Philadelphia, and Dallas–Ft. Worth. Outside of the top 20 markets, there are station groups that own a large number of stations (e.g. group owner Nexstar owns or operates many stations through local **management service agreements** [MSAs]), because the markets are small the national reach is considerably less than the eight stations owned by ABC. In 2014, the FCC began to increase its scrutiny of SMAs to prevent their use as a loophole to circumvent FCC ownership rules.

FYI: AFFILIATION SWITCHING

When the Fox network helped New World Communications buy a group of television stations in 1994, it did so because it had a plan to get its own strong affiliates in major markets: The deal specified that all stations in the New World purchase would become Fox affiliates.

This touched off a mad scramble for network affiliations among stations in some major markets.

The stations in Phoenix, Arizona, underwent these interesting changes:

- KTSP (channel 10), the CBS affiliate, became KSAZ, a new Fox affiliate.
- KNXV (channel 15), the Fox affiliate, became the ABC affiliate.
- KTVK (channel 3), the ABC affiliate, became an independent station.
- KPHO (channel 5), an independent station, became the CBS affiliate.

Fox wanted to have a VHF affiliate in the Phoenix market and achieved this goal when KTSP was purchased and switched its affiliation from CBS to Fox.

The number of stations owned by large media companies changes rapidly as new deals are made to strengthen overall market position. Large media companies own groups of stations and are referred to as group owners. The biggest are called **supergroups**. Supergroups own many properties across a spectrum of media and related industries. Most non-network group owners have stations affiliated with several different networks because a diversity of network affiliations lowers the group's business risk. When one is down, another will probably be up.

Companies are buying local television stations for retransmission consent. Local stations receive money from cable companies to have their station's programming 'retransmitted' over cable. In some stations, retransmission consent brings significant revenue to the station without additional work or expense.

FYI: ADVANTAGES AND DISADVANTAGES OF CONSOLIDATION OF BROADCAST STATIONS

ADVANTAGES:

- More resources are available for program production and creation.
- There is a guaranteed distribution of programs to other stations in the same group.
- Programming can be repurposed easily among stations.
- Advertising packages can be sold for more than one station or for a combination of radio, television, or other media.
- Radio station groups can offer a wide variety of program formats.
- Economies of scale and efficiency help groups and their stations stay economically healthy.
- Career opportunities available to employees can be favorable. Employees of a group are often hired internally for similar or better jobs at other stations in the group. There are both horizontal (similar job at another station) and vertical (moving up to more responsibilities) job opportunities in a large group.
- A group of stations has more leverage than individual stations when negotiating with syndicators and cable companies.

DISADVANTAGES:

- Fewer voices are heard, and therefore there is less diversity of opinion and programming.
- Formulaic programming is often used, even though many groups have the resources to produce unique and innovative programming.
- Large groups can undercut the advertising prices of smaller groups and thus create financial hardship for competition.
- Career opportunities are reduced because the stations can function with fewer people running them; this downsizing also creates an overall reduction in the number of jobs available to media professionals.
- There is less local programming, especially in news.
- In some companies, there is anti-union sentiment, lower wages, and less job security.

Networks and Station Compensation and Retransmission Fees

The income for network-affiliated television stations traditionally has been comprised primarily of two revenue streams. One is selling advertising, both to local merchants or services and to national advertisers. The second has been the **station compensation** given to the local station in exchange for airing network programs and commercials. A third component, in practice since the 1990s, is **retransmission consent**.

The compensation model of the networks paying their affiliates for airtime is complicated, depending on which network, which market, and which stations, as well as the economic situation at the time.

Relationships between the networks and local affiliates continue to evolve. Local stations want to make more money from the networks through station compensation for airing network shows, and they want more advertising time available to them during primetime. Networks are reluctant to pay station compensation and now demand **reverse compensation** from affiliates who receive network programs.

For example, in 1994, after the Fox network had negotiated a contract with the National Football League to air NFC games, Fox faced having to pay its affiliated stations for clearing airtime to show the games. Fox's affiliates agreed to lower their compensation to help the network. The NFL contract helped establish Fox as a viable fourth network, and the affiliates benefitted from carrying the games because of the high viewership numbers. The higher ratings translated to higher advertising revenue. This network-affiliate arrangement opened the door to future shifts in this station–network relationship. Since that arrangement, network compensation to its affiliates has changed, with stations getting little or no compensation.

Another important revenue source for television stations is the retransmission consent payments made by cable companies to television stations for the privilege of carrying the station on their cable systems. Local cable providers are required by the FCC to include all over-the-air television channels on the cable company's basic programming line-up, the **must-carry rule**. This rule pre-dates the Web, when local television stations only worried about competition from other affiliates in distant markets. Must-carry helped small networks, like The CW or ION, to reach large audiences delivered by cable. As the number of cable channels increased, they began to demand payment from the cable and satellite companies (ultimately from the subscribers) for the exclusive programming created by that channel, such as ESPN.

This model led local television stations to demand similar payment from cable companies, known as *retransmission consent*. Broadcast stations argue their local content – particularly news and sports – is valuable to consumers. Threatening to remove their signal from local cable systems has helped broadcast groups successfully negotiate retransmission compensation. In 2008, retransmission fees accounted for less than 5% of the revenue for a Texas-based group owner's stations. In 2012, those fees had nearly tripled. Retransmission fees provide a third income stream to go with advertising and network compensation. Negotiating these fees can be contentious, resulting in station signal blackouts on local cable systems. In 2019, the retransmission fees that all U.S. television station owners collected from cable, direct broadcast satellite and telco video providers amounted to approximately $11.7 billion.

Retransmission fees provide needed revenue for station groups, but complicate the relationship between affiliates and networks. Networks claim they are due a share of the retransmission revenue because their prime-time programs keep local stations in business. Local stations contend that the unique character of local news, weather, and event coverage is what keeps viewers tuning in. In 2019, the so-called *reverse retransmission fees* sent from local stations and groups to the networks amounted to more than $3.5 billion. This was roughly 50% of the total sent to major networks (ABC, CBS, NBC, FOX, MyNetworkTV, The CW, Telemundo, Univision, and UniMás).

Multichannel Television Delivery

In 2007, the FCC extended the rule requiring multichannel delivery system owners (e.g. cable and satellite television) who own or provide programming for their own delivery systems to make their channels available to the rest of the multichannel industry at a reasonable price. Without this rule, a company with large interests in broadcast and cable television, could avoid making its programming available to the direct broadcast satellite (DBS) industry. Without access to a full array of programming, satellite services cannot compete with cable.

Cable Television

The Telecommunications Act of 1996 allowed telephone companies to compete with cable and offer video programming services. This brought upheaval and consolidation to the cable industry. The cable industry enjoyed several years of growth and development but has had difficulty adjusting to the new online realities. Consumers are moving away from cable to adopt any number of online streaming channels from whoever will provide the service.

In 1970, about 20% of U.S. households in the top five markets were served by a multiple-system operator (MSO), and in the top 50 markets that number was 63.5%. By 1999, almost 60% of households in the top five markets were served by MSOs, and almost 92% in the top 50 markets. Over the past 25 years, cable's market share of television subscribers has dropped significantly. In the mid-1990s, cable had about 95% of all television subscribers. Now, because of competition from satellite services and telcos, cable's market share is about 50%.

Satellite Television

Since the late 1990s, satellite television has been offered to audiences primarily from two companies, DirecTV and DishNetwork (DISH). After experiencing times of growth and success, currently, both DISH and DirecTV may go the same way as satellite radio when XM and Sirius merged to stay alive. By late 2019, falling subscriber numbers at both DISH and DirecTV led to speculation that merging is only a matter of time. In fact, DirecTV parent AT&T announced it would no longer promote DirecTV as the prime means of delivering video in the home. The future with DISH might allow them to at least survive by providing television to customers who can't get a streaming service due to poor Internet.

The inescapable trend is cord cutting and dish cutting. In 2018, DISH lost over one million subscribers, and another 500,000 in 2019. Even their Sling TV shed roughly 94,000 subscribers in the fourth quarter of 2019.

Streaming Video Services

Streaming provides consumers several inexpensive options for bringing television to the consumer. Some of the most popular brands include Netflix, Hulu, Sling, Amazon Fire, Apple TV, Roku, and Disney+. New services are being launched regularly, as this emerging market appears to be replacing cable for many consumers. While cable managed prices by bundling channels, video streaming services are capitalizing on inexpensive services with targeted program specialization. The key element is giving the user the desired video-on-demand (VOD), and there are at least four revenue-generation models driving video streaming services.

First, there is **Subscription Video-on-Demand (SVOD)**. This service resembles cable service, but is not as expensive. SVOD provides more targeted, consumer-specific content. Users subscribe to a monthly or yearly membership to access the service. Subscribers typically can watch as often or little as they like on any device. Examples of SVOD include Netflix, Amazon Prime, HBO Go, and Hulu. This type of streaming is probably the most common model to date.

A second approach is **Advertising-Supported Video-on-Demand (AVOD)**. Probably the most popular approach, the traditional broadcast model where users can freely see whatever is offered, but they have to watch commercial ads in return. YouTube has used this approach, where advertising is placed either before the video (pre-roll), within the program (mid-roll), or after the program ends (post-roll).

A third approach is **Transactional Video-on-Demand (TVOD)**. This approach is commonly known as pay per view, where the user pays only for what they watch. This model works well for live sports events, concerts, and other limited-time offerings. For example, look to the WWE streaming app.

A fourth approach is a **hybrid model** which may combine any of the first three. It is common to subscribe (SVOD) and still have to watch commercial ads before or after the program (AVOD). Services also try to combine approaches where, for example, for a higher monthly rate (SVOD), a user is provided added value in the form of free access to HD or free access to special programs (TVOD). YouTube Premium and, arguably, Amazon Prime offer elements of the hybrid approach.

As new entries are launched, they will fit into one of the above-mentioned categories. As video streaming competes for viewers and subscribers continue to migrate away from cable, the industry will watch closely to see which are most lucrative and most popular with consumers.

SEE IT LATER

The FCC Looks to the Future

The FCC oversees the overarching media and communication trends for the United States. The FCC implements the nation's broadband strategy, building the infrastructure to ensure there is sustained growth and investment. The FCC is working on the following ongoing initiatives:

- **5G**. As the United States rolls out the next generation of wireless connectivity, new issues will emerge. New networks and technologies will enable faster speeds and low-latency wireless broadband services, cultivating the Internet of Things and innovations not yet imagined. The FCC oversees the implementation of the Facilitate America's Superiority in 5G Technology Plan (the 5G FAST Plan). Three key components are (1) creating more space in the electronic spectrum for commercial use; (2) updating infrastructure policy; and (3) modernizing outdated regulations.
- **Digital divide**. Broadband Internet is critical to economic opportunity, job creation, education, and civic engagement. The FCC works to deliver high-speed Internet for all. Some places do not have access to important services available through the Internet, putting some Americans at a disadvantage in their employment, health, and well-being.
- **Accessible communications for everyone**. The FCC is looking at how to improve connectivity for disabled Americans, particularly the millions of Americans who are deaf, hard of hearing, speech disabled and deaf-blind. The FCC effort Accessible Communications for Everyone, seeks to break down telephone access barriers for these populations using innovative technologies of the 21st century.
- **Robocalls and spoofing**. Combatting unlawful robocalls and malicious caller ID spoofing is a top consumer protection priority. The FCC seeks better policies that allow stronger enforcement actions.
- **Telehealth**. The FCC is helping to expand how health services can be extended through broadband Internet. The FCC is looking into how consumers, particularly those living in rural areas, lack access to affordable broadband and might not be able to receive critical health services.

Now Media Universe

Radio

The radio industry continues to be dominated by a few large group owners. For a century, most of the focus was on radio's local voice in communities. Today, radio is guided by its financial bottom line. Stations that are profitable make few if any changes. Stations losing money are sold to others who implement whatever format and structure appear potentially successful. The key, as always, is the audience. If audience measurement reports indicate the audience is big enough for sales to generate enough income, the station will continue its operation. But storm clouds persist, and overall radio listenership inches downward. Online music services continue to creep, and younger demographics in particular show less loyalty to radio in an era of iPods, iPads, iPhones, smartphones, Spotify, and Shazam. Napster is ancient history to today's teens, but their attitude toward music is that it is ubiquitous, it is digital, and it is best when freely shared. Radio may survive as a music delivery platform, but future success may be tied to doing what radio has always done best – be the voice of the local community.

Television

Television continues to change, as does the meaning of the phrase 'watching TV.' We may be referring to broadcast, cable, online delivery, general video (digital or analog), a medium, a platform, a technology, a type of video recording, a particular program, a format, and a catch-all reply when asked 'What are you doing?' as you are lying on the couch reading.

Television has transitioned from analog to digital, and television stations now split their signal to broadcast programming over more than one channel. Millions of households watch local television signals provided by cable, satellite, and increasingly, the Internet. The migration away from cable has so far shifted more than 20% of cable subscribers from a few years ago to à la carte streaming channels. Netflix, Disney+, Hulu, and even YouTube are seen by consumers as competition and alternatives to 'television.'

Stations will continue to operate, and expect ownership changes to continue as economic instability and market forces bring change to media industries. Changes in the rules governing ownership caps and duopoly rules may help television stations remain competitive even though it may bring more consolidation. The fragmenting of television audiences across platforms presents a serious threat to stations without signature local news broadcasts or some other local program brand.

The flow of programming continues to change. The old system where programs are pitched to the networks for broadcast now considers running those programs on network streaming channels such as CBS All Access or NBC's Peacock. Although programs still progress from the networks to services like Hulu and then to syndication, other paths have also emerged.

The relationship between cable companies, broadcast groups, and networks will continue to change, reflecting the new models brought about by streaming services. The terms of retransmission consent deals and the network-affiliate relationship will also change according to the new market forces brought about by the Internet.

Several media groups are looking to use – if not already using – the Internet to directly reach the consumer. Disney+ gained 28 million subscribers in the first quarter of 2020. Of course, Disney has brand and synergies between its properties that most other media companies don't have.

Local broadcast television stations will survive and continue to change. As the networks all launch their own Internet streaming services online, the role of the local station will undoubtedly change. In several markets, stations will increase local news programming as a way to re-brand themselves and retain their visibility and influence in the community. The question is how many local news organizations can a market support in this changing environment? As noted elsewhere, there must be enough advertising support which is directly related to audience size. More viewers equal more advertising, and vice-versa.

As the national networks reach out to audiences through the Internet, local stations feel left by the wayside. In the traditional model, the networks supplied programs to local stations as a means of reaching and holding the local audience. The local station gave up the time (program and advertising) in order to be paid back later through the affiliate agreement. These relationships will evolve as the networks offer their programs directly through the Internet. For example, if a couple wants to watch NBC's *Saturday Night Live*, they can go to the NBC streaming channel PeacockTV.com, or watch the local NBC affiliate KPNX-TV channel 12. If they wish to watch parts of it later, they can visit various Web sites, such as NBC.com, or see what's uploaded onto YouTube.

The most important point is that viewers today have choice. 'Any screen any size' has empowered the consumer to watch any program on whatever device he or she wants. You can watch *Saturday Night Live* in your living room on your 47-inch smart television, on your 13-inch computer monitor in your bedroom, or your 5-inch smartphone if you're out with friends. You can also choose to watch something else and watch the show the next day.

The implications for advertising are profound. Advertisers must embrace all the platforms using all forms of digital advertising in order to reach their intended audience.

> ### FYI: DISNEY LAUNCHES DISNEY+
>
> Disney jumped into the online streaming video wars in late 2019 and became an immediate hit. After five months the channel had surpassed 50 million subscribers, reaching far beyond what the company had hoped. The original goal was 60 million subscribers in five years. Disney's success is widely seen as the future of home entertainment, taking on Netflix as well as other hopefuls like Peacock, Apple TV Plus and HBO Max, and unconventional short-form mobile services like Quibi. Disney+ streams shows and movies from Disney's franchises including Star Wars, Marvel, and Pixar titles, and all the family-friendly movies and animation from Disney itself, plus new originals and programming it acquired by taking over Fox, such as *The Simpsons*. At a rate of $7 a month ($70 if prepaid for a full year), Disney+ is well below competitors like HBO Now and HBO Max, and Netflix's cheapest tier, at $9 a month.
>
> In the wake of the coronoavirus pandemic, Disney+ took on added importance for families in self-isolation looking for video entertainment. Disney+ used the channel to release new big-screen films like *Artemis Fowl*, a sci-fi fantasy based on a popular series of young-adult books. Although originally scheduled to open in theaters, the pandemic made that impossible so the company turned it into a Disney+ original film, skipping theater release entirely. This is the power of horizontal and vertical integration.
>
> Source: Sorrentino and Solsman, 2020

> ### FYI: LEAST-LIKED COMPANIES: SUBSCRIPTION TELEVISION
>
> The University of Michigan's annual American Consumer Satisfaction Index polls Americans to rate household brands and services. In 2019, media industries once again were rated the lowest of 400 brands and services including gas stations, grocery stores, airlines, and even the U.S. Postal Service. In particular, Subscription television services and ISPs both tied at the bottom at a score of 62. Consumers cited high prices, unreliable service, and poor customer service leading to low ratings.
>
> Among the *Subscription television companies*, AT&T's U-verse TV (69) scored the highest, followed by Verizon's Fios (68), DISH Network (67), and DirecTV (66). Far below was Altice's Optimum (61), Charter's Spectrum (59), and Cox Communications (59), followed by Frontier Communications, Comcast's Xfinity (57), Mediacom (56) and Altice's Suddenlink (55).
>
> The category of *Internet Service Providers* also appeared at the bottom of the ACSI rankings and was widely criticized as slow and unreliable, providing poor customer service, and too expensive. Rated highest was Verizon's Fios (70) followed by AT&T Internet (69), Altice's Optimum (63), Comcast's Xfinity (61), Cox Communications (60), Altice's Suddenlink (60), Charter's Spectrum (59) and CenturyLink (59). The lowest-scoring ISPs were Mediacom (56), Windstream, (57) and Frontier Communications (55). As in the subscription television industry, ISP firms scoring lowest here continue to be some of the worst performers among the 400+ companies in the ACSI.
>
> Source: Johnston, 2019

Cable Television

The cable industry is struggling to find a strategy to stay competitive in the online jungle where only the fittest survive. Streaming television is easy and relatively inexpensive, and cable continues to look for a workable business model.

For years, cable companies were some of the most mistrusted and maligned companies in the country. That legacy explains why many young people have no problem with cutting the cord. Digital natives expect to consume 'now media' using whatever platform suits them in that moment, be it television, computer, game console, table, or smartphone. Price is more important than loyalty. Older viewers may be willing to stay with cable because they enjoy the older programs in syndication.

The trend of going over the top to get programs to viewers has attracted some of the biggest technology companies like Apple (Apple TV), Google (Chromecast), and Amazon (FireTV). As such, cable's most attractive service may be Internet connectivity rather than bundled video tiers.

Cable companies hope to maintain subscriber numbers, and it remains to be seen if 'à la carte' can work. Consumers dislike paying for channels they never watch and seem increasingly willing to pay less for fewer channels as long as they stay connected to the Internet.

Meanwhile, both DISH and DirecTV will need to consider whether they will go the way of satellite radio in the wake of decreasing audience subscribership.

Satellite Radio

SiriusXM announced increased earnings in early 2020, but the outlook is complex. At the end of March, 2020, it reported more than 34.8 million subscribers, including more than 30 million self-pay users. But the coronavirus pandemic brought the economy to a halt and the company reported it lost 143,000 net subscribers in the first quarter of 2020. The company pointed to drops in auto sales, advertising, and responses to marketing campaigns during that time.

Most satellite radio listening takes place in the automobile, and the company is positioning itself for a more competitive environment. New automobiles are being equipped with a full range of Internet-connected entertainment functions, giving drivers and passengers more options. SiriusXM purchased music service Pandora in 2018, and the music-streaming service announced it had added 51,000 net self-pay subscribers by the end of March, 2020. Pandora announced it had reached 6.3 million subscribers. Ad revenue at Pandora brought in an additional $241 million in the first quarter of 2020, up 4% from the previous year.

Telcos

Telephone companies have all jumped into the fray to be full-service providers like cable companies. Major players such as Sprint/T-Mobile, AT&T, and Verizon all offer what they bill as free video packages to go with their broadband services. Video streaming channels are thrown in to sweeten the offer to provide the old cable 'triple play' of broadband Internet, wireless (telephone), and video services. Each telco offers a variety of plans for providing multichannel streaming subscriptions for Netflix, Hulu, Amazon Prime, HBO, and/or Disney+. Like cable, additional streaming services can be included for a price. Video streaming is offered on top of standard broadband Internet and wireless data plans, which vary with each company. The bottom line is still for consumers to be careful and compare packages and prices for the one that fits best for their household.

The Internet and Wireless

The Internet continues to grow and extend itself into every aspect of our lives. As the network is upgraded from 4G to 5G, so also the applications will expand. Expect the Internet to continue being an integrated provider of sound, visuals, information, entertainment, home security, and more. In terms of where people get their television, expect streaming to continue to grow. In February 2020, Nielsen reported that people are spending more of their television time with streamed content. In 2018, households reported 10% of television time was streamed and now that has grown to almost 20% of total television time. The report also identified that 60% of Americans subscribe to more than one paid video streaming service. The most popular services were Netflix (31%), followed by YouTube (21%), Hulu (12%), Amazon (8%), and several others filling out the remaining 28%. Since that time Disney+ launched and outperformed by picking up 28.6 million subscribers in less than three months.

SUMMARY

The businesses that comprise 'now media' are in a fast-paced and highly competitive environment. They employ new technologies as they are made available to the public, and use them in sometimes novel and exciting ways. Across the spectrum of businesses involved in electronic media, they all look for ways to generate enough revenue to stay profitable.

One hundred years ago, broadcasters experimented with various business models and employed various strategies to generate revenue. The new technology of AM radio was used in ways that galvanized communities and changed our views about communication forever. Broadcasting flourished and policymakers were guided by the scarcity principle because there were limited frequencies for radio (and then television) stations. Today that world has been turned upside down. 'Now media' has leapfrogged over broadcasting, cable, and telephony, integrating their strengths while giving the consumer almost unlimited choice.

The federal government encourages diversity and consumers can express their freedom of choice through the wealth of options available in the current 'now media' environment. The days of a few broadcast voices in a community are long gone, as are the days of one cable television provider offering bundled channels despite the objections of consumers. The convergence of television, computers, telephony, and broadband has created the 'now media,' an expanded marketplace of ideas, and all forms of electronic media will evolve and find their function in modern society.

BIBLIOGRAPHY

ACSI. (2019, May 21). The ACSI Telecommunications Report 2018–2019. *Acsi.org.* Retrieved from: www.theacsi.org/news-and-resources/related-research/journal-articles

Albarran, A. (2010). *Management of electronic media* (4th ed.). Boston, MA: Wadsworth.

Albiniak, P. (2009, April 19). NAB 2009: Top 25 station groups. *Broadcasting & Cable.* Retrieved from: www.broadcastingcable.com/article/209430-NAB_2009_Top_25_Station_Groups.php

Aswad, J. (2020, April 28). SiriusXM revenues and earnings rise, despite plunging auto sales and advertising. *Variety.com.* Retrieved from: https://variety.com/2020/digital/news/siriusxm-revenues-earnings-rise-plunging-auto-advertising-1234591665/

Baig, E. (2015, January 11). *Cord-cutting TV fans can rejoice.* New York: USA Today.

Bannerman, N. (2020, March 13). FCC completes large scale 5G spectrum auction. *Capacitymedia.com.* Retrieved from: www.capacitymedia.com/articles/3825104/fcc-completes-large-scale-5g-spectrum-auction

Berman, M., & Hayes, K. (2019, March 21). NATPE 2019: A sudden surge in new first-run syndicated series is forefront. *American Target Network.* Retrieved from: http://americantargetnetwork.com/natpe-2019-a-sudden-surge-in-new-first-run-syndicated/

Bouma, L. (2020, March 20). AT&T is ending marketing of DIRECTV & U-Verse TV. *Cordcuttersnews.com.* Retrieved from: www.cordcuttersnews.com/att-is-ending-marketing-of-directv-u-verse-tv/

Brodkin, J. (2015, September 4). Verizon sale of Fios and DSL network in three states clears FCC hurdle. *ARS Technica.* Retrieved from: http://arstechnica.com/business/2015/09/verizon-gets-approval-to-sell-part-of-fios-and-dsl-territory-to-frontier/

Cole, G. (2019, July 24). Massachusetts close to enacting economic nexus, marketplace facilitator law. *Avalara.com.* Retrieved from: www.avalara.com/us/en/blog/2019/07/massachusetts-to-require-remote-sellers-and-marketplaces-to-collect-sales-tax.html

Day, L. A. (2006). *Ethics in media communications* (5th ed). Boston, MA: Wadsworth.

Eile, E. (2014, August 6). Gannett to spin off publishing business. *USA Today.* Retrieved from: www.usatoday.com/story/money/business/2014/08/05/gannett-carscom-deal/13611915/

Ernst, M. (2003). Cybersmear: Supreme Court 'chats' over libelous internet messages … trust the leaders. Issue 3, Spring 2003. Retrieved from: www.sgrlaw.com/ttl-articles/861/

FCC's review of the broadcast ownership rules. (2020). *Federal Communications Commission.* Retrieved from: www.fcc.org

Friedlander, J., & Bass, M. (2020, February). Year-end 2019 RIAA music revenues report. *RIAA.com.* Retrieved from: www.riaa.com/wp-content/uploads/2020/02/RIAA-2019-Year-End-Music-Industry-Revenue-Report.pdf

Fu, H., Mou, Y., & Atkin, D. (2011). The impact of the Telecommunications Act of 1996 in the broadband age. In A. Stavros (Ed.), *Advances in Communications*, Volume 8, pp. 117–136. © 2011 Nova Science Publishers, Inc. Retrieved from: www.researchgate.net/publication/281389845_The_Impact_of_the_Telecommunications_Act_of_1996_in_the_Broadband_Age

Future is bright for powerline broadband [press release]. (2004, July 14). *FCC.* Retrieved from: www.fcc.gov [July 29, 2004]

Gartenberg, C. (2020, February 19). Dish Network floats merger with DirecTV over pace of cord-cutting. *TheVerge.com.* Retrieved from: www.theverge.com/2020/2/19/21144345/dish-network-merger-directv-streaming-services-subscribers-q4-2019

Goldman, D. (2010, February 2). Music's lost decade: Sales cut in half. *CNN.money.com.* Retrieved from: www.money.cnn.com/2010/02/02/news/companies/napster_music_industry/

Greppi, M. (2001, November 19). NBC, Young and Granite roll the dice. *Electronic Media.* Retrieved from: www.craini2i.com/em/archive.mv?count53&story5em178785427096220069 [February 12, 2004]

Hess, A., & McIntyre, D. (2015, January 17). 10 Most hated companies in America. *24/7 Wall St.* Retrieved from: http://247wallst.com/special-report/2015/01/14/americas-most-hated-companies/#ixzz3P7NDElCY

History of cable. (2016). *California Cable & Telecommunications Association.* Retrieved from: www.calcable.org/learn/history-of-cable/

Inside Radio. (n.d.). Who owns what. *Insideradio.com.* Retrieved from: www.insideradio.com/resources/who_owns_what/

Jessell, H. (2004, January 5). Dotcom redux. *Broadcasting & Cable.* Retrieved from: www.broadcastingcable.com/archives

Johnston, K. (2019, May 21). Netflix and video streaming widen lead over subscription TV in customer satisfaction, according to the ACSI. *ACSI.org.* Retrieved from: www.theacsi.org/news-and-resources/press-releases/press-2019/press-release-telecommunications-2018–2019

Knee, A., Greenwald, B., & Seave, A. (2009). *The Curse of the Mogul: What's wrong with the world's leading media companies.* London: Portfolio.

Kurtz, P. (2019, August 27). Number of U.S. TV homes grows to 120.6 million, says Nielsen. *TVTechology.com.* Retrieved from: www.tvtechnology.com/news/number-of-u-s-tv-homes-grows-to-120-6-million-says-nielsen

Lind, R., & Medoff, N. (1999). Radio stations and the World Wide Web. *Journal of Radio Studies*, 6(2), 203–221.

Nakum, V. (2019, June 14). 7 online video streaming businesses having awesome revenue model. *Trootech.com*. Retrieved from: www.trootech.com/video-streaming-businesses-and-how-they-work/

National Cable Television Association. (2020, March). Retrieved from: www.ncta.com/industry-data

NBC v. *United States*. (1943). 319 U.S. 190.

Neiger, C. (2014, October 22). The future of cable: Bleak in the face of more cord cutters. *The Motley Fool*. Retrieved from: www.fool.com/investing/general/2014/10/22/the-future-of-cable-bleak-in-the-face-of-more-cord.aspx

Oxenford, D. (2017, February 24). FCC approves for the first time 100% foreign ownership of US broadcast stations. *Broadcastlawblog.com*. Retrieved from: www.broadcastlawblog.com/2017/02/articles/fcc-approves-for-the-first-time-100-foreign-ownership-of-us-broadcast-stations/

Parsons, P., & Frieden, R. (1998). *The cable and satellite television industries*. Boston, MA: Allyn & Bacon.

Performance royalties: The state of play. (2014, July 30). *Intellectual Property*. Retrieved from: www.commlawblog.com/2014/07/articles/intellectual-property/performance-royalties-the-state-of-play/

Retransmission fee race poses questions for TV viewers. (2013, August 2). *USA: USA Today*. Retrieved from: www.usatoday.com/story/money/business/2013/07/14/tv

Schechner, S., & Dana, R. (2009). Local TV stations face a fuzzy future. *Wall Street Journal*. p. A1.

Seward, Z. (2014, January 21). The race is on to launch an internet TV service in the US. *Over the Top*. Retrieved from: http://qz.com/169249/the-race-is-on-to-launch-an-internet-tv-service-in-the-us/

Sorrentino, M., & Solsman, J. E. (2020, April 17). Disney Plus: Everything to know about Disney's service during coronavirus. *Cnet.com*. Retrieved from: www.cnet.com/news/disney-plus-everything-to-know-coronavirus-artemis-fowl-onward-pixar/

Spangler, T. (2020, February 19). Traditional pay-TV operators lost record 6 million subscribers in 2019 as cord-cutting picks up speed. *Variety.com*. Retrieved from: https://variety.com/2020/biz/news/cable-satellite-tv-2019-cord-cutting-6-million-1203507695/

Sterling, C., & Kittross, J. (2002, February 10). *Stay tuned: A history of American broadcasting*. Mahwah, NJ: Erlbaum.

Stiles, G. (2016, July 25). Meet the cable-cutters. *Mail Tribune*. Retrieved from: www.mailtribune.com/news/20160725/meet-cablecutters

Swaine, J. (2018, December 12). National Enquirer owner admits to 'catch and kill' payment to ex-Playmate. *TheGuardian.com*. Retrieved from: www.theguardian.com/us-news/2018/dec/12/national-enquirer-trump-payments-david-pecker-catch-and-kill

Television isn't really dying – but there is a war over it. (2014, July 2). *Business Insider*. Retrieved from: www.businessinsider.com/what-people-get-wrong-about-the-death-of-tv-2014-7

Trick, R. (2003, May). Powell: Newspapers to 'fare well' in cross-ownership decision. *Presstime*, p. 8.

ViacomCBS. (2020). Making connections around the world. Retrieved from: www.viacbs.com/brands

Weinschenk, C. (2019, July 30). Retransmission fee forecast calls for another steep rise: $11.72 billion in 2019. *Telecompetitor.com*. Retrieved from: www.telecompetitor.com/retransmission-fee-forecast-calls-for-another-steep-rise-11-72-billion-in-2019/

Wharton, D. (2019, November 14). 200 house members now support Local Radio Freedom Act. *NAB.org*. Retrieved from: www.nab.org/documents/newsRoom/pressRelease.asp?id=5201

Wiquist, W. (2020, April 2). FCC proposes $6m fine against TracFone in lifeline case. *FCC.gov*. Retrieved from: www.fcc.gov/document/fcc-proposes-6m-fine-against-tracfone-lifeline-case-0

Chapter twelve

Working Behind the Scenes in Media

The authors thank Glenn T. Hubbard, Ph.D. (East Carolina University), for his contributions to this chapter.

In the 1950s, television won the public's hearts and eyes, much to the chagrin of the radio industry. But through station formatting and focusing on music, radio found a way to live side by side with television. As new types of music have hit the airwaves, formatting options have expanded, which has led to more competition in many radio markets. Digital technology made radio production more efficient and, in some cases, improved sound quality, but otherwise radio station operations have remained basically the same.

The same can be said for television. Though technology has improved picture and sound quality, new types of programs have been created, and old types circle in and out of favor, television stations still operate similarly to how they have for decades. Major improvements came about as improved technology led to new production techniques and digital delivery became possible.

Internet delivery of broadcast radio and television fare further complicates the role of traditional media and how they operate, especially as what were once strictly considered radio or television companies have merged with Internet, film, newspaper, and other types of media corporations.

This chapter looks at how the operation, production, and distribution of electronic media programming has changed through the years, especially with the advent of cable, satellite, and Internet as program delivery systems.

SEE IT THEN AND SEE IT NOW

Radio Stations

From radio's origin in the early 1920s, six basic functions have kept broadcast stations in business: General management, engineering, production, programming, sales, and promotions. These operations and business functions were carried out in much the same way in all stations. General management oversaw all business needs including personnel. Engineering was in charge of production equipment such as recording devices, microphones, cameras, lighting, and the studio, as well as transmitters and other equipment necessary to broadcast over the air. Programming executives decided what shows and music should air. Production created and recorded the shows, as well as commercials in many cases. Sales generated revenue by filling airtime with commercials. And promotions got the word out about the station.

From the early 1950s until the mid-1990s, the radio audience grew along with revenues. The number of stations increased dramatically, and most stations realized enough of a profit to stay in business. Although a good part of the audience had shifted from AM to FM, most stations had a management structure that stood the test of time and worked well in both the small, independently owned stations and the stations that belonged to corporate groups.

The Telecommunications Act of 1996 kick-started ownership consolidation and convergence in both radio and television. Radio stations changed their organizational charts and operations, and consolidating station jobs among group personnel became common.

Before 1996, a radio station often had a full array of managers and staff in each of its departments. But since 1996, some of the managers working for radio station groups perform managerial tasks for more than one station. This allocation of duties was especially common after an acquisition. For example, suppose an owner of a group of stations like iHeartMedia acquires four stations in a market. Before the acquisition, each of those four stations might have had its own program director. After the acquisition, one *regional* or *market program director* would make the programming decisions for all four stations, or one sales director and one assistant director would manage sales for all four stations. It is also common practice for salespeople to sell for more than one station and possibly for all stations in a group; thus fewer sales reps are needed after consolidation. Although selling for more than one station presents some conflicts, such as competition among stations within the same group, these problems are small when compared to the value of reducing expenses by decreasing personnel.

In some cases, station ownership groups save money by combining the studios of several stations into one facility. Operating one large facility with several studios and on-air control rooms is often less expensive than operating several fully equipped studios at different locations. Consolidation also reduces overhead costs such as bookkeeping, billing, subscriptions, professional membership fees, and insurance.

General Management

The general management of a radio station, sometimes one **general manager** in smaller stations but often a management team, is responsible for policy planning, hiring and firing, payroll and accounting, purchasing, contract administration and fulfillment, and the maintenance of offices, studios, and workplaces. In addition, general management provides leadership, promotes communication among station departments, and makes all final decisions that affect the station and its employees. In many cases, the general manager supervises more than one station; often the same general manager is in charge of the AM and FM operations or clusters of stations owned by the same company.

Usually working under the general manager, the **station manager** typically oversees the operations of one station. Overall, the station manager has four important roles: Make the station function efficiently, hire and fire station personnel, provide the maximum return to the owners or shareholders, and adequately serve the commitments of the station's license. The station manager must be familiar with media labor unions, if there are any in the given market, and must be knowledgeable about legal issues that may affect the station. Knowledge of FCC rules and regulations, as well as knowledge of local, state, and federal laws that impact the station and its operation, is crucial.

Engineering

Since the beginning of broadcasting, the technical functions of a station included the construction, maintenance, and supervision of a station's broadcast equipment. The chief engineer oversees the engineering department of the station, installs and maintains the transmitter and antenna and other broadcast equipment, manages the archiving of programs, supports operations, and is the liaison with the FCC on technical issues, and in some cases deals with labor unions representing the engineering staff.

The engineering department, which often includes information technology (IT) is responsible for the operation, care, and installation of technical broadcast equipment. In some stations, this department is now split into an engineering department, which maintains and installs the transmitter and all technical equipment, and an operations department, which runs the equipment and makes sure that there is appropriate workflow from one area of the station to another. A station's **workflow** is the movement and completion of tasks within and between departments, like making sure that a new commercial spot flows from the copywriting desk, to the production studio, and on to the airwaves by deadline.

The chief engineer hires personnel to install and maintain the equipment and also makes recommendations to the station manager or general manager about new technologies and equipment needed by the station. Knowledge about computers and computer networks has become much more important because all broadcast stations now rely heavily on computer technology for daily operation.

Production

Typically, a basic early radio production system consists of microphones to pick up sounds or phonograph records and magnetic recordings of prerecorded shows. The audio signal then travels to an audio console or audio board, which contain a series of volume controls and switches to determine what signals are sent to the transmitter and over the airwaves. The radio station transmitter converts the audio signal into fluctuations of a beam of electromagnetic radiation, which is then sent to the transmitting antenna. The combined signal is broadcast from the transmitting antenna in the form of invisible electromagnetic waves vibrating (or modulating) at a speed (frequency) assigned to the station.

CAREER TRACKS: DAVE ZORN, NEWS/SPORTS DIRECTOR, GREAT CIRCLE MEDIA

WHAT IS YOUR JOB? WHAT DO YOU DO?

I am currently the news/sports director for Great Circle Media in Flagstaff, Arizona. I do daily newscasts for our stations: 92.9 KAFF Country, 93.9 The Mountain, Hits 106, Arizona Shine and KAFF Country 93.5/AM 930 in Flagstaff, as well as Magic 99.1 and Arizona Shine in Prescott. I also do play-by-play for high school football and basketball in Flagstaff and I'm the radio halftime host on the NAU Lumberjacks Radio Network.

WHAT WAS YOUR FIRST JOB IN ELECTRONIC MEDIA?

My first job was about six months out of high school at 1270 AM KDJI in Holbrook, Arizona 2020 is my 30th year in radio and it's been a great ride so far.

WHAT LED YOU TO THE JOB?

I was working at Burger King in Holbrook and one of my co-workers, who was working at the radio station, asked me to go with him to call a game in Round Valley, so I did. The next day the owner-GM called me and offered me a job. I said yes and started learning the art of radio in a small little studio off I-40. I learned everything from newscasts, production, remotes, etc. I was very lucky to have someone throw me into the fire like that. My dad was in radio for years and always tried to keep me out of it. But when he saw me excited, he knew he couldn't fight it and started being my mentor.

Figure 12.1 Dave Zorn.

WHAT ADVICE WOULD YOU HAVE FOR STUDENTS WHO MIGHT WANT A JOB LIKE YOURS?

First: Two Ps: Preparation and Patience. Always be ready and be prepared, be overly prepared. The second you're not prepared, you will get stuck, not only in this business, but in life. As for patience, in most cases this industry doesn't go as fast as you want it. Sometimes you need to stay at a job, not only to gain experience but gain what I like to call 'resumé years.' If people see you at a job for a few years, you may get the job over someone else. Also, learn everything, don't fall into just one category. The more well-rounded you are when you hit the market, the more attractive you will be to companies looking to possibly hire you. Finally: STAY HUMBLE. No one likes a braggart, or a 'look-at-me' person. It's a career-killer. Be confident, but always be open to criticism. You will get some. Learn from it.

Programming

The programming functions of early radio included securing programming from a network and creating enough programming to fill the rest of the broadcast transmission schedule. The programming department, led by the program director, was responsible for planning ahead and implementing (or executing) the programming plan.

In addition to network programs, local radio stations produced their own shows. Sometimes musicians and singers would play live from the station's studio. Talk shows, news shows, and live readings were also broadcast from the station. Later, as broadcast technology improved, a station could air music from a remote location, such as a concert hall. Live music was sent via telephone wires to the radio station studio, and from there to the station's transmitter, which broadcast the signal over the airwaves. Later, live remotes covered sports matches and other events. These 'live remotes' were often sponsored by the concert hall, hotel, ballroom, arena, or wherever the action was taking place. Once all the microphones and telephone lines were in place, the station engineer merely had to flip a switch to get the remote on the air. Some stations relied heavily on live remotes. For example, in the 1920s, station WBT in Charlotte, North Carolina, played live concert remotes from the Charlotte Hotel every day from 12:30 p.m. to 2:00 p.m.

Musicians were often on a station's payroll. It was not unusual for some radio stations to have ten or more musicians on the full-time staff. The studio musicians performed live numbers, announcers' background music, or even a sponsor's jingle. The musicians performed in the station's sound studio, which was equipped with microphones that were connected to an audio console that controlled the sound level and combined or mixed the sound from the various microphones in the studio.

Radio relied on live programs. If a performer suddenly got sick, or otherwise did not get to the station by airtime, no recording was available to fill time until a substitute performer could be found. When these kinds of problems came up, the station often relied on the studio musicians to perform on short notice.

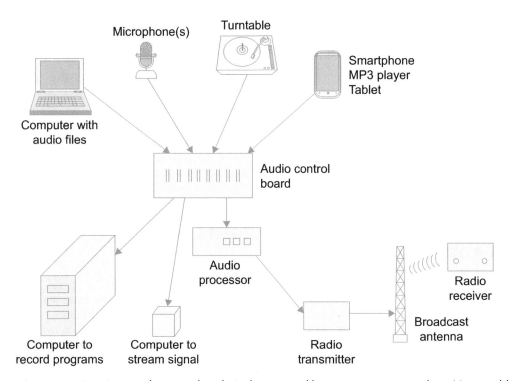

Diagram 12.1 How radio is produced. Audio sources like a computer, microphone(s), turntable, and audio player (smartphone, MP3 player, or tablet) supply signals to the audio control board. The control board or audio console mixes and controls the audio signals and controls the level of each signal. Then the audio control board (or console) combines the signal(s) and sends it out to an audio processor or a recording device, which then sends the signal on to a streaming server and the radio transmitter. The transmitter creates a radio wave on a specific frequency (e.g. 107.1 mHz) and sends the signal to an antenna. The signal is then broadcast out from the antenna to radio receivers in homes and cars.

Beginning in the 1920s, local station audio programs were sometimes recorded on phonograph records, referred to as **electrical transcriptions**, also known as ETs, using the 78 RPM (revolutions per minute) speed for music or the slower 33⅓ RPM speed for voice. At either speed, the sound quality was crackly and poor. By the late 1930s, recording equipment used steel tape or wire, and though the sound was better, it still was not very clear.

Program Director In a modern station of today, the **program director (PD)** is responsible for everything that goes out over the air. He or she usually works closely with a **music director** if the station has a music format, and a **news director** if the station produces its own **newscasts**. The PD often supervises a **director of production**, who in turn oversees the creation of commercials, promotions, and other prerecorded messages.

The PD has always been responsible for music selection and scheduling. In addition, in many stations, the news operation is part of the programming department. Stations with a large news component, such as those that are news–talk or sports, often have a separate news department.

The news director is responsible for writing, scheduling, and delivering newscasts throughout the day. If the station has a syndication service or a network that supplies some or all of the news, then the news director is in charge of the contractual agreements with those businesses. In addition, the news director is responsible for the entire staff, which may include the following:

- **News producers**, who put together news stories and news programs.
- **Reporters**, who go out of the studio to gather news stories.

ZOOM IN 12.1

BING CROSBY AND AUDIO RECORDING

During the Golden Age of Radio in the 1930s and 1940s, radio networks wanted only the best-quality shows on the air and did not allow their affiliate stations to air recorded programing. Big stars, like crooner Bing Crosby, often had to perform their live shows twice: Once for the East Coast and once for the West Coast.

Until the late 1940s, the radio networks did not have much interest in recording methods because, at the time, the quality of recorded programs was noticeably worse than live programs. Network programs produced in New York or elsewhere on the East Coast were delivered to local stations across the country by way of high-bandwidth telephone lines with better sound quality than regular phone lines, but a two- or three-hour time difference made it impossible for programs to entertain evening audiences in the Mountain and Pacific zones if aired live at the same time they were broadcast for evening audiences in the East. They would have aired before people were home from work on the West Coast. Therefore, famous entertainers like Bing Crosby, who worked for NBC, had to perform each show twice – once for a network feed to stations in the Eastern and Central time zones, and again for the Mountain and Pacific time zones. Crosby disliked such repetition and tried to figure out a work-around. While performing in Europe during World War II, he noticed that many performances were recorded on audiotape, and felt that high-quality audio recordings could simplify the lives of performers and the networks in the U.S.

Crosby started the Crosby Research Foundation to seek patents on equipment that would improve audio recording. He left NBC, where there was no interest in recorded shows, and started working for ABC, which was accepting both of Crosby himself as a star performer but also of his ideas about audio recording. Crosby invested money in the Ampex Corporation that created the first commercial reel-to-reel audio recorder for the U.S. market. Crosby gave one of the recorders to a musician friend, Les Paul, a famous guitar player, who pioneered multitrack audio recording and is often credited as being the 'father' of the electric guitar.

By the early 1950s, plastic-based magnetic tape was being used to produce audio recordings that were practical, affordable, and of high quality, and Crosby became the first national performer to prerecord his shows onto audio magnetic tape. A taped live show would be fed electronically over phone lines to stations in different time zones, with the high quality of the sound intact, thus eliminating the need for a second performance. Bing Crosby produced and recorded his radio programs using the same procedures (rehearsing, doing a 'take' with high-quality recording, and editing) used in motion picture production. Crosby's style of producing radio became the industry standard.

Figure 12.2 Bing Crosby ad.

- **Newscasters (anchors)**, who deliver the news to the audience in front of the camera or microphone.
- **Specialized news personnel**, like play-by-play and color announcers for sports events, and traffic reporters.
- **Weather specialists and meteorologists**, who deliver the current weather and forecasts.
- **Special reporters**, who work as business analysts, environmental reporters, and other special topic journalists.

Sales

Since 1922 when WEAF in New York broadcast the first 'toll broadcasting' message, commercials have provided broadcast stations' most significant revenue stream. Businesses paid broadcasters for the privilege of reaching consumers with messages about products and services and commercial spots became part of the basic structure of commercial radio.

Program sponsorship was the standard practice for advertising on radio in the 1930s and 1940s. Advertisers enjoyed the local notoriety of being associated with a program and its host. Sponsorships were sold for a variety of programs and program lengths. Some programs were only 15 minutes in length, while others, such as baseball games, would last for several hours. The title of the show may have included the sponsor's name, such as *The Eveready Hour* (sponsored by Eveready batteries), or a musical show would feature a band named after the product, such as the Lucky Strike Orchestra.

Spot advertising was also sold, but it was not very popular until the 1950s because it was believed that the direct association of the sponsor with a show was more effective than buying one or more of many spots included in or between shows. Since the 1950s, spot advertising has dominated the local radio sales activities as sales has become an increasingly sophisticated endeavor.

As the principal contact with advertisers, **account representatives** must be good communicators and knowledgeable about local businesses. In addition, a salesperson must understand marketing strategies and messaging, and know how to interpret business clients' needs and translate them into commercial campaigns. Account representatives sell the radio audience (airtime) to the advertisers. Sales goals are typically established for a given month based on last year's sales plus a certain percentage increase. Therefore, sales requires a drive and an almost constant effort to make more sales and bring in more money, and hence a larger commission. As stated in *Broadcast Management* by Quaal and Brown (1976, p. 245), sales people should have

> *an extroverted personality, a high degree of intelligence, a gregarious nature, signs of perseverance, an ability to get along with people, a good appearance, correct manners, sincerity, a true interest in sales, and, frankly, a strong desire to make money.*

In most large markets, the salespeople do all of the selling but one or more employees also help with research and ratings, sometimes even conducting audience surveys to inform the sales effort.

In smaller markets, announcers and other station personnel often sell commercials in addition to their regular station responsibilities. In small markets, the on-air personalities often become household names and are typically welcomed to visit businesses in the market, even though the visits are sales calls. Small market sales are more relationship driven than a hard-core 'by the numbers' decision.

Sales representatives need to know about the business they are calling on – its seasonal fluctuations, products, image, and general business needs. Based on what they learn about the business, sales representatives then match the business's needs to the station, present a sales package consisting of the number of spots, air times, and costs, and try to convince the business to advertise on the station. If successful, the salesperson might write and produce the spots, schedule them at the best times to reach the advertiser's customers, and sometimes even has to collect the bills.

Account reps typically answer to a **sales manager**, usually an experienced sales person who also is very familiar with radio broadcasting and marketing. The sales manager hires and trains the sales staff and sets sales goals. In larger stations, the sales manager's position is often split into two positions: **Local sales manager** and **national** (or **regional**) **sales manager** to handle national sales (e.g. Coca-Cola) and local sales (e.g. Ralph's Barber Shop).

In some stations, the sales department also schedules commercials. An employee with the somewhat confusing title of **traffic manager** is responsible for scheduling commercials and other program elements. The term 'traffic' in this context has nothing to do with reporting on the flow of cars during rush hour, but with the flow of commercials spots on the air.

Promotions

Before the introduction of television, there was less need for radio promotion than there is now. Radio was unique and drew an audience as soon as radio receivers became available. It was the only electronic medium before World War II, and there were few stations on the air. The audience did not have television, cable, satellite, or the Internet competing for its attention. Nonetheless, stations did promote themselves to stand out from competing stations, but to nowhere near the degree that they do now.

Radio promotions are usually short messages that tell of upcoming programs or otherwise try to get listeners to stay tuned. On-air promotion includes simple promotional announcements such as the station call letters and slogans ('KRCK radio: RoCK Radio at its best!'), and sometimes preproduced spots about upcoming programs, giveaways, contests, and special events initiated by the station. Promotions are very effective if they are aired at the same time or within the same show each day. Because air time is expensive, promotions usually run during unsold commercial time. But station managers realized the benefits of regularly scheduled self-promotion and started setting aside time for promotions when they were most likely to be heard.

The promotions department also creates promotional spots to run on television, in newspapers, on billboards, Web sites, or some other medium, a

CAREER TRACKS: RYAN KLOBERDANZ, NATIONAL IMAGING DIRECTOR, NEWS/TALK RADIO, SALEM COMMUNICATIONS

WHAT IS YOUR JOB? WHAT DO YOU DO?

I work with all of Salem's news/talk stations and help them develop and produce their on-air image.

HOW LONG HAVE YOU BEEN DOING THIS JOB?

Thirteen years.

WHAT WAS YOUR FIRST JOB IN ELECTRONIC MEDIA?

Nights/weekend board operator for a small station.

WHAT LED YOU TO THIS JOB?

Figure 12.3 Ryan Kloberdanz.

I always loved radio and was approached by a man that worked at the local radio station in my hometown who asked me if I wanted a part-time job. I took that part-time job, and here I am.

WHAT ADVICE WOULD YOU HAVE FOR A STUDENT WHO MIGHT WANT A JOB LIKE YOURS?

Listen. Students have the ability to listen to stations all around the country and the world via the Internet. Listen to the production on the stations, learn different styles, and develop your own style based on what you have learned from the styles of others.

practice known as **cross-promotion** or external promotion. Cross-promotions also include sponsoring community events such as 5k races, or local art fairs, and even includes imprinting the station's call letters on pens, coffee mugs, and t-shirts. Cross-promotion is done primarily to reach non-listeners who are watching television, or reading a paper, driving down the interstate, or running a 5k, and convince them to tune in.

Internal and external promotions are designed with different goals in mind:

- **Audience acquisition (external):** To entice people who do not listen to sample the station.
- **Audience maintenance (external):** To urge listeners to continue to listen to the station.
- **Audience recycling (external):** To give listeners a reason to listen to the station on various days and at different times of the day (vertical recycling persuades the audience to return later in the same day, and horizontal recycling persuades the audience to listen during the same time period each day of the week).
- **Sales promotion (external):** Promotions aimed at businesses (e.g. discount plans) so they will buy advertising on the station.
- **Morale building (internal):** Stations set up in-station promotions (e.g. sales contests) to generate and maintain the sales staff's energy and self-motivation.

Promotions are expected to generate ratings, revenue, and goodwill (enhance the image of the station). These goals have to be accomplished while making sure that the promotional materials are in good taste, congruent with the overall station image, and realistic and factual enough not to create false expectations in the minds of the audience or advertisers.

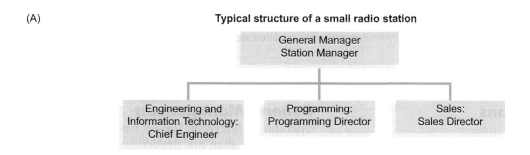

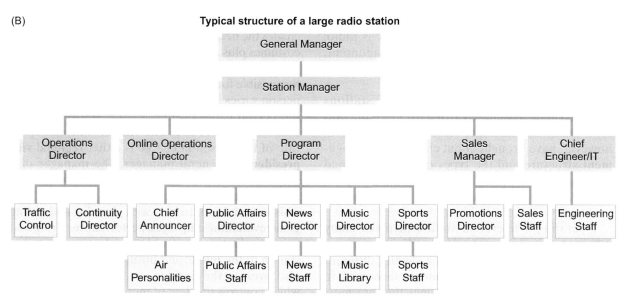

Diagram 12.2A and 12.2B These flowcharts show the departments commonly found in radio stations.
Source: Sherman, 1995

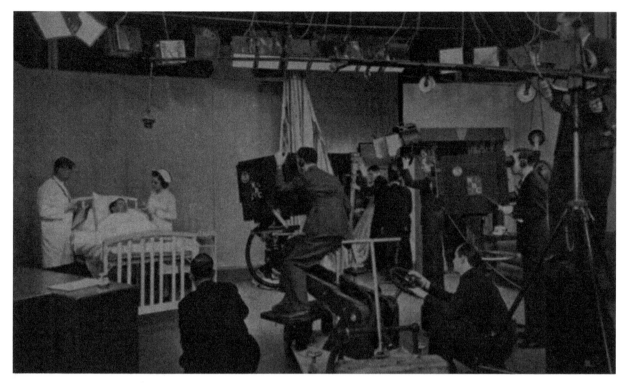

Figure 12.4 In the early days, a television studio camera often required many people to operate it.
Source: Photo courtesy of MZTV Museum

Television Stations

Television adopted the network radio model for business and operations and became immediately successful. Early television stations looked to radio operations to set up their enterprises. They adopted the same seven basic functions as radio: General management, business management, engineering, production, programming, sales, and promotions.

Some of the basic ways television stations operated in the past have carried over into the present. Technological improvement and innovation, however, have brought about changes in management structure and the types of jobs performed in modern stations. Production techniques, however, are very different than they were ten years ago.

Operating a television station involves many different organizational schemes. Each group or station in a group might have a slightly different organization, but the departments discussed in the following sections are the ones that are most common to television stations today.

General Management

Operating a television station has always required more people than are needed to operate a radio station. Moreover, television personnel perform a variety of tasks, many of which are different from those in radio. In particular, the visual element of television creates the need for make-up, hairstyling, lighting, and costumes plus a wide range of sets and props.

The **general manager** of a television station is responsible for all of its departments. The title of this person varies by company and depends on whether the person is responsible for one station or more than one. If more than one station is involved, the title of the person in charge is either **regional vice president** or general manager. The manager of a single station is usually called the **station manager**.

Whatever their title, managers typically have assistants and staff to help with planning and setting goals, evaluating employee performance, hiring and firing, leading and motivating, and representing the station to the public and business community. They also oversee the business or finance department, which controls the flow of money both into and out of the station.

FYI: TELEVISION STATION DEPARTMENTS AND JOB TITLES: 1950s VERSUS 2020

Note the similarities and differences between the jobs and personnel of 1950s' television production and contemporary studio functions.

1950s' jobs		Modern-day jobs
Department	Personnel	General manager
Executive Offices	Station manager	Sales manager
		• Sales associates
	Sales manager	
	Program manager	News director
		• Executive producers
		• Producers
		• Writers
		• Editors
		• Graphic designers
		• Assignment editors
		• Photographers
		• Reporters/Anchors
		• Sports
	Engineering	
Program Production	Writers	
	Directors	
Engineering	Engineers	
	Operating maintenance	
Sales and Service	Salespersons	
Scenic	Designers	Director of engineering
	Artists	• Managers
Carpentry	Carpenters	• Maintenance engineers
Property Shop	Property workers	• Operating engineers
Electrical	Electricians	• Building maintenance
Visual and Sound	Effects specialists	engineer/Carpenter
Effects		
Paint Shop	Artists	
Painters	Program development manager	
Wardrobe	• Producers	
Control Room	• Directors	
Studio	• Editors	
Actors	• Promotions editors/Photographers	
Dressing Rooms	Make-up artists	
Hairdressers	Programming director	
Film Studio	Projectionist	
Operating Engineer	Marketing/Community relations director	
Librarian	Station promoters	
Master Control	Operating engineers	
Transmitter	Operating engineers	
	Maintenance engineers	
	Interactive director	
	Web site designers/editors	

Business Management

The head of the business department is known by a number of titles, such as **business manager**, **chief financial officer (CFO)**, or **controller**. Other personnel are accountants and bookkeepers, who record transactions and debit or credit the transactions to the appropriate station accounts. CFOs also create budgets and track expenses.

The business department takes care of billing accounts receivable, such as payments that advertisers owe the station, and then collecting the money. When the station purchases equipment, such as new cameras, the business department is usually consulted beforehand to see if the station can afford it. The business department also produces reports required by the government and by general management to determine the station's financial situation.

Human Resources

Television stations with a large number of employees typically have a human resource department. Human resource managers recruit and hire new employees and provide and explain the benefits and services available to employees. In non-union stations, human resources often act as the liaison between employees and their bosses. They also conduct exit interviews and sometimes offer placement help to employees who are leaving the station.

Engineering

The engineering department has the responsibility of installing and maintaining all broadcast equipment used by the station, which includes not only the audio and video production equipment but also the signal transmitting equipment. Engineers are often classified as: Maintenance engineers, who keep production equipment working; operating engineers, who keep the station on the air; building/maintenance engineers, who take care of the facility; and IT engineers, who deal with computers and telephone networking.

Production

Television production in the 1940s and 1950s was quite a bit different than it is now. Cameras were huge, expensive, and limited in terms of capability. Because the early broadcasts were all in black and white, little consideration was given to color. Clothing or objects in the colors of pink and blue often appeared as identical tones of gray on a black-and-white screen, and colors like brown, purple, and even dark green yielded the same shade of dark gray. Moreover, to ensure that the performers' facial features would be apparent onscreen, white face make-up and black or dark green lipstick were used. Lighting equipment was heavy and lights had to be bright enough for the cameras to capture images. The lights generated quite a bit of heat, often raising the temperature of a small television studio set to an uncomfortable level. Visual effects were primitive in these early days. Miniature sets made of cardboard were common. Titles were drawn or printed on cards and then placed in front of the camera.

Shooting outside the studio meant taking a heavy studio camera into the field and supplying it with power from an electrical outlet (cameras were not battery operated until the 1970s). On-location shooting was thus limited to those programs that justified the huge outlay of equipment, personnel, and vehicles to transport equipment, sets, and props, such as sporting events, concerts, parades, and special events like political party conventions.

Television production personnel work in a variety of departments in the station. For example, those who operate television production equipment often work in the programming department or sometimes in the engineering department. Production personnel create the **station promotional announcements** (also known as SPAs). Sales managers sometimes oversee the production of commercials. In a station that produces some of its own programs, production personnel work in the program development division of the programming department, and staff members who shoot news video on location, known as **news photographers** or **video journalists** (and now often referred to as **multimedia journalists**), are usually part of the news department.

Rehearsal Time The production of a television drama or comedy requires the actors to know their lines. Such memorization was not needed in radio, where actors could voice their lines with a script in hand. As such, television is more like live theater, requiring quite a bit of rehearsing for actors to learn their lines as well as their movements.

A rehearsal requires everything that is needed for the final taping: Lighting, props/scenery, cameras, camera operators, engineers and other personnel, and studio space. Also, since one program is rehearsing while another program is airing, the station needs to have at least two studios. For this reason, the typical broadcast schedule includes out-of-studio productions, such as sporting events (e.g. boxing, baseball), as well as in-studio productions.

CAREER TRACKS: KYLE MAJORS, DIRECTOR OF TECHNOLOGY, FOX5 SAN DIEGO

WHAT IS YOUR JOB? WHAT DO YOU DO?

Director of Technology. I oversee the technical operations and engineering of a Fox affiliate in San Diego, California. In this role, I am responsible for the station's technical operations, including studio, news, commercial, syndicated, network, and remote productions.

I am also responsible for general building maintenance like HVAC systems, electrical power, plumbing, and all safety and emergency systems for the building and staff.

In other words, my team makes sure anything that plugs into an outlet or takes batteries at a TV station works properly.

My role includes making yearly operating plans and budgets and accomplishing capital projects.

Figure 12.5 Kyle Majors.

I keep the station in compliance with policies and regulations set by the Federal Communications Commission in order to be able to broadcast on public airways. I make sure we adhere to all local, state, and federal laws, including the Sarbanes-Oxley Act and Occupational Safety & Health Administration or OSHA regulations.

I report to the station's general manager, who is the head of the TV station.

HOW LONG HAVE YOU BEEN DOING THIS JOB?

Since 2011. Before then I was the station's operations manager and oversaw the personnel associated with the production of the news and operations of the station.

WHAT WAS YOUR FIRST JOB IN MEDIA?

My first job was at KNAZ in Flagstaff, AZ where I ran cameras, VTRs, audio mixers, and video switchers in the studio. I also worked as an announcer at an FM radio station.

WHAT LED YOU TO THE JOB?

Starting as a camera operator at a small station, I worked through many different positions at different television stations until I understood how a television station operates. Television production requires a team of people who work in unison using different systems to put on live television and cover local news. At one point in my career, the general manager of the station thought I understood how the people and systems worked together and promoted me to this position.

WHAT ADVICE WOULD YOU HAVE FOR STUDENTS WHO MIGHT WANT A JOB LIKE YOURS?

While jobs in television are collective and specialized, computer skills are necessary in all roles. Most people know how to use a computer, but we are always looking for people who understand how a computer works.

Television is part craft, part technical. It is important to know what makes good television as well as how the specialized equipment used to create television works. Those good at the what and how go far in this industry.

Camcorders The dominant trend in camcorders is for small, lightweight cameras that store images on memory cards. Two styles of cameras are common: The traditional video camera, which is often used with a shoulder mount, and the **DSLR (digital single-lens reflex)** camera, which is similar in style and shape to the 35 mm still film cameras that have been around since the 1960s but shoot both excellent video and still pictures and record audio as well. Studio cameras have typically been larger and heavier, because a reduced need for portability inside a studio environment allows for the use of higher quality, heavier and larger components, but even studio cameras have decreased in size in recent years, and it is not uncommon for studios to integrate small camcorders into productions alongside larger cameras.

Figure 12.7a This typical portable video camera can be used in the field or in a television studio. The quality of the video it produces approaches the quality of a cinema (film) camera.
Source: Courtesy Panasonic

Figure 12.6 Early television production included the use of: (a) cardboard cards for prompting the talent with their lines; (b) cardboard miniature sets like this harbor scene; (c) other simple devices like a drum showing program credits.

Figure 12.7b This camera resembles the 35 mm film cameras used for many years for still photography. This digital single-lens reflex or DSLR camera is small and lightweight but can shoot very high-quality video.
Source: Courtesy Canon

Studio Production Television programs are produced either in a studio or outside the studio in the field. Studio television production requires large, specially designed spaces, lights, and vast amounts of electrical cable, portable cameras, and a crew able to take on many different functions.

A television studio is a large, open space designed to control several aspects of the production

CAREER TRACKS: ERIN BIGELOW, DEPUTY HEAD OF STUDIOS FOR VERIZON MEDIA

WHAT IS YOUR JOB? WHAT DO YOU DO?

I am the Deputy Head of Studios for Verizon Media. I build and manage television studios around the country. I manage live production operations for large-scale digital television news studios. I manage field operations for video production around the world. I manage Live Event Operations for multiple Verizon Brands around the world. My team functions as an in-house production company for Verizon Media with clients such as Yahoo Finance, Yahoo Sports, Huffington Post, Tech Crunch, In the Know, and many others. My favorite part of the job is working on construction projects to build new TV studios in especially challenging space environments like so much of New York City!

WHAT WAS YOUR FIRST JOB IN MEDIA?

Figure 12.8 Erin Bigelow.

My very very first TV job was as an overnight Broadcast Operations Technician with NAU Television Services. I then started my career in 2004 when I was still a Junior at Northern Arizona University (NAU) as a full-time Director/TD at KNAZ-TV in Flagstaff, AZ.

WHAT LED YOU TO THE JOB?

The work I did as the Head of Production with NAU *Live!* (the daily student-produced newscast on campus) led me to the position of Director/TD and eventually Production Supervisor at KNAZ-TV. I had a role very similar to my current role (on a much smaller scale) and worked very hard to train the crew members in the control room, on post-production workflows and field operations. My work was recognized and appreciated as I managed the schedules and trainings for 60+ students at the height of NAU *Live!*

WHAT ADVICE WOULD YOU HAVE FOR STUDENTS WHO MIGHT WANT A JOB LIKE YOURS?

I made the MOST of my time at Northern Arizona University. I stayed active in extracurricular clubs, I made the live television production team my on-campus family, and the newsroom was my second home. I loved learning new things and was driven by the ever-changing live news environment. Everything I did on-campus, I treated as if it were my paid job; from classes to extracurricular activities. I would suggest to anyone in college now to treat the time at school as practice for a career. How you utilize your todays will shape how your tomorrows look!

environment. The lighting and sound, for instance, are under complete control of the television crew and manipulated to fit the production. The television studio is temperature controlled so performers are comfortable under the hot lights. The studio is as soundproof as possible; that is, little or no noise from the outside world can be heard inside it. Microphones are placed in key locations around the set to maximize audio quality. Television studios are windowless to prevent unwanted light infiltrating a scene. Lighting is supplied by special video lights, most of which are attached to a system of pipes, called the lighting grid, that is attached to the ceiling. The lights are controlled from a centralized lighting board, which allows a member of the lighting crew to connect numerous lights and control their intensities for use in a production.

Studio cameras are mounted on large, heavy, roll-around devices called **pedestals**. Using a pedestal, the camera operator can make smooth and easy camera movements in any direction. The pictures from the cameras are fed through cables to a small room near the studio called the **control room**. Inside the control room, the camera cables are connected to a camera control unit (CCU), which is used to control the color and brightness of each camera. The video signal travels to a device called a **switcher**, which is similar to the audio board that processes the audio signal. The switcher combines signals from other cameras, videotape, or other video storage devices.

Character generators are like word processors. They add titles to pictures, often called 'lower thirds' because it's common to place the names and titles of people appearing on camera toward the bottom of the screen. Character generators also create a **credit roll** at the end of a program, showing the names of the people who made the program, or a **crawl** across the bottom of the screen, as in the case of a severe weather watch or special bulletin. **Special effects generators (SEGs)** can make a multitude of creative changes to the video, such as slow motion,

ZOOM IN 12.2

The Tricaster is a versatile but affordable video switcher that has allowed many studios to upgrade from standard-definition to high-definition video. The switcher relies heavily on software to accomplish the many functions it serves in the television production studio.

Figure 12.9 A Tricaster TV switcher.
Source: Courtesy New Tek

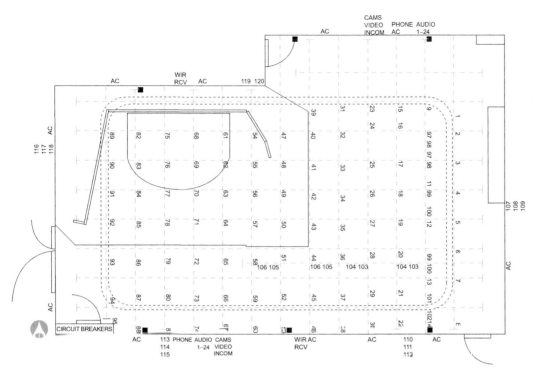

Diagram 12.3 Typical television studio floor plan. A news set is in the upper-left corner. The numbers along the dotted lines indicate the locations of lights. The connections for cameras, audio, telephone, and power are shown along the perimeter of the drawing.

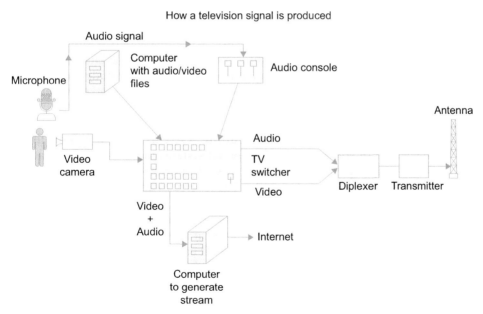

Diagram 12.4 A video camera captures the image of the subject and a microphone captures the sound. The video signal is sent to the TV switcher, while the audio signal is sent to the audio console. The video is then combined with other video signals that are generated in computers to add titles or graphic material to the signal and then sent to the diplexer. The audio can be combined with other audio sources at the audio console and then sent to the switcher or directly to the diplexer. The diplexer combines the audio and video signals, and the combined signal is then sent to the transmitter and the broadcast antenna. For non-broadcast program sources for cable, satellite, or the Internet, instead of an antenna, the signal is sent to a satellite uplink and on to a satellite that either sends directly to the audience or to cable head ends to be distributed via cable systems. Video programs for streaming are sent to the Internet via a streaming computer that is connected to the Internet (or cloud).

color variations, or a strobe effect. Many of these functions are the work of computers with special software.

The switcher also produces the changes between cameras or video sources that keep the program moving. The change can be a **dissolve**, in which one shot slowly changes to another; a **fade**, or a slow change from black to a picture or from a picture to black; a **cut**, or an instantaneous change from one camera to another; or a **wipe**, as when one video image pushes another off the screen.

In addition to the **switcher**, the video signal goes to one or more monitors so the director, the person in charge of the actual production, can see what each camera sees, what pictures are available via other sources (e.g. digital video recorder or SEGs), and which picture is actually going on the air. The director can make changes as the production is happening. For example, he or she might suggest that one of the camera operators get a **close-up shot (CU)** instead of a **wide shot (WS)**. The director might also tell the **technical director (TD)**, the person who runs the switcher, to change from one camera to another to get a **medium shot (MS)** of the host of the program.

Shot changes are accomplished by using a set of commands that tell the TD what video picture to use next. The director might tell the TD, "Dissolve to camera 2," which means that the TD should use a dissolve transition from the current video source to camera 2. The TD will push the appropriate button or move the appropriate lever on the switcher to accomplish the change requested by the director.

The video signal is then either stored on some type of storage device, like a DVR, DVD, or solid-state memory unit, or the signal can be sent directly to the transmitter for broadcasting. In some productions, such as when ESPN is covering a sports event, the signal is sent to a satellite uplink so the program can be transmitted to cable systems and direct broadcast satellite providers such as DISH or DirecTV. Some productions use the Internet to bring a remote signal to the studio.

In many larger markets, the operation of studio television has changed because of technology and budgets. Studios often had three or more cameras for news shows or studio shows, and each camera had its own operator. In addition, a crew member would act as a **floor director** and direct the talent's attention to the correct camera. Since budgets have been tight and camera-control technology has become more sophisticated and cost-effective, some studios are now automated. In other words, one person controls three or more studio cameras and performs all necessary camera shots and camera movements via one remote-control unit.

Numerous types of programs are produced in television studios: News, interview, talk, game, and quiz shows, along with dramas (mostly soap operas), although many of these types of programs are national, usually not produced in-house at local stations. Local news shows, however, are almost always delivered live from the local station's studio. Other programs are prerecorded and broadcast at a later time.

Portable/Field Production After battery-powered cameras became available for news and general production in the late 1970s, news stories and entertainment programs could readily be shot outside the studio. Field production added a sense of realism to television that was missing in the earlier days of studio television. Instead of constructing a set for each scene or program, a portable video crew could go to an appropriate location.

In addition, portable video has allowed news photographers to shoot breaking news and have it aired almost immediately at the station in a process known as **ENG**, or **electronic news gathering**. Stations regularly use field crews to do live shots during newscasts, which are sent to the stations via microwave. After the video is shot, a videotape editor prepares it for airing. Due to advances in technology and decreases in budgets, reporters are now often expected to shoot and edit their own stories.

Editing Editing systems are now **non-linear** computer-based systems that allow random access to any video shot or scene without having to fast forward or fast reverse to find it. Non-linear systems can create an array of special effects, such as slow motion, wipes, and dissolves. Another highlight of a digital non-linear system is its random access process that makes it easy for an editor to find desired shots or scenes without having to spend time fast forwarding or rewinding. With non-linear editing, shots or scenes can be easily added or removed anywhere in the program, and the computer adjusts the program length automatically. Linear editing was like composing a paper on a typewriter. If a mistake was made or new information needed to be added, the whole piece had to be retyped. Non-linear editing, on the other hand, is like using a word

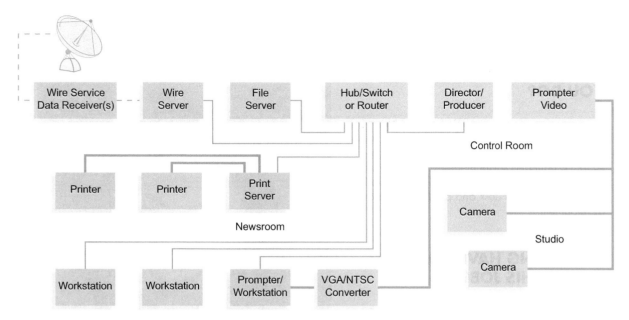

Diagram 12.5 Wired TV newsroom. In a wired newsroom, many people can access news information for editing and scriptwriting. The news information is sent from a wire receiver to a wire server. The files are then sent through a router to a file server and to a news director or producer. Other producers or writers can access the information from the file server at workstations in the newsroom. After the information is edited and put into script form, it can be sent directly to the prompters on the cameras in the news studio. The newscaster reads the story from the script in front of the camera.

processing program. If a mistake is made, it is relatively easy to correct.

Programming

Many network radio programs made a smooth transition to television, solving at least part of the question of what to show to audiences. The television networks delivered entertainment shows the same way that radio did: Through stations interconnected by telephone wires. The networks provided the programming, and the stations merely broadcast it.

Television stations did not get all of their programs from the networks, however. To fill non-network times, stations either produced their own programs or found other program sources. Television programming was more complicated and expensive to produce than radio programming, requiring more rehearsal time and involving more people and more equipment than radio programs.

The programming department in a television station is managed by the program director (PD) and deals primarily with program acquisition and scheduling rather than program production. Because most television stations are either network owned and operated or network affiliates, they get most of their programs from the networks. The PD must fill the remaining airtime, usually by obtaining programming that has been produced from an outside source, like a *syndicator*. Stations use outside sources for programming, primarily because the expense of producing local television entertainment shows is very high and requires time, equipment, and often studio space. However, the station almost always produces some local programs, such as local public affairs and news shows.

Most network-affiliated television stations and independent television stations in medium and large markets produce at least one local news program each day. Local news is often very profitable and draws an audience that may keep watching other programs on the same channel. Local television news departments often have large staffs and receive strong support from general management. Stations often air newscasts at several times during the day, requiring many hours of news gathering and production, both in the studio and in the field.

The **news director (ND)** is responsible for all newscasts and personnel in the department. In addition to the ND and staff, there are producers of newscasts and news stories, as well as writers, story and script editors, assignment editors, video photographers, reporters, and anchors. Most stations also employ several weathercasters and several sportscasters.

CAREER TRACKS: MARIA HECHANOVA, REPORTER

WHAT IS YOUR JOB? WHAT DO YOU DO?

I'm a morning reporter for 3TV & CBS 5 News in Phoenix. My job duties include presenting stories on air and writing for digital platforms including AZfamily.com and social media. I'm also required to research and pitch stories. In addition, I work with the assignment desk and producers on story development.

HOW LONG HAVE YOU BEEN DOING THIS JOB?

I've been a 3TV/CBS 5 reporter since summer 2017. I landed my first TV job in 2009.

WHAT WAS YOUR FIRST JOB IN ELECTRONIC MEDIA?

Producer/multimedia journalist (MMJ).

WHAT LED YOU TO THE JOB?

I love to write. I love to talk. I love to ask questions. I took a journalism class in high school out of curiosity and discovered I could get paid for doing all three skills.

Figure 12.10 Maria Hechanova.

WHAT ADVICE WOULD YOU HAVE FOR STUDENTS WHO MIGHT WANT A JOB LIKE YOURS?

Be a sponge. Watch the news. Study it. Apply for and earn a TV news internship. Ask questions. Find a mentor. Prove to your mentor you're worth his or her time. Don't get into the business just because you want to be on TV. Be persistent. Don't leave college without a resumé reel.

CAREER TRACKS: RAJAH MAPLES

WHAT IS YOUR JOB?

I am a TV news anchor, news manager and multimedia journalist for KHQA (CBS and ABC) in Quincy, Illinois. Our television station serves viewers in a Tri-State area of West Central Illinois, Northeast Missouri, and Southeast Iowa.

WHAT DO YOU DO?

I anchor KHQA's 5 p.m., 6 p.m., and 10 p.m. newscasts. I also help provide guidance to produce those three newscasts. I also serve as a writing coach for journalists in our newsroom; provide guidance on news selection for the day, and report on various news stories in our Tri-State region, including features, 'hard news,' crime and court reporting, presidential campaign visits to Southeast Iowa leading up to the Iowa Caucuses, working with CBS Newspath and CNN Newsource on story retrieval and uploading, working with our three statehouse reporters in our region's three state capitols, communicate with viewers via Facebook, Twitter, our YouTube channel, and upload and proof stories on our Web site.

Figure 12.11 Rajah Maples.

HOW LONG HAVE YOU BEEN DOING THIS JOB?

I've been serving in my current job for 15 years.

WHAT WAS YOUR FIRST JOB IN MEDIA?

Working on my college newspaper during my junior year at Stephens College, a historic all-women's college in Columbia, Missouri, that has been educating women since 1833.

WHAT LED YOU TO THE JOB?

Fulfilling a college requirement in order to graduate with my degree. Working on my college newspaper was not something I would've chosen. However, that's where I discovered (perhaps through serendipity?) that I thoroughly enjoyed the art of storytelling. I discovered how much I loved interviewing people, listening to their stories, and doing my best to tell their stories through the art of writing. After my semester of working on my college newspaper, I decided to 'branch out' and intern at a television station to see if I'd enjoy that form of storytelling as well. The rest is history. I'm not sure I would've ended up in television news had it not been for discovering my passion of storytelling while working on my college newspaper.

WHAT ADVICE WOULD YOU HAVE FOR STUDENTS WHO MIGHT WANT A JOB LIKE YOURS?

Try it through an internship before making a career decision. In fact, that's my advice for students pursuing any job. I knocked on a lot of doors (on purpose and not) and tried many different jobs in the process of finding a job and career that I feel I was 'destined' to do. Many careers, jobs, and projects sound terrific on paper. Many of them don't sound exciting at all (such as my college newspaper seemed to me before I actually tried it). However, you'll never know what you'll like or won't until you actually try anything and everything. You never know what you'll discover along the way. Enjoy the journey. Enjoy the adventures. Prepare but don't stress about the future. Follow your gut, instinct, and the many successful people who have come before you. Their wisdom is invaluable. Last but not least, enjoy the storytelling. Listening to other people's stories and accurately telling them with fairness, respect, and accuracy is one of the greatest gifts that one can ever do for humankind. And it's one that absolutely feeds my soul. I can't believe I actually get paid in dollars and cents for a job that pays me rewards and wisdom of a lifetime.

CAREER TRACKS: DOUG DREW, EXECUTIVE DIRECTOR, NEWS DIVISION, 602 COMMUNICATIONS

WHAT IS YOUR JOB? WHAT DO YOU DO?

Executive director, News Division, 602 Communications. Conduct training for television stations and networks in reporting, producing, and writing. I decided to focus my consulting on being a Breakfast Television Consultant. Morning news is really my passion, so I spent the last ten years creating this very specific niche. Ninety percent of my time was spent helping stations improve their morning news: Content, production, format, style, personality, and talent development. The other 10% of my time was spent doing international consulting that was more about general news.

Figure 12.12 Doug Drew.

HOW LONG HAVE YOU BEEN DOING THIS JOB?

More than ten years.

WHAT WAS YOUR FIRST JOB IN MEDIA?

I was a reporter for the NBC affiliate in Flagstaff, Arizona.

WHAT LED YOU TO THE JOB?

I spent four years as a reporter for the television station at Ohio University, where I went to school. OU allows students to produce and present a half-hour newscast each night. A few of the students hold paid positions, and I was fortunate enough to obtain one of them. In addition, while I was at Ohio University, I also worked for the Associated Press Broadcast Division. With the opportunities to work at the television station and with AP while in school, in addition to my classwork in the radio-television program, I was more than ready to enter the real world of broadcast news.

WHAT ADVICE WOULD YOU HAVE FOR STUDENTS WHO MIGHT WANT A JOB LIKE YOURS?

Get involved ASAP as an intern, or work in a broadcast newsroom while still in school. The hands-on experience is invaluable. When you're applying for jobs, employers will look beyond your education to see what experience you have.

Sales

Television sales departments are similar to radio sales departments in that they are divided by national and regional/local categories. However, television stations often have more salespeople, more assistants, and more people involved in audience research than do radio stations. And because of television's larger audience, television advertising is usually more expensive and generates more dollars for the station than radio advertising does.

The sales department typically consists of a general sales manager, a national sales manager, and a local sales manager. It also employs account executives to sell spots; a traffic manager to schedule commercials; and researchers to collect, interpret, and prepare audience ratings information for use in sales. One of the easiest ways for young television professionals (especially those with management career objectives) to enter the business is through sales. The broadcast industry seeks energetic people to sell advertising time. The sales department is often the quickest path to upper-level management and has the side benefit of paying successful salespeople well.

Promotions

A television station often has a community relations department that promotes the station and participates in community events. Another title for this department is **marketing**, as this department markets the station's product (i.e. its programs and personalities) to the audience and advertisers.

The promotions department has developed an increasingly important role since the 1970s due to industry changes – primarily the decreasing dominance of the networks, the rise in importance of local news, and the need to establish station identity among the numerous channels available from cable and satellite. The heavy competition for viewers (and, of course, advertisers) has created a promotions effort in stations that is based on consumer research, competitive positioning, long-range strategizing, and targeting specific audience segments.

Cable Companies

Cable started merely as a way to extend the reach of broadcast television stations beyond the ranges of their over-the-air signals but eventually changed the face of the television industry, expanding channel offerings from fewer than a dozen to hundreds. Most of us take for granted in modern times that watching television means choosing from a variety of options targeted to most every imaginable demographic. Despite its simpler origins, the cable industry is now a complex maze of technical operations, advertising, programming, production, and a big-money corporate business structure.

Operations

Cable television was a small business from 1948 until the early 1970s. Most early ventures were operated by a handful of owners in a small town. Large cities had not yet been wired for cable, because franchise deals between cable companies and local governments were not in place, and free broadcast television signals were available over the air. Early cable company owners took on many roles and often supervised many overlapping departments, including engineering, sales, and business. But since the 1980s, cable services have become well-organized businesses with functions similar to those of radio and television stations as well as other functions unique to the cable business.

As the cable business became lucrative, it drew the attention of large media companies, which began buying up smaller cable systems in the 1970s, a trend that was later accelerated by the change brought on by satellite technology. To meet the demands of subscribers, in the late 1970s cable systems had to become 'satellite capable' to receive the new premium channels such as HBO.

Larger companies had an advantage in acquiring franchises. Small cable companies could not afford to bid on large-city franchises because the cost to build such a system was too high. Therefore, the larger franchises made deals with the bigger companies, who could raise the large amount of money needed for financing. Ultimately, many small systems had to sell out because they could not afford to upgrade the systems they owned or expand to larger, more profitable markets. Since the 1970s, most small systems have been acquired by large **MSOs (multiple-system operators)**, such as Comcast, Charter, Cox, and Altice.

General Management

Similar to broadcast groups that have stations in multiple markets, most cable companies are comprised of multiple cable systems. Therefore, general management must be able to manage distant systems with varying technological, personnel, and business needs. General management of a system is responsible to regional or national management to adhere

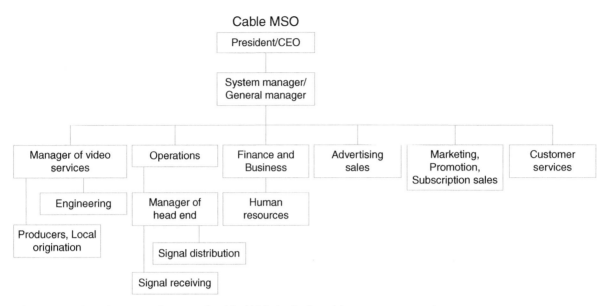

Diagram 12.6 Structure of a typical cable MSO (multiple cable system operator).

to the principles of good business and the brand of the company that owns the systems. Most MSOs use the power and economies of scale to leverage deals with the various channels that are shown on their system. Management goals include getting the most desirable channels for the audience (ESPN, CNN, Fox News, HBO) at the best price from the channel provider and negotiating with local broadcasters to get an affordable retransmission fee plan in place.

Engineering

A cable service's primary function is to distribute programming to cable subscribers via cable wires, therefore engineering takes the driver's seat in cable operations. The engineering department receives television signals from both local and distant stations in a facility called a *head end*; the physical location where signals are assigned to particular channels.

Once the various television signals arrive at the head end and are placed on channels, all of the channel signals are sent out through a system of shielded **coaxial cable wire** (a conductor with a metal sheath around a common axis or center). The system for distributing the signals to subscribers resembles a tree with a trunk and branches. Amplifiers are set up at intervals along the path to enhance the signals. Newer systems are commonly wired with the use of fiber optic cable made of strands of optically pure glass as thin as a human hair that carries digital information over long distances. Compared to copper-based coaxial cable, fiber optic cable carries much more information.

Until the 1970s, most cable systems had 12 or fewer channels, and the television signals were placed on the 12 VHF channels (numbered 2 through 13), regardless of what channel each signal was originally assigned. After HBO's satellite delivery debut in 1975 the number of cable networks available to local cable systems boomed, making it necessary for cable systems to provide more channels. The engineering solution to the need for more channels was to provide subscribers with set-top converters that expanded the number of channels that could be received on their television sets. Eventually, newer sets with built-in tuners capable of receiving hundreds of channels overtook the market, making channel-expanding converter boxes unnecessary.

Since the advent of high-definition (HD) television, most cable programs are typically received through a digital converter box (called a 'receiver') or a DVR (digital video recorder). Although a direct cable wire to a television connection will work, the picture quality may not be up to digital standards nor as clear as the picture from an **HDMI (high-definition multimedia interface)** cable connection between the digital box and the television. Many of the channels distributed through a digital box are not available with a direct cable-to-TV connection.

Production and Programming

In the 1970s, cable services began producing their own shows after the FCC mandated local-originated programming and **public-access channels**. Many systems built small television studios with sufficient equipment to make local talk or talent shows. Engineering personnel were hired to keep equipment working properly. The channels provided to the local programs are referred to as **local-origination** channels. Although making local shows was an exciting idea to many, viewing the shows was not very popular. The idea that local audiences would want to watch local shows was just that ... an idea. The reality was that local programs simply were too amateurish to garner a loyal audience large enough to justify the expense, space, and equipment needed to sustain a consistent local-origination operation. Most cable systems abandoned the local-origination studio as soon as they were legally able to do so. The requirements changed from public-access channels to **PEG channels**. PEG stands for public, educational, and government. The channels were designated to serve the local communities, providing channels for public information (e.g. libraries), education (local schools), and government (city council meetings, etc.).

ZOOM IN 12.3

The Federal Communications Commission issued its *Third Report and Order* in 1972, which launched hundreds of public-access television production facilities. The report contained a rule that required all cable systems in the top 100 U.S. television markets to offer three access channels, one each for public, educational, and local government use. The rule was amended in 1976 to require that cable systems in communities with 3,500 or more subscribers set aside up to four cable television channels and provide access to equipment and studios for use by the public.

Most local production in cable systems involves commercial and system promotional announcements rather than entertainment and information programs, although some MSOs with cable systems serving very large markets have a local channel for news. Budgets, audience size, and competition from local broadcasters make cable a risky endeavor. In addition, some systems that do have a local news channel often partner with traditional non-cable businesses, typically broadcast stations or newspapers, that provide some news content.

Sales

The task of a cable service's sales department was straightforward during the small-system era. The only product offered was a subscription to the cable system's television channels. A system did not insert its own locally sold commercials, and it did not have premium channels or pay-per-view programs to offer. Since only one cable system was franchised in a market, there were not any competitors for subscribers to sell against. As the small-system era ended and the MSO era began, local advertising sales became more important and more competitive with other local media outlets, such as the local television and radio stations and the local newspapers.

Today's cable world is much more competitive. Since direct broadcast satellite television became available, cable companies compete vigorously for subscribers. In addition, some media markets have two or more cable companies that serve the same region and compete for the same subscribers. Probably the biggest change in cable sales, however, comes not from subscriptions but from advertising. There are now hundreds of advertising-supported cable networks. When a cable service picks up a cable network, for example Comcast in Knoxville putting Food Network on its line-up, local sales reps sell commercial time on Food Network's programs to businesses in the Knoxville market. Comcast's sales reps, therefore, compete with sales reps from the local television affiliates who are also out trying to sell local businesses commercial time on television. Local-insertion cable advertising is a good buy for businesses because cable systems tend to cover smaller areas than broadcast television stations, so their ads cost less and are more targetable. Similarly, because cable systems have multiple channels available for ad insertion, businesses can choose which networks to advertise on based on their target demographics.

Promotions

Cable systems promote special television channel events or the addition of a new or desirable channel to their channel line-up to encourage subscriptions. Since competition for subscribers changed with the entry of satellite television in the mid-1990s, cable companies spend much more time and energy promoting their offerings and their 'bundles' to potential subscribers. Cable systems want customers to subscribe not only to cable television but also to their Internet service and their telephone service (VoIP). Marketing this array of services requires cross-promotion involving other media, such as newspapers, radio, and direct mail, but also a nearly constant barrage of promotional announcements on the system's channels, filling unsold commercial time with the company's package deals for television, Internet, and telephone service. As cord cutting becomes more common, cable companies have stepped up their promotional efforts to get more people to sign on in an effort to regain lost subscribers and attract new subscribers.

Satellite Delivery – Satellite Radio

The satellite radio industry paid the FCC more than $80 million in 1997 for the use of the frequencies for distributing satellite signals. The major companies XM Satellite Radio and Sirius merged in 2008 to become SiriusXM, thereby reducing consumer choice to subscribe to satellite or not subscribe. Although experiencing financial difficulties and slow growth in subscribers at first, SiriusXM has continued to operate and adapt to the current media environment, with nearly 35 million subscribers and a recently completed acquisition of the streaming service Pandora strengthening its position as a popular delivery medium for audio content.

General Management
The satellite radio industry relies on an organization or structure similar to that of radio broadcasting. At the top of the organization chart is senior management, including a chief executive officer (CEO), an executive vice president, and a chief financial officer. This top layer of management has five areas of responsibility: Engineering, production, programming, sales, and promotions. Each division or department has numerous subdivisions. For example, in the programming department, a satellite radio station will have separate subdivisions for music, news, and sports. Engineering has a satellite uplink subdivision and a studio subdivision. In satellite radio, the provider, SiriusXM, programs its music channels. Some news and sports channels are provided by other services such as CNN, ESPN, and National Public Radio.

Engineering
The engineering department in satellite radio is responsible for maintaining the satellite uplink and downlink operations, construction and maintenance of studios (both on-air and production), and installation and maintenance of all production equipment. Engineering is also responsible for IT, and owns and operates studios in New York City, Washington, DC, Los Angeles, Nashville, and several other locations.

Engineering is also responsible for updating the satellite radio receivers and the associated software. Unlike satellite television, satellite radio receivers often are part of in-dash radios, and reception of the satellite radio signal does not require installation of special dish antennas. SiriusXM does not manufacture receivers but works closely with equipment manufacturers like Alpine, Audiovox, Cambridge Audio, Denon, JVC, and Clarion to make sure that equipment is available to the audience to receive a high-quality audio signal, in a home or in a car.

Production
Production work in satellite radio is similar to any audio production facility. Live recording takes place in an audio studio and involves a studio with sound-absorbent treatment to reduce echo and is isolated from sound generated outside the studio, similar to all production and broadcast station facilities. Many independent studios supply programming for SiriusXM's hundreds of channels of programming. All of its programs are digital and sent to a central location for uplinking to the SiriusXM satellites. Each type of show, whether talk, specialty music, news, or sports requires producers who specialize in that genre and individual producers who help produce each program. All audio sources such as prerecorded clips and microphone feeds are combined at an audio console with which an on-air personality, an audio engineer, or a producer determines what goes on the air.

Figure 12.13 This satellite radio will receive SiriusXM satellite signals throughout the United States. On many newer model cars the standard AM/FM radio has become an AM/FM/HD/XM radio with an additional 'band' that receives the SiriusXM signal.

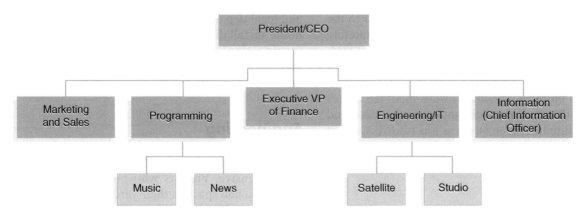

Diagram 12.7 Structure for satellite radio distribution company.

Programming

A recent alignment of channels on SiriusXM included at least 78 channels of commercial-free music, 18 sports talk and play-by-play sports channels, 22 talk channels, nine comedy channels, 15 news channels, nine traffic and weather stations, 18 Latin stations, plus 14 'other' channels. The diversity of programming requires PDs for each of the various formats. In addition to the satellite channels that are similar to terrestrial radio station formats, such as alternative or jazz, there are numerous channels that carry play-by-play coverage of sports events, and channels dedicated to a particular artist such as The Beatles, Elvis Presley, Pearl Jam, and Phish. The number and specific format of the many channels in satellite radio is in constant flux. Channels change seasonally; for example, holiday music in December. As musical styles evolve, channels are added and deleted to suit subscribers' preferences.

Sales and Promotion

Satellite radio requires two kinds of selling. One type involves marketing to the general public to gain and keep subscribers. Satellite radio subscribers not only commit to a monthly subscription fee but also rent or own a satellite receiver. This subscription purchase is different from a subscription to a video service like Netflix or audio service like Spotify that merely requires an account and an app to begin receiving the programming, as opposed to the purchase of special equipment.

With tens of millions of subscribers, SiriusXM is a good way for national advertisers to reach a large audience with a single buy. However, the audience is splintered among the many commercial channels offered. Advertising is placed either by format (e.g. all of the comedy channels), or on individual channels. The commercials are read live by on-air talent or produced by high-profile personalities and customized for either single spot placement or a customized segment sponsorship. Sales to advertisers requires a high level of sophistication because of the large number of channels available.

Satellite Delivery – Television

Satellite television has existed since the 1970s, but the industry was very different than it is now. When HBO began distributing its programming to cable systems using satellites in 1975, enterprising companies began selling satellite dishes capable of receiving the HBO signal. Other cable channels followed HBO, and soon satellite **television receive only** (TVRO) dishes became somewhat common. These large dishes were popular in rural areas that were not being served by the local cable companies. At first the dishes were expensive, but by 1984 the price had dropped to about $1,000 per installation, and more than 600,000 TVRO dishes were installed in 1985. The installation charge (including the cost of the equipment) was the only charge to the customer. Once installed, the dishes received television signals direct from satellite for free. Getting television programs for free was similar to later peer-to-peer sharing of copyrighted music and video – essentially pirating the signals.

The TVRO business changed dramatically beginning in January 1986 when HBO and other satellite-distributed channels put a stop to program pirating by scrambling their signals, so dish owners were no longer able to view the highly desirable pay channels. Within a year of HBO's scrambling, half of the TVRO companies went out of business. Although some channels were not scrambled, the lure of buying and installing a TVRO to get free premium channels was gone. The growth of cable into the suburbs and rural areas also contributed to the eventual demise of the TVRO business. The satellite television industry changed once again in the mid-1990s when direct broadcast satellite (DBS) companies began a subscription multichannel television delivery business. The dishes were small (18 inches across), the selection of channels was extensive, and the satellite television business was resurrected, but with a new business model.

ZOOM IN 12.4

Telecommunications company AT&T, a provider of multichannel video, Internet service, and telephone service, added satellite television to its services. In July of 2015, the FCC gave its approval for AT&T to merge with satellite television provider DirecTV. Amid fears of the lessening of competition in the multichannel video delivery market, the FCC and the Department of Justice felt that the combination of AT&T's land-based Internet and video business with DirecTV's satellite delivery of video does not pose a significant threat to competition and should not harm consumers.

At first the biggest disadvantage of direct satellite delivery was that the companies were restricted from delivering the local television stations' programs. The satellite industry lobbied Congress and eventually was given legal permission to deliver local station signals to satellite-subscribing homes through the Satellite Home Viewer Improvement Act of 1999. Cable and satellite still compete for subscribers, but cable has the edge because it can offer high-speed Internet service and other multichannel and audio services.

General Management

The general management of the satellite television companies has similarities with management of cable companies. The business of satellite television is collecting television channels from a variety of sources and securing the rights to include the channels in the various bundles to market to individual homes and businesses. For example, gaining the right to send ESPN to the homes of subscribers, management must negotiate with Disney, the owner of ESPN, regarding which ESPN channels (e.g. ESPN, ESPN2, ESPN Classic, ESPNews, ESPN Deportes, ESPNU) it will carry and for how long before renegotiations are necessary. General management oversees the acquisition of content (programming and legal departments), distribution of content and home installation and maintenance of equipment (engineering), marketing of subscriptions to the audience (promotions and sales), and business and finance (business).

Engineering

Although there was quite a bit of interest in satellite dish delivery, the early TVRO dishes were large and expensive. Viewers who bought a dish were those who had space in their yards, could afford the dish, and could not or did not want to receive cable programming.

The two main responsibilities for satellite engineering have been to maintain in orbit the satellites that receive signals from the program channels and send them back down to earth, and to provide and install satellite receiving equipment in subscribers' homes. Now satellite television channels are expected to be online and available on all mobile devices.

Production and Promotion

Satellite television companies are primarily multichannel program delivery services and currently do not produce their own entertainment programming. Their production is limited to producing informational programs about how to use the equipment, pay a bill, troubleshoot problems, or simply promotional programs about the channels offered, special free trials of channels, or information about video on demand. Otherwise, all programs are licensed from independent production companies.

Sales

When satellite television first started competing with cable, sales reps had the formidable task of selling an unfamiliar service. They had to explain satellite delivery, inform a potential client of its benefits, and get them to buy a huge, unsightly satellite dish. For a while, reps could offer free premium cable programs as an incentive to buy a dish, but premium cable channels soon caught on and began scrambling their signals in such a way that they could only be decoded by a special box.

The business changed when DBS satellite services began. Sales people were no longer selling equipment but instead were selling multichannel program delivery subscriptions. Also helping in their sales efforts was a growing distrust and dislike of cable companies' customer service.

DISH offers advertising to tens of millions of subscribers, but its advertising is addressable through the satellite receiving equipment. In other words, DISH knows who its subscribers are and can target them through the receiving equipment. For example, if an advertiser wants to reach women, DISH can place ads in Lifetime Movie Network, MTV, E!, BBC, and other channels that have strong female viewership. Another aspect of satellite television advertising is the ability to place a 'trigger' during an ad or program. A message is placed on the screen that tells the audience to 'press the select button now' to get more information about the product or service being advertised.

Internet – Streaming Operations

Most media outlets have been streaming video content to their viewers since the 1990s. But streaming did not take hold with Internet users until major technological improvements in the late 2000s. Now all that is required to stream media is a computer that accepts the audio and/or video signal and translates the information digitally. This digital information is then sent out from the 'server' computer to all 'client' computers that request the digital information via a stream. In this way, streaming functions as both a one-to-one media source and a one-to-many media source.

At first streaming media presented challenges to the audience. Because of slow download speeds, audience members often had to face many starts and stops in the stream while the computer displayed a 'buffering' message, during which the audio and/or video would stop and then start. Now, for most Internet users with broadband connections, buffering is much less common, and programs are usually streamed smoothly without interruption.

Streaming has become easy and dependable enough to allow millions to stream media from Netflix, Amazon Prime, Hulu, or HBO Go with few if any delays or problems. Now that streaming is nearly problem free, it can be said that streaming is replacing downloading media. Streaming allows an audience member to 'rent' media – watch or listen in real time – as opposed to having to buy it. Downloading takes time for the information to be stored on a computer and requires storage space on that device as well. Streaming takes up very little storage space, and that space is not needed once the listening or viewing of the streamed media ends.

Competition in video-streaming services has grown substantially in recent years, but the first major subscription-based service, Netflix, remained the largest into the 2020s. In 1997 Netflix (started by Reed Hastings and Marc Randolph) allowed customers to order physical DVDs online to be delivered by mail to their homes. The company introduced a monthly subscription model in 1999, allowing customers a certain number of movies per month with no late fees. At that point Netflix was a relatively small company with only around 100 employees, most of whose jobs were to fill orders by packaging DVDs and putting them in the mail. The company had fewer than 1,000 movie titles available for order in its early days, but this expanded rapidly as the company negotiated deals with movie studios and independent production companies. By 2002, Netflix had 670,000 monthly subscribers and mailed 190,000 DVDs, but the high costs of acquiring physical product and shipping kept the company from becoming profitable for several years. In the mid-2000s, the company began pursuing technological developments that would allow online streaming instead of selling physical product, and first launched a streaming service in 2007 as Internet bandwidth and digital video technology had advanced enough to make streaming high-quality video affordable and technologically feasible. The company eventually began funding its own shows and movies. As a studio, Netflix makes its own deals with directors and producers and takes a more active role in the production of films and television shows. By 2020, the company had about 183 million subscribers around the world and 8,600 employees, in the areas of content and talent acquisition, product development, technical operations, business, finance, legal affairs, and communication, with executive officers overseeing each of these areas. In addition to business and technical staff, the company hires researchers, media producers, marketers, public relations operatives, and producers of corporate media such as podcasts and video promoting the company itself.

Digital delivery online has some real advantages as compared to broadcast, cable, or satellite delivery:

- Streaming begins when the user requests the stream, any time, day or night.
- Users can tap into interactive applications like creating a personal playlist.
- The content deliverer (e.g., the station) can monitor the size of the audience and the length of the audience's listening/viewing.
- Users can multitask while listening.
- Streamed content does not stay on a user's computer or device, thus protecting the copyright holder.

- Broadcast stations and cable channels can stream live media as it is being broadcast or cablecast, but can also give access to a program library of recorded material.
- Once streaming media is set up for a station, it requires very little supervision or maintenance.

Streaming media technology has greatly improved in the 20 years it has been technologically feasible. The weak link in this delivery system is the user's connection speed.

SEE IT LATER

Radio Stations

Radio has changed dramatically through the years but is still a valuable source of information and entertainment for millions. Continued change is inevitable, but as long as people drive cars and engage in other activities that command their visual attention, audio media will likely remain an important utility and companion. Radio's demise has been predicted repeatedly through the years, and there may be reason to question whether its business model is sustainable, but its popularity with audience remains strong.

A recent study by Nielsen Media Research showed that although online listening is a mainstream activity, radio still reaches an overwhelming majority of the American public. Radio reaches 89% of Americans age 12 and older each week, down only slightly from 91% a few years earlier, a reach that has not changed much in the past ten years despite all of the new ways to listen to audio. It is difficult to track Internet radio listenership, but one 2015 study showed that about 53% of people older than age 12 listened to online radio at least once a month. This percentage represented nearly twice the listenership from five years earlier (2010). Overall, AM/FM radio revenue for advertising declined slightly, but revenue from digital advertising showed strong gains. Anecdotally, some stations, including two whose managers were interviewed for this chapter, report thousands of weekly streams, whereas the number was often less than 100 streams ten years ago. This increase is likely because of improvements in available Internet bandwidth – especially from Wi-Fi availability – along with the explosion in popularity of smart phones as a media consumption tool.

The broadcast radio industry is working on new ways to improve audio quality and offerings as a way to compete with online and satellite delivery. **HD radio** has seen slow adoption by audiences. As of 2019, roughly 18% of cars on the road were equipped with HD radio receivers, including about half of new cars, but the vast majority of vehicle owners are sticking with over-the-air analog stations.

Radio's distribution model is evolving to keep up with online delivery and radio satellite services. Despite competing services, traditional radio will still maintain its advantage of being local. Local radio plays music that appeals to its market, runs commercials for hometown businesses, and airs local weather, traffic, and event information. Radio's future depends not so much on new technology like HD radio but on staying competitive with its ability to appeal to its market, thus making it attractive to local audiences and advertisers. Sports and news, along with personality-driven programs about topics of local interest, are among traditional radio's greatest strengths. Local stations must maintain an online presence to help stations stay close to their audience and provide the opportunity for audience members to stream the station's signal on computers, tablets, and smartphones.

Production styles are changing because of smaller budgets, less advertising support, and increased competition from a variety of sources, but will largely remain the same because so much of radio relies on prerecorded music.

The consolidation and merger rage of the 1990s and early 2000s has slowed considerably. Although radio station consolidation increases efficiency and profitability, mostly because of the need for fewer employees, mergers can be very expensive, and station groups can incur huge financial debt. Even after consolidation, station groups found that some stations were now worth less than when they were purchased.

Even contemporary station groups are digging themselves out of their financial holes. For example, in an effort to save itself, Clear Channel rebranded itself as iHeart Radio and heavily promotes the service for online listening. Station groups are banking that in the long run consolidation will streamline operations and increase efficiency and profitability, mostly because of the need for fewer employees. If a group owns 20 stations and six of them are of the classic rock format, buying programming for all six can be accomplished with one 'classic rock package' buy. Buying programming for many stations gives group owners buying power, which translates into a lower cost per station, because of volume discounts. Such economic leverage gives group owners a strong advantage over owners who buy programming for only one station.

In the near future, it does not appear that radio stations will be changing their operations. Audiences are still listening to radio, and revenue to stations has only decreased slightly. With that in mind, the organization of radio stations will most likely remain stable, at least in the immediate future. The radio industry will continue to employ general managers, PDs, engineers, IT personnel, announcers, and perhaps most importantly, salespeople. An excellent radio station with a terrific format, great personalities, and a loyal audience will only stay in business if its sales operation is effective.

Television Stations

Television stations have been facing financial pressures for years now that have forced consolidation, contraction in the number of employees, and the need to keep expenses as low as possible.

High-quality audio and video production that in the past could have been produced only by expensive equipment manufactured for that purpose is now being created by less expensive equipment that can do a variety of tasks with remarkably high quality that rival film cameras. For example, digital camcorders that cost less than $2,000 are now available in various HD formats and are capable of producing images better than more expensive camcorders did only a few years ago. Some camcorders come with digital storage devices similar to the removable 'jump' or 'flash' card devices. Feature films and network programs and commercials that were once shot only on film are now being shot on HD or UHD (4K) video. Even smartphones are capable of good video quality, and many small- and medium-market television stations now use them for some of their news gathering. And during the coronavirus pandemic, some reporters without access to field cameras relied solely on their iPhones to record events or shoot a package.

Video editing is one area of television production that has changed dramatically in the past ten years. Almost all computers are now equipped with some basic video editing software. Costing only about $100, new video editing programs permit sophisticated editing that is better than software that cost thousands of dollars a few decades ago.

In the past, a news crew consisted of a producer, reporter, videographer, and sound person.

Better equipment that is easy to use, lighter, and more compact enables a smaller crew to cover news stories in the field. Instead of a four-person crew, many stations now expect a single multimedia journalist to set up the equipment (tripod, camera, sound, and lights), do a stand-up introduction to the story in front of the camera, conduct an interview, shoot video of the scene or location of the story, edit the story, and send it back to the station for airing. In some cases, the location shot is shown live on air. The video may also be posted online. Added responsibilities of reporters have changed the job preparation necessary for entry-level video journalism jobs. Television reporters are expected to be able to shoot video, be on camera, write the story, edit the video, and send the video back to the station. This new 'one-man band' or 'multimedia' style of reporting requires a wide range of production, journalistic, and multimedia skills.

Even in-studio jobs are being cut. Studio cameras can now be robotically controlled, obviating the need for a camera operator for each studio camera. Instead of three camera operators, one person can sit in the control room and remotely control all camera functions and camera moves needed for a live news show.

Conceptually, multicasting held the promise of increasing advertising revenue by increasing advertising availabilities on newly created subchannels, but the reality has not met expectations. Advertisers may be lured into trying the various digital subchannels but at the trade-off of removing dollars spent on the main channel.

Although television stations do offer some localism, primarily in local news, much of what they broadcast is also available online. In addition, streaming services might lessen the need for local television stations.

The biggest challenge to television stations lies in viewing habits. Many people are cutting the cord by canceling their cable and satellite television subscriptions, and even those who have maintained such subscriptions are streaming more of their programs. Although the broadcast networks are adapting by creating subscription opportunities for streaming directly to the audience, local stations may not be able to do so. When audiences can get their favorite network prime-time programming directly from the network, local affiliates will need to provide other more compelling content to keep their audience.

ZOOM IN 12.5

HOW MANY VIDEO CAMERAS?

The idea of individuals creating television 'programs' has been around since camcorders became available at reasonable prices to the general public in the 1980s. The show *America's Funniest Home Videos* has been on network television since 1989 and has depended upon people sending in their funny home videos. Although not every household had a camcorder, the show never lacked for material. In 1999, Nokia cell phones had a feature that allowed sending video messages. At the time, there were probably a few million video cameras available, most stored in closets waiting for children's birthday parties and some in the hands of video professionals who shot video regularly. At that time, none of the cameras was connected to the Internet.

Technology has dramatically changed the availability of video cameras among the population. It is estimated that now there are almost 3.5 billion video cameras in use in the world. Obviously, this is because smartphones can also shoot video. Also, these phones can be easily connected to the Internet. The Internet is a network with billions of potential sources of video information. How does this change our picture of the world?

Cable Television

Due to the nature of the equipment, cable systems must have many local employees to provide service for the head end, maintain wiring and servicing amplifiers, and to satisfy customer needs. For example, a subscription to a broadband service involves a cable modem installation, a service that requires local personnel who have computer technology knowledge and experience and who can physically install the equipment. MSO-owned systems rely upon upper management to decide about programming packages, special deals to encourage people to subscribe, and other promotions that involve convincing subscribers to add premium channels or Internet and telephone (VoIP) service to their subscriptions.

The biggest challenge for cable systems is cord cutting in favor of streaming. This trend changes the cable business from one with an emphasis on being a multichannel program provider that also provides Internet service and telephone service to a business that may have Internet service as its main selling point. If the cord-cutting trend continues, which appears highly likely, cable companies will hire fewer personnel involved in programming and fewer engineers and technical people and more people to work in Internet and telephone service.

One strategy to stave off dramatic declines in the number of subscribers to cable television service is a change in how channels are bundled, or grouped in tiers. Some companies, such as Verizon, are heeding complaints of television subscribers who dislike paying for hundreds of channels to get the ten channels they really want. Cable viewers in the future will want more of an 'à la carte' or 'skinny bundle' menu for selecting channels. The cable companies are taking notice, primarily since cable's reach into homes in the U.S. is shrinking. The future may see slimming down of the size of the bundles that cable companies offer their subscribers. Unless cable companies can stop or at least slow down the cord-cutting trend, the operation of systems will change, based upon less revenue which will lead to hiring fewer personnel.

Satellite Radio

The satellite industry – including radio and television delivery – reaps the advantage that, unlike broadcast radio, its signals cover huge geographic areas. In fact, two satellites in geostationary orbit can cover the entire contiguous 48 states.

Satellite radio has an advantage over terrestrial radio of nearly continuous coverage for subscribers while traveling. Terrestrial radio has geographic limitations that satellite radio does not have. A traveler can drive across the country with uninterrupted satellite radio service. FM broadcast stations are limited to a 75- to 100-mile radius around the transmitter. AM stations have a broader reach that increases at night, but AM signals do not have the same high quality of audio fidelity as FM. In other words, music played on an AM station does not sound nearly as good as it would on FM. Satellite radio has excellent sound fidelity on all of its channels. Satellite radio requires a special receiver preinstalled in newer model cars. For out of car use, a subscriber must buy or rent a receiver.

Many satellite radio programs are licensed from independent producers. The larger variety of program options are selected by a programmer. The studios also require personnel to help produce programming, maintain the technical operation of the facility, and schedule the activities in the studio. If revenue shrinks for SiriusXM, the number of studios delivering independently produced programs to them, as well as the number of personnel operating these independent production companies, will also decrease.

Satellite Television

Unlike cable and its ability to provide two-way communication through an infrastructure already in place, satellites are restricted in the two-way arena: That is, consumers cannot send or uplink content to a satellite. This has changed somewhat, as satellite Internet has become available and is offered by DBS providers, but it is not as fast or reliable as other forms of Internet delivery. Nevertheless, satellite television delivery will most likely continue to operate as it has for the past 20 years, as a multichannel provider of entertainment and information that competes directly with cable for subscribers and will continue to improve its Internet service. The merger of AT&T with DirecTV changes the bundling opportunities for satellite television in the future, and both DirectTV and DISH offer Internet bundles using either DSL or separate companies providing the emerging technology of satellite Internet.

Because most of the revenue for both satellite radio and satellite television comes from subscriptions, most of the promotional and sales activity deals with marketing to individual subscribers. The satellite industry faces the same threat as cable television: Video streaming. As people find other, more economical ways to obtain television entertainment, satellite television will have to find solutions to the problem of a shrinking subscription base. Although some of the loss in revenue because of decreased subscribers has been offset by higher prices and added revenue from movies on demand, this may be a short-term solution to a long-term problem.

Both satellite radio and satellite television have added an online streaming component to their business model; both businesses must support a sophisticated and expensive satellite uplink and downlink system. The radio and television satellite industries are buoyed by millions of subscribers. The satellite industries are middlemen in the entertainment business; they deliver programming from content creators to an audience whose viewing habits favor a passive audience accustomed to paying high fees for many channels that are never listened to or watched. Both industries have an "antiquated business model, and suffer from ... a lack of innovation" (Gerber, 2013). Some investors seem to think that the satellite business is in danger of being overtaken by a company or companies that will create content, simplify the viewing experience, and reduce the number of remote controls the audience must use to access the various programming services. In other words, to be successful in the future, satellite distribution companies need to change their operation to become innovators that produce content (similarly to Netflix and Amazon Video) and innovate technology to simplify the listening and viewing experience.

SUMMARY

Radio became a dominant medium in the 1920s. Programming was live, for the most part, and included mainly musical performances. Programming was mostly live music performances, broadcast either from the studio or a remote location.

Television inherited radio's network system and then it proceeded to steal its programming ideas and its audience. By the early 1950s, it had pushed radio out of the way as the dominant electronic entertainment and news medium, leaving it to fend for itself. Radio hung in there and eventually bounced back into public consciousness when it ceded narrative programs to television and focused on what an audio-only medium does best – music. Rock 'n' roll led the way to radio's recovery. Once management learned that formatting a station by type of music was the best way to compete with other stations, radio boomed.

Meanwhile, the television industry had its own unique challenges to face. As viewers demanded more sophisticated and live-action shows, stations and networks had to find a way to produce them. Higher-quality shows meant larger studios, more rehearsal time, and additional personnel such as directors.

Radio and television production has not changed much over the years. Program formats are similar, and production techniques have changed gradually with technology. The signal flow in both audio production and television production is much the same today as it has been for the past 50 years. The most significant change in production has come about with the introduction of digital equipment. Compared to its analog counterparts, digital equipment produces a better-quality product and gives production personnel more flexibility to experiment.

Distribution patterns have changed over the past 50 years. Although over-the-air broadcast radio transmission is still the dominant form of audio transmission, satellite radio and Internet-based audio services have eroded the listening audience a bit. SiriusXM is a subscription service that provides an array of audio channels, some of which are commercial-free. Satellite's biggest advantage is in its diverse program offerings that can be heard in almost any location in the continental U.S. Internet audio services provide user-customized listening.

Television program distribution has changed more than radio. The network-to-local station program distribution model that has existed since television's inception is no longer the only way to distribute programs.

Technology will continue to change the radio, television, cable, and satellite businesses and how they operate. As consumers rely more upon the use of in-home broadband connections and the mobile Web, the media will need to reconsider their business models and their method of distribution to keep up with the demands of technology and the audience.

BIBLIOGRAPHY

Armstrong, J. (2007). Constructing television communities: The FCC, signals and cities, 1948–1957. *Journal of Broadcasting & Electronic Media*, 51(1), 129–146.

Bond, P. (2009, September 25). DVRs in 36% of households. *The Hollywood Reporter*. Retrieved from: www.hollywoodreporter.com/hr/content_display/technology/news/e3i470b0d4b3627285705f84f75e2ab3902

Cheng, R. (2009, April 20). Telcos, satellite join cable's push to build pay wall on Web. *Wall Street Journal*. Cited in Benton Foundation Newsletter (April 21, 2009).

Eastman, S., & Klein, R. (1991). *Promotion and marketing for broadcasting and cable* (2nd ed.). Prospect Heights, IL: Waveland Press.

Gerber, R. (2013, September 18). Investing in the future of television. *Forbes*. Retrieved from: www.forbes.com/sites/greatspeculations/2013/09/18/investing-in-the-future-of-television/?sh=6484624e3eba

Hampp, A., & Learmonth, M. (2009, March 2). TV everywhere – as long as you pay for it. *Advertising Age*. Cited in Benton Foundation Newsletter (March 2, 2009).

Head, S., Sterling, C., & Schofield, L. (1994). *Broadcasting in America*. Boston, MA: Houghton Mifflin.

Hilliard, R., & Keith, M. (2001). *The broadcast century and beyond* (3rd ed.). Boston, MA: Focal Press.

Hutchinson, T. (1950). *Here is television*. New York: Hastings House.

Insider Audio. (2019). More than half of new cars now equipped with HD Radio. Retrieved from: www.insideradio.com/more-than-half-of-new-cars-now-equipped-with-hd-radio/article_055842a0-3f18-11e9-af44-abb5c736f701.html [May 24, 2020]

Johnson, B. (2009, October 5). Media revenue set for historic 2009 decline. *Advertising Age*. Cited in Benton Foundation Newsletter.

Kaye, B., & Medoff, N. (2001). *The World Wide Web: A mass communication perspective*. Mountain View, CA: Mayfield.

Kittross, J. (1976). A fair and equitable service or, a modest proposal to restructure American television to have all the advantages claimed for cable and UHF without using either. *Federal Communications Bar Journal*, 29, 91–116.

Learmonth, M. (2010, January 18). Thinking outside the box: Web TVs skirt cable giants. *Advertising Age*.

Napoli, P., & Yan, M. (2007). Media ownership regulations and local news programming on broadcast television: An empirical analysis. *Journal of Broadcasting & Electronic Media*, 51(1), 39–57.

Palmer, S. (2015, May 9). Up periscope: I see the future of video. *Shelly Palmer*. Retrieved from: www.shellypalmer.com/2015/05/up-periscope-i-see-the-future-of-video/

Parsons, R., & Frieden, R. (1998). *The cable and satellite television industries*. Boston, MA: Allyn & Bacon.

Quaal, W., & Brown, J. (1976). *Broadcast management*. New York: Hastings House.

Romano, A. (2002, September). 'Dearth of women' in top spots. *Broadcasting & Cable*, p. 9.

Settel, I. (1960). *A pictorial history of radio*. New York: Citadel Press.

Sherman, B. (1995). *Telecommunications management*. New York: McGraw-Hill.

Thottam, G. (2001). Cable. In E. Thomas & B. Carpenter (Eds.), *Mass media in 2025* (pp. 15–26). Westport, CT: Greenwood Press.

Turner, E., & Briggs, P. (2001). Radio. In E. Thomas & B. Carpenter (Eds.), *Mass media in 2025* (pp. 75–84). Westport, CT: Greenwood Press.

TV basics: Alternate delivery systems-national. (2009). *Television Bureau of Advertising*. Retrieved from: www.tvb.org/rcentral/mediatrendstrack/tvbasics/12_ADS-Natl.asp [October 10, 2009]

Vogt, N. (2015, April 29). State of the news media 2015. *Pew Research Center*. Retrieved from: www.journalism.org/2015/04/29/audio-fact-sheet/

Willey, G. (1961). End of an era: The daytime radio serial. *Journal of Broadcasting*, 5, 97–115.

Chapter thirteen

Feature Films: 'The Movies'

Ross Helford and Paul Helford (Northern Arizona University)

The instinct to tell a story visually is a fundamental human characteristic that dates back to the cave paintings of our earliest ancestors. In the millennia that followed, language was developed, which allowed stories to be passed orally from generation to generation. But, unlike the cave paintings, these stories lasted only as long as the civilizations that passed them on. It was only with the innovation of written language that these stories were given a similar sense of permanence to those ancient cave paintings.

Words and images became intertwined with storytelling. The innovation of theater gave a new sense of dramatic realism to some of these stories. Whether the content of these stories, artworks, and plays pertained to the boundless limits inherent in the human imagination or if they strove to represent reality with as much accuracy as the best technology of the time would allow, it was not until the innovation of photography in the mid-19th century that the dynamic film industry we know today was born.

SEE IT THEN

In film's infancy, during the latter half of the 19th century, audiences were captivated merely by the novelty of moving images. But it was in those days, with exhibitors charging audiences pennies to view these moving pictures, that the first seeds of the film industry were planted. What had begun as a series of photographs spun on a zoetrope (a cylinder with vertical slits along the side that, when spun, made the images appear as if they were moving), had evolved into a montage of moving images interconnected to tell a rudimentary story or anecdote. A few short decades later, movies would evolve to become not only a fully realized storytelling art form, but ultimately an industry that has proven as mighty and profitable as any the world has ever seen.

Although most people think of Hollywood as the birthplace of the film industry, its origin actually begins in Fort Lee, New Jersey. There, in 1893, Thomas Edison was experimenting with motion pictures. He built a studio named the Black Maria. It was a tar-paper-covered dark studio room with a retractable roof where short films of various topics were shot. The motion picture business in the area around Fort Lee prospered for about 20 years, until around 1911, when the first studio was built in Hollywood. California's climate, with moderate temperatures, wide-open spaces, and year-round sunshine proved to be cost-effective for the fledgling film industry. Another reason southern California attracted filmmakers in those days had to do with Edison's overly litigious nature. Because he held the patents on most filmmaking equipment, the long distance between the east and west coasts made enforcement by Edison much more difficult. As a result, filmmakers flourished in California, creating the origins of the studio-dominated film industry that endures to this day.

An Entertainment Industry

In the early decades of the film industry, America was not the center of the entertainment world it is today. Yes, the United States was a major player in the evolution of the early industry, with luminaries such as Charlie Chaplin and D. W. Griffith creating enduring and controversial works that continue to engage, inspire, infuriate, and educate, but a great many of the artistic and technological leaps and bounds were happening an ocean away in France, Russia, and, most notably, Germany. The unrest and

Figure 13.1 This device, the zoetrope, was used for visual entertainment in the 19th century.

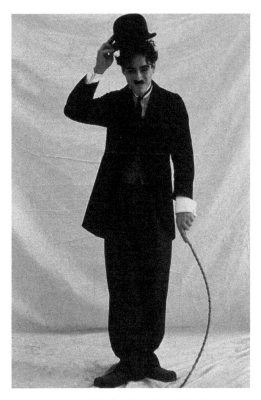

Figure 13.2 Charlie Chaplin's films featuring the Little Tramp character were enormously popular in the early days of the U.S. film industry.
Source: Courtesy TriStar Pictures/Photofest. © TriStar Pictures

violence that swept through Europe during World Wars I and II devastated economies and film industries, allowing America to emerge as the worldwide film industry leader.

Although the history of cinema worldwide remains a vibrant story of invention and innovation, this chapter will primarily focus on the American film industry, as well as its evolution into the multimedia, multiplatform, multi-billion-dollar entertainment juggernaut that is its present and future.

As America rose to the top of the film industry, California – and, more specifically, sunny Los Angeles – became the film capital of the world. The 1920s gave rise to the studio system. In those early days, a silent film would be screened in an auditorium, often with an accompanying live organ that would provide both the film's soundtrack and sound effects. To tell a story (aside from a sparse usage of title cards to convey essential dialogue and descriptions), silent cinema was entirely reliant on continuity editing and the non-verbal expressiveness of its actors. Silent movies spoke a universal language that was immediately accessible to non-English speakers – such as America's growing immigrant population, as well as a global audience.

The Production Code

In its early years, the film industry was largely unregulated for content, and the variety of cinematic offerings during the silent era was not all that different from what it is today. There were family films, comedies, action-adventure, sprawling epics, and no shortage of sex and violence.

It was not until 1934 that guidelines for cinematic content began being enforced. The Motion Picture Production Code adopted by the Motion Picture Producers and Distributors Association (MPPDA, later known as the Motion Picture Association of America, or MPAA) set forth general standards of 'good taste' and specific do's and don'ts about what could and could not be shown in movies. The Production Code was informally known as the Hays Code, named for its creator, Will H. Hays, who served as chairman of the Republican National Committee and then U.S. Postmaster General before becoming the first chairman of the MPPDA, serving from 1922 to 1945. The justification for Hays' Production Code was the "moral importance of entertainment." These guidelines varied from the prohibition of foul language and overt sexuality (which is why in black-and-white movies and, later, television shows, married couples were shown sleeping in separate beds) to forbidding mixed-race romances, and requiring that all bad deeds be punished in the end. The Hays Code lasted for more than 30 years, ending in 1968 when the industry adopted the MPAA film-rating system that endures to this day.

Hollywood studios enforced the Hays Code, just as they would later become complicit in the anti-communist hysteria of the 1950s, blacklisting a great many of its creative talents. The Blacklist was literally a list of names of people who were suspected of being communists. It did not matter that *being* a communist was not against the law. An accusation was all that was needed to get people fired and blacklisted from the industry. Careers and the lives of innocent people were ruined. Writers who were blacklisted often used pseudonyms or 'fronts' through which their work would be submitted. Screenwriter Dalton Trumbo won Oscars in 1953 (*Roman Holiday*) and 1956 (*The Brave One*) under different 'fronts' while blacklisted. Trumbo also became the symbol for the end of this shameful chapter in Hollywood's history when producer-star Kirk Douglas proved even more heroic than the title character Trumbo penned for him, insisting that he be hired to write the 1960 film *Spartacus*.

The Business of Film

At the beginning of the Great Depression, from 1930 to 1933, weekly movie attendance dropped from 110 million to 60 million. In an attempt to lure back audiences, Hollywood studios introduced the double feature. The first feature would be the A list production, with its big budget and big stars, and the second would be the low-budget B genre flick – thrillers, westerns, gangster, horror, and science fiction. The double feature proved a marketing masterstroke, and during the Depression years, Hollywood entered its Golden Age. This was the era of the Hollywood dream factory, in which American success stories and happy endings were mass produced for a population

ZOOM IN 13.2

HAYS PRODUCTION CODE

Listen to the NPR story about the Hays Product at: www.npr.org/templates/story/story.php?storyId=93301189

ZOOM IN 13.1

THE MOTION PICTURE PRODUCTION CODE GENERAL PRINCIPLES

1. No picture shall be produced that will lower the moral standards of those who see it. Hence the sympathy of the audience should never be thrown to the side of crime, wrongdoing, evil, or sin.
2. Correct standards of life, subject only to the requirements of drama and entertainment, shall be presented.
3. Law, natural or human, shall not be ridiculed, nor shall sympathy be created for its violation.

Figure 13.3 *Gone with the Wind*: set during the Civil War; number 6 on the American Film Institute's (AFI's) 100 Greatest Movies of All Time.
Source: Photo courtesy of MGM/Photofest. © MGM

Figure 13.4 *The Wizard of Oz*. This 1939 film used extensive special effects, was shot in glorious Technicolor, and remains one of the most beloved movies ever made; number 10 on the AFI's 100 Greatest Movies of all Time. Source: Photo courtesy of MGM/Photofest. © MGM

in desperate need of temporary relief from their real-life woes.

The year 1939 is often looked at as the pinnacle of the Golden Age. An abbreviated list of the classic films released this year include *Gone with the Wind* (Academy Award for Best Picture), *The Wizard of Oz, Gunga Din, Mr. Smith Goes to Washington, Stagecoach, Dark Victory, Love Affair, Wuthering Heights, Beau Geste, Intermezzo, The Adventures of Sherlock Holmes, Destry Rides Again, Another Thin Man, The Hunchback of Notre Dame, Ninotchka,* and *Only Angels Have Wings.*

In the late 1940s, another innovation would profoundly affect the movie industry: Television. People stayed home to watch television, because in their newly built suburbs, there often was not a movie theater nearby. Television, not movies, became the entertainment of choice, and theater audiences declined.

The Hollywood Empire Fights Back

If there is one constant in the film industry, it is the whirring speed of technological innovation.

Silent movies were quickly phased out after the unprecedented success of the first 'talkie', *The Jazz Singer* (1927). Sound created a seismic shift in the industry, forcing a re-evaluation of what constituted a movie star. An actor gifted in the exaggerated melodramatics so integral to silent cinema was often unable to adapt to the more nuanced storytelling that could be revealed through dialogue. Some of the greats, like Charlie Chaplin and his golden voice, eventually went on to make influential movies in this new era – though even Chaplin resisted for over a decade before releasing his first talkie, *The Great Dictator* (1940), which is widely considered his masterpiece. A great many other silent stars of 1920s cinema rapidly faded into obscurity. By the late

1930s, Technicolor movies would slowly supplant black-and-white films, and in the 1960s, movies in full color would become the norm.

One of the ways in which Hollywood responded to declining attendance caused by television was to develop widescreen formats, changing the size of the screen (for the first time in film's half-century history) from the standard, nearly square 1.33:1 aspect ratio to a widescreen ratio that was more than twice as wide as it was high. On this new widescreen, Hollywood gave audiences epics that were too big and colorful for small, black-and-white televisions at home. Westerns, war sagas, biblical and costume films such as *The Ten Commandments* and *Lawrence of Arabia*, and literary adaptations like *From Here to Eternity* and *Moby Dick*, would come to define the early years of the widescreen era.

FYI:

The first image is from *The Artist*, about a silent superstar who cannot make the switch to 'talkies,' which won the Oscar in 2012. The second image is from the 1950 classic *Sunset Boulevard*, about a faded silent star desperate to reclaim her rightful place in Hollywood.

Figure 13.5a *The Artist*.

Figure 13.5b *Sunset Boulevard*.

By the late 1960s, movie censorship virtually ended, with a new rating system that would finally replace the unpopular Hays Code. Rating films at that time from G to X based upon suitability for children, the new system allowed filmmakers to explore themes of violence, sex, and liberation from societal constraints and norms in films like *The Graduate*, *Bonnie and Clyde*, *Easy Rider*, *Midnight Cowboy*, and *Rosemary's Baby*.

The 1970s were a new cinematic Golden Age in which movies such as *The Godfather*, *Jaws*, and *Star Wars* were to become the epitome of creative and popular success, leading to the first increase in movie attendance since the 1940s. A movie generation had been born. The first film school-educated movie-makers emerged, and the public was more sophisticated, informed, and interested in film as an art than it had ever been before. The 1970s was the epoch of director-driven films. Francis Ford Coppola, Steven Spielberg, Martin Scorsese, Robert Altman, George Lucas, Brian De Palma, Roman Polanski, Woody Allen, Mel Brooks, and Clint Eastwood were a few of the director *auteurs* (directors so distinctive that they are perceived as a film's primary creative influence) whose films defined the era.

Forty years prior, *Jaws* and *Star Wars* would have been the low-budget B movie at the tail end of a double feature. Filmmakers like Spielberg and Lucas grew up loving these films, and their influential touch led the way in redefining how Hollywood made genre films. Ultimately, these movies would come to define the modern-day blockbuster.

Stars and Heroes

Whereas in decades past, a megawatt star like Cary Grant or Clark Gable would be called on to headline any big-budget venture, the 1980s saw the rise of the action hero, making superstars of the likes of Arnold Schwarzenegger and his lesser ilk: Actors of limited range whose bulging biceps, menacing scowls, and well-placed one-liners sold tickets to a worldwide

Figure 13.6 *The Graduate*, a 1967 comedy drama, is listed as number 71 on AFI's 100 Greatest Movies of all Time, and was selected for preservation by the National Film Registry because of its cultural and aesthetic significance.
Source: Photo courtesy of Embassy Pictures Corporation/Photofest. © Embassy Pictures Corporation

Figure 13.7 *Jaws*, number 56 on AFI's 100 Greatest Movies of all Time, directed by Stephen Spielberg, is the story of an existential threat that becomes infinitely worse after being ignored and denied by the authorities. It was released on June 20, 1975, and is considered the first 'summer blockbuster.'
Source: Photo courtesy of Universal Pictures/Photofest. © Universal Pictures

Figure 13.8 Big movie stars like Cary Grant became icons in our culture. He even got a postage stamp with his picture on it in 2002.

audience to feed its seemingly limitless appetite for action. It might be said that action movies, in the tradition of the silent movies of old, had become cinema's newest international language.

By the 1980s, special effects in Hollywood had come a long way from the stop-motion wonders of *King Kong* (1933). It should be noted, however, that until CGI (computer-generated imagery) effects became commonplace in the 1990s after the visual triumphs of movies such as *Terminator 2* and *Jurassic Park*, stop motion was used extensively in special effects – as were model miniatures; rudimentary computer graphics; advances in animatronics and puppetry from such effects masters as Jim Henson and Stan Winston; and optical special effects, which were used to extraordinary effect in movies like *Superman* (1978), *Star Wars*, and *Close Encounters of the Third Kind*. In special effects-laden *Ghostbusters* (1984), the apartment building where the ghosts converge in the climactic sequence is a real 20-story building onto which models and a matte painting were added to make it appear as if it stretches into the stratosphere.

By the mid-1990s, CGI was on its way to becoming a fully integrated cinematic technology, as well as a viable art form and industry. Pixar Studios led the way, with mega-hits and critical darlings like *Toy Story* (1995) and *A Bug's Life* (1998). Before long, every major studio was churning out CGI features. This began a precipitous decline in cel animation, which had been how animated films were made for nearly a century before computers were able to do it faster and cheaper.

It should go without saying that CGI was a game-changer, giving audiences a glimpse into a visual world that in the past could have been realized only in animated films, fantasy and sci-fi novels, and comic books. Today, it is integrated so seamlessly, it is barely noticed as a special effect at all. However, in the 1990s and first decade of the 2000s, when the technology was much newer and rapidly evolving, filmmakers had to strike a careful balance with this fun, new toy.

Looking back on the early years of CGI, the filmmakers who used the technology most judiciously – as James Cameron did in *Terminator 2* and Spielberg in *Jurassic Park* – created movies that were both hits in their time while remaining visually impressive for decades to come. This is contrasted with the great many movies that relied all too heavily on computer effects, only to have their films look like outdated videogames after a few short years (have a look at *Terminator 3*). Perhaps there is no better example of this than Peter Jackson's *Lord of the Rings* franchise (2001–2003), which looks nearly as visually impressive today as it did at the turn of the millennium. Like Cameron and Spielberg, Jackson used CGI when necessary, such as for the computer-generated character Gollum or for the epic battle sequences, but in other instances, he relied on decades-old tried-and-true cinematic 'cheats' like forced perspective, use of body doubles, and make-up to get the desired visual effect.

ZOOM IN 13.3

A still photo from a cel drawn movie. While CGI has largely supplanted cel animation, there is still something irreplaceable about the hand drawn animated masterpieces of yesteryear, like Disney's *Sleeping Beauty* (1959).

https://artinsights.com/sleeping-beauty-production-cels-why-is-original-art-from-sleeping-beauty-is-so-beautiful/

Figure 13.9a *Sleeping Beauty.*

SEE IT NOW AND SEE IT LATER

Feature Films Today and Tomorrow: The Entertainment Industry's Radical 21st-Century Transformation

In the relatively short span between the first decade of the 21st century and today, it simply cannot be overstated how much has changed about the way in which entertainment is distributed and consumed. In 2001 (the year Arthur C. Clarke and Stanley Kubrick predicted there would be colonies on the moon – to say nothing of a giant Star Baby hanging out somewhere beyond Jupiter's orbit), media was consumed by going to the movies, watching television, or playing a VHS or DVD. Although there was limited streaming available, most people were still using painfully slow dial-up to connect (YouTube would not become a thing until 2005). As the decade progressed, however, it became apparent that the next phase in the entertainment industry would be delivered through a high-speed modem directly to a computer or television screen.

There are, perhaps, two crucial events that brought us from there to here. The first was the Writers Guild strike that virtually shut down Hollywood for three months beginning in November, 2007. At the time, few people fully understood the implications, much less the *meaning*, of what was then being referred to as 'new media.' But what was fully understood by Hollywood's creative unions was that the last time a game-changing entertainment delivery platform was introduced into the marketplace (home video in the 1980s), writers saw only a small fraction of the gargantuan profits reaped by studios.

In the wake of the strike, many surmised that, much like following the resolution of the previous Writers Guild strike in 1988, the industry would pick up with a furious buying frenzy as it worked to make up for lost time. Instead, studios had done some realigning of their own, revamping their business model to be more economically efficient, buying and developing fewer original and speculative scripts, and relying even more on previously existing properties such as remakes, sequels, novels, and comic books.

Then, in 2010, a company that was founded in 1997 as a mail-based DVD rental hub, began

Figure 13.9b A still from *Humorous Phases of Funny Faces*, the first animated movie from 1906. Watch this short at: www.youtube.com/watch?v=wGh6maN4l2I

ZOOM IN 13.4

View animation from *Humorous Phases of Funny Faces* at this site.

www.youtube.com/watch?v=wGh6maN4l2I

offering streaming services, and the brave new world of Netflix brought the art of passive media consumption to a whole new level. As streaming services have become increasingly ubiquitous, movie theaters have had to operate with increasing nimbleness and creativity to keep crowds coming. While the adage that it is always better to see it on the big screen still perhaps holds true, when many homes have a flat panel widescreen television, high-quality sound system, and access to more hours of media entertainment than can be consumed in a hundred lifetimes, getting people out to a theater has never been more challenging.

By the end of the first decade of the 2000s, studios found great success revamping, with the benefit of modern technology, a gimmick that was first tried nearly a century earlier. 3D (or stereoscopic) movies – the concept of adding a depth dimension to the flat screen – first appeared in the United States during the Great Depression, in the early 1930s. The technology came roaring back in the 1950s with the horror blockbuster *House of Wax* (1953). Periodically since then, 3D films have appeared with mixed results. The 1969 soft-core porn film *The Stewardesses* made lots of money; Andy Warhol gave 3D a try with *Andy Warhol's Frankenstein* in 1973; and there was *Jaws 3D*, among others, in the heyday of the 1980s slasher film craze.

FYI:

3D movies have been tried for many years and had a resurgence of popularity from 2005 on. Although the interest even reached the White House, 3D popularity has waned and by 2020 almost no television sets with 3D capability are still being sold.

Beginning in 2005, 3D was being shot on digital video, and the results were far more compelling than previous attempts. Instead of using two 65-mm 150-lb film cameras to shoot each scene, filmmakers could use two 13-lb digital video cameras to get the 3D effect. These days, it is expected that any major popcorn flick will be given the 3D treatment. While it is still fun to catch *The Rise of Skywalker* in 3D, by the 2020s, the novelty has somewhat worn thin. Still, at the peak of the 3D resurgence, there were movies, like Cameron's *Avatar* (2009) and Alfonso Cuarón's *Gravity* (2013), whose revolutionary and

Figure 13.10 *2001: A Space Odyssey* was a visionary masterpiece so ahead of its time that the actual 2001 looked nothing like it.

Figure 13.11 The Obamas watching a 3D movie.

Figure 13.12 George Clooney and Sandra Bullock in space, from the film *Gravity*.

very deliberate use of 3D, elevated the technology as a legitimate art form.

Theaters have added additional perks to the moviegoing experience, including the ability to order food from your seat – even alcohol for the over-21 crowd (though best of luck trying to make it through the full 160-minutes of *Once Upon in Hollywood* after downing a pint or two of craft IPA on tap). New theaters have also revamped their seating to be more comfortable and roomier; and some theaters offer couches and recliners.

Then, there is assigned seating. These days, if you want to see the biggest movie on opening night, all you need is to buy tickets online, pick your seats, and show up. From a convenience standpoint, this is invariably less troublesome than first come, first served. And so, much like the double feature of yesteryear, the palpable thrill of lining up around the block at Mann's Chinese Theater in Hollywood to see the first midnight showing of *Harry Potter and the Chamber of Secrets* or *Fellowship of the Ring* or *The Phantom Menace* is now a part of history.

Of course, no gimmick no matter how innovative or consumer-friendly, is possible when the very act of attending a movie might put one's health, or indeed one's life, in jeopardy. As this chapter is being revised, the coronavirus pandemic has shut down much of the country, crashed our economy, and transformed life as we know it. Movie theaters, much like all non-essential businesses, have had no choice but to close their doors, lay off employees, and fret for their futures.

A *Los Angeles Times* article written by Ryan Faughnder (2020) laid out the predicament theaters were in, noting, "Movie theaters have gone from a $15-billion-a year business – about $11 billion in domestic box office grosses and $4 billion in concessions – to effectively no revenue."

FYI: PREPARE FOR THE DEATH & REBIRTH OF HOLLYWOOD

In an article for *Medium* entitled "Prepare for the Death & Rebirth of Hollywood" (2020), Richard James offered a dire assessment of the post-pandemic state of movie theaters, predicting mass bankruptcy by the major chains. In this vacuum, the demands for a theatergoing experience will not simply vanish. Instead, Janes foresees a theatrical resurgence in which "distributors such as Warner Bros., Disney, Lionsgate, and yes, Netflix, Amazon, and Apple … [will] actually own movie theaters." For this to happen, however, a landmark 1948 Supreme Court antitrust case known as the Paramount Consent Decrees, "making it illegal for studios and distributors to own movie theaters," will have to be changed. When these rules were first enacted, there was a very real fear that studios would create monopolies in which only their films would be able to be seen. But, Janes argues, with the multitude of platforms in which media can now be consumed, such fears are hopelessly antiquated. The United States Department of Justice agreed with this assessment and, on November 22, 2019, filed to terminate the Paramount Consent Decrees.

Should the Paramount Consent Decrees become no more, Janes predicts this radical redefinition of the theatergoing experience will mean "movie theaters are going to turn into entertainment centers with big money pumped into them, creating premium experiences in a way that the current owners never could."

Until such a time, the entertainment business as a whole is more prepared than ever before to deliver its products so that anyone with an Internet connection can consume content to their heart's fulfillment.

Most distributors, however, are not jumping at the prospect of offering streaming content in lieu of theatrical distribution. According to Julia Alexander in an article for *The Verge*

> [S]tudios ... earn an impressive return on investment when they release big tentpole, event movies in theaters. The global box office in 2019 alone saw $42.2 billion, up 1% from 2018. That is the kind of money that no amount of streaming subscriptions can easily match.

So, despite the increasingly lucrative returns on streaming, there is nothing currently in place that can make up for the profitability of theatrical blockbusters.

Still, for smaller movies, synchronous streaming/theatrical releases have been making economic sense for some time. The groundwork for how we got to this point was laid most prominently by Steven Soderbergh. In the first iteration of Soderbergh's career, he helped usher in the independent film revolution of the 1990s with his groundbreaking indie hit *Sex, Lies, and Videotape* (1989). A decade later, he found himself squarely in the A-list mainstream, with critical and commercial hits like *Erin Brockovich* (2000), *Traffic* (2000), and *Ocean's Eleven* (2001). But that maverick spirit remained. In 2006, when YouTube was in its infancy and high-speed Internet was just becoming widespread, Soderbergh became the first mainstream filmmaker to simultaneously release a film theatrically and stream it online with the largely improvised, low-budget *Bubble*.

It took some time for this release strategy to catch on with smaller films that might not otherwise have had the opportunities to gain a wider audience. However, by the early part of the 2010s, with movies such as the 2013 Richard Gere thriller *Arbitrage*, it was proving to be a winning business model.

FYI: THEATRICAL RELEASES DURING THE PANDEMIC

Many blockbusters set to be released during 2020 either had their release dates pushed back or were simply released digitally. In late 2020, Warner Bros. announced its entire 2021 slate would be released on HBO Max. Disney, similarly, has made its biggest releases, such as *Hamilton*, *Mulan*, and *Soul*, available on its streaming platform, Disney+. While the stubborn endurance of a badly mismanaged pandemic has undoubtedly forced studios' hands in this regard, offering blockbusters as streaming content has been far from financially disastrous. The family film sequel *Trolls World Tour* proved this in the early weeks of lockdown, when it earned $30 million in its opening week, ten times that of the previous digital rental debut. It would go on to earn more in its first three weeks of digital release than the original had in its first five months in theaters. This prompted a headline from *The Wall Street Journal*: "'Trolls World Tour' Breaks Digital Records and Charts a New Path for Hollywood."

Figure 13.13 *Trolls World Tour*.

Then came *The Interview* (2014), about two bumbling media personalities tasked with assassinating North Korean dictator Kim Jong-un. In another time, this movie surely would have been little more than another silly Seth Rogan–James Franco comedy striving to overcome critical disinterest on its way to box office success – or maybe not. No one will ever know, because in a stunning example of life imitating art, the film's content so outraged the hermit republic that Sony – which was distributing the movie – suffered a crippling cyber-attack that the FBI linked to North Korea. It might never be fully known whether this attack was primarily motivated by Kim Jong-un's documented antipathy toward *The Interview* or if there were other forces at work. However, the fall-out from the hack was monumental indeed. Fearful of threats of violence, movie theaters across the country refused to screen *The Interview*. Sony, in turn, pulled its December 2014 release, which led to much criticism that the studio was caving to censorship. This ultimately led to *The Interview* being released as a streaming download on iTunes and Google Play before getting a limited release on fewer than 100 screens a month later.

Since then, cyber warfare has become an increasingly dreadful part of everyday life, affecting the world far beyond the entertainment industry, and permeating everything from the small-scale ransom of a person's private data to presidential elections. As we make our way through the third decade of the 21st century, the possibilities for global disaster on par with the devastating toll of the COVID-19 pandemic, through cyber warfare, might very well jump from the imagination and permeate our daily lives.

At the dawn of television, studios fretted it would harm their bottom line. At the dawn of home video, studios fretted it would harm their bottom line. At the dawn of streaming, studios fretted it would harm their bottom line. Instead, each of these groundbreaking technologies followed their own unique trajectories to enhance, not detract from, the business model. Adjustment can be painful, and the unwritten future can feel frightening, but the spirit of human innovation, creativity, entrepreneurship, and,

Figure 13.14 *The Interview*.
Source: Photo courtesy of Getty Image #46083152

Figure 13.15 The outbreak of the coronavirus brought an increasing relevance to every outbreak movie ever made – especially the disturbingly prescient *Contagion* (2011) directed by Steven Soderbergh.

FYI: NETFLIX WINS OSCARS

Slowly, the Hollywood mainstream has been coming around to accepting the inevitability of streaming first-run big movies. According to Adam Epstein, writing for QZ,

> Hollywood has not been pleased with Netflix's decision to release most of its films to subscribers online the same day that they're put in theaters, which challenges the century-old relationship between distributors and theater owners. Nor are they happy with the small number of theaters that Netflix does allow its movies to be screened in.

And yet, in the six years beginning in 2015, the number of Oscar-nominated Netflix properties has grown from one in 2015 to 24 in 2020, the most nominations of any studio.

yes, the pure, raw power of an industry behemoth whose influence permeates the lives of billions, has, time and again, seen its way through to a brighter, more lucrative tomorrow.

The Studio System, Branding, and the Bloated Disney Empire

An argument could be made that the studio system of today is not all that different from in the past. The major studios, after all, continue to finance, market, and distribute feature films that are intended – in their first run – to be viewed in a theater.

Moreover, much like today, in decades past, 'branding' was very much a part of how each studio positioned itself in the marketplace. For example, in the 1930s, at the dawn of the talkie, Warner Brothers was known for producing low-budget, gritty films, Paramount for its sophisticated comedies, MGM for its glossy productions, RKO for its special effects, Fox for its biographies and musicals, Universal for

Figure 13.16 The Netflix-produced *The Irishman* was released theatrically on November 1, 2019, and was streaming less than a month later. It was nominated for ten Academy Awards.

its horror films, Republic for its westerns, Disney for its animation. Columbia had been a straight-up B-movie studio until the legendary Frank Capra directed films like *Mr. Deeds Goes to Town*, *Mr. Smith Goes to Washington*, and, after World War II, *It's a Wonderful Life* (which was produced by Capra's own independent production company). Capra earned the studio multiple Oscars, propelling it into a major studio player.

Fast forward to the present day, and RKO, Republic, and even the once-grand MGM have alternately folded, declared bankruptcy, or had their libraries sold and acquired. Fox splintered, and its entertainment division, 21st Century Fox, was sold to Disney in 2019. Warner Brothers, Paramount, and Universal (now Comcast NBC Universal) remain major players, as does Columbia – though under another name, as it was acquired by Sony in 1989.

Of course, in other ways, the studio system is virtually indistinguishable from what it once was. Starting in the 1980s, studios (that in the past were owned by movie moguls) had become profitable arms of the multinational corporations that own them today: Sony Corporation, Time WarnerMedia, Comcast, ViacomCBS, and the Walt Disney Company.

And that is before we get into all the streaming studios that are not only very much in the game, but asserting themselves as major players moving forward, giving the long-time studio giants a run for their money.

But even if the game has more players, its rules were largely written decades ago by those mega-national corporations. A case could be made that this new influx of corporate capital in the 1980s was one of the primary factors leading to ever-increasing budgets, to the point where today a movie that costs $100 million (a sum that just 30 years ago was virtually unthinkable) is considered on the low end of a blockbuster budget. Why would studios and their corporate parents be willing to spend so much to make and market a single movie? Perhaps this can best be answered by James Cameron's *Titanic* (1996), a movie whose budget kept going up and up, whose release date kept getting pushed back, and whose pundits speculated financial disaster on par with a certain infamous collision between ship and berg. Instead, *Titanic* became the highest-grossing movie ever made – until Cameron broke his own record with 2009's *Avatar* (which was then broken in 2019 by *Avengers: Endgame*). For the first time, the notion that a single film could gross more than a billion dollars in global profits had become a reality.

The concept of branding has also come a long way from the 1930s. Nowadays, it is hard to see much of a difference in what type of movie one studio makes as compared with another. A perusal of the top-grossing movies year in and year out, for example,

Figure 13.17 The success of *Easy Rider* helped stimulate American filmmaking starting in the late 1960s. It is considered part of the 'American New Wave' of filmmaking that flourished from the late 1960s until the early 1980s.
Source: Photo courtesy of Columbia Pictures/Photofest. © Columbia Pictures; Photographer: Peter Sorel

reveals multiple studios making big-budget forays into comic book adaptations, CGI family features, sequels, and remakes.

Branding for the modern era, then, is more about studios being able to generate properties with franchise potential. For example, Universal has the *Fast and Furious* series, Paramount *Transformers*, and Warner Brothers the D.C. comics franchise. Then there is Disney, which has for years been swallowing up studios and franchises, beginning with Pixar in 2006, then Marvel Entertainment in 2009, Lucasfilm in 2012, and 21st Century Fox in 2019. Disney's dominance atop the box office charts, to say nothing of the runaway success of its streaming app Disney+ (launched in late 2019), has been staggering. As a point of comparison, 2014 saw Warner Brothers, Fox, and Disney neck and neck in terms of international box office dominance. In 2019, just five short years later, *eight* of the top-ten highest-grossing movies of the year were Disney properties.

Speaking of Star Wars

It is hard to believe that a few short decades ago, *Star Wars* fans were desperate for more. *Return of the Jedi* had been released in 1983, and as for another *Star Wars* trilogy, there were but rumors, innuendo, and a long, long wait.

When *The Phantom Menace* (1999) was finally released, many who had waited their entire childhoods for this singular moment of euphoria entered the theater with impossible expectations, only to walk out two and a half hours later feeling disappointed. For a great many, that disappointment would turn to disdain, and neither *Attack of the Clones* (2002) nor *Revenge of the Sith* (2005) would do much to improve their standing – though, one can make a case that flawed though they may be, there is still a lot of cool stuff in all three movies (the Darth Maul lightsaber battle, CGI Yoda going mano-a-mano with Count Dooku, the chilling image of Dark Side Anakin approaching the Jedi temple with murderous intent, to name but a few).

Seven years after *Revenge of the Sith*, Disney purchased Lucasfilms for an astronomical $4.05 billion, and immediately set to work reinvigorating the franchise. Few were surprised that Disney's first go at it, *The Force Awakens* (2015), was destined to become the highest-grossing domestic movie of all time. Director J. J. Abrams and *Empire Strikes Back* scribe Lawrence Kasden (who wrote the script along with Abrams and Michael Arndt) were even mostly able to overcome the critical apathy and fan vitriol that plagued the previous three releases.

One *Star Wars* fan, however, was not particularly enamored with the newest chapter. This was none other than its creator, George Lucas. In an interview with Charlie Rose, a grumpy Lucas compared Disney to "white slavers." While Lucas later apologized for the remark, it does not mean he did not have a point (however hyperbolic and arguably offensive the statement might have been). With its media and merchandising muscle seeming to permeate virtually every corner of the globe, a compelling argument could be made that Disney does take on certain characteristics of the kind of imperialistic planet-crushing agent of the Dark Side commonly associated with *Star Wars* villains.

Meanwhile, fans have little need to hunger for more *Star Wars* content. Along with the completion of the most recent trilogy, as well as two one-offs – *Rogue One* (2016) and *Solo* (2018) – and the series *The Mandalorian* that kicked off Disney+, there is a staggering multitude of *Star Wars* features and series in various stages of development.

Perhaps these days, instead of hungering for the next *Star Wars* movie, fans are instead looking back at a simpler time when the *Star Wars* universe felt less crowded. Through the rose-tinted lenses of nostalgia, fans long for the version where Han shoots first in the original *Star Wars: A New Hope* (1977) that would go on to become an unexpected sensation – an homage to the matinee B-movie genre flicks the film school-educated Lucas grew up loving, coupled with inspiration ranging from Akira Kurasawa's 1958 masterpiece *Hidden Fortress* to Joseph Campbell's seminal book on comparative mythology, *The Hero with a Thousand Faces* (1949).

By comparison, there is a certain focus-grouped-to-death formula to this multi-billion-dollar franchise that make Lucas's original indeed feel as if it came from a long, long time ago and a galaxy far, far away.

The Independent Revolution

In the 1980s, the formative years of another influential movement in American film became evident. This movement had its roots in the renegade filmmaking of the 1960s and 1970s from the likes of Dennis Hopper and John Cassavettes. Hopper starred in, co-wrote, and directed *Easy Rider* (1969), which, from drug-dealing protagonists to a shocking burst of violence seconds before the end credits, bucked all Hollywood conventions and became a megahit. Cassavettes wrote and directed a slew of off-beat films in the 1970s, including *Minnie and Moskowitz*, *A Woman Under the Influence*, and *The Killing of a Chinese Bookie*. The artists inspired by Hopper, Cassavetes, and other visionary provocateurs of the 1970s quickly found blockbuster-minded Hollywood a hostile environment to make their small, personal, and often twisted, violent, and highly stylized films. So, filmmakers such as John Sayles, Jim Jarmusch, Spike Lee, David Cronenberg, Joel and Ethan Coen, Sam Raimi, David Lynch, and Steven Soderbergh would blaze the trail for the independent revolution that would reach its maturity in the mid-to-late 1990s.

Independent (indie) films are, by definition, movies that raise their money outside of the studio system, though connotatively, the term tends to make people think of movies that are low-budget, serious, quirky, experimental, and/or non-mainstream.

In fact, because indies experienced so much success in the 1990s, every major studio soon had an 'indie' wing – finding, acquiring (and occasionally developing) the kinds of challenging, edgy movies that tend to be released at the end of each year in hopes of garnering an Oscar nomination or two.

Unions and Making a Living in the 'Biz'

Careers in the entertainment industry, even at the lowest rung of the ladder, are incredibly difficult to come by. When a raging pandemic is not keeping everyone at home, young people stream to Southern California by the scores, nursing the ambition to make their dreams come true.

Rare, however, is the breakthrough talent who takes Hollywood by storm. The truth of the matter is, despite those who luck into a job as a production assistant or intern, the overwhelming majority of Hollywood dreamers will give up, go home, and do something less soul-crushing with their lives. While it remains true that one can make a great living in Hollywood, it is important to note that the vast majority of people working in the entertainment industry enjoy little to no job security.

But the exceptional few who, through luck, perseverance, talent, connections, and other intangibles, do stay, these are the people who will form the backbone for the entertainment industry's next generation. The production assistant today could, in ten years, be a studio executive, staff writer, producer, editor, or camera operator, or one of hundreds of different careers that keep the entertainment industry running.

Hollywood is a union town, and virtually every job that goes into the producing of a film (from camera operators to grips to sound engineers to transportation) is done through organized labor. Likewise, the 'creative' fields, such as writers, actors, directors, and editors, are also organized into unions. With all the money that flows through the entertainment industry, members of union film crews enjoy high wages. Although it is difficult to join these unions, once a person is in, it is possible to enjoy steady work. This is not always the case with those in the creative unions, where separate guilds representing the interests of writers, actors, and directors certainly do a lot toward guaranteeing fair (and handsome) wages, as well as royalties and protections from exploitation. However, these guilds do not help the individual creative talents find work, and it is not uncommon for a writer, director, or actor to enjoy a brief string of success before washing out and never working again.

It might seem antithetical for a movie studio to spend $20 million on a movie star and then penny pinch at the margins; however, this practice not only happens, it is, essentially, business as usual. Money people, after all, are always trying to maximize profits and minimize costs. In the past, the studio system worked well because everything was done 'in house.' Movie studios were essentially mini-cities. Every aspect, from the initiation of the screenplay to the final cut, from the wardrobes and make-up to the props and sets, was created on the studio lot. Even the biggest stars (as well as those rising hopefuls) would be contractually attached to a single studio.

While studio lots are still in business, the studio system as previously organized is no more. Thanks to the globalization of the entertainment industry and particularly the digitization of video, audio, and special effects, production need not all be done in the same place, and it is often much cheaper when done elsewhere.

The phenomenon of 'runaway production,' for example, has become increasingly commonplace. Ever concerned about the bottom line, studios and production companies will move their productions away from Los Angeles to places in the country and across the globe where the labor laws are more relaxed (or non-existent), where the dollar is stronger, and costs are lower.

Diversity in Film

In spite of South Korea's *Parasite* winning the Oscar for Best Picture, the 2020 Academy Awards were a historically bad year for diversity. A month earlier, right after Oscar nominations were announced, literary giant (and voting Academy member) Stephen King tweeted, "I would never consider diversity in matters of art. Only quality. It seems to me that to do otherwise would be wrong." This caused something of a social media uproar, which was contextualized well by Aja Romano, writing for Vox:

Many people who responded to King's surprising statement were quick to point out that all too frequently, the systems that produce art are designed to shut out minority artists – meaning that many of them never even get the chance to be judged by the quality of their work.

King went on to write an editorial for *The Washington Post* ("The Oscars are still rigged in favor of white people"), clarifying his comments:

I ... said, in essence, that those judging creative excellence should be blind to questions of race, gender or sexual orientation.
　I did not *say that was the case today, because nothing could be further from the truth. Nor did I say that films, novels, plays and music focusing on diversity and/or inequality cannot be works of creative genius. They can be, and often are.*

Indeed, while there have certainly been strides made in improving diversity, women, people of color, and the LGBTQIA community are still severely under-represented on both sides of the camera – you can look no further than this very chapter, in which the overwhelming majority of people mentioned in the past, present, and future of cinema happen to be white males.

CAREER TRACKS: MATTHEW FEDERMAN, EXECUTIVE PRODUCER/SHOWRUNNER

WHAT IS YOUR JOB? WHAT DO YOU DO?

I'm an Executive Producer/Showrunner in television. The term Showrunner doesn't appear in the credits but it designates the Executive Producer who is the head writer and has final say on creative decisions. My writing partner is also a Showrunner and runs the set while I run the Writers room. We come up with the story as a group, then writers go off to write their script. I give them notes. They rewrite, and then I do the final rewrite.

HOW LONG HAVE YOU BEEN DOING THIS JOB?

I've been an Executive Producer for three years. I've been a professional writer for 16 years.

Figure 13.18 Matthew Federman.

WHAT WAS YOUR FIRST JOB IN THE ENTERTAINMENT INDUSTRY?

Production Assistant, which is the lowest level job. You do things like making copies of scripts, distributing those scripts around the office and the city, getting lunches, etc. It is low paying, a lot of hours, everyone who does it has a college degree and there would be a line of people waiting to take the job if you didn't want it so you can't complain.

WHAT LED YOU TO THE JOB?

I always wanted to be a writer, decided in college I wanted to write for movies or TV, found out after college that the best way to get into TV was through the assistant route, so I made a plan for how to get an assistant job and worked my way up.

WHAT ADVICE WOULD YOU HAVE FOR STUDENTS WHO MIGHT WANT A JOB LIKE YOURS?

There are two parallel paths that have to converge: Your writing needs to be at a high level and you need to know people in order for your writing to be seen. So first, always be working on your craft, develop a group of writers you trust to give and get notes from, and get good at implementing notes (while maintaining your vision) because that's a lot of what the job is. You'll have to move to LA at some point, and then you need to work on the other track of meeting good, like-minded people and building your camp with them, which means you all help each other whenever you can and rise together. But never stop writing and never be waiting for the phone to ring.

Figure 13.19 A still image from the Oscar-winning Korean film, *Parasite*.

ZOOM IN 13.4

Every year, the Ralph J. Bunche Center for African American Studies at UCLA publishes a diversity report. See https://socialsciences.ucla.edu/wp-content/uploads/2018/02/UCLA-Hollywood-Diversity-Report-2018-2-27-18.pdf

Some statistics from their 2018 edition that pop out include:

- In 2016, women leading the billing in top theatrical films was at 31.2%, a small increase from 25.3% in 2013.
- While people of color make up 38.7% of the U.S. population, as of 2015–16, they comprised 18.7% of broadcast scripted content. This paltry number, in fact, is a marked improvement over where we were in 2011–2012, when people of color only comprised a minuscule 5.1%.
- On a downward trend, the report noted: "The share of the top films with casts that were 10 percent minority or less decreased from more than half in 2011 (51.2 percent) to 37 percent in 2016."
- The percentage of people of color directing top movies has been virtually unchanged between 2011 (12.2%) and 2016 (12.6%), with only one slight outlier in 2013 (17.8%).
- Meanwhile, show creators by gender has shown a small downward trajectory, with women comprising 26.5% in 2011–2012, and only 22.1% in 2015–2016.
- The percentages have remained equally dismal for people of color writing top theatrical films. In 2011, that number was 7.6%. In 2016, it was 8.1%.

Figure 13.20 In 2016, the hit Amazon series *Transparent* was hailed as being 'the most trans-inclusive show in Hollywood history.'

Beyond the binary male/female designation, the UCLA diversity report does not deal with issues of gender and sexuality. However, GLAAD (formerly the Gay & Lesbian Alliance Against Defamation) publishes its own report. In its 2018 overview, GLAAD found that of 109 films released by major studios, 12.8% featured characters who identified as lesbian, gay, bisexual, transgender, and/or queer. It also found that gay men were by far the most represented (64%), and that racial diversity of LGBTQ characters increased from 20% in 2016 to 57% in 2017.

> **ZOOM IN 13.5**
>
> GLAAD: The GLAAD Studio Responsibility Index maps the quantity, quality, and diversity of LGBTQ people in films released by the seven major motion picture studios.
> Learn more at: www.glaad.org/sri/2018

So, even today, diversity in the entertainment industry is a mixed bag. On the one hand, there are so many more opportunities for a multiplicity of divergent voices to be seen and heard. On the other hand, as Stephen King noted, "[A]s with justice, judgments of creative excellence should be blind. But that would be the case in a perfect world, one where the game isn't rigged in favor of the white folks."

Feature Films Today and Tomorrow

While there have been many hard lessons humanity has had to learn as we emerge from the COVID-19 pandemic, so too have certain long-held assumptions held up quite well. One of these, that the film industry – which is now, of course, an overly quaint and inaccurate term to describe the largely digital, sprawling passive entertainment hydra that is 21st-century entertainment media – remains recession proof (and Depression proof). While planet Earth shut down, flat panel smart TVs across the globe streamed desperately needed escapist content while the sirens blared and the death count soared.

It is precisely because this industry has adapted so adeptly to every kind of change humanity has undergone in the centuries-spanning epoch since the zoetrope was all the rage, that it will continue to thrive into the brave, uncertain, unwritten beyond. This industry, after all, has been around so long, COVID–19 was not even the first global pandemic it has had to weather. It has thrived through hot wars and cold wars, prosperity and peace, reflecting who we are, who we aspire to be, the best of ourselves, and the worst.

The future of this mighty, resilient, and enduring industry is being written every day. Although there are many unknowns as we gaze into the future, of one thing we can be certain: just as it always has, the entertainment industry will continue to change, adapt, and evolve.

SUMMARY

The instinct to tell a story is a fundamental characteristic that dates back to humanity's sentient awakening. Visual storytelling has evolved from ancient cave paintings to live theater to – in the latter half of the 19th century – moving photo collages, which would become the starting point for cinema, an enterprise that has endured and evolved to become the trillion-dollar industry it is today.

Although most think of Hollywood as film's birthplace, its story in fact began in Fort Lee, New Jersey, where, in 1893, Thomas Edison began experimenting with motion pictures. It was in part to escape the litigation-happy Edison (who held the patents to most filmmaking equipment) that filmmakers moved west to California, where the first movie studio was built around 1911.

In the early decades of the film industry, America was one of many countries (along with France, Russia, and Germany) leading the way in terms of artistic, technical, and commercial achievement. However, the unrest and violence that swept through Europe during World Wars I and II devastated those economies, allowing America to emerge as the worldwide film industry leader.

The 1920s gave rise to the studio system. Silent cinema, with its reliance on continuity editing and the non-verbal expressiveness of its actors, spoke a universal language that was immediately accessible to English speakers and non-English speakers alike, helping to forge a shared American identity that incorporated both the growing immigrant population as well as those who had been living in the country for generations.

In 1934, the Production Code (or Hays Code) was enforced by the studios, setting forth standards of 'good taste' in which, among other things, there could be no foul language or overt sexuality. Mixed-race romances were also forbidden, and all bad deeds had to be punished in the end. The Hays Code lasted until 1968, when the industry adopted the MPAA film-rating system.

Hollywood studios enforced the Hays Code, just as they would later become shamefully complicit in the anti-communist hysteria of the 1950s, blacklisting a great many of its creative talents, and ruining lives and careers.

If there is one constant to the story of cinema, it is the speed with which technology either becomes incorporated into the product or else triggers other kinds of innovation. In the late 1920s, 'talkies' gradually replaced silent cinema. By the late 1930s, color began to supplant black-and-white films. When television began to threaten box office receipts, cinema's then nearly square aspect ratio was transformed to the widescreen format we know today. In the 21st century, with increasingly impressive home entertainment setups and streaming technology, theaters have adapted with 3D, IMAX, luxury seating, and serving food and drinks – but thanks to the coronavirus pandemic, the movie theater experience as we know it is in danger of extinction. Meanwhile, CGI would redefine what could be possible in a live-action movie, beginning in the 1990s with movies such as *Jurassic Park*, *Terminator 2*, and *A Toy Story*, until becoming commonplace and increasingly seamless here in the third decade of the 21st century.

In the 1970s, the first generation of film school-educated filmmakers gained prominence, bringing about a new epoch of director-driven films whose works defined the era. Whereas 40 years earlier, movies like *Jaws* and *Star Wars* would have been the low-budget B movie at the tail end of a double feature, in this new era, these would be the precursors to the modern-day blockbuster.

The 1980s saw the formative years of another important movement: Independent cinema (or indies). Independent films are, by definition, movies that raise their money outside of the studio system, though connotatively, the term tends to make people think of movies that are low-budget, serious, quirky, experimental, and/or non-mainstream. In the 1990s, thanks in no small part to movies like *Pulp Fiction* and *Clerks*, indies experienced so much success that every major studio soon had its own 'indie' wing – finding, acquiring (and occasionally developing) the kinds of challenging, edgy movies that tend to be released at the end of each year in hopes of garnering an Oscar nomination or two.

In the early years of Hollywood, each studio was characterized by the types of movies it made. Warner Brothers was known for low-budget, gritty films, Paramount for sophisticated comedies, MGM for glossy productions, RKO for special effects, Fox for biopics and musicals, Universal for horror, Republic for westerns, and Disney for animation, while Columbia began as a low-budget studio before Frank Capra helped transform it into a major player with a string of Oscar-winning hits. These days, the remaining movie studios are profitable arms of giant, multinational corporations. As such, today's studios tend to brand themselves less by the types of movies they make and more by the types of franchises they bank on to deliver massive profits for their corporate overlords. All the while, they are being pushed from an increasing number of streaming services whose tendrils are digging ever deeper into every aspect of the entertainment mainstream, up to and including the Academy Awards.

As for careers in the entertainment industry, there are the exceptional few who become the big stars, directors, producers, and writers. There are also legions of employees who make the movie industry run. Hollywood is a strong union town, and in so being, it guarantees crew members high wages; but it has also contributed to the phenomenon of 'runaway productions,' in which bottom line-conscious producers and studios seek less expensive incentives elsewhere in the country or overseas.

The film industry has evolved by leaps and bounds from those early days of Zoetropes and nickelodeons. However, that initial attraction to experiencing storytelling through moving images remains as captivating today as it was nearly a century and a half ago. The film industry is a global force that continues to permeate more markets, and the fact that today anyone can make a film on a cell phone means anyone with a story to tell is not only able to do so but also able to make it available for anyone to see.

Glimpsing into the future, while there are certain to be plenty of surprises along the way, we can expect a movie industry that will continue to adapt and evolve while attracting new generations of fans and talent.

BIBLIOGRAPHY

Alexander, J. (2020, March 13). Blockbuster movies delayed by coronavirus precautions likely won't become streaming exclusives. *The Verge*. Retrieved from: www.theverge.com/2020/3/13/21178200/blockbuster-movies-delayed-mulan-mutants-f9-quiet-place-theaters-streaming

Bryant, J. (2015, December 30). George Lucas says he sold Star Wars to 'White Slavers.' *Variety*. Retrieved from: http://variety.com/2015/film/news/star-wars-george-lucas-disney-white-slavers-1201669959/

Epstein, A. (2020, January 13). Netflix has taken over the Oscars. *Quartz*. Retrieved from: https://qz.com/1784161/netflixs-leads-all-2020-oscar-nominations-thanks-to-the-irishman-and-marriage-story/

Faughnder, R. (2020, March 25). Will movie theaters – and moviegoing – survive coronavirus closures? *Los Angeles Times*. Retrieved from: www.latimes.com/entertainment-arts/business/story/2020-03-25/movie-theaters-recover-coronavirus-closures

GLAAD. (2018). *2018 GLAAD studio responsibility index*. Retrieved from: www.glaad.org/sri/2018

Harris, M. (2009). *Pictures at a revolution*. London: Penguin Books.

Hunt, D., Ramón, A. C., Tran, M., Sargent, A., & Roychoudhury, D. (2018). *Hollywood Diversity Report 2018: Five years of progress and missed opportunities*. Retrieved from: https://socialsciences.ucla.edu/wp-content/uploads/2018/02/UCLA-Hollywood-Diversity-Report-2018-2-27-18.pdf

Janes, R. (2020). *Medium*. Retrieved from: https://medium.com/swlh/prepare-for-the-death-rebirth-of-hollywood-f3853aacbee0

King, S. (2020, January 27). The Oscars are still rigged in favor of white people. *The Washington Post*. Retrieved from: www.washingtonpost.com/opinions/stephen-king-oped-the-oscars-are-rigged-for-white-people/2020/01/27/ad29c4e8-407c-11ea-aa6a-083D01b3ed18_story.html

Krikowa, N. (2019). Intervention as activism: Advocating queer female representation through independent film production. *Refractory: A Journal of Entertainment Media*.

Mast, G., & Kawin, B. (2007). *A short history of the movies*. New York: Pearson.

Romano, Aja. (2020, January 14). Stephen King's confusing tweets about diversity missed a larger point about the Oscars. *Vox*. Retrieved from: www.vox.com/culture/2020/1/14/21065850/stephen-king-diversity-art-tweets-oscars

Schwartzel, E. (2020, April 28). 'Trolls World Tour' breaks digital records and charts a new path for Hollywood. *The Wall Street Journal*. Retrieved from: www.wsj.com/articles/trolls-world-tour-breaks-digital-records-and-charts-a-new-path-for-hollywood-11588066202

Shapiro, A. (2016, September 13.) How *Transparent* became the most trans-inclusive show in Hollywood history. *Out*. Retrieved from: www.out.com/out-exclusives/2016/9/13/how-transparent-became-most-trans-inclusive-show-hollywood-history

Wolk, A. (2015, February 19). Networks as an anachronism. Retrieved from: http://tdgresearch.com/networks-as-an-anachronism

Chapter fourteen

The Personal and Social Influence of Media

The authors thank Jason Stamm, Ph.D. (University of Tennessee) and Mary Beadle, Ph.D. (John Carroll University) for their contributions to this chapter

Although television and other media have positive effects, it seems as though most of the attention is focused on the negative effects. The debate about the effects of media content on behaviors, values, and attitudes is complex because humans are complex. There are no simple answers to the many questions concerning how mediated content influences media consumers. As Dick Cavett, long-time humorist and talk show host, once commented, "There's so much comedy on television. Does that cause comedy in the streets?"

This chapter begins with a historical look at concerns about mediated messages and the development of theories that help explain the connection between mediated messages and human behaviors and attitudes. The chapter provides an overview of how media effects are studied. Contemporary issues concerning violence on television and the effects of viewing such material are examined next. The chapter ends with a look at the implications of the effects of mediated content and the implementation of rating systems, V-chips, and other tools that screen offensive content.

SEE IT THEN

Strong Effects

Concern about the negative influence of the media is not a modern-day phenomenon. Religious and government organizations have attempted to suppress printed works that they deem ideologically contrary and unfit for consumption since the advent of the printing press in the mid- to late 1400s. In addition to religious and government elites, members of the upper social strata have long tried to stifle the written word and keep new ideas from the masses, because knowledge is power, and they wanted to keep the power among themselves. With increased literacy and access to mass-produced writings came an even greater need to keep information out of the hands of the general public to silence opposition.

In the 1800s, the penny press sizzled with sensationalized accounts of criminal activity, sexual exploits, scandals, and domestic problems, which prompted community outcry about the negative social effects of reading such scintillating material. In the 1920s, parents protested violent content in motion pictures, and in the 1930s and 1940s, they focused on comic book violence.

The years between the end of World War I and the onset of the Great Depression were a time of growth in the United States. People were migrating from their rural homes into the industrialized cities, leaving behind networks of friends and family and old ways of life in which traditions, behaviors, and attitudes were passed along from one generation to the next. Moving to the city disconnected individuals from family and social ties and left them to assimilate into a new culture. Without family and friends to depend on, newcomers to a city turned to newspapers, magazines, and books to learn about new ways of life and to keep up with current events. The mass media became a central and influential part of everyday life.

Social scientists sought to explain the social and cultural changes brought on by migration to the cities, industrialization, and increased dependence on the media. They tied personal behavioral, attitudinal, and cognitive changes to the media and later developed theories to help explain how certain aspects of the media affect our lives.

A **theory** is basically an explanation of observed phenomena. Researchers offer varying definitions of theory. "Theories are stories about how and why events occur" (c.f. Baran & Davis, 2000, p. 29). "Theories are sets of statements asserting relationships among classes of variables" (Baran & Davis, 2000, p. 30). A theory must contain four basic criteria: conceptual definitions, domain limitations (application of conceptual definitions), explained relationships, and predictions ("what could, should, and would happen" when the concepts are measured) (Wacker, 2004, p. 629). Mass communication theories are explanations of the relationship between the media and the audience and how this relationship influences or affects audience members' everyday lives.

Magic Bullet Theory

In the 1920s, the United States was still struggling with the effects of World War I and rebuilding its economic and social structures. Newspapers were the main source of information, but radio was beginning to build an audience.

Along with these new information and entertainment providers came the fear that the media could take over people's minds and control the way they thought and behaved. Many thought of the media as a '**magic bullet**' or '**hypodermic needle**' that could penetrate people's bodies and minds and cause them to all react the same way to a mediated message. This concept of an all-powerful media was a widely held and frightening belief. These fears were not unfounded when considering the successful propaganda campaigns waged during World War I, the newness of the mass media, and the move to the cities that left many people without close social networks.

Propaganda and Persuasion Theories

During World War I, propaganda was used to spread hatred across nations, to concoct lies to justify the war, and to mobilize armies. Although propaganda was used more intensely in Europe, it quickly spread to the United States. Starting in the 1920s, the world followed Adolf Hitler's rise to power, which was aided by his domination over radio and carefully crafted propaganda campaigns. During this time, there was also a broad range of social movements in the United States. Radio was an especially powerful tool for spreading propaganda and persuasive messages.

Definitions of propaganda vary slightly and often overlap with definitions of persuasion, but psychologist Harold Brown has distinguished between the two concepts. According to Brown, propaganda and persuasive techniques are the same, but their outcomes differ. **Propaganda** is "when someone judges that the action which is the goal of the persuasive effort will be advantageous to the persuader but not in the best interests of the persuadee," whereas **persuasion** is when the goal is perceived to have greater benefits to the receiver than to the source of the message (Severin & Tankard, 1992, p. 91).

ZOOM IN 14.1

Examples of World War I and World War II propaganda can be found at these sites:

- German Propaganda Archive (speeches, posters, writings): www.calvin.edu/academic/cas/gpa

- Snapshots of the Past – World War I and II posters: http://bir.brandeis.edu/handle/10192/23520

Figure 14.1 A Nazi Party election poster, urging German workers to vote for Adolf Hitler.
Source: © Lebrecht Music & Arts/Corbis

Limited Effects

Imagine that it is the night before Halloween in 1938. You live in a rural farmhouse in New Jersey. You do not have a telephone or a television, your nearest neighbor is half a mile away, and you depend on your radio to link you to the outside world. The radio airwaves are filled with news about Hitler's rise to power in Germany, and rumors abound that outsiders are infiltrating the United States to initiate the fall of democracy. It is a scary world.

You settle in after dinner and tune your radio to the *Mercury Star Theater* program. You hear that tonight's program is a recreation of H. G. Wells's book *The War of the Worlds*. But if you were not listening carefully or had tuned in a little late, you would not have heard that announcement. In that case, what would you have done when dance music playing as part of the program was interrupted with a 'news report' that a spaceship had landed in New Jersey and that we were at war with Martians? Would you have tuned to another radio station to find out if the report was true?

Many radio receivers in those days could pick up only one or two stations. You would not have had a television to turn on or a telephone to call your friends. Even people with telephones could not call out, because the phone lines were jammed. Would you have just laughed off the report, or would you have panicked?

When radio was still a relatively new medium, many listeners relied on it as their primary news source and believed what they heard. When a 'news report' interrupted the music, many listeners believed that Martians were indeed invading Earth. Mass panic ensued. People jumped in their cars and headed somewhere, anywhere. They hid in closets and under beds. Some even thought of committing suicide. Other listeners, however, realized the broadcast was merely a rendition of *The War of the Worlds*. They listened with bemusement and thought the program was quite clever and entertaining.

Considering what was thought about the magic bullet theory at the time, it would make sense to conclude that everyone who heard *The War of the Worlds* broadcast panicked. But that is not what happened. Scholars at Princeton University took the lead by conducting many studies about audience reaction to the broadcast. Basically, the scientists found that listeners' reactions to the show were influenced by several factors, such as education, religious beliefs, socioeconomic status, political beliefs, whether listeners tuned to the broadcast at the beginning of the program or sometime during the show, where listeners were during the broadcast (e.g. rural home, city apartment), and whether they were alone or with others when listening.

Research about *The War of the Worlds* panic and other studies show that the media are not as all powerful as once thought. Audience members do not all react in the same way to the same mediated stimulus, because everyone's life experiences are different, and thus people filter messages such that they interpret them in their own ways. The findings from this line of research led to the **limited-effects perspective**, which is the contention that media have the power to influence beliefs, attitudes, and behaviors, but that influence is not as strong as once thought. Moreover, the media are not just evil political instruments but have positive effects as well. That the media's influence is limited by personal characteristics, group membership, and existing values and attitudes makes us less vulnerable and not easily manipulated by what we see and hear.

Figure 14.2 An alien spacecraft opens fire in a scene from the 1953 movie *The War of the Worlds*.
Source: Photo courtesy of Paramount/the Kobal Collection

ZOOM IN 14.2

To learn more about *The War of the Worlds*, go to:

- www.museumofhoaxes.com/hoax/archive/permalink/the_war_of_the_worlds/

Listen to the entire 1938 broadcast:

- www.archive.org/details/OrsonWellesMrBruns
  www.youtube.com/watch?v=XsOK4ApWl4g

Several other perspectives came out of limited-effects research. After conducting a series of studies, researchers Paul Lazersfeld and Elihu Katz discovered that rather than media directly influencing the audience, messages are filtered through a two-step flow process. **Two-step flow** theory explains that messages flow from the media to opinion leaders and then to opinion followers. The process starts with gatekeepers, such as news producers, newspaper editors, and others who filter or edit media messages. The messages are passed on to opinion leaders, or influential members of a community, who then pass on the messages to opinion followers, or the people the gatekeepers and opinion leaders are trying to influence. For example, suppose that a neighborhood group opposes building a new road through its community. Rather than just send anti-road messages to the mass public, the group will be more effective if it persuades a smaller number of influential homeowners (opinion leaders) that the road will harm the neighborhood and then has the opinion leaders influence the larger group of neighbors (opinion followers) to rally against the road. The two-step flow theory supports limited effects by demonstrating that opinion leaders are often more influential than the media.

Further support for the limited-effects perspective emerged with the identification of selective processes. **Selective processes** are "defense mechanisms that we routinely use to protect ourselves (and our egos) from information that would threaten us. Others argue that they are merely routinized procedures for coping with the enormous quantity of sensory information constantly bombarding us" (Baran & Davis, 2000, p. 139).

There are three basic ways in which selective processes operate:

1. **Selective exposure** is the tendency to expose ourselves to media messages that we already agree with and that are consistent with our own values and beliefs.
2. **Selective perception** is the tendency to change the meaning of a message in our own mind so it is consistent with our existing attitudes and beliefs.
3. **Selective retention** is the tendency to remember those messages that have the most meaning to us.

Selective processes support the limited-effects model by explaining how we filter mediated messages so they do not affect us directly. Rather, we choose what messages to expose ourselves to and then screen and alter the meanings of those messages so they are consistent with our current attitudes and beliefs. Long-lasting effects are further limited because we remember only the messages that had meaning to us in the first place.

Moderate Effects

The mid- to late 1960s were marked with social unrest, instability, riots, and protests. Concerns arose about the effects of watching news footage of real-life violence. It was commonly believed that the viewing public, and children in particular, could not discern between fictionalized violent content and violent news content, and that viewing aggressive behavior caused people to act more aggressively in real life.

On the one hand, the media claimed that there was no relationship between viewing violence and increased aggression; in other words, the media claimed they had a limited effect on the viewing public. Yet on the other hand, the media claimed to their advertisers that they could indeed persuade the public to purchase certain products; in other words, the media had a strong effect on viewers. The limited-effects perspective was questioned in light of this inconsistency, and new research indicated that media effects were not as limited as previously believed but rather had a **moderate effect**.

FYI: VIOLENCE IN THE 1960S

Although the decade began peacefully, many factors contributed to violence in American society that sent shock waves throughout the country and the government. The civil rights movement raised awareness about racial inequality, President John F. Kennedy was assassinated in 1963, and civil unrest and even civil disobedience resulted from U.S. involvement in the Vietnam War. As Americans turned against the war, antiwar and civil rights demonstrations turned violent in cities across the country. The Reverend Dr. Martin Luther King Jr., a civil rights advocate and leader, and Senator Robert F. Kennedy, a presidential candidate and brother of the late president, were both assassinated in 1968.

Congress and the public turned their attention to the causes of violence and social unrest. The media, and especially violent television programs, were blamed for contributing to social ills. Near the end of the decade, Congress started gathering scientific information about the causes of violence in society. The U.S. Surgeon General supported numerous studies that investigated the relationship between violence in the media and violence in society.

In the early 1960s, Stanford University psychologist Albert Bandura was studying the effects of filmed violence on children. For example, in one variation of Bandura's 'Bobo doll' experiments, children watched a short film of other children playing with Bobo dolls. A Bobo doll is an air-filled plastic doll, usually about three feet tall. The doll is weighted at the bottom, so when it is punched, it bounces back and forth. One group of children was shown a film of kids punching and kicking their Bobo dolls and yelling angry, nonsensical words while doing so. A second group of children was shown a film of children playing nicely with their Bobo dolls. After viewing one of the films, each child was given his or her own Bobo doll to play with. As it turns out, the children who viewed the film of the kids playing nicely with their dolls were also gentle with their own dolls, but the children who watched the Bobo dolls being subjected to violent play also kicked and punched their dolls and spouted nonsensical words while doing so.

Bandura and other researchers demonstrated that children and adults learn from observation and model their own behavior after what they see, whether in real life or in films or on television.

Figure 14.3 Police tangled with demonstrators at the 1968 Democratic National Convention, one of many violent incidents that year.
Source: © JP Laffont/Sygma/CORBIS

Figure 14.4 In one study, children imitated film-mediated aggression by beating up Bobo dolls.
Source: LearningSpace at the Open University

Further, the media teach people how to behave in certain situations, how to solve problems and cope in certain situations, and in general present a wide range of options upon which to model their own behavior, thus lending support for media as a strong influence. Violence and aggression are very complex, and despite thousands of studies, researchers are still grappling with the role media play in the development of these human characteristics.

ZOOM IN 14.3

To see Dr. Albert Bandura and original clips from the experiments, visit www.youtube.com/watch?v=dmBwWlJg8U

Powerful Effects

While the debate about media effects continued, studies were commissioned to specifically test the influence that television content (especially violence) had on viewers, particularly children and young adults. In 1969, the U.S. Surgeon General's Scientific Advisory Committee on Television and Social Behavior was created to conduct research on television's effect on children's behavior. After two years of extensive study, the committee concluded that there was enough evidence to suggest **powerful effects** in terms of a strong link between viewing televised violence and engaging in antisocial behavior and that the link was not limited just to children who were already predisposed to aggressive behavior.

The report stirred up much controversy. On one side, parents, medical associations, social groups, teachers, and mental health specialists called for restrictions on televised violence. On the other side, the television industry pointed to research on limited and moderate effects and lobbied hard against any new Federal Communications Commission (FCC) regulations on television. The television industry finally bowed to pressure and agreed to limit the extent of violence in children's programs and to times when children would be less likely to watch. Violence in television programs in general, however, has not decreased by much, and so the struggle to limit such content continued.

More evidence supporting the **powerful-effects perspective** came to light in the 1980s, when several major violence studies concluded that viewing violence is strongly related to aggressive behavior. Additionally, *The Great American Values Test* demonstrated that television could influence beliefs and values. This half-hour television program special was actually a research project in which viewers first assessed their own values, and then the program pointed out inconsistencies to get viewers to question their values and change their minds. The researchers were amazed that a half-hour-long program was influential enough to change viewers' attitudes and values. Even more important, the researchers found that those viewers who were more dependent on television were more likely to change their attitudes. *The Great American Values Test* and other similar research lent further credence to the idea that the media do indeed have a powerful influence on the viewing public.

Although everyone is out to protect his or her own interests, there is a general consensus that viewing too much violent or objectionable content might, under some circumstances, have serious social and personal consequences. Thus, the argument centers on what, if anything, should be done to curb violent images.

Radio Effects

Television is not the only medium that has come under fire. In late 1973, a father driving with his 15-year-old son was shocked to hear George Carlin's 1973 radio routine, *Filthy Words*. Carlin's expletive-filled monologue was broadcast in the middle of the day on a New York radio station owned by the Pacifica Foundation. The father complained to the FCC, demanding that the Pacifica-owned station lose its license for airing the offensive routine. The ensuing lawsuit and public outcry spurred the FCC to bar the broadcast of words that are indecent, or "language that describes, in terms patently offensive as measured by contemporary community standards for the broadcast medium, sexual or excretory activities or organs" (*FCC v. Pacifica Foundation*, 1978), and it established seven words too 'dirty' ever to say on the airwaves.

Figure 14.5 Comedian George Carlin became well known for a monologue that featured seven 'dirty' words.
Source: Photo courtesy of Photofest

Ten years after the hoopla surrounding George Carlin's on-air comedic dirty-words routine, attention focused on indecent words found within song lyrics. After being offended by the lyrics in the Prince song 'Darling Nikki,' Tipper Gore, wife of then-Senator Al Gore, founded the now defunct Parents Music Resource Center to protect children and young adults from the influences of indecent and violent lyrics. This group believed that young people would do what the song lyrics suggested and would be inclined to accept deviant behavior as normal.

Hearings before the Senate Commerce Committee in 1985 led to the record industry voluntarily putting parental advisory stickers on record albums and CDs that contained "profanity, violent or sexually explicit lyrics, including topics of fornication, sado-masochism, incest, homosexuality, bestiality and necrophilia."

SEE IT NOW

Powerful Effects Today

The **powerful-effects** perspective dominates today, but it is very different from the magic bullet theory of yesterday. Enough is known about the influence of the mass media to know that not all people respond in the same way to the same message. The powerful-effects model is complex, and the circumstances must be right for certain effects to occur. Most young viewers today have grown up watching an enormous amount of violent television, yet they are not all aggressive and violent, as suggested by the magic bullet theory. Rather, some viewers may be influenced and might become more aggressive than others under some circumstances.

In this chapter, the effect magnitudes are neatly delineated, but this has been done to simplify a very complex issue. Although there has been a dominant perspective at any given time, research findings have not always been consistent. Within each era of thought, numerous studies have demonstrated different outcomes.

Agenda Setting

Agenda setting refers to the mass media's power to influence the importance of certain news events. In other words, the more airtime and Web space that is devoted to an event, the more important it seems to the audience. Agenda setting is a function of the **gatekeeping** process that news media practice daily. Because of the time and space constraints of radio and television, news producers select (or 'gatekeep') stories and events to cover and then present them on the air and on their Web sites. These stories and events become increasingly important to the audience with repetition and consistency across various media. When an event is the lead story on the broadcast and cable networks' news segments, people believe it to be of utmost importance, when, in fact, other events may be more noteworthy but for various reasons (e.g. political pressure, lack of compelling video) are largely ignored. The repetition of certain themes leads to a **cumulative effects model** that if the media do not air a story, people do not think it is important. In this sense, the news media do not tell us what to think but rather what to think about.

As the number of independently owned news outlets continues to decrease because of aggressive consolidation, there will be fewer individual 'voices' (or media owners), and thus, more agenda setting. Consumer advocates, civil rights groups, religious groups, independent broadcasters, writers, and concerned citizens fear that instead of getting several different perspectives about a certain issue, they will get one perspective – likely that of the corporate owners. Powerful publishers and editors can give prominence and positive 'spin' to issues that may help them (such as tax breaks) and squelch stories that may hurt them (such as recalls on products with which the media corporation has financial ties).

Research on the Mass Media

The discussion up to this point has shown that understanding the effects of media content is not a simple matter. For example, about 4,000 research studies have been conducted that look just at the effects of mediated violence. These studies are not conducted casually but use strict procedures and methods for learning about how people use the mass media and how the mass media influence people socially and culturally.

Survey Research

Gathering information about the media audience involves a systematic method of observation that results in data that can be measured, quantified (counted), tested, and verified. **Survey research** is one of the oldest research techniques and perhaps the most frequently used method of measuring the radio, television, and 'now media' audience. Surveys are used to explain and describe human behaviors, attitudes, beliefs, and opinions. Survey research usually entails some sort of questionnaire or observation. Individuals may be asked questions about what they think about new television programs, what programs they watch, what Web sites they depend on most heavily, and how many hours per day they listen to the radio. Survey research is designed to be as objective and unbiased as possible.

Content Analysis

Content analysis is a research method used to study the content of television programs, song lyrics, Web sites, and other mediated messages. For example, content analysis reveals the number of times indecent language is used on television, the number of times violence is promoted in song lyrics, and the number of times the nation's economy is mentioned as the lead story on network news programs or their Web sites. Although content analysis does not tell about media effects or audience use of media, it does tell about media content.

Laboratory Experiments

Experimental methods allow researchers to isolate certain elements they want to study. Laboratory experiments usually involve a **test group**, which is exposed to a variable or condition under study, and a **control group**, which is not exposed to the variable or condition. For example, researchers may be interested in knowing how many times cable viewers change channels. In a laboratory setting, individuals in the test group would watch digital cable, and those in the control group would watch television without digital cable. The number of times the viewers in the test group changed channels would be compared to channel changes made by those in the control group. Researchers would then know whether having digital cable service results in more channel switching.

The biggest drawback to laboratory experiments is that people may not behave or react in a lab as they do in real life. For instance, viewers at home might switch channels less often than in a lab, where perhaps factors such as boredom and knowing they are part of an experiment might change their behavior.

Field Experiments

The purpose of a **field experiment** is to study people in their natural environment instead of in an artificial laboratory setting. Field researchers do not have much control over outside influences, but the trade-off is that they get to observe real-life behaviors. It may be more valuable to observe viewers switching channels in their own homes, where they are more likely to behave as they normally do, than in a lab, where they may behave differently.

Effects of Mediated Words and Images

The question now turns from the strength of media effects, to the media effects themselves. Generally, media content influences the way people behave, the way they think, and the way they react emotionally. People react differently to mediated content, and just how they are affected depends on many factors. Sometimes the effects are long lasting and sometimes they are fleeting. Sometimes they are more intense than at other times, depending on social, psychological, and situational factors. After watching a violent show with a friend, a viewer could feel fine, but the friend might be too hyped up to sleep.

We all like to think that we are immune to the influences of the mass media, and we have probably all heard someone say something like, "I watched a lot of TV when I was growing up, and I'm not a murderer, so it didn't hurt me." This **third-person effect**, in which individuals claim that they are not

Figure 14.6 Researchers are very interested in how and why viewers watch television.
Source: Photo courtesy of iStockphoto. © mikkelwilliam, image #0116545

as susceptible to mediated messages as others, often leads to an emphasis on others' viewing habits rather than on our own. But researchers have shown that there are three basic commonalities in the ways we react to various mediated content.

Behavioral Effects

We learn to behave by watching what others do and then following those examples. The same principle applies to the media. Viewers take behavioral cues from the media and apply them to their own lives. The following types of behavioral effects are most common.

Imitation There is particular concern that people may imitate the behavior they see on screen. Imitation means that a viewer will behave in the same way or take a similar action as a character on television or in a videogame. In a 1993 lawsuit, a parent alleged that her 5-year-old son set fire to a house after witnessing the cartoon characters Beavis and Butthead commit the same act. Although there has been much publicity about cases like this, in which children have engaged in extreme behavior that they claim to have imitated from television, these situations are rare. It is also very difficult to say whether a viewer directly imitated what he or she saw or already had a predisposition to the behavior or aggressive actions and television was merely a reminder.

Identification It could be that viewers, especially children and young adults, identify with media figures in a broader sense. For example, young viewers may want to be like their favorite television characters and so take on similar characteristics without really imitating their behaviors. Viewers might wear T-shirts, for instance, imprinted with pictures or names of their favorite television personalities or characters or even walk or talk the same as their idols, but they will not usually imitate taboo behaviors.

Inhibition/Disinhibition Punishments and rewards influence the likelihood of modeling and imitating mediated behavior. Viewing a character being punished or dealing with negative consequences creates an **inhibitory effect** in viewers. Viewers do not want to experience the same punishment as the character and will therefore refrain from engaging in the same negative behavior. On the other hand, positive reinforcement creates a **disinhibitory effect**. When a negative action is rewarded, inhibition decreases, and so the likelihood of repeating the bad behavior increases. Disinhibition also occurs when an authority figure directs a person to behave badly, therefore displacing responsibility from the perpetrator ("He told me to do it").

Arousal Viewing action-filled scenes and televised violence arouses emotions. If a viewer is feeling somewhat aggressive or stressed and then watches an action-packed show or depictions of violence, she may become more stimulated, and her initial aggressiveness or stress may intensify and lead to aggressive or violent behavior. Some athletic team coaches believe that showing aggressive sports films to players shortly before a game heightens levels of arousal and makes it more likely that they will play more aggressively.

Catharsis Although most evidence supports the contention that viewing aggression leads to increased arousal, there is some support for the opposite perspective: That viewing violence leads to catharsis or the release of aggressive feelings; watching others act out feelings of anger relieves aggressive feelings, so that we may behave more passively after watching onscreen violence. Although some empirical evidence points to a catharsis effect, the validity is largely based on tradition rather than scientific observation and testing.

Desensitization Viewing certain types of content may also lead to desensitization, or the dulling of natural responses due to repeated exposure. In other words, the shock value of seeing violence, uncomfortable situations, or sexual acts is diminished the more times they are repeated. For example, when car alarms were still a novelty, everyone would stop and look around and wonder if a vehicle was being broken into when they heard one go off. Now when an alarm shrieks, most people do not pay any attention because car alarms have become a nuisance. Similarly, repeated exposure to intense television images and offensive online graphics has lessened their effect. And just as reactions to mediated violence have dulled, so, too, have reactions to real-life situations. When we are desensitized to violence in the media, we are less likely in real life to help someone in trouble, or call the police, or feel aroused or upset when witnessing a violent act.

Affective (Emotional) Effects

When viewers see a character being mutilated, shot, thrown over a cliff, stabbed, beaten up, run over by a train, mangled in a car crash, or harmed or killed in other horrendous ways, they cannot help but react emotionally. Viewers also react emotionally to positive images, such as characters getting married, performing acts of kindness, and showing physical affection.

Everyone experiences some sort of emotional reaction to a mediated image, even if that reaction has become blunted due to overexposure. But it is the negative images that get the most attention. Most Americans were glued to their television sets and the Web as they watched the horrors of September 11, 2001. Viewing the intense violence of the day increased levels of posttraumatic stress symptoms and anxiety, especially among heavy television viewers. About 18% of viewers who watched more than 12 hours of television per day reported increased distress levels after 9/11, compared to 7.5% of those who watched television less than four hours per day.

Cognitive Effects

The media also influence how and what viewers think about the world. Portrayals of minorities, women, families, and relationships influence social and cultural attitudes and perceptions of real life. For example, more than just counting the *number* of television programs in which women and minorities appear, studies are concerned with how these groups are *portrayed*, and the mediated *effects* on viewers. The mediated portrayals of women, minorities, gays, and others, reinforce societal norms and values. This socialization process is life-long but is especially powerful on young people. Even though not all **stereotypical depictions** are negative, they are nonetheless harmful because they objectify, depersonalize, and even deny individuality. In addition, television and the other mass media could provide the dominant or perhaps the only view of certain groups in society, and shape the way viewers think about the world.

FYI: TELEVISION BOOSTS WOMEN'S RIGHTS INTERNATIONALLY

Although much attention is focused on the negative influence of television, new evidence points to the positive effects on women in developing nations. Studies show that exposure to television is associated with higher school enrollment for girls and greater autonomy and rights for women, along with a lower birth rate, which frees women from the home.

Programs such as *Baywatch* – which is the most-watched program around the world and has been seen by more than one billion people – and soap operas often portray women as having equal rights to men, as being equal partners in marriages, as having an education and a career, and as standing up to men and challenging the traditional roles. Further, girls are often named after strong, independent female television characters, which could further signify a desire to increase women's power in societies dominated by men.

Source: Soap operas boost rights, 2009

Using the Media

So far, this chapter has examined the ways in which the media, especially television, influence their audiences – or, as some would say, how the media *use* their audiences. The chapter now focuses on how audiences use the media.

Uses and Gratifications of the Mass Media and 'Now' Media

The uses and gratifications approach examines how audiences use the media and the gratifications derived from this use. This perspective is based on the premise that people have certain needs and desires that are fulfilled by their media choices, either through use of the medium itself or through exposure to specific content. The uses and gratifications model is used to answer such media use questions as these: Why do some people prefer watching television news to reading the newspaper? Under what conditions is an individual more likely to watch a sitcom rather than a violent police drama? What satisfactions are derived from watching soap operas? The model is based on these assumptions: (1) the audience actively and freely chooses media and content; (2) individuals select media and content with specific purposes in mind; (3) using the media and exposure to content fulfills many gratifications; and (4) media and content choice are influenced by needs, values, and other personal and social factors.

The more strongly a particular medium or content gratifies a viewer's needs, the more likely the viewer is to continue using that medium or to depend on that content. For example, if a viewer's need for feeling smart is fulfilled by a particular game show, he or she will probably continue watching that show. Audiences watch particular shows, listen to particular music, and even select specific Web sites for many reasons, such as to escape, to pass time, to unwind, to relax, to feel less lonely, and to learn new ideas and perspectives.

Sometimes it is not just the content that is gratifying but rather the medium itself. The act of watching television, regardless of what is on, may be relaxing, or the act of surfing the Internet, regardless of the sites being accessed, may gratify the need to feel productive.

Television is watched in two primary ways: Instrumentally and ritualistically. **Instrumental viewing** tends to be goal oriented and content based; viewers watch television with a certain type of program in mind. Conversely, **ritualistic viewing** is less goal oriented and more habitual in nature; viewers watch television for the act of watching, without regard to program content. Research suggests that perhaps television-viewing behavior should not be thought of as either instrumental or ritualistic but as falling along a continuum. In other words, sometimes a viewer might watch instrumentally, sometimes ritualistically, or sometimes a bit of both.

The uses and gratifications approach is important in understanding how audiences use the mass media and their reasons for doing so. The theory connects media use to the audience's psychological needs and attitudes toward the media, and it also explains how personal factors influence media use. Additionally, the uses and gratifications model tracks how new communication technologies change media use habits and how social and cultural changes influence content selection.

ZOOM IN 14.4

Next time you watch television, think of why you turned on the set and why you chose the particular program you are watching. Think of the last time you watched television. Did you watch ritualistically, or did you turn on the television to watch a particular program?

Source: Time Flies, 2018

Multitasking/Task-Switching

There was a time when computers were touted as time savers. It was prognosticated that when computers became integrated into daily life, there would be more time for other activities. But that does not seem to be the case. To have enough time to sleep, work, watch television, use the Internet, listen to the radio, read books, and fit in other activities, we need 38 hours per day. It all gets fit in by **multitasking**.

Humans have always been able to multitask – singing while showering, walking while chewing gum – without decreasing their concentration. But that is not the case with electronic devices because they command a higher level of cognition than, say, cooking several different dishes at the same time. Media multitasking becomes problematic depending on the number of different actions being taken simultaneously and the media being used – some media are more conducive to multitasking than others. For example, listening to music while surfing the Web or reading a novel go together more readily than watching television while reading a chemistry textbook.

When we engage in several activities at the same time, we are more easily distracted, less organized, and generally perform poorly as compared to those attending to one task at a time. It takes too much prefrontal brainpower to focus on various stimuli at the same time. Even though humans can perceive two stimuli at the same time, we cannot process them simultaneously – thus we experience a delay in our responses. When presented with new stimuli, physiologic brain processes that drive concentration are stretched to accommodate attention to several different tasks. A person has finite attention and concentration abilities, and the greater the number of tasks being done at once, the less able a person is to concentrate on any one of them. Moreover, it can take up to 20 minutes for the brain to 'reboot' after an interruption. The more distractions when learning, the less we are able to remember. The more tasks that are performed at the same time, the higher the error rate. Students often study with the television on in the background, thinking that it really does not interfere with learning, but their brains are processing the images, which leaves less power to concentrate on their schoolwork.

If multitasking reduces productivity and performance, then why do we keep doing it? Because not only do we get social rewards through the misperception that we must be very hardworking and organized to juggle it all, but we also get a kick of dopamine, a neurotransmitter "that some believe is the master molecule of addiction" (Quittner, 1999). Every time we complete a task, our brains give us a shot of the chemical dopamine, which enhances our mood. The more we multitask, the more hits of dopamine, the better we feel.

Talking on a cell phone is particularly disruptive. Psychologists assert that the act of talking on a cell phone decreases cognition and awareness and leads to preoccupation or 'inattention blindness.' Cell phone use decreases auditory as well as visual functions – talkers look right at an object but do not really see it. Although a little stimulation, such as music while exercising, might boost performance, too much induces stress, which puts the mind in overactive mode, which decreases focus.

In our fast-paced society, multitasking has almost become a badge of honor. Young people grow up thinking that multitasking is the best way to accomplish their work. Just over eight in ten young people aged eight to 18 frequently multitask.

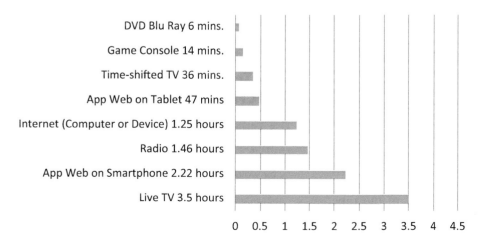

Diagram 14.1 Media use per day.

Although many people multitask, only 2% can do it well. Multitasking reduces work productivity by 40%, and it takes multitaskers 50% longer to complete a task and they make 50% more errors. The strangest part of all is that those who think they are good at multitasking actually perform poorly when compared to those who acknowledge that multitasking is difficult.

Task-switching is a form of multitasking but involves doing one task at a time but rapidly switching back and forth between several tasks. For example, starting to write an email, but getting up to pour a cup of coffee, then looking at the weather forecast online, then writing another line of the email, then looking at a Facebook notification, then writing two more sentences of the email, is an example of task-switching. Multitasking and task-switching are often thought of as the same activity and although task-switching is cognitively a bit different from multitasking, it has the same negative effects as doing two or more activities at once.

FYI: MULTITASKING

Nine in ten Americans multitask while watching TV. Here's what they do.

Surf the Internet on a computer	32%
Read email	28%
Text message	28%
Use a social network	26%
Talk on the phone	19%
Write email	17%
Shop online	16%
Play video games	16%
Microblog (e.g. tweet)	14%
Just watch TV	10%

Source: 9 in 10 Americans, 2015

Effects of Television

Effects of Television on Lifestyle

Since its widespread adoption in the 1950s, television has changed the way we live our lives – sometimes for the better, other times for the worse. As the saying goes 'too much of a good thing, is a bad thing.' It is difficult, however, to determine overuse. Residents of West Virginia, Alabama, and Arkansas spend the most time watching television at almost four hours per day, whereas residents of Alaska, Utah, and New Mexico are in front of the tube only about two hours per day. What is considered too much or too little viewing varies by person and the types of programs viewed. For example, of the 5.5 hours per day people devote to leisure activity including sports, about 3.5 hours are spent watching television, DVDs, and videos.

As more and more hours of the day are taken up with television and other forms of video, other activities such as socializing and reading suffer. From 2005 to 2018, the time spent reading books for pleasure has decreased from about 38 minutes per day to 26 minutes, while television viewing increased from 2 hours 36 minutes per day to around 3.5 hours per day. And senior citizens are especially at risk as they shift time from reading and socializing to watching television, which as a passive and solo activity could increase cognitive decline, unhappiness, isolation, and early death.

Although the consequences are focused on more intently than the benefits, television can be a positive influence. Television has psychological effects; it can make us laugh, lighten our mood, make us feel less lonely, strengthen bonds with family and friends through group viewing, reduce stress, and increase relaxation. Television is also educational; it teaches us how to behave, informs us of social norms, keeps us current with news, teaches us about science, and enlightens us about other countries and cultures.

Effects of Binge Watching

Binge watching has become a new concern for parents, physicians, and mental health experts. Sitting in front of a television set or hunched over a laptop or tablet watching episode after episode of a program is physically, mentally, and emotionally unhealthy. The escapist pleasure of binge watching can increase feelings of loneliness and isolation, depress mood, increase anxiety, disrupt sleep, and keep people from doing what they should be doing, such as studying or working or completing chores and errands. Like video game playing, sitting in one spot for long stretches of time binge watching can lead to addiction and other long-term problems.

HEALTHY WAYS TO BINGE WATCH

- *Set limits*: Set a timer or ask a friend to text you.
- *Watch together*: Create a shared connection by watching with a friend.
- *Take five*: Pause, stretch, get a drink of water, and take a bathroom break.

- *Watch mindfully*: Give the show your full attention.
- *Guard your sleep*: Don't watch before sleeping. Mental stimulation from the blue light can make it hard to fall asleep.
- *Combine good habits*: Exercise – lift weights or use a treadmill while watching.
- *Be honest*: Take a truthful look at your TV habits; not so much the content but the habit of avoiding other things you do not want to do.

Source: Cleveland Clinic, n.d.

Effects of Televised Violence

Arguably, the most troubling content on television is violence. Violence comes in many forms – shooting, beating, rioting, arguing, cursing, and car, train, and plane crashes.

Television violence and its effects on viewers is an ongoing problem. Six decades of research on media violence consistently shows that exposure to violent acts on television is related to aggression, such as bullying, peer-directed and intimate partner violence, and desensitization to violence in real life. Exposure to television violence can also cause viewers to become more fearful of becoming victims of violence, warp perceptions of real life, induce feelings of anger, and even increase heart rate and blood pressure.

The number of violent acts during primetime increased 75% to 4.41 acts per hour between 1998 and 2005 and then decreased to 3.7 in 2010. Additionally, the Parents Television Council found that nearly one-half of prime-time broadcast shows that aired in early 2013 contained some form of violence, and almost one-third depicted gun violence. Despite calls to reduce violence on television, the Parents Television Council calculated 28% more violent acts from 2008 to 2018.

FYI: PERCEPTIONS OF MEDIA'S CONTRIBUTION TO VIOLENCE

When Americans were asked if various media contribute to crime in the United States:

- 92% said television contributes to crime.
- 91% said local television news contributes to crime.
- 82% said video games contribute to crime.
- 80% worry about Internet predators.

Sources: Common Sense Media, 2009; Parents, children and media, 2007; Potter, 2003

Cultivation Researcher George Gerbner and associates spent many years examining how television cultivates a worldview. In particular, they conducted many content analyses that compared television content to real-life situations and to viewers' perceptions of actual life. Cultivation researchers first count how many times a certain action occurs on television, for example, the number of characters who die as a result of a gunshot. Next, they ask viewers how often they think gunshot deaths occur in real life. Then they research public records to find out how many people actually died from gunshots. Following this method of data collection to investigate cultivation effects, comparisons can be made between real-life occurrences, viewers' perceptions of real life, and what is shown on television.

Many viewers say that because much of what they see on television is fictional, it does not influence their perceptions because they know what they are seeing is not real. However, research does indeed show that our cognitive perceptions are shaped by television. Heavy television viewers' estimations of real life violence and crime levels are more closely in accord with those portrayed on television. For example, compared to light viewers of violent television, heavy viewers are more likely to fear being a victim of violent crime, to exaggerate their chance of being victimized, and to go out of their way to secure their homes and buy guns for protection. In sum, heavy viewers believe the world is a scarier place than it really is. Additionally, cultivation may lead viewers to accept violent acts as a normal part of life and to think of violence as an acceptable way to solve conflicts. Television cultivates a '**mean-world syndrome**,' in which viewers believe that the world is a mean and scary place to live and alter their behaviors accordingly, all based on television's warped depiction of real life.

It is especially difficult to dismiss violence and not believe that the world is a scary place when the news media blow crime coverage way out of proportion. For instance, the national homicide rate decreased 33% between 1990 and 1998, yet television coverage of murders increased 473%. Television news is filled with child abduction stories or reports on how to protect children from what seems to be a spike in kidnappings. But in reality, the number of child abductions has actually decreased since the 1980s. According to the FBI, 332 children were abducted by a stranger in 2014. That number is out of about 50 million school-age children in the United States. In comparison, about 1,000 children die in car accidents and 1,300 are killed by firearms each year, but these deaths receive much less media coverage than kidnappings. The FBI reports that violent crime rates in the U.S. have steadily declined from 758.2

per 100,000 population in 1992, to 429 per 100,000 in 2009, to 382 per 100,000 in 2017, yet two-thirds of the public believe that crime is on the increase.

Effects of Televised Profanity

Cursing is a type of verbal aggression that attempts to degrade the concept of self or the opinion of another person with the intent of causing psychological harm. People learn how to behave by observing others in real life and on television. Behaviors learned on television are often imitated in real life, and profanity is easier to mimic than physically aggressive actions.

Television writers might script expletives to give emotionally charged scenes and situations a sense of reality, but critics claim that a reflection of reality should not be the trade-off for exposure to offensive language, especially if, as feared, repeated exposure leads to desensitization to profanity, and thus to greater acceptance and use of cursing and insults in real life.

A Parents Television Council study discovered that the amount of **profane language**, which is considered verbal aggression, increased 69.3% from 2005 to 2010, and by 43.5% from 2008 to 2018. Academic studies found that on programs aired by broadcast networks during primetime, profanities were uttered 5.5 times per hour in 1990, 7.2 times in 2001, and by 2005, curse words were said 9.8 times per hour. When looking at types of words, expletives that are considered mild have decreased, while strong curse words have increased, as evidenced by a 2,409% jump in the frequency of the f-word (partially bleeped or partially muted).

Because of FCC oversight, broadcast programs have long contained fewer incidents of cursing than cable programs, which are not under the FCC's purview. For several decades, those numbers held true. For example, in 2005, cable shows contained 15.37 occurrences of crude language per hour compared to 9.8 on broadcast, a 44.2% difference. However, 2013 research by the Parents Television Council indicates that cable programs contain only 6% more incidents of cursing than broadcast shows. The research further indicates that the gap is narrowing because cursing in broadcast programs is increasing, not because cursing on cable is decreasing.

Curbing Obscenity, Indecency, and Profanity on the Airwaves

***Miller v. California* (1973)** The U.S. Supreme Court set new standards for obscenity and indecency on the airwaves in 1973 with its *Miller* v. *California* ruling. The ruling restricts obscenity, which has stricter legal standards than indecency. Whereas obscenity cannot be shown, indecency, in some circumstances, is allowable. Because the Internet and subscription services – like cable companies and satellite television providers – are not regulated by the FCC, they are free to air indecency. But broadcast networks and stations are regulated by the FCC and thus must follow its restrictions on indecency as imposed by the *Pacifica* ruling of 1978.

The Court ruled that **obscenity** was not allowed on any broadcast medium, and it established the following criteria for determining obscenity:

- Whether the average person, applying contemporary community standards, would find that the work, taken as a whole, appeals to the prurient interest, and
- Whether the work depicts or describes, in a patently offensive way, sexual conduct specifically defined by the applicable state law, and
- Whether the work taken as a whole, lacks serious literary, artistic, political, or scientific value.

Telecommunications Act of 1996 In an attempt to appease concerned parents, legislators, and others who rallied to clean up the airwaves, Congress passed the **Telecommunications Act of 1996**. The Act set parameters for protecting the public from objectionable content. For example, it mandated that all television sets with screens larger than 13 inches be equipped with V-chips or other means of blocking objectionable content and required a rating system to alert viewers to violent and sexual content.

The television industry first adopted an age-based rating system similar to the one used by the movie industry, but the system was criticized for inadequate protection from offensive content. The age-based system was enhanced with a system of content warnings. Current television ratings are a blend of both age and content indicators.

Age Ratings

TV-Y: Appropriate for all ages
TV-Y7: Appropriate for age 7 and above
TV-G: Most parents would find content appropriate for all ages
TV-PG: Contains material that most parents would find unsuitable for younger children
TV-14: Contains material that most parents would find unsuitable for children under 14
TV-MA: For mature audiences, unsuitable for children under 17

Content Ratings

V: Violent content
S: Sexual content
L: Coarse language
D: Sexually suggestive dialogue
FV: Fantasy violence (children's programming only)

These age- and content-based ratings are shown on the screen as easy-to-understand icons that appear at the beginning of most television programs.

Ratings opponents fear that the system might increase the number of violent and sexual images, because producers can now justify such content because viewers are warned and given the opportunity to avoid such programming. Some broadcasters do not comply with the content ratings system and object to its use. One executive producer expressed the fear that content ratings will jeopardize the success of certain programs. For example, the now canceled hospital drama *ER* could have been be considered violent because it showed car accidents and victims of gunshots.

Although the ratings system is hailed as a way to protect children from age-inappropriate content, many parents do not rely on it, claiming it is not very helpful in identifying specific types of content or in limiting children's exposure to sensitive content. One study showed that ten years after the implementation, only 8% of adults could correctly identify the content descriptors, and only 30% of parents with 2- to 6-year-old children could do so.

Although eight in ten parents are aware that televisions come with a built-in **V-chip**, only 16.5% use it. Instead of relying on the V-chip to block certain shows, parents prefer to limit viewing time (63%), restrict content (53%), and watch television with their children (49%).

Other means of blocking objectionable content are available in the marketplace. TVGuardian and other similar devices mute more than 150 offensive words and phrases from television programs and edit profanities out of closed-captioned scripts.

Broadcast Decency Act of 2005 Interest in indecent content surged after Cher exclaimed, "f*** 'em," during the 2002 live Billboard Music Awards program, and singer Bono uttered, "this is really, really f***ing brilliant" during a broadcast of the Golden Globe Awards in 2003. After the infamous 'wardrobe malfunction' that partially exposed Janet Jackson's breast during the 2004 Super Bowl halftime show, the FCC and CBS were immediately flooded with phone calls, letters, and emails from viewers who were fed up with indecency, especially on a program watched by so many young viewers. Network executives were forced to explain their

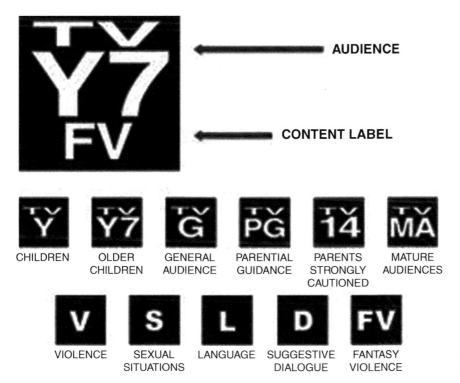

Figure 14.7 Example of TV content and age rating system.

programming standards in special congressional hearings prompted by the post-Super Bowl uproar.

Stemming from the Super Bowl and other incidents and viewer disgust and frustration with raunchy television in general, Congress passed the **Broadcast Decency Enforcement Act** in 2005. The bill allowed the FCC to increase fines for indecency violations from $32,500 to $325,000. Additionally, rather than fining a broadcast television station on a per-program basis, the FCC could issue fines for *each* indecent incident.

Threatened with heavy fines, the broadcast television industry fought back. It claimed that FCC indecency guidelines were unclear and arbitrary and that there was no way of knowing whether programs were in compliance until after they aired. In 2012, the Supreme Court threw out sanctions against broadcasting companies ruling that the "commission failed to give FOX or ABC fair notice prior to the broadcasts in question that fleeting expletives and momentary nudity could be found actionably indecent" (Liptak, 2012, B1). In the wake of the ruling, the FCC is free to modify its indecency policy, and networks, at least for now, are free to self-regulate. But the ruling has left parents and watchdog groups concerned that the amount of cursing and indecent depictions on television will increase and become more explicit.

Effects of the Internet and Mobile Devices

Overall, people are spending slightly less time with traditional media and more time online. In fact, some people are actually spending *too* much time on the Web. Although the Internet has improved our lives in many ways, it has also created many problems. Internet addiction, cyberbullying and harassment, deep fake videos, workplace stress, and decreased leisure time contribute to emotional distress.

Effects of Digital Addiction

Many people are addicted to the Web, social media, texting, email, chat rooms and newsgroups, and other online venues and activities. About 81% of the U.S. population goes online daily. Younger adults (age 18–29) are setting a faster pace than other age groups with almost one-half constantly online. Moreover, about three-quarters of online users would rather give up chocolate, coffee, or alcohol than the Internet, and 20% would give up sex. College students, who enjoy freedom from parental supervision and unlimited access to the Internet through wireless and campus broadband connections, are at particular risk for Internet addiction, with about 71% saying they could not function without the Internet. They sometimes find that the stress of school, work, and social situations leads them to seek a safer, less demanding environment on the Internet. There, they can escape from real-life problems and socialize with new and perhaps anonymous social networked friends.

Overuse of digital technology greatly affects intimate relationships, friendships, and family connections. Whether by computer, tablet, or smartphone, digital connections have become more important than face-to-face ones. Phone calls are quickly giving way to texting as a more comfortable and convenient way of conversing, yet excessive texting also leads to failing grades, social anxiety, dependence, stress, and other psychological disorders.

In 2019, American adults spent about three hours per day using a smartphone for any reason – calling, texting, social media, Internet. For addicts, being without a cell phone or tablet is paralyzing. Almost 25% of younger adults exhibit addictive behavior, such as becoming upset or panicky if denied constant access to their phone, the inability to limit time they spend on the phone, and using the phone instead of studying or doing tasks.

About one-half of cell phone users cannot leave home without their phone even if just for a walk around the block or a 15-minute trip to a convenience store. About 84% of Americans said they cannot go a single day without their mobile device in hand, almost three-quarters would feel panicked if they lost their cell phone, 20% would rather go without shoes for a week than a cell phone, and 14% would feel totally desperate. Further, 63% of cell phone users check their phone every hour, 20% every 30 minutes, and 9% look at it every five minutes. Moreover, 93% of mobile phone users sleep with their phone within arm's reach, and 10% sleep with their phones under their pillows.

Mobile device addiction is called **nomophobia** (for no-mobile-phone phobia), the fear of being disconnected from the virtual world. It is estimated that about three-quarters of 14- to 24-year-olds and about 70% of 25- to 34-year-olds are nomophobic. College students are very susceptible to mobile device addiction. The average college student compulsively checks his or her cell phone 60 times and sends almost 110 texts each day.

According to a poll conducted by the National Sleep Foundation, more than half of 15- to 17-year-olds sleep about seven hours a night, 90 minutes less than the minimum recommendation. Tech-addicted teenagers are getting even fewer hours of sleep. Some young people even have a term for it: Vamping, a reference to those other legendary creatures of the night. They document their all-nighters by posting selfies on Instagram from bed, with the hashtags #teen and #vamping.

FYI: INTERNET ADDICTION

About one in eight Americans experience signs of Internet addiction.

- 90% of Millennials check their smartphone before getting out of bed.
- 65% use the Internet to escape their problems.
- 33% of university students spend five hours a day online
- 21% of women wake up in the middle of the night to check Facebook.
- When students disconnect for 24 hours, they feel:
  - "like some sort of withdrawal"
  - "inferior"
  - "panicked"
  - "Media is like a drug"

Source: ansonalex.com

FYI: SMARTPHONE ADDICTION

- Teens who spend five or more hours a day on electronic devices are 71% more likely to exhibit suicide risk factors.
- 36% of Millennials spend two or more hours per workday looking at their phones for personal reasons.
- 41% of teenagers feel overwhelmed by the quantity of notifications they receive on a daily basis.
- 33% of teens spend more time socializing with close friends online instead of in-person.
- 52% of teens sit for long periods of time in silence, on their smartphones, while hanging out with friends.

Source: SlickText, 2019

ZOOM IN 14.5

Take this test to find out if you are a smartphone junkie: www.huffpost.com/entry/nomophobia-smartphone-sep_n_7266468

The overuse of mobile devices has had other unintended and dangerous consequences. Mobile communications are also linked to a significant increase in distracted driving, resulting in injury and loss of life. The National Highway Traffic Safety Administration reported that in 2018, driver distraction claimed 2,841 lives – a distraction is any activity that takes attention away from driving, such as texting or using a navigation system. Almost 80% of Millennials have used a phone while driving, 79% text, and 72% glance at their phone while behind the wheel. Gen X and Baby Boomer drivers are less likely to let their phones distract their driving. As of 2019, 20 states prohibit the use of handheld cellphones while driving, and 48 states ban text messaging.

FYI: DRIVING AND TEXTING

- 38% of teens text while driving.
- 56% of those 18 and older report that they sometimes or frequently text.
- 26% of accidents involving automobiles are caused by cell use while driving.
- Texting while driving is associated with other risky behaviors, such as not regularly wearing a seatbelt and drinking alcohol while driving.

Source: Survey Finds, 2018

Effects of Cyberbullying and Digital Harassment

The digital sphere makes it easy to go after people we do not like. Many people have been emotionally traumatized by online trolls, bullies, and liars.

With social media platforms largely unregulated for hate speech and other aggressive postings, and having become more mainstream, vulnerable users are targets for slings and arrows that can ruin their reputations and lives. Online culture has also changed from issue-oriented to personality-driven. Online users, then, are personally attacked for voicing an opinion, and the attackers are not always anonymous – bloggers and other cyberpersonalities are willing to go after anyone who does not agree with them. Hateful words are deemed 'opinions' and thus the authors are largely immune from repercussion.

From emotional abuse to death threats, online users are subjected to hoaxes, misinformation, disinformation, racist, sexist and misogynistic memes, deep fakes, and outright lies, all of which negatively affect the psychological well-being of online users. Especially damaging is when online images are stolen and superimposed in deep fake videos, especially pornographic ones. Such videos can harm the victim's job prospects, ruin their reputation, and cause deep emotional distress to the point of suicide. Once the video circulates in cyberspace it is impossible to 'take it down,' or even inform the millions of viewers that it is fake. The bigger problems, however, are what to do about online users who know a video is fake, but watch and share it anyway, and what to do about users who watch a video, then learn it is fake, but still believe it.

The gaming industry took a negative, very public, turn when in 2013 independent game designer Zoe Quinn released a free game online called, *Depression Quest*. One of Quinn's former boyfriends wrote on the gaming message board 4Chan that Quinn had sexual relations with a gaming journalist in order to further her career with positive reports. Quinn then received so many misogynistic attacks and death threats via 4Chan, Twitter, and Reddit that the FBI had to step in. The online assault on Quinn became known as '**gamergate**,' when actor Adam Baldwin used the hashtag in multiple Twitter posts. Other women who condemned the virtual attacks also became targets for abuse and threats, including racism, neo-Nazism, homophobia, and anti-Semitism. In some cases, the women's home addresses and exact times of planned attacks were posted, which drew an investigation from the FBI for criminally punishable threats.

The series of events that comprise 'gamergate' have been largely called a 'cultural war' between a traditional group of young, male gamers and a more diverse group that includes female game developers who call for greater inclusion. No arrests or charges have been brought by the FBI, though in response to the incident many game developers condemned the attacks made and encouraged future reviews of how women and minorities are treated in the industry. Despite the efforts, cyberbullying seems to have escalated since gamergate in 2014, with about 20% of online users saying they have been bullied or harassed online.

Effects of Technology-Induced Workplace Stress

Workplace managers are beginning to recognize that email is one of the main sources of employee stress and dissatisfaction. The typical office employee spends 2.5 to 3 hours per workday (12.5 to 15 hours per workweek) reading, responding to, and managing about 120 daily work-related emails, accounting for about 70% of all email traffic. About one-half of U.S. workers feel pressured to reply to a message within an hour of receiving it. All this checking and responding makes employees less efficient, less productive, and more stressed out. Companies are struggling with ways to help employees balance their work responsibilities with the distractions of email and the Internet.

With the goal of increasing productivity and satisfaction, some businesses hire outside consultants to help employees manage their time. Other employers have gone so far as to turn off their company's email servers for several hours each day. Germany is calling for anti-stress regulations to restrict bosses from emailing employees during non-working hours. Volkswagen in Germany turns on the email servers 30 minutes before the work day and turns them off 30 minutes after work hours. Daimler has gone a step further by automatically deleting all emails sent to an employee while on vacation. In 2017, France enacted legislation that requires employers to negotiate with employees their obligations to respond after hours. Savvy managers are beginning to recognize a significant reduction in employees' stress levels and an increase in productivity when they check email less frequently.

Effects of Digital Media on Lifestyle

Digital media have changed what we do in our leisure time, and not always for the better. Blue light screens, animation, colorful links, graphics, video, and notifications stimulate our brains and perk us up. For the sake of sound mental health, happiness, and general well-being, people need to power down the digital devices and television, yet the opposite is happening.

The Internet has trained our minds to be on the constant lookout for more stimulus and stronger dopamine rushes, which bring on the inability to concentrate and read. Much to our personal and social detriment, reading has been greatly affected by the Internet. We have become so accustomed to visual stimulation that we quickly become restless when reading a printed book or newspaper. After a few minutes of quiet reading, we get the urge to look online, and think that we can comprehend online words as well as printed words. Yet, reading an in-depth article is almost impossible as our eyes are drawn to graphics, photos, and video, often not even related to what we are reading.

People often lament not having enough time to read for pleasure, or even to read a newspaper. We do have the time, but we have turned to social media as a substitute for other media. Reading at a pace of 400 words per minute, it would take 417 hours to read 200 books, far less than the 608 hours per year spent on social media or the 1,600 hours watching television. The point is that when concentrated reading decreases, so does our understanding of the world and our ability to learn and create knowledge.

Effects of Video Games

Video games have become a part of everyday in-home and mobile entertainment. Some people are hardcore players, and others just play to pass the time, for instance when they are waiting for a bus or in a boring meeting or class. But no matter who plays and how the games are played – arcade, console, computer, portable/handheld, or mobile modes – by 2019, video gaming had become an almost $130 billion industry worldwide, with the U.S. accounting for about $45 billion.

In 2000, about 49 million people in the United States played some type of online computer game. By 2009, that number had jumped to 114 million, by 2013 to about 183 million (about 58% of the U.S. population), and by 2019 to 205 million (about 66% of the population). Further, about two-thirds of U.S. households own a video game system.

Playing video games, like watching television, browsing the Internet, and using social media comes with benefits as well as consequences. The type of video game, the devices used to play them, how often and for how long they are played, and why they are played, all contribute to the effects on players.

Types of Video Games

Video games are played in arcade, console, computer, portable/handheld, and mobile modes. **Arcade gaming** refers to a coin-operated gaming machine played in a public place like a bar or casual restaurant. Early arcade videogames such as *Pac-Man* and *Space Invaders* were very popular and set the gaming industry on a successful trajectory.

One of the first home console **gaming systems** was the Magnavox Odyssey. The Odyssey, released in 1971, was the forerunner to Sony's PlayStation and Microsoft's Xbox. After using the Odyssey system, Nolan Bushnell, cofounder of Atari, came up with the very popular *Pong* arcade game, which was later released as a console version. When Nintendo introduced *Game Boy* in 1989, it added an 'on-the-go' way to play video games, and diminished the need for arcade games. Today's **console games** have come a long way from the soundless Odyssey system and feature an array of interactive multimedia features displayed on a connected television screen and manipulated by a joystick or some other type of controller. PlayStation and Xbox are two of the most popular video console gaming devices.

Computer games are those that are played directly on a computer. Computer games these days are sold through the Internet and downloaded onto a computer. A big draw to computer gaming is playing with others online. With a fast enough broadband connection, players compete in real time with little lag between moves.

Gaming systems such as Game Boy Advance and PlayStation Portable are dedicated portable/handheld devices, which can be played on smartphones or tablets. *Tetris*, introduced in 1994, was the first game that could be played on a mobile phone, and three years later Nokia brought on the game *Snake*. Mobile games are sometimes embedded in the device, but most games these days are downloaded as apps. Games are played either solo within the device or with others through a mobile cloud.

In keeping with the First Amendment, the U.S. government does not restrict video game content. Video game retailers, however, abide by a voluntary ratings system. The Entertainment Software Rating Board (ESRB) has established the following age-based ratings system:

Figure 14.8 Examples of video game rating icons.

C: Content intended for young children
E: Everyone. Suitable for all ages
E10+: Suitable for viewers older than 10
T: Suitable for ages 13 and older
M: Suitable for ages 17 and older
AO: Adults only. Not suitable for players under the age of 17. These games may contain high levels of violence, profanity, or sexual content.

ZOOM IN 14.6

MOST POPULAR STREAMED GAMES THROUGH TIME (2015–2019)

Check out this YouTube video that takes you through a moving timeline of the most popular streamed games 2015–2019: www.youtube.com/watch?v=PRmhnfwyLNY

There are three primary types of online shared game playing: (1) identity and **self-presentation**, (2) **collective identity**, and (3) **phatic communication**. With *self-presentation*, gamers determine how they want to be seen by others. For example, filling out a personality test; posting their horoscope or tarot readings; sharing movie preferences; or playing games in which they reveal their personality sets a public persona. It is a way to let friends know who you are, or at least who you want them to think you are.

With *collective identity* formation tools, friends define their friends with questions such as: "What's the best way to make [friend's name] happy?" "How would you describe [friend's name] sense of style?" "How would [friend's name] occupy him or herself if he or she was thrown in jail?"

Social network interactions are also a form of *phatic communication*, which is a linguistic term that defines a type of expression that is used only for social reasons instead of for the purpose of sharing information. For example, "How are you?" and "Fine, thanks" is phatic communication, because it is mostly used as a polite recognition and not as a conversation starter or information exchange. The same can be said for some types of games and online interactions, such as sending someone a heart icon or a hug or updating one's latest game accomplishments through Facebook.

Social Effects of Video Gaming

As video games have become more popular, concern has been growing about the social effects, especially increased aggression, that stem from playing such games. Studies suggest that video games may be more influential than television, for several reasons: (1) they are more interactive, which increases involvement; (2) a large percentage of games involve violence as the main activity, and players are encouraged to 'kill' and 'injure' as many of the 'enemy' as possible to win; and (3) the games' portability makes them somewhat of a companion, as they can be played almost anywhere on a mobile device.

The most popular types of game genres are action, shooter, role-playing, sport, and adventure, and a smartphone is the most common way to play, followed by computer, and dedicated game console. Gamers play about 12 hours per week. The average age of a U.S. gamer is 33. Further, 21% people under the age of 18, 40% of 18–35 year olds, 18% of 36–49 year olds, and 21% of those older than 50 play video games. Back in 2006, the ratio of male to female gamers was 62:38 but by 2019 the ratio was a more even 54:46, with about 9% playing every day.

About 70% of college students play on a regular basis. College professors are unhappy to know that about one-third of college students have played games during class sessions and that the average college freshman has spent twice as much time playing video games (10,000 hours) as reading (5,000 hours).

Addiction to Video Games

In 2013 'Internet-Use Disorder' became a full-fledged mental health diagnosis within the *Diagnostic and Statistical Manual of Mental Disorders*, and online gaming is a subcategory of Internet-Use Disorder. Playing violent video games stimulates psycho-neurological receptors that give the player a 'high,' producing symptoms similar to those induced by drugs and other pleasurable activities. High levels of body and brain involvement also lead to the production of dopamine, which can make us giddy and less able to concentrate.

About 9–10% of all players show signs of video game addiction. Problem gamers (about 4% of all gamers) spend a whopping 50 to 80 hours per week playing video games, with roughly 40% doing so to escape reality. Many organizations list possible warning signs of video game addiction. Some of these signs include a preoccupation with games, negative physical and emotional symptoms when trying to stop playing, spending less time with friends and family, curtailing other enjoyable activities, lying about the amount of time spent gaming, falling GPA, job loss, depression, and increased use of drugs and alcohol.

Effects of Digital Media on Children and Young Adults

The typical child of today spends an average of 53 hours per week using various media – television, Internet, video games, and music. Children younger than five years old typically spend their media time watching television; about 2.5 hours per day. Children and young adults may be particularly susceptible to mediated words and images, both positive and negative.

Educational and non-violent programs, such as *Sesame Street*, help children learn positive social behaviors, enhance their imaginative powers, and even develop problem-solving skills. Studies conducted over the years have shown that children who watched *Sesame Street* when they were young later demonstrated higher academic achievement and better reading skills than children who did not watch the program. This effect lasted through grade school and even into high school. Whether watching television shows or playing video games, "Children identify quite closely with electronic characters of all sorts, and … these identifications may have important implications for their emotional well-being as well as for the development of their personality" (McDonald & Kim, 2001, p. 241).

Conversely, negative images, especially depictions of violence, can cause short- and long-term harm. Given that about 20 to 25 acts of violence are shown in children's television programs each hour, most children have witnessed about 8,000 killings before the age of 12 and 200,000 acts of violence before the age of 18.

FYI: CHILDREN (0–8) AND MEDIA EXPOSURE PER DAY 2017

- Watching TV/DVDs/videos: 1 hour 40 minutes
- Listening to music/audio: 18 minutes
- Reading (being read to): 29 minutes
- Playing video, computer or mobile games: 25 minutes
- General computer/digital use: 10 minutes
- Video chatting: 1 minute

Source: The Common Sense Census, 2017

FYI: TWEENS (8–12 YEARS OLD) AND MEDIA EXPOSURE PER DAY 2019

- Watching TV/DVDs/videos: 2 hours 30 minutes
- Playing video, computer or mobile games: 1 hour 28 minutes
- Listening to music/audio: 43 minutes
- Reading (print/digital): 29 minutes
- Browsing Web sites: 14 minutes
- Using social media: 10 minutes
- General computer/mobile use: 7 minutes
- Video chatting: 5 minutes

Source: The Common Sense Census, 2019

Figure 14.9 Some feel that playing video games may be addictive given the high levels of body and brain involvement.

FYI: TEENAGERS (13–18 YEARS OLD) AND MEDIA EXPOSURE PER DAY 2019

- Watching TV/DVDs/videos: 2 hours 52 minutes
- Listening to music/audio: 2 hours and 5 minutes
- Playing video, computer, or mobile games: 1 hour 36 minutes
- Using social media: 1 hour and 10 minutes
- Browsing Web sites: 37 minutes
- Reading (print/digital): 29 minutes
- General computer/mobile use: 28 minutes
- Video chatting: 19 minutes

Source: The Common Sense Census, 2019

Figure 14.10 College students are at particular risk for Internet addiction.
Source: Photo courtesy of iStockphoto. © mdmilliman, image #4528806

Sexploitation

Children and young adults are vulnerable in many ways to television and online advertising and content, but none more sinister than online sexual exploitation (sexploitation). Sexual predators often gain access to youths through online gaming sites. Predators establish online relationships with young people by disguising themselves as teenagers and joining in the games. The predators gain a child's confidence through gaming chat platforms and other gaming virtual communities. Once a predator

Figure 14.11 Television addiction.
Source: Photo courtesy iStockphoto. © stray_cat, image #2397753

has established a friendship they swoop in and threaten to tell the child's parents and their social media friends about their 'relationship' unless the child posts nude photos of him/herself or even of their brothers or sisters. Predators also talk teens into showing their genitals while playing a game in exchange for online gift cards.

Such extortion (sextortion) has increased dramatically from about 50 reports in 2013 to about 1,500 in 2019. The Justice Department considers sextortion a significant and growing threat to children and teens. More than 25% of sextortion cases have led to threat of or attempted suicide. Young victims believe there is something wrong with them, they are ashamed, they withdraw socially, and their grades suffer. Police departments, the FBI, and other law enforcement agencies do their best to zero-in on and arrest online pedophiles but face many obstacles in their pursuits. Gamers themselves resist giving up their anonymity and because the gaming culture thrives on group play, eliminating audio and video interactions would take away the heart of online gaming. In the meantime, authorities and some gaming companies themselves have put out tips for parents to protect their children online.

SEE IT LATER

In today's world, television, the Internet, and mobile communication are the central focus of media use. The remainder of this chapter covers the implications of media effects and how television and Web content is shaping our future.

Objectionable Content

Many people believe that the media are responsible for the content they air and should curb violent and negative images for the good of society as a whole. Concerned viewers, parents, and policymakers have aligned themselves with the more conservative members of the FCC, who have proposed designating one hour each evening as family viewing time with wholesome programming. They have also asked television networks to voluntarily cut back on violent, sexual, and offensive programs.

On the other hand, many viewers and the television industry have countered that V-chips, language-muting devices, and a content- and age-based ratings system already protect children from objectionable content, and that such content should not be curtailed because of the possibility of negative effects or because some viewers are offended. Many strongly believe that it is the parents' responsibility to monitor what their children watch on television, what music they listen to and buy, and what Internet sites they visit. Moreover, if viewers are offended by certain content, they should stop watching or listening to it. In other words, viewers are responsible for their own exposure to offensive content.

It is difficult to sift through all the arguments, especially when parents are leaning on legislators to clean up television while, at the same time, they are taking their kids to movies that contain violent and sexual scenes, and buying them video games

FYI: CHILDREN AND ONLINE CONTENT

- 73% of parents say they know a lot about what their children do online.
- 65% of parents say they closely monitor their children's media use.
- 59% said the Internet is mostly a positive influence on their children.
- 55% think online sexual content contributes a lot to inappropriate sexual behaviors.
- 7% of parents say they know little or nothing about what their children do online.

Sources: Common sense media poll, 2009; Parents, children and media, 2007; Potter, 2003

Figure 14.12 By the time a child graduates from high school, he or she will have spent about 33,000 hours watching television, as compared to about 13,000 hours in school.
Source: Photo courtesy of iStockphoto. © mdmilliman, image #4528806

containing mature and violent images. Given these and other contradictory behaviors, it is understandable why the television industry is reluctant to voluntarily censor its content but instead rallies hard against further regulations on violent, sexual, and verbal content.

Combating Sexploitation

To combat online sexual abuse and sextortion of children and adults, companies such as Roblox and Microsoft are developing systems to detect and block explicit language and contact information from their gaming chat sites. Microsoft is developing more effective anti-sextortion software that it plans to offer to other tech businesses for free. Other companies are working on systems to analyze users' selfies and estimate their ages, and on ways to detect gamers from supplying false information about their age (grown-ups purporting to be teenagers) by infiltrating their social media accounts and looking at their friends and activities.

Now Media

New communication technologies have given rise to new uses of media. 'Now' media such as television remote control devices, fast-forward buttons, DVRs, DVD players, streaming television, music services, and satellite radio, are changing existing media use habits by giving users more choice of what, when, where, and how to watch and listen to programming.

The Internet has already changed and will continue to change uses of radio and television. Broadcast television, cable television, newspapers, news magazines, and radio news are already taking a backseat to the Internet for political information. Yet, when it comes to hearing about breaking news, such as when Osama bin Laden was killed, U.S. adults still turn to television first. Television news is still king and it is the preferred news medium for 41% of Americans.

But the biggest worry is the gradual disappearance of newspapers. In 1996, about half of adults read a printed daily newspaper, with 2% reading online. By 2019 news-consumption habits had changed dramatically, with 37% of Americans preferring to read news online either on social media or a news Web site or app. Unfortunately, advertising revenue and digital subscriptions have not made up for the move to online. Ad revenue has dropped precipitously from $65 billion in 2000 to $19 billion in 2016. With few exceptions, such as *The New York Times* and *The Boston Globe*, digital subscription and revenue have not yet been enough to help the print product. Since 2004, about 1,800 U.S. newspapers (about 20% of the national total) have ceased operations, leaving about 200 counties and three million people without a newspaper. For thousands of other newspapers, especially those bought out by hedge funds and other non-news companies, the future is bleak as newsroom staffs have been decimated by severe budget cuts. Independent media are slowly being taken over by major corporations, slashing the number of independent voices down to a few. Just 25 companies own two-thirds of U.S. daily newspapers, and five companies own one-third of 1,400 television stations. As the promise of free news, albeit shallow and assembled by algorithms, seduces people online, revenue for in-depth, professional news organizations falls, making them vulnerable to hedge funds and other deep pocket corporations who swoop down and buy up them up, shut them down, sell the real estate, make a fortune, and walk away happy. With the craven slaughter of newspapers, it is little wonder that only 13% of Americans read a printed daily – there are few papers left to read and skeleton staffs are only able to provide little original content.

The newspaper industry and avid readers hope that digital paywalls will boost revenue enough to keep dailies and weeklies afloat, and that the downturn in print will slow as people increase their reliance on professional journalism. It could be that local news saves the day as people are more inclined to turn to area television stations and newspapers to find out about what is happening in their communities, information that is hard to find online or on social media.

Swimming Against the Tide of Technology

Whether they are called digital dissenters, techno-skeptics, or luddites, there is a growing number of activists who are resisting new communication technologies. Digital critics question the benefits and fear the consequences of technologies that are moving forward at warp speed. In many ways, the computer and later the Internet were looked to as saviors that would usher in a utopian society. Although in many ways they have lived up to expectations, in other ways they have disappointed and led to a world of limited privacy, techno-elites, big data collection, and a gig economy of contract workers without benefits and whose offices are their laptops. In a machine-versus-human society, what is good for machines takes precedent over what is good for humans. Personal preferences, purchases, hobbies, everyday movements, and habits are tracked and recorded and processed by algorithms that know more about an individual than she knows about herself. Digital dissenters are attempting to rein in runaway data and help shape a digital world in which man and machine can live side by side but with humans as first priority.

SUMMARY

Various perspectives are offered to explain the effects of media content individually, socially, and culturally. Those perspectives include the strong-effects model and magic bullet theory; the limited-effects model and the research that stemmed from the broadcast of *The War of the Worlds*; the moderate-effects model; and the powerful-effects model, which takes many factors into consideration when examining media influence.

Viewing media violence and playing violent video games can affect people behaviorally, emotionally, and cognitively. Viewers, especially children, may imitate aggressive behaviors, identify with unsavory characters, become anxious and fearful, and become desensitized to violence in real life. Repeated exposure to mediated violence breaks down social barriers and may influence anti-social behaviors.

The media, especially television, socializes and shapes our attitudes, values, and beliefs about the world around us. Program content and commercials both strongly influence the way we think about ourselves and others, and life in general. The Internet also strongly influences our behavior, especially socially. Further, some individuals are addicted to watching television and using the Internet, and as a consequence, they forget about the world and fail to meet their responsibilities. Teens and young adults are often vilified on social media sites causing mental anguish and trauma that could take years to overcome.

The television, Internet, and video game industries, parents, psychologists, educators, legislators, activists, and other interested parties are struggling with the many social and cultural issues surrounding 'now media' use and content. In this struggle, rights to freedom of speech go head-to-head with the desire to protect viewers from objectionable words and images. Arguments include whether viewers need protection and what negative effects mediated content may have. Some worry that our fascination with television and the Internet is turning us into media junkies who live in darkened rooms, transfixed by our screens.

These concerns will not likely be settled in the near future and, in fact, are more likely to grow as 'now' media become more ubiquitous and our dependency deepens.

BIBLIOGRAPHY

7 ways to binge-watch TV without harming your health. (n.d.). Retrieved from: https://health.clevelandclinic.org/7-ways-to-binge-watch-tv-without-harming-your-health/

9 in 10 Americans multitask while watching TV. (2015). *Statista*. Retrieved from: www.statista.com/chart/3485/tv-multitasking/

The 16th annual study of the impact of digital technology on Americans. (2018). *Center for the Digital Future*. Retrieved from: https://digitalcenter.org/wp-content/uploads/2018/12/2018-Digital-Future-Report.pdf

About three-in-10 U.S. adults say they are almost 'constantly' online. (2019, July 25). *Pew Research Center*. Retrieved from: www.pewresearch.org/fact-tank/2019/07/25/americans-going-online-almost-constantly/

Achenbach, J. (2006, June 25). Dropping the f-bomb. *The Washington Post*, pp. B1, B4.

Achenbach, J. (2015, December 27). The resistance. *The Washington Post*, pp. A1, A16.

The adult video gamer market in the U.S.: Tapping into the new diversity of video-game players. (2009, January 1). *Market Research.com*. Retrieved from: www.marketresearch.com/search/results.asp?sid538903034–458950752–420123802&query5video1game&submit15Go [September 28, 2009]

Age breakdown of video game players in the United States in 2019. (2020). *Statista*. Retrieved from: www.statista.com/statistics/189582/age-of-us-video-game-players-since-2010/

Ahrens, F. (2006, June 8). The price for on-air decency goes up. *The Washington Post*, pp. D1, D6.

Alarming video game addiction statistics. (2018, September 14). *Addictions.com*. Retrieved from: www.addictions.com/video-games/alarming-video-game-addiction-statistics/

Andersen, R. E., Crespo, C. J., Bartlett, S. J., Cheskin, L. J., & Pratt, M. (1998). Relationship of physical activity and television watching with body weight and level of fatness among children. *Journal of the American Medical Association*, *279*, 938–942.

Anderson, C. A., Suzuki, K., Swing, E. L., Groves, C. L., Gentile, D. A., Prot, S., et al. (2017). Media violence and other aggression risk factors in seven nations. *Personality and Social Psychology Bulletin*, *43*(7), 986–998.

Angier, N. (2005, September 20). Almost before we spoke, we swore. *The New York Times*. Retrieved from: www.nytimes.com/2005/09/20/science/almost-before-we-spoke-we-swore.html

Are Alexa and Siri considered AI? (2019). *Bernard Marr & Co*. Retrieved from: https://bernardmarr.com/default.asp?contentID=1830

Baker, P. (2006, June 16). Bush signs legislation on broadcast decency. *The Washington Post*, p. A6.

Bandura, A., Bryant, J., & Zillmann, D. (1994). Media effects: Advances in theory and research. *Social Cognitive Theory of Mass Communication*. Hillsdale, NJ: Lawrence Erlbaum Associates.

Baran, S. J., & Davis, D. K. (2000). *Mass communication theory* (2nd ed.). Belmont, CA: Wadsworth.

Barclay, E. (2016, January 29). Scientists are building a case for how food ads make us overeat. *NPR*. Retrieved from: www.npr.org/sections/thesalt/2016/01/29/462838153/food-ads-make-us-eat-more-and-should-be-regulated

Barnes, R. (2009, April 29). Supreme Court rules that government can fine for 'fleeting expletives.' *The Washington Post*. Retrieved from: www.washingtonpost.com/wp-dyn/content/article/2009/04/28/AR2009042801283.html

Barnes, R. (2012, June 21). Supreme Court overturns FCC sanctions on network, sidesteps larger issue. *The Washington Post*. Retrieved from: www.washingtonpost.com/politics/supreme-court-overturns-fcc-sanctions-on-networks-sidesteps-larger-issue/2012/06/21/gJQAwffxsV_story.html

Baxter, L. A., & Kaplan, S. J. (1983). Context factors in the analysis of prosocial and antisocial behavior on prime time television. *Journal of Broadcasting & Electronic Media*, *27*(1), 25–36.

Bella, T. (2012, May 24). The '7 dirty words' turn 40, but they're still dirty. *The Atlantic*. Retrieved from: www.theatlantic.com/entertainment/archive/2012/05/the-7-dirty-words-turn-40-but-theyre-still-dirty/257374/

Benton, J. (2019, May 21). Another milestone passed for newspaper: The Boston Globe is the first local newspaper to have more digital subscribers than print. *NiemanLab*. Retrieved from: www.niemanlab.org/2019/05/another-milestone-passed-for-newspapers-the-boston-globe-is-the-first-local-newspaper-to-have-more-digital-subscribers-than-print/

Birch, J. (2019, June 3). How binge watching is hazardous to your health. *The Washington Post*. Retrieved from: www.washingtonpost.com/lifestyle/wellness/how-binge-watching-is-hazardous-to-your-health/2019/05/31/03b0d70a-8220-11e9-bce7-40b4105f7ca0_story.html

Bond, P. (2008, December 18). Study: Young people watch less TV. *Reuters*. Retrieved from: www.reuters.com/article/us-study-idUSTRE4BH10Y20081218 [October 10, 2009]

Bowles, N., & Keller, M. H. (2019, December 18). While they play online, children may be the prey. *The New York Times*, pp. 1A, 24–25A.

Breus, M. J. (2018, January 18). Binge watching and its effects on your sleep. *Psychology Today*. Retrieved from: www.psychologytoday.com/us/blog/sleep-newzzz/201801/binge-watching-and-its-effects-your-sleep

Bureau of Labor Statistics. (2018). Time spent in leisure and sports activities. Retrieved from: www.bls.gov/news.release/atus.t11A.htm

Bushman, B. J. (2018). Teaching students about violent media effects. *Teaching of Psychology*, *45*(2), 200–206.

Canary, D. J., & Spitzberg, B. H. (1993). Loneliness and media gratifications. *Communication Research*, *20*(6), 800–821.

Cantor, J. (1998). *'Mommy I'm scared': How TV and movies frighten children and what we can do to protect them*. San Diego, CA: Harcourt Brace.

Carey, B. (2013, February 11). Shooting in the dark. *The New York Times*. Retrieved from: www.nytimes.com/2013/02/12/science/studying-the-effects-of-playing-violent-video-games.html?_r=0 [December 12, 2014]

Cellular phone use and texting while driving laws. (2019, May 29). *NSCL*. Retrieved from: www.ncsl.org/research/transportation/cellular-phone-use-and-texting-while-driving-laws.aspx

Chai, C. (2017, December 15). Ad bans lead to less fast food eaten in Quebec, study says. *Global News*. Retrieved from: https://globalnews.ca/news/209938/ad-bans-lead-to-less-fast-food-eating-in-quebec-study-says/

The changing news landscape. (2009). *Media Literacy Clearinghouse*. Retrieved from: www.frankwbaker.com/mediause.htm [October 1, 2009]

Child obesity facts. (2019, June 24). *Centers for Disease Control and Prevention*. Retrieved from: www.cdc.gov/obesity/data/childhood.html

Children and media violence. (2009). *National Institute on Media 1 Family*. Retrieved from: www.mediafamily.org/facts/facts_vlent.shtml [October 1, 2009]

Clancey, M. (1994). The television audience examined. *Journal of Advertising Research*, 34(4), special insert.

Commission reports to Congress on kids, content and protection. (2009, September 2). *Zogby International*. Retrieved from: www.zogby.com/Soundbites/ReadClips.cfm?ID519055 [September 28, 2009]

Common Sense Media. (2009). Is social networking changing childhood? Retrieved from: www.commonsensemedia.org/teen-social-media

Common Sense Media. (2017). Census: Media use by kids age zero to eight. *Common Sense*. Retrieved from: www.commonsensemedia.org/sites/default/files/uploads/research/csm_zerotoeight_fullreport_release_2.pdf

Common Sense Media. (2019). Census: Media use by tweens and teens. *Common Sense*. Retrieved from: www.commonsensemedia.org/sites/default/files/uploads/research/2019-census-8-to-18-full-report-updated.pdf

Connley, C. (2017, August 17). This company has an ingenious way to free employees from email on vacation. *CNBC*. Retrieved from: www.cnbc.com/2017/08/17/one-companys-genius-way-to-free-employees-from-email-on-vacation.html

Coughlan, S. (2019, November 29). Smartphone 'addiction': Young people 'panicky' when denied mobiles. Retrieved from: www.bbc.com/news/education-50593971

Crecente, B. (2018, September 11). Nearly 70% of Americans play video games on smartphones (study). *Variety*. Retrieved from: https://variety.com/2018/gaming/news/how-many-people-play-games-in-the-u-s-1202936332/

Crespo, C. J., Smit, E., Troiano, R. P., Bartlett, Susan J., Macera, C. A., & Anderson, R. E. (2001). Television watching, energy intake, and obesity in US children. *Archives of Pediatric and Adolescent Medicine*, 155, 360–365.

Crime in the United States. (2017). *FBI*. Retrieved from: https://ucr.fbi.gov/crime-in-the-u.s/2017/crime-in-the-u.s.-2017/topic-pages/tables/table-1

A decade of deceit. (2018). *Parents Television Council*. Retrieved from: https://go.parentstv.org/decades-report/documents/Decades-Report.pdf

The deadline team. (2014, June 18). New study: TV violence makes people more afraid of crime, but not afraid there is more crime. *Deadline*. Retrieved from: http://deadline.com/2014/06/new-study-tv-violence-makes-people-more-afraid-of-crime-but-not-afraid-there-is-more-crime-792399/

Dennison, B. A., Erb, T. A., & Jenkins, P. L. (2002). Television viewing and television in bedroom associated with overweight risk among low-income preschool children. *Pediatrics*, 109, 1028–1035.

Dickinson, K. (2018, April 5). Multitasking – the good, the bad, and the ugly. *The Washington Post*. Retrieved from: https://jobs.washingtonpost.com/article/multitasking-the-good-the-bad-and-the-ugly/

Dietz, W. H. (1990). You are what you eat – what you eat is what you are. *Journal of Adolescent Health Care*, 11, 76–81.

Distracted driving. (2017). *NHTSA*. Retrieved from: www.nhtsa.gov/risky-driving/distracted-driving

Distribution of computer and video gamers in the United States from 2006 to 2019, by gender. (2020). *Statista*. Retrieved from: www.statista.com/statistics/232383/gender-split-of-us-computer-and-video-gamers/

ESRB ratings guide. (n.d.). *Entertainment Software Rating Board*. Retrieved from: www.esrb.org/ratings/ratings_guide.jsp#rating_categories [December 12, 2014]

Essential facts about the computer and video game industry. (2019). *Entertainment Software Association*. Retrieved from: www.theesa.com/wp-content/uploads/2019/05/ESA_Essential_facts_2019_final.pdf

Federal Communications Commission. (2003). *Complaints Against Various Broadcast Licensees Regarding their Airing of the 'Golden Globe Awards' Program 1*. File No. EB-03-IH-0110.

Federal Communications Commission. (2006). *Frequently Asked Questions*. Retrieved from: www.fcc.gov/eb/oip/FAQ.html [September 14, 2006]

Federal Communications Commission. (2012a). *Regulation of Obscenity, Indecency, and Profanity*. Retrieved from: http://transition.fcc.gov/eb/oip/Welcome.html

Federal Communications Commission. (2012b). *V-Chip – Putting Restrictions on What Your Children Watch*. Retrieved from: www.fcc.gov/guides/v-chip-putting-restrictions-what-your-children-watch

Federal Communications Commission v. *Pacifica Foundation.* 438 U.S. 726 (1978).

Federman, J. (1998). National television violence study, Vol. 3, Executive summary. *Center for Communication and Social Policy – University of California, Santa Barbara.* Retrieved from: www.ccsp.ucsb.edu/execsum.pdf

For local news, Americans embrace digital but still want strong community connection. (2019, March 26). *Pew Research Center.* Retrieved from: www.journalism.org/2019/03/26/nearly-as-many-americans-prefer-to-get-their-local-news-online-as-prefer-the-tv-set/

Fry, R. (2019, October 1). The number of people in the average U.S. household is going up for the first time in over 160 years. *Pew Research Center.* Retrieved from: www.pewresearch.org/fact-tank/2019/10/01/the-number-of-people-in-the-average-u-s-household-is-going-up-for-the-first-time-in-over-160-years/

Genre breakdown of the most popular U.S. video game genres by sales in 2018. (2020). *Statista.* Retrieved from: www.statista.com/statistics/189592/breakdown-of-us-video-game-sales-2009-by-genre/

Gerbner, G., & Gross, L. (1976). Living with television: The violence profile. *Journal of Communication, 26,* 173–199.

Gerbner, G., Gross, L., Jackson-Beeck, M., Jeffries-Fox, S., & Signorielli, N. (1978). Cultural indicators: Violence profile no. 9. *Journal of Communication, 27,* 171–180.

Gervis, Z. (2018, August 15). Phones turn bedroom into a no-sex zone. *New York Post.* Retrieved from: https://nypost.com/2018/08/15/phones-turn-bedrooms-a-no-sex-zone/

Graft, K. (2009, August 7). 40 percent of U.S. homes have a gaming console, as HDTV adoption rises. *Gamasutra.* Retrieved from: www.gamasutra.com/php-bin/news_index.php?story524757 [September 28, 2009]

Graham, M. (2020, February 12). Unilever will stop advertising ice cream to kids over obesity concerns. *CNBC.* Retrieved from: www.cnbc.com/2020/02/12/unilever-to-stop-targeting-food-ads-to-kids-over-childhood-obesity-concerns.html

Greenya, J. (2005, November). Can they say that on the air? The FCC and indecency. *DC Bar for Lawyers.* Retrieved from: www.dcbar.org/for_lawyers/

Gregoire, C. (2015, May 18). This scientific test will tell you how addicted you are to your smartphone. *HuffPost.* Retrieved from: www.huffpost.com/entry/nomophobia-smartphone-sep_n_7266468

Griffiths, M. D., & Shuckford, G. L. G. (1989). Desensitization to television violence: A new model. *New Ideas in Psychology, 7(1),* 85–89.

Guta, M. (2019, September 22). Hey marketers, Americans still spend 5 hours a day on email. *Small Business Trends.* Retrieved from: https://smallbiztrends.com/2019/09/email-usage-statistics.html

Haridy, R. (2018, August 14). The right to disconnect: The new laws banning after-hours work emails. *New Atlas.* Retrieved from: https://newatlas.com/right-to-disconnect-after-hours-work-emails/55879/

Harwell, D. (2019, January 1). 'Deepfake' porn becomes a high-tech tool of harassment. *The Washington Post,* A1, A2.

Harwell, D. (2020, February 9). U.S. radio's top player blames AI for layoffs. DJ's say that's spin. *The Washington Post,* G1, G2.

Haughney, C. (2013, April 30). Newspapers post gains in digital circulation. *The New York Times.* Retrieved from: www.nytimes.com/2013/05/01/business/media/digital-subscribers-buoy-newspaper-circulation.html?_r=0 [December 12, 2014]

Hilliard, R. L., & Keith, M. C. (2007). *Dirty Discourse.* Malden, ME: Blackwell.

Ho, D. (2006, April 15). Networks appeal TV indecency rulings. *The Atlanta Journal-Constitution,* pp. b1, b5.

Holley, P. (2019, August 30). Tech firms strive to extend life beyond the last breath. *The Washington Post,* A1, A6.

Hopkinson, N. (2003, October 23). For media-savvy tots, TV and DVD compete with ABCs. *The Washington Post.* Retrieved from: www.washingtonpost.com

How face recognition evolved using artificial intelligence. (2020, January 7). *Facefirst.* Retrieved from: www.facefirst.com/blog/how-face-recognition-evolved-using-artificial-intelligence/

The impact of food advertising on childhood obesity. (n.d.). *American Psychological Association.* Retrieved from: www.apa.org/topics/kids-media/food.aspx [December 11, 2014]

Infante, D. A., Riddle, B. L., Horvath, C. L., & Tumlin, S. A. (1992). Verbal aggression. *Communication Quarterly, 40(2),* 116–126.

Ingraham, C. (2019, June 30). As Americans tune in, they tune out books. *The Washington Post,* G3.

Is there more crime in the U.S. than there was a year ago, or less? (2019). *Statista.* Retrieved from: www.statista.com/statistics/205525/public-perception-of-trend-in-crime-problem-in-the-usa/

The issues. (2009). *Mc Spot Light.* Retrieved from: www.mcspotlight.org/issues/advertising/index.html [October 1, 2009]

Jamieson, P. E., & Romer, D. (2014). Violence in popular US prime time TV dramas and the cultivation of fear: A time series analysis. *Media and Communication, 2(2),* 31.

Jeong, S. (2019, August 18). When the online mob comes after you. *The New York Times,* SR6.

Kaye, B. K., & Johnson, T. J. (2003). From here to obscurity: The Internet and media substitution theory. *Journal of the American Society for Information Science and Technology, 54*(3), 260–273.

Kaye, B. K., & Johnson, T. J. (2014). The shot heard around the World Wide Web: Who heard what where about Osama bin Laden's death. *Journal of Computer-Mediated Communication, 19*(3), 643–662.

Kaye, B. K., & Sapolsky, B. S. (2009). Taboo or not taboo? That is the question: Offensive language on prime time broadcast and cable programming. *Journal of Broadcasting & Electronic Media, 53*(1), 22–37.

Kellogg, C. (2013, July 2). Hours spent reading books around the world. *Los Angeles Times*. Retrieved from: http://articles.latimes.com/2013/jul/02/entertainment/la-et-jc-hours-reading-books-around-the-world-20130702

Kessler, G. (2019, June 30). The Fact Checker tackles the spread of false and misleading videos with a guide for reviewers. *The Washington Post*, A6.

Kite-Powell, J. (2018, September 30). Making facial recognition smarter with artificial intelligence. *Forbes*. Retrieved from: www.forbes.com/sites/jenniferhicks/2018/09/30/making-facial-recognition-smarter-with-artificial-intelligence/#202893adc8f1

Kushlev, K., & Dunn, E. W. (2015, January 11). Stop checking email so often. *The New York Times*, p. 12.

Labaton, S. (2006, April 18). U.S. networks turn to the courts in search of clearer rules. *International News Herald*, p. 16.

Lafayette, J. (2018, March 16). Nielsen launches TV product placement measurement system. Broadcasting+Cable. Retrieved from: www.broadcastingcable.com/news/nielsen-launches-tv-product-placement-measurement-system-171571

Lavers, D. (2002, May 13). The verdict on media violence: It's ugly and getting uglier. *The Washington Post*, pp. 28–29.

Leadem, R. (2017, July 14). How you respond to emails matters more than you think. *Entrepreneur*. Retrieved from: www.entrepreneur.com/slideshow/297223

Li, L., Schultz, M. P. H., Andridge, R., Yellman, M., Xiang, H., & Zhu, M. (2018). Texting/emailing while driving among high school students in 35 states, United States, 2015. *Journal of Adolescent Health, 63*, 701–708. Retrieved from: www.jahonline.org/article/S1054-139X(18)30250-7/pdf

Liptak, A. (2012, June 22). Supreme Court rejects F.C.C. fines for indecency. *The New York Times*. Retrieved from: www.nytimes.com/2012/06/22/business/media/justices-reject-indecency-fines-on-narrow-grounds.html

Livingston, G. (2002, July 28). Sense and nonsense in the child abduction scares. *San Francisco Chronicle*, p. 6, D.

Lohr, S. (2019, October 3). Murkiness of online ads spurs a quest for clarity. *The New York Times*, p. B4.

Losing the news: The decimation of local journalism and the search for solutions. (2019, November 20). *PEN America*. Retrieved from: https://pen.org/wp-content/uploads/2019/12/Losing-the-News-The-Decimation-of-Local-Journalism-and-the-Search-for-Solutions-Report.pdf

Makuch, E. (2016, February 16). U.S. video game industry generated 23.5 billion in 2015, increasing 5%. *GameSpot*. Retrieved from: www.gamespot.com/articles/us-video-game-industry-generated-235-billion-in-20/1100–6434834/

Marino-Nachison, D. (2014, December 6). Engineer became the father of video-game industry. *The New York Times*, p. B4.

Martin, M. M., Anderson, Carolyn M., & Cos, G. C. (1997). Verbal aggression: A study of the relationship between communication traits and feelings about a verbally aggressive television show. *Communication Research Reports, 14*(2), 195–202.

Martin, N. (2019, September 25). The major concerns around facial recognition technology. *Forbes*. Retrieved from: www.forbes.com/sites/nicolemartin1/2019/09/25/the-major-concerns-around-facial-recognition-technology/#4c14cca94fe3

McDonald, D. G., & Kim, H. (2001). When I die, I feel small: Electronic game characters and the social self. *Journal of Broadcasting & Electronic Media, 45*(2), 241–258.

McGregor, J. (2014, September 14). A scatterbrain? The neuroscience of getting organized. *The New York Times*, p. G7.

McLennan, D., & Miles, J. (2018, March 21). A once unimaginable scenario: No more newspapers. *The Washington Post*. Retrieved from: www.washingtonpost.com/news/theworldpost/wp/2018/03/21/newspapers/

McQuail, D. (2000). *Mass communication theory* (4th ed.). London: Sage Publications.

Media violence study. (2013). *Parents Television Council*. Retrieved from: http://w2.parentstv.org/main/Research/Studies/CableViolence/cableviolence2013.aspx [December 10, 2014]

Millennials drive while using cellphones at worse rate than Gen X or Boomer: Liberty Mutual. (2019, August 18). *Carrier Management*. Retrieved from: www.carriermanagement.com/news/2019/08/18/196692.htm

Mukherjee, S. (2017, June 20). Cause of child death in America. *Forbes*. Retrieved from: https://fortune.com/2017/06/20/cdc-suicide-teen-gun-deaths/

National TV Violence Study. (1998). Executive summary. Retrieved from: www.academia.edu/944389/National_Television_Violence_Study_Executive_Summary_Editor_University_of_California_Santa_Barbara_

NCIC missing person and unidentified person statistics for 2014. (2014). *FBI*. Retrieved from: https://archives.fbi.gov/archives/about-us/cjis/ncic/ncic-missing-person-and-unidentified-person-statistics-for-2014

Nelson, S. S. (2014, December 1). German government may say 'nein' to after work emails. *National Public Radio*. Retrieved from: www.npr.org/blogs/parallels/2014/12/01/366806938/german-government-may-say-nein-to-work-emails-after-six [December 12, 2014]

New media study discovers Americans need 38 hours per day to complete their tasks. (n.d.). *Earthtimes.org*. Retrieved from: www.earthtimes.org/articles [September 28, 2009]

The New York Times v. *Sullivan*. (1964). 376 U.S. 254.

Nomophobia, the fear of not having a mobile phone, hits record numbers. (2013, June 2). *news.com.au*. Retrieved from: www.news.com.au/technology/nomophobia-the-fear-of-not-having-a-mobile-phone-hits-record-numbers/story-e6frfro0-1226655033189 [December 12, 2014]

Online and digital news. (2012, September 27). *Pew Research Center for the People & the Press*. Retrieved from: www.people-press.org/2012/09/27/section-2-online-and-digital-news-2/ [December 12, 2014]

Paik, H., & Comstock, G. (1994). The effects of television violence on antisocial behavior: A meta analysis. *Communication Research, 21*(4), 516–546.

Palmgreen, P., Wenner, L. A., & Rosengren, K. E. (1985). Uses and gratifications research: The past ten years. In K. E. Rosengren, L. A. Wenner, & P. Palmgreen (Eds.), *Media Gratifications Research* (pp. 11–37). Beverly Hills, CA: Sage Publications.

Parents, children and media. (2007). *Kaiser Family Foundation*. Retrieved from: www.frankwbaker.com/mediause.htm [February 6, 2010]

Parents Television Council. (2011). *Habitat for Profanity*. Retrieved from: www.parentstv.org/PTC/news/release/2010/1109.asp

Pember, D. (2001). *Mass media law*. Boston, MA: McGraw-Hill.

Pennebaker, R. (2009, August 30). The mediocre multitasker. *The New York Times*, p. A5.

Penetration rate of gamers among the general population in the United States from 2013 to 2018. (2020). *Statista*. Retrieved from: www.statista.com/statistics/748835/us-gamers-penetration-rate/

Peoples, G. (2019, September 12). Average music listening time is down: How much does this matter? *Billboard*. Retrieved from: www.billboard.com/articles/business/streaming/8529828/average-music-listening-time-down

Poovey, B. (2002, April 15). Device lets viewers bleep at home. *Knoxville News Sentinel*, p. C3.

Potter, W. J. (2003). *The eleven myths of media violence*. Thousand Oaks, CA: Sage.

Quittner, J. (1999, May 10). Are video games really so bad? *Time*, pp. 50–59.

Radcliffe, S. (2016, July 6). Junk food ads may cause kids to overeat. *Parenthood*. Retrieved from: www.healthline.com/health-news/junk-food-ads-cause-kids-to-overeat

Roberts, D. F., & Foehr, U. G. (2008). The future of children. *Children and Electronic Media, 18*(1), 11–37.

Rosengren, K. E., Wenner, L. E., & Palmgreen, P. (1985). *Media gratifications research*. Beverly Hills, CA: Sage Publications.

Rubin, A. M. (1981). An examination of television viewing motives. *Communication Research, 8*(2), 141–165.

Rubin, A. M. (1984). Ritualized versus instrumental viewing. *Journal of Communication, 34*, 67–77.

Severin, W. J., & Tankard, J. W., Jr. (1992). *Communication theories: Origins, methods, and uses in the mass media* (3rd ed.). New York: Longman.

Sherman, E. (2017, May 24). Advertising that targets kids makes them eat more fast food. *Food & Wine*. Retrieved from: www.foodandwine.com/news/advertising-targeting-kids-makes-them-eat-more-fast-food

Shute, N. (2013, January 24). If you think you're good at multitasking, you probably aren't. *National Public Radio*. Retrieved from: www.npr.org/blogs/health/2013/01/24/170160105/if-you-think-youre-good-at-multitasking-you-probably-arent [December 10, 2014]

Signorielli, N. (1990). Television's mean and dangerous world: A continuation of the cultural indicators perspective. In N. Signorielli & M. Morgan (Eds.), *Cultivation analysis* (pp. 85–106). Beverly Hills, CA: Sage Publications.

Signorielli, N. (2005). Age-based ratings, content designations, and television content: Is there a problem? *Mass Communication & Society, 8*(4), 277–298.

SlickText. (2019). 37 addiction statistics for 2019. Retrieved from: www.slicktext.com/blog/2019/10/smartphone-addiction-statistics/

Soap operas boost rights, global economist says. (2009). *National Public Radio*. Retrieved from: www.npr.org/templates/story/story.php?storyId5113870313 [October 21, 2009]

The state of online gaming. (2019). *Limelight Networks*. Retrieved from: www.limelight.com/resources/white-paper/state-of-online-gaming-2019/

Sterling, C., & Kittross, J. (2002). *Stay tuned: A history of American broadcasting*. Mahwah, NJ: Erlbaum.

Steyer, J. P. (2002). *The other parent*. New York: Atria.

Stone, M. (2014, July 31). Smartphone addiction now has a clinical name. *Business Insider*. Retrieved from: www.businessinsider.com/what-is-nomophobia-2014-7

Sundem, G. (2012, February 24). This is your brain on multitasking. *Psychology Today*. Retrieved from: www.psychologytoday.com/us/blog/brain-trust/201202/is-your-brain-multitasking

Survey finds one in three U.S. teens text while driving. *AAFP*. Retrieved from: www.aafp.org/news/health-of-the-public/20180928textndrive.html [September 28, 2018]

Taggart, J., Eisen, S., & Lillard, A. S. (2019). The current landscape of U.S. children's television: Violent, pro-social, educational, and fantastical content. *Journal of Children and Media, 13*(3), 276–294.

Takahashi, D. (2009, December 14). Gamer population surges – consoles in 60% of households. *Games Beat*. Retrieved from: www.games.venturebeat.com/2009/12/14/video-gamer-population-surges-as-60-percent-of-households-now-have-game-consoles

Telford, T., Heath, T., & O'Connell, J. (2020, February 13). Newspaper giant McClatchy files for bankruptcy, hobbled by debt and declining print revenue. *The Washington Post*. Retrieved from: www.washingtonpost.com/business/2020/02/13/newspaper-giant-mcclatchy-files-bankruptcy-hobbled-by-debt-declining-print-revenue/

Time flies: U.S. adults now spend nearly half a day interacting with media. (2018, July 31). *Nielsen*. Retrieved from: www.nielsen.com/us/en/insights/article/2018/time-flies-us-adults-now-spend-nearly-half-a-day-interacting-with-media/

Tracy, M. (2019, August 7). The New York Times up to 4.7 million subscribers as profits dip. *The New York Times*. Retrieved from: www.nytimes.com/2019/08/07/business/media/new-york-times-earnings.html

Tsukayama, H. (2015, November 14). Young minds immersed in media. *The Washington Post*, p. A12.

Turkle, S. (2012, April 21). The flight from conversation. *The New York Times, Sunday Review*, pp. 1, 8.

U.S. video game sales reach record-breaking $43.4 billion in 2018. (2019, January 22). *Entertainment Software Association*. Retrieved from: www.theesa.com/press-releases/u-s-video-game-sales-reach-record-breaking-43-4-billion-in-2018/

Vaala, S. E., Bleakley, A., Castonguay, J., & Jordan, A. B. (2017). Parents' use of the V-chip and perceptions of television ratings: The role of family characteristics and the home media environment. *Journal of Broadcasting & Electronic Media, 61*(3), 518–537.

Value of the global video game market from 2012 to 2021. (2020). *Statista*. Retrieved from: www.statista.com/statistics/246888/value-of-the-global-video-game-market/

Violent video games. (n.d.). *ProCon*. Retrieved from: http://videogames.procon.org/view.resource.php?resourceID=003627 [December 12, 2014]

Vranica, S. (2015, October 13). Marketers may be falling out of love with product placement. *The Wall Street Journal*. Retrieved from: www.wsj.com/articles/marketers-may-be-falling-out-of-love-with-product-placement-1444771858

Wacker, J. G. (2004). A theory of formal conceptual definitions: Developing theory-building measurement instruments. *Journal of Operations Management, 22*(6), 629–650.

Wallis, C. (2009, March 19). The multitasking generation. *Time*. Retrieved from: www.time.com/magazine/ [January 11, 2010]

Watson, B. (2014, February 24). The tricky business of advertising to children. *The Guardian*. Retrieved from: www.theguardian.com/sustainable-business/advertising-to-children-tricky-business-subway

Warzel, C. (2019, August 18). Gamergate gave us the post-truth information war. *The New York Times*, SR6.

Wu, B. (2019, August 18). Why was there no reckoning? *The New York Times*, SR6.

Wurmser, Y. (2019, May 30). U.S. time spent with mobile 2019. *eMarketer*. Retrieved from: www.emarketer.com/content/us-time-spent-with-mobile-2019

Yancey, P. (2017, July 29). The death of reading is threatening the soul. *The Washington Post*, B2.

Zill, N. (2001). 'Does *Sesame Street* enhance school readiness?' Evidence from a national survey of children. In S. M. Fisch & R. T. Truglio (Eds.), *'G' is for 'growing': Thirty years of research on children and Sesame Street* (pp. 115–130). Mahwah, NJ: Erlbaum.

Index

Note: Entries in *italics* indicate diagrams, figures and tables.

3D television and movies 79, 213, 216, 376, *377*
4chan 410
4G wireless technology 135–6
5G FAST Plan 326
5G wireless technology 136, 198, 229, 231, 326
6DoF (six degrees of freedom) 222
8MK-WWJ Detroit 31
9/11 terrorist attacks 245–6, 401
20th Century Fox 74, 109, 201, 382–3
360° video 218, 220, 222, 228–9
2001: A Space Odyssey 375, *377*

ABC (American Broadcasting Company) 8–9; market share 8, 60; news on 106; origins of 9, 309; ownership of 310
Abrams, J. J. 384
accelerating change 19
Accessible Communications for Everyone 326
account representatives 339
Acosta, Jim 202
ACSI (American Consumer Satisfaction Index) 328
ACT (Action for Children's Television) 68
action cameras *195*
action heroes 372
ad auctions 298–9
ad exchanges 298–9
ad-blocker software 267
addressable geofencing *262*
advertising xi, 8, 237, 272; from business perspective 245; on cable television 357; from consumer perspective 246; creative strategies 244; criticisms of 268–73; history of 237–8; on radio 34–5, 238–40, 243, 248–9, *250*, 339; residuals for music in 45; sales of 292–4, 298–9; on television 68–9, 74, 240–4, 249–54, *250*, 363 (*see also* television commercials); trust in media for *271*; on YouTube 18
advertising agencies 63, 237–9, 242, 246–7, 272

advertising campaigns 247
advertising revenue: for broadcast television 74, 241; in Great Recession 318; by medium 245; for radio 240, 316; for streaming services 47
advertorials 257–8
Aereo 194
affective effects of media 401
affiliated stations: radio 35, 38; relationship with network 110, 310; revenue streams 324; in star model 312; switching affiliations 323; television 56, 60, 74, 312, 324
African Americans 40, 84–5, 87, 95, 198, 387
agenda setting 398–9
Agha-Soltan, Neda 147
AirPort Base Station 135
Alexa 2, 147, 195, 232
Alexander, Julia 379
algorithms 1, 15, 144, 202, 267, 271, 416
All-Channel Receiver Act 1962 57–8
all-news format 89
AM radio 8, 48; 1980s resurgence of 10; bandwidth of 54; in cars 11; evolution of 40; expansion of 41; and FM radio *42*; news on 89
Amazon: advertising on 202, *246*, *258*; cloud storage from 14; and cooperative advertising 258–9; and Kindle 194; and USPS 127
Amazon Echo 147, 195
Amazon Music 138
Amazon Prime ix, 18, 80, 82, 114
ambient awareness 171–2, 176
Ambisonics 229
AMC 116, 118, 224, 287
American Bandstand 86, *87*
An American Family 101–2
American Honda 42
American Idol 20, 103, 290, 302; price of commercial on 249
American Marconi 31–2
America's Funniest Home Videos 101–2, 364
Amos 'n' Andy 84, *85*

Andreessen, Marc 126, 203–4
Android devices 144
Android TV 192
anonymity, online 145
anonymous sources 170, 321
answering machines 184
antenna, transmitting 64, 334
anthologies 91–2, 118
AOL (America Online) 128
AP (Associated Press) 24, 106, 354
Apple: and encryption 175; home speakers from 12; and music consumption 314; smart glasses from 230–1; and television 76, 81; and Wi-Fi 135; *see also* iPads; iPhones; iPods
Apple HomePod 195
Apple Music 138, 302
Apple Newton 184–5
Apple Pay system 191
Apple TV 78, 192, 263, 325, 328
Apple Watch 196
apps *see* mobile apps
AQH (average quarter hour) 286–7
AR (Augmented Reality) 211–12, 218; and Apple technology 222–3, 231; Google Glass as 220; rise of 231–2; Sword of Damocles as 214
Arab Spring 164, 170
ARB (American Research Bureau) 280
Arbitron 280
ARF (Advertising Research Foundation) 298
Arnaz, Desi 97, 112
arousal 401
ARPAnet 123
artificial intelligence 176, 195, 201–3, 233
The Artist 371
ASCAP (American Society of Composers, Authors, and Publishers) 35, 44–5, 314, 317
aspect ratio 75
asynchronous media 12
ATSC (Advanced Television Systems Committee) 75–6, 79
AT&T 5; ad spending 246; breakup of 308; and DirecTV 80, 360, 365; media

425

ownership by 311; mergers with other telcos 198; and radio 32, 35; and telephone technology 183–4; video packages 329
attention span 176
audience fragmentation 150, 245, 249, 252–3
audience measurement 17, 279; and advertising sales 291–2; calculations of 285–6; challenges of 283–4; new developments in 301–3; reporting 286–90
audiences: demographic analysis 290–1; demographic changes in 17; for mass media 12
audimeter 280
audio console 334, 336, 348–9, 358
audio recording 12, 202, 337
audion tube 26, *28*, 83
automobiles: accidents and mobile devices 409; Internet audio in 45, 147; radios in 38, 43, 46
AV (Augmented Virtuality) 211–12, 218, 227–8
Avatar 376, 382
AVOD (advertising-supported video-on-demand) 326
Ayer, Francis 238

The Bachelor/Bachelorette 103, 302
Baird, John Logie 51
Baird disc *52*
balance sheet 318, *319*
Baldwin, Adam 410
Ball, Lucille 95, 97, 112
Bandura, Albert 395–7
bandwidth: in electromagnetic spectrum 41, 43, 54; on Internet 124
banner display advertising 254–6; clicks on 296; mobile 256–7
Barney 109
bartering 294
baseball *104*, 109, 344
Batman (1966 television program) 116
BBC (British Broadcasting Corporation) 51, 308, 361
BBS (Bulletin Board System) 157
The Beatles 11, 98, *99*, 359
behavioral effects of media 11, 400–1
Bell, Alexander Graham 4, 25, 183, 308
Benny, Jack 84–5
Berners-Lee, Tim 123, 130, 312
BET 70, 320
Betty and Bob 84
Beulah 84–5, 95
bicycle networks 68
The Big Bang Theory 93, 95, 112–13, 194, 251, 289; price of commercial on 249
big data 17, 149, 196, 417
big-box advertising 255
Bigelow, Erin *347*
Billboard 302
billboards, online 255
Biltmore Agreement 39
bin Laden, Osama 169

binge watching *193*, 194, 404–5
Black Lives Matter 169
Black Mirror 233–4
blackface 84, *85*
blacklisting 63–4, 369, 389
blocking 117
blogs 131–3, 147, 167–9
blue light 405, 410–11
Blue's Clues 109
Bluetooth 136, 147, 196
BMI (Broadcast Music, Inc.) 44, 314, 317
Bobo doll 395, *396*
Bochco Productions 110
bonus programming 251
books, printed and electronic 15
Boston Marathon bombings 164, 169, 174, 198
Bozo the Clown 109
brand awareness 142
brand image 80, 244–5, 259
brand loyalty 259
branded content 171, 257–8
branding, in movie industry 381–3
Breaking Bad 118
bridging 117
broadband 124; access to 198, *199*, 326, 329
Broadcast Decency Act 2005 407–8
broadcast networks: advertising on 243–5, 249, 252; audience measurement for 287, 289–90; beginnings of 56; educational *see* educational television; news on 106–7; owners of content 81–2; ownership by 309–10; and program production 109–10, 112–13; relationship with affiliates 60–3, 74; streaming platforms of 18, 114; and Telecommunications Act 1996 316; websites of 79, 142
broadcast radio: beginnings of 31–3; commercial 1, 32, 34–6; news on 38–9; and other radio formats 17–18, 45; program selection for 14–15; regulation of 36–8, 46
broadcast television 8; beginnings of 54–7; and cable television 66, 74; characteristics of 12–13; in color 59–60; in coronavirus lockdown ix–x; future of 79–80; in McCarthy era 63–4; news on 105–7, 245; and primetime 113; signal path for 58, 64; via computer 78
broadcasting, deregulation of 311
Brown, Harold 392
Browning King company 239
building/maintenance engineers 344
Bullock, Sandra *378*
The Burns and Allen Show 85
Bush, George W. 186
business management, in television stations 342, 344
buzz marketing 246, 259
BuzzFeed 142, 258

Cable Act 1992 316
cable channels 8, 15; advertising-supported 70; bundling 364; buying advertising on 294; definition 71; distribution of 16; growth and ownership of 311; independent producers for 110; and localism 16; news-based 106–7; programming for 18–19, 113–15; in vertical integration 15; viewership of 116; *see also* premium channels
cable companies: and advertising sales 294; definition 71; and home satellite dishes 72–3; operations of 355, *356*; renting citizens 67; and streaming 76–8; VOD from 15; *see also* MSOs
cable networks *see* cable channels
cable nevers 80
cable systems 8; definition 71; regulation of 66, 71
cable television 64–7, 81; advertising on 244–5, 252–3; audience measurement for 287, 289–90; future of 80, 328, 364; market share of 325; profanity on 406; programming for 117–18; and satellite transmission 69–70; signal path *67*, *71*; subscriptions to 316; and Telecommunications Act 1996 316; terms for 71
caching 296
California: legal restrictions on data collection 266; movie industry in 367–8; net neutrality act 149
call letters 31, 34, 36, 55, 281, 341
call-in programs 89
Cambridge Analytica 19
camcorders 346, 363–4
Cameron, James 382
Candid Camera 101
CAN-SPAM (Controlling the Assault of Non-Solicited Pornography and Marketing Act) 2003 268
Capra, Frank 382, 389
Captain Kangaroo 107–9
Carlin, George 397, *398*
Carmack, John 218–19
Carney, Art *96*
The Carol Burnett Show 97–8
Caruso, Enrico 83
Carvin, Andy 170
Cassavettes, John 384
catch and kill 321
category contextual targeting 261
catharsis 401
CAVE (Cave Automatic Virtual Environment) 215–16
cave paintings 2
CBS (Columbia Broadcast System): beginnings of 36; and The CW 15; merger with Viacom 16; news on 106; ownership of 310; radio network 8, 38–40; radio news programs 40; streaming service 114; television affiliates 60
CBS All Access 313, 327
CBS color system 59
CCPA (California Consumer Privacy Act) 266

CCU (camera control unit) 348
CDA (Communications Decency Act) 11, 316
CDs (Compact Disks) 18, 137–9, 147, 314
celebrities, surveying opinions of 291
cell phones xi, 21, 184; addiction to 409; in citizen journalism 147; cognitive effects of use 404; data plans 308; demographics of ownership *187*; etiquette for 188; function of 185; generations of technology 136; interfering with other appliances 189; Internet connections via 13, 148; radio via 45; social impact of 186–8; and texting 127; video shot on 364; *see also* iPhones; smartphones
CEO (chief executive officer) 195, 226, 313, 321, 358
Cerf, Vincent 123
CERN (European Laboratory of Particle Physics) 123, 200
CET (Carnegie Commission on Educational Television) 68
CFO (chief financial officer) 344
CGI (computer-generated imagery) 374, 389
chain broadcasting 35, 309
channel surfing 252–3
Chaplin, Charlie 367, *368*, 370
Chapman, Rex 173
character generators 348
Charter Communications 246, 311
chat rooms 127–8, 408
Cheers 112, 173, 288
Cherry Productions 110
children, media exposure of 413
children's television 81, 84, 107–8, *109*; advertising on 270; and rating system 406–8; social effects of 413; violence in 397
Children's Television Act 1990 68
Christensen, Clayton 17
Chromecast 192, 328
CIMM (Coalition for Innovative Media Measurement) 298
Cinemax 114
CineVR 218, 228–9, 231
citizen journalism 169
Citizen Protect the World 198
Clark, Dick 86, *87*
Clark, Jim 126
Clear Channel 42, 315, 362
click-throughs *see* CTRs
Clinton, Bill 34
Clinton, Hillary 165
Clio Awards 247–8
Clooney, George *378*
cloud computing 14, 195
CNBC ix, 107
CNN (Cable News Network) 10, 69–70, 74, 107
cognitive effects of media 11, 401
Colbert, Stephen *105*, 169
college radio: first 31; limits of signal 16; music on 90

color television *59*, 60, 73, 81
Columbia Phonograph Company 36
Comcast: ad spending 246; and vertical integration 15
comedies: on radio 84–5; on television *see* situation comedies
commercial radio: programming 87; rise of 34
commercial time 88, 112; presold 114
Commission on the Causes and Effects of Violence 68
communication models 5–6
communication technology: development of 2–8; timeline of *9*
Communications Act 1934 36–8, 238–9, 315
communism 63–4, 369
community antenna television 64–6; *see also* cable television
competition, and ethics 321–2
computer-generated 217–18, 224, 234; *see also* CGI
computers: personal 17–18, 312; and television sets 76
comScore 290, 296–7
Conrad, Frank 31–2, 47
consolidation 16, 21, 323–5, 334, 362; advantages and disadvantages 324; and agenda setting 398; in radio 48; in television 81
Contagion 381
content analysis 399, 405
convergence 15, 21, 136, 197, 204, 315
conversion zones 262
cookies 143–4, 266
Cooper, Martin *184*
cooperative advertising 258–9, 294
Cop Watch 198
COPS 102
Copyright Act 1907 35
Copyright Act 1976 44
copyright law 44–5, 47, 312
cord cutting 78, 80, 82, 117, 302, 325, 358, 363–4
cord nevers 117
corner peel advertising 255, *256*
coronavirus ix–x, 127, 169; collaboration on finding a cure 201; impact on advertising 17, 245; Internet misinformation on 19, 317; and satellite radio 329; and social media 169, 171, 173; and streaming services 76, 192, 328; theatrical releases during 379; and webcams 194–5
Counterattack 63
counterprogramming 117, 253
CPB (Corporation for Public Broadcasting) 68–9, 90, 108
CPC (cost per click) 279, 296, 301
CPI (cost per impression) 297, 301
CPM (cost per thousand) 292, 297–8, 303
CPP (cost per point) 292, 303
CPT (cost per transaction) 292, 296, 301, 303
Crawford, Cindy *244*

creative boutiques 246–7, 272
credit card numbers 143
crisis communication 173
Cronkite, Walter 40, 104, 106
Crosby, Bing 337, *338*
Crossley, Archibald M. 279–80
cross-media ownership 16, 318
cross-promotion 340
Crusader Rabbit 109
CTRL+Labs 230
CTRs (click-through rates) 296, 298–9, 303
CTV (Connected TV) 261–2, *263*
CU (close-up shot) 350
cultivation 405
cumulative effects model 398
cumulative persons (cumes) 286–7
Cumulus Media 248, 315
The CW Network 15, 113, 320
cyber warfare 380
cyberbullying 408, 409–10
cybercasting 136, 138

Dancing with the Stars 20, 103, 302
dating shows 103
Davis, H. P. 34
dayparts 293, 303
DBS (direct broadcast satellite) 10, 13, 81, 357, 360; beginnings of 72–3; and cord cutting 77–8; digital conversion 75; future of 80; and geosynchronicity 70
de Forest, Lee 26, *28*, 30, 47, 83
death threats 174, 410
debt service 318
deceptive advertising 269, 271
deconvergence 16, 21
deep fakes 133, 202, 233, 408, 410
defamation 322–3
Dell 76, 81, 221
demo discs 88
desensitization 401
desktop production 21
Dexter 114
Di Bona, Vin 102
Dickson, W. K. L. 213
digital addiction 408–9
digital assistants 13, 184, 195, 202
digital devices 1, 17; location-based 134; social impact of 197–203; varieties of 184–97
digital dissenters 417
digital equipment 75, 279, 365
digital media: effects on children and young adults 413–16; lifestyle effects of 410–11
digital natives 176, 328
digital radio 14, 42–3
digital subchannels 43, 363
digital targeting 267
digital television 60, 74–6, 192, 347; signal stream 77
digitization 14–16, 21, 41, 385
diodes 26
DirecTV 42, 72, 80, 325, 328, 360, 365

dirty tricks 321
dish cutting 325
Dish Network 73, 325, 328, 365; advertising on 361; and Nielsen ratings 290
Disney+ 1, 15, 327–9; and coronavirus lockdowns ix; new releases on 379; program library of 19; *Star Wars* content on 384; subscribers to 8
Disney: buying *Star Wars* 384; dominance of 382–3; and Hulu 312; vertical and horizontal integration 15, 310; and XR 216, 224
Disney Channel 108, 116
Disney v. Sony 39
display, of mass media 12–13
display banner targeting 259–61
disruptive technologies 17
distance, of mass media 13–14
distribution, of mass media 13
DJs (disc jockeys): hosting music formats 48; and payola scandal 90; and rock 'n' roll 86; use of term 40
DMAs (designated market areas) 280, 284
DMCA (Digital Millennium Copyright Act) 45
DNS (Domain Name System) 125
Domino, Fats 88
dopamine 403, 413
Dotto 99
double features 369, 372, 378, 389
Douglas, Kirk 369
download sites 76
downloading, illegal 137–8, 322
Downton Abbey 115, 118
dramas: primetime 113; on radio 84; on television 92
Dreams (film) 211, *212*
Dreamscape 224, 231
Drew, Doug 354
driving, and mobile devices 409
DSL (digital subscriber line) 13, 365
DuMont network 60, 63, 310
duopoly 318
DVDs: sales of 76; storage on 14
DVRs (digital video recorders) xi, 12, 14, 141, 204; and audience measurement 287–9; avoiding advertising with 193, 250, 252
dynamic inventory retargeting 263–4

Easy Rider 372, 383–4
e-books 194
The Ed Sullivan Show 97–8, *99*
Edison, Thomas 213, 367, 389
editing systems 350
educational television 58–9, 68–9, 81
The Electric Company 109
electromagnetic spectrum *10*, *27*; regulation of 9–10, 28, 30, 36, 74, 197; television frequencies on 54
electronic media: business models for 307–9, 312, 329; effects of 11, 21; finances of 318; regulation of 10–11; trends and terminology in 14–16; *see also* convergence
Ellen 112
The Ellen Degeneres Show 112
Ellul, Jacques 204
email 17, 127–8, 174
Emmy Awards 94, 96, 108, 113–14, 118, 251
emoji 162, *172*
emoticons 172
emotional effects 11
encryption technology 145, 175
ENG (electronic news gathering) 350
engineering department: at cable companies 356; in radio stations 333–4; in satellite radio 358; in satellite television 360; in television stations 342, 344
Epstein, Adam 381
equal time provision 311
ER 92, 407
ESPN 16, 69–70, 74; advertising on 253; and Sling TV 193; sports news on 107
ESRB (Entertainment Software Rating Board) 412–13
ethical dilemmas 233, 321–2
ethics 2, 177, 318–21; codes of 311, 320–1
ETs (electrical transcriptions) 337
extramercials 255–6, 272

Fabacher, Trey *295*, 296
Facebook 133, *163*; advertising on 259, 301; Beacon program 266; in coronavirus lockdown ix; friends on 171; history of 162–3; instant messaging on 128; and other SNSs 164–5, 176; television presence on 76; user numbers 161–2; user-generated content on 19; and XR 217, 219, 221–2, 228–30, 234
Facebook reactions 162
Facetime 190
facial recognition technologies 202–3, 232
Fahlman, Scott 172
fair use 44
fairness doctrine 10, 311
fake news 169
Farnsworth, Philo T. 51–4, *55*, 80
Farook, Syed 175
Farrow, Mia 92
Faughnder, Ryan 378
Faulk, John henry 64
fax machines 183–4
FCC (Federal Communications Commission) 317–18; allowing cross-media ownership 16; and broadcast television 54–5, 57, 59–60, 81, 309; and cable television 66, 324, 356–7; and cell phones 184–5; and deregulation 311; and digital television 74–5; and educational television 69; electromagnetic spectrum auction 197–8; and ethics 318–20; fin/syn rules 112; foundation of 38; Internet Policy Statement 148; and MSAs 323; and multichannel delivery 325; and objectionable content 397, 406–8, 415; ongoing initiatives 326; primetime access rule 112; and product placement 251; regulation of radio 41, 43, 45–6, 48; and satellite radio 358; and satellite television 69, 73, 360; and station operations 334; subscription-based and broadcast media 10; and telecommunications equipment 316; utility ruling 148–9
Federman, Matthew *386*
feedback circuit 28
Felix the Cat 57
Fessenden, Reginald 26, 30–1
fiber optic cable 312, 356
field experiments 399
field production 111, 350
file sharing 44–5, 137–8
files 137
film, television programs recorded on 63
film cameras 63, 346, 376
films *see* movies
filter bubbles 142, 203
fin/syn rules 112–13
Firefox 126, 130, 144
FireTV 328
first-run syndicated programs 112
Fitbit Flex 196
fixed buys 293
flat-screen monitors 191–2
Fleming, John 26
Flickr 167
flight simulators 213, 216
floating ads 255, *257*
FM radio 48; and AM radio *42*; invention of 28, 41; music formats on 89; non-commercial 59, 90
Food Network 114, 253, 357
Fort Lee, NJ 367, 389
Fox Network 74, 81, 113, 310, 323–4
Fox News 10, 107, 356
Fox News Radio 248
franchises: in cable television 355; film 383–4, 389
Frawley, William 97
FRC (Federal Radio Commission) 36, 38, 47
free speech 8–9, 316, 323
Freed, Alan 86
frequency allocation 14, 36
frequency discounts 293, 298, 303
Friendster 158, *160*, 176
FTC (Federal Trade Commission) 259, 266, 268–9, 271
full-service advertising agencies 238, 246–7, 272
Funt, Allen 101
Fyre Festival 176

Game of Thrones 93–4, 114, 194, 289, 302
game shows 98–100; reality 103; timeslots for 112
Gamergate 410

Gannett Company 16
gatekeeping 398
Gates, Bill 169
GE (General Electric) 26, 32; and early television 54; and NBC 35; ownership of broadcasting 310
general management: in cable companies 355; in radio stations 333–4; at satellite radio 358; of satellite television 360; in television stations 342
General Mills 84, 238
geofencing 261, *262*
geosynchronous satellites *70*, 71–2
geotargeting 261
Ghostbusters 374
GLAAD 388
Gleason, Jackie *96*, 97
GM (General Motors): ad spending 246; and DirecTV 72; and satellite radio 42
Gofundme 173, 308
Golden Globe Awards 94, 96, 114, 407
Golf Channel 244, 253
Gone with the Wind 95, 98, *369*, 370
The Good Doctor 92–3, 118, 288
Google+ 158
Google: cloud storage from 14; and home speakers 12–13; instant messaging on 128; and shared endorsements 266
Google Ads 299
Google Cardboard *220*
Google Chrome 126, 130
Google Chromecast *78*
Google Expeditions 225
Google Glass 196, 220, 224–5
Google Home 195
Google Play 138, 189, 380
GoPro cameras *18*, 195
Gore, Tipper 398
government subsidy or ownership model 308
GPS (global positioning system) 70, 271, 301
The Graduate 372–3
Grant, Cary 372, *374*
Gravity 376, 378
gravity ads 255
Great American Values Test 397
Great Recession 127, 245
Greeley, Horace 135
Grey's Anatomy 93, *94*, 113–14, 118, 194, 302; price of commercial on 249
grids 200–2, 204
Griffith, Andy *243*
Griffith, D. W. 367
GUI (graphical user interface) 214
Gupta, Sanjay *105*
Gutenberg, Johannes 3–4, 23, 238

Hadid, Bella 176
hammocking 117
haptics 215–16, 229–30, 232
Harding, Warren G. 33
hashtags 164–5, 409–10
Hays Production Code 368–9, 372, 389

HBO (Home Box Office) 18, 69–70, 74; award-winning programming 118; financial relationship with networks 113; movies on 97; as premium channel 114; satellite provision of 69, 311, 356, 359; *Sesame Street* on 108
HBO Max 76, 115, 192, 328, 379
HBO Now 114, 141, 192
HD (high-definition): radio 14, 43, 362, 366; television 1, 74–5, 79, 82, 199, 326, 356, 363
HDMI (high-definition multimedia interface) 356
headphones 211, 213, 224, 229
head-to-head programming 117
healthcare, XR in 226–7
Hear It Now 104
Hechanova, Maria *352*
Heilig, Morton 214
Herrold, Charles D. 30–2
Hertz, Heinrich 25–6
HGTV (Home and Garden Television) 116, 253, 290
hieroglyphics 2, *3*
higher education, online delivery of x
Hitler, Adolf 392, *393*
hits 296
HMD (head-mounted displays) 213, 215, 218, 220–1; *see also* VR
Hollywood xi, 60; beginnings of film industry in 367; blacklisting in 369; unionization of 385; *see also* movie industry
Holzhauer, James 100
home entertainment systems 124, *199*
home improvement shows 103
homepage takeover ads 256
homuncular flexibility 215, 227
The Honeymooners 96
Hooper, C. E. 279–80
Hoover, Herbert 35, 238
Hope, Bob 85, 97
Hopper, Dennis 384
horizontal integration 15
hot clocks *see* program clocks
HouseParty ix
hover ads 255
The Howdy Doody Show 107–9
HSS (hypersonic sound system) 199–200
HTC Vive 221–2, 231
HTML (hypertext markup language) 126, *129*
Hughes, Howard 95
Hulu xi, 15, 78–9, 114, 141; and coronavirus lockdowns ix; launch of 312; networks releasing first episodes on 116
human resources 79, 344, 356
Humorous Phases of Funny Faces 376
HUT (Households Using Television) 285
hypertext 126

I Love Lucy 95, *97*, 112
IAB (Interactive Advertising Bureau) 298

IBOC (in-band on-channel) 14, 43
ICANN (Internet Corporation for Assigned Names and Numbers) 125
identification, as media effect 400
IFPI (International Federation of the Phonographic Industry) 314
iHeartmedia 248, 315
iHeartRadio 13, 138, 202, 303, 362
image advertising 244
imitation, of media 400
immersion 211, 217–18, 223–4, 229, 231–2, 234
immersive audio 229
implantables 196–7
inattention blindness 403
in-banner audio/video advertising 255
indecency 397–8, 406–8
independent television stations 60, 81; and cable systems 71; and satellite uplinks 70; in star model 312
Industrial Revolution 23, 238
influencers 20, 176
infomercials 238, 269–70, 308
inhibition/disinhibition 400–1
Instagram 15, 158, 164–5, 223
instant messaging 128, 157
Instapundit 132
integration, vertical and horizontal 15
intellectual property 45, 224
interactive/cyber Agencies 247
interactivity: definition 217; of mass media 13; and satellite television 80; and XR 212–13, 224, 228, 234
interconnects 253, 294
International News Service 39
Internet x, 149–50; audio via 136–40; becoming mass medium 123–4, 312; characteristics as medium 11–13, 21; credibility of information on 145, *146*; defamation on 322; development of 123; FCC regulation of 148–9; First Amendment protections of 10–11; future of 329; home access to *127*; how it works 124, *125*; measuring audiences on 294–6; mobile access to 185; most popular activities on *134*; as new media 1; obscenity and violence on 316; popularization of 34; privacy, surveillance and trust on 143–7; resources on 127–31, 133–4; as system of interruption 170; users of 134–5; via cable companies 77–8; via satellite 80
Internet addiction 408–9, *414*
Internet advertising 253–5; advantages and disadvantages 265–6; attitudes towards formats *260*; company revenue shares *258*; format revenue shares *254*; formats of 254–9; industry revenue share *260*; measuring exposure 296–8; and privacy 266–8; sales of 298–9; targeting strategies 259–65
Internet Explorer 130
Internet radio 17, 43–5, 136–8, 147, 149–50, 191; highlights of *137*

Internet Tax Freedom Act 1998 317
Internet telephony 190
Internet television 78–9, 82, *140*, 141–2, 191–2, 194; impact on industry 18; and news 150; and second screening 147; in star model 315; *see also* streaming services
Internet-Use Disorder 413
interstitial advertising 255, 272
The Interview 380
interviews, paying for 321
inverted pyramid format 24
ION Television 113
IoT (Internet of Things) 17, 19, 195–7, 201
IP addresses 125
iPads 1, 15, 78, 188–9
iPhones: and AR 223; films shot with 18; news gathering with 363
iPods 45, 314, 327
Iran, 2009 protests in 147
IRC (Internet Relay Chat) 157
The Irishman 382
ISPs (Internet service providers) 123, 312, 328; cable companies as 78; and copyright law 45
IT (information technology) 334, 344, 358

Jackson, Janet 407
Jackson, Michael: on *The Dating Game* 103; death of 142, 147; funeral of 192–3, 290; as virtual performer 233
Jackson, Peter 374
Janes, Richard 378
Jaws 372, *373*, 376, 389
JCET (Joint Committee on Educational Television) 68
Jennings, Ken 100
Jeopardy! 100, *101*, 112–13, 378
JOA (joint operating agreement) 318
Johnson, Lyndon B. 106
journalists, blogs of 169
JSA (Joint Sales Agreement) 318
Judge Judy 112–13, 287
Julia 95
Jurassic Park 374, 389

KABC Los Angeles 40
Katz, Elihu 394
KDKA Pittsburgh 31–4, 84, 238
Keeping Up with the Kardashians 102
Kennedy, John F. 67, 81, 395
Kennedy, Robert F. 68, 81, 395
keyword contextual targeting 261
KFAX San Francisco 40
KHQA Quincy 353
kill switches 198
Kim Jong-un 380
Kindle 15, 194
Kinescope film 63, 68, 112
King, Larry 38
King, Martin Luther Jr. 68, 81, 395
King, Stephen 385, 388
Kloberdanz, Ryan *340*
Korean War 63

KQED San Francisco 115
Kraft Television Theater 92, 241
Kurosawa, Akira 211, 217

laboratory experiments 399
Lanier, Jaron 215–16
laptop computers 1; editing video on 18; screen size 12; viewing television on 78
The Larry Sanders Show 118
Lasswell, Harold 5
late-night talk shows 104
Law & Order: SVU 92, 113
Lazersfeld, Paul 394
leaderboard advertising 255
leading in and out 117
legacy media 8, 16
LGBTQ community 388, 401
LHC (Large Hadron Collider) 200
Library of Congress 138, 170
Limbaugh, Rush 10, 310
limited-effects models 393–5
Link, Edwin 213
LinkedIn 166–7, 301
liquor, advertising 270
list retargeting 264
listservs 127, 157
live maps 230
live product demonstrations 241
live remotes 336
LiveJournal 158
LMA (local marketing agreement) 318
local discounts 294
local news: Internet input into 76; on radio 46, 89, 139; on social media 157; on television 80, 106–7, 325, 327, 345, 350–1, 363, 405
localism 16
Lodge, Oliver 26
The Lone Ranger (radio show) 38, 84
loneliness 171
lookalikes 264
Lord of the Rings films 374
Lott, Trent 169
low-orbit satellites 70
LTE (long-term evolution) 136
Lucas, George 372, 384
Lucasfilms 384
Luckey, Palmer 217–21, *219*, 225, 229, 234
Lumière brothers 213

machine learning 15, 195, 201
Mad Men 93, 116, 118, 244
made-for-television movies 96–7, 113
magic bullet theory 392–3, 398
Magic Leap 220, 222–4
mailing lists, electronic 127; *see also* listservs
maintenance engineers 343–4
Major League Baseball 38
Majors, Kyle *345*
Maples, Rajah *353*, 354
Marconi, Guglielmo 8, 26, *28*, 47
marketing, four Ps of 237
marketplace of ideas 9

mass communication: development of 4–8; models of *5–6*
mass media 4; audience use of 402, *403*; characteristics of 11–14; effects of 399–401; regulation of 8–11, 14, 28; research on 399; social influence of 391–9
Mathis v. Cannon 322
The Matrix 234
Matsuda, Keiichi 232
Maxwell, James Clerk 25
MBS (Mutual Broadcasting System) 38, 46, 84, 310
McCain, John 165
Meadows, Audrey *96*
mean-world syndrome 405
media adoption rates *124*
media companies: print and electronic operations 16; profitability of 16–17; vertical and horizontal integration 15; *see also* consolidation
media groups, ownership of 323–4
media literacy 18–19
media theories 392–7, 417
media-buying services 247
Meet the Press 105
Megan Wants a Millionaire 101
Melomind 199–200
MeTV 113
MGM 381–2, 389
The Mickey Mouse Club 109
microblogs 131; *see also* Twitter
Microsoft, and television 76, 81
Microsoft HoloLens 220, 222–5, *223*, 231
Microsoft Mixed Reality system 221–2, 231
microwave transmission 38, 66, 69
middle-earth orbiters 70
mid-season replacements 116
Mighty Morphin Power Rangers 109
Millennials: as cord nevers 117; media use 168; private information of 145; and *Saturday Night Live* 98; sources of news 107, 143, 163, 170, 177; use of technology *135*
Milton, John 9
miniseries 95–7, 107
Minow, Newton 67
MMDS (Multichannel Multipoint Distribution Systems) 71–2
mobile advertising 254, 257, 271–2, 296, 301
mobile apps 15, 19, 134, 188–9; for messaging 145; for monitoring police 198; music and video via 13, 45–6, 116; and privacy 144–5; statistics 189
mobile devices 19, 148; addiction to 409–10; and citizen journalism 169; radio via 13, 45; social impact of 198; viewing television on 76
mobile games 257, 402, 412
mobile phones *see* cell phones
moderate-effects models 395–7
Modern Family 95, *98*, 113, 251
monetization 17–18, 194, 307, 313

Montuori, John C. *300*, 301
Moore's Law 19, 230
Morse code 4, 23–6, 183
Mosaic 126, 130
motion sickness 216, 229–30
motion tracking 214, 216, 220–1, 229–30
The Movie Channel 97
movie industry: 21st century transformation of 375–81; beginnings of 367–8; and copyright 44; future of 388; golden age of 369–70; and streaming 76; technological innovation 370–1, 389; and television viewing 97; unions in 384–5
movie production companies 109
movie stars xi, 20, 370, 372–4, *374*, 385
movie studios 8, 19, 81, 361, 367, 389; *see also* studio system
movie theaters: 8K resolution in 79; and coronavirus 76, 378–9; movies released to 96–7; recent trends in 376–8; VR facilities in 224–5
movies: director-driven 372; diversity in 385–8; downloading 322; on DVD 12, 14; impact on culture of 20; independent 384; and interactivity 228–9; on premium channels 69–70, 114; product placement in 258; social media discussion of 158, 164; on streaming services 13, 76, 141, 191, 328, 361, 378–80; on television 95–7; VR in 234
MP3 format 137
MPAA (Motion Picture Association of America) 368–9, 372, 389
MR (Mixed Reality) 211–12, 218; Microsoft and 221, 223–4; multiplayer entertainment 224–5
Mr. Rogers 109
MS (medium shot) 350
MSA (metro survey areas) 280, *281*, 323
MSNBC 10, 107
MSOs (multiple-system operators) 71, 311, 325, 355–7
MTV 70, 74, 81; advertising on 253; reality programming on 102
MUDs (multi-user dungeons) 157, 176
multi-channel delivery systems 325; *see also* cable television; satellite television
multimedia skills 363
multiple-camera filming 95, 103
multitasking 20, 403–4
Murdoch, Rupert 310
Murrow, Edward R. x, 39–40, 64, 104–5, 203
music: and copyright law 44–5, 137; effects of 11; live 8, 33, 307, 336; new consumption patterns of 314; in now media 1; offensive lyrics in 398; over radio 31, 83–5, 87–8, 317; playlists of 90; on television 86–7; top digital download sales *139*
music preference research 290–1
music services 17

music streaming 43, 45, 314; measuring audiences for 302–3, 327; revenues of 18; royalties from 47
must-carry rule 324
Muybridge, Eadward 213
MyNetworkTV 94, 113, 325
MySpace 158, 176

NAB (National Association of Broadcasters): code of ethics 311, 320–1; and liquor advertising 270; and Performance Tax 317
Napster 44, 137–8, 312, 314
narrative programs 91–7
NASA, use of VR 215
native ads 257–8
NATPE (National Association of Television Program Executives) 117–18, 313
NBC (National Broadcasting Company): divestment of stations 9, 309; and FM radio 41; foundation 35; market share 8, news on 106, 140, 142; ownership of 15; streaming service 114; television affiliates 60
NBC Blue and Red networks 35, 38, 48, 83, 309
NBC v. United States 309
NCIS 92–3, 288
ND (news director) 337, 343, 351
nerve tracking technology 230
NET (National Educational Television) 68
net neutrality 148–9, 151
Netflix 18, 78–80, 82, 114; and binge watching 194; and coronavirus lockdowns ix, 192; development of 361, 375–6; program library of 15, 19; television sets connecting to 141; and theatrical releases 381, *382*; watching movies on 141
Netscape Navigator 126, 130
network O&O (owned and operated) 312
network programming 35; ability to reject 309; advertising on 248, 310; cooperative 38; creation of 313; radio 88; television 60
networks: beginnings of 35–6, 56; *see also* broadcast networks; radio networks
neuroplasticity 227
New York City: early radio in 29, 33; origins of cable television in 64; television headquarters in 91
New York Sun 8
New York Times, first issue of 7
New York Times v. Sullivan 323
news: depictions of violence on 405–6; on the Internet 142–3, 147–8; national 24, 107; over radio 38–40, 89, 104; on social media 143, 168–70, 176–7; on television 105–6, 351; *see also* local news
News Corporation 74, 158, 312
news crews 363

news department 337–9, 344
newscasters 38–40, 203, 339, 351
newscasts: in news/talk format 89; video packages for 80
newsgroups 128
newspapers: disappearance of 416–17; origins of 3–8; and radio 38–9; and television news 107; websites of ix, 142
news/talk format 87, 89
Newsy 107
Newton, Isaac 69
NFL (National Football League) 38, 324
niche audiences 11, 114
Nickelodeon 108, 291, 320
Nielsen, A. C. 280
Nielsen diaries: for radio 281, *282*; for television 282–3
Nielsen Media 17–18
Nielsen meters *283*, *284*
Nielsen Music 302
Nielsen ratings system 116, 280–3, 290, 303; reporting from 286–8
Nielsen Social 302
NielsenAudio 280–1
Nintendo: Game Boy 411; Power Glove 215–16; Virtual Boy *216*; and VR 222
Nixon, Richard 67, 69, 107
Noble, Edward J. 309
non-commercial radio 59, 89–90
non-commercial television 68, 108, 115
non-linear editing 350
non-narrative programs 91, 97
non-primetime 113
now media 1–2; advertising on 272; in coronavirus lockdowns ix; future of 416–17; use of term 1
now television 118, 191
NPR (National Public Radio) 90, 310; founding of 68; funding for 68–9; programming on 87
NSI (Nielsen Station Index) 287; ratings book *287*
NTI (Nielsen Television Index) 287
NTSC system 54, *55*, 60, 79
Nuñez, Armando 313

Obama, Barack 34, 165, *377*
Obama, Michelle 270, 377
objectionable content 397, 406–8, 415–16
obscenity 133, 315–16, 406
Occupy Wall Street 170
Oculus Rift 12, *13*, 219, 221, 234
Oculus system 12, 215, 217–22, 230–2, 234
The Office 112–14, 197
off-network syndication 112–13, 313
Olympic Games, televising 103–4
O'Neal, Ryan 92
one-to-many communication 4
one-to-one distance communication 4
online product ratings systems 271
operations department 334
opportunistic market 250
organic (OLED) television sets 79
The Osbournes 102
Oscars 113, 288, 369, 371, 381, 384–5, 389

OTT (Over-the-Top content) 78, 114, 191–3; advertising on 262, *263*; *see also* streaming services
out-of-home viewing 284, 288
over-the-air radio: continuing use of 45, 139; disadvantages of 42; royalty payments 47

Pacifica Foundation 397
packet-switching 123–4, 190, 197
Pai, Ajit 149
Paley, William S. 36, 240, 242
Palm Pilot 185
Palmer, Shelly 19
Palmer, Volney B. 238
Pandora xi, 13, 43–4, 138; audience for 302; revenues of 47; SiriusXM acquiring 329, 358
paradoxes of technology 204
paramount ads 255
Paramount consent decrees 378
Parasite 385, *387*
Parents Music Resource Center 398
Parents Television Council 405–6
Parkins, Barbara *92*
Parsons, L. E. 66
patent pooling 32, 35
Pattiz, Norm *46*
Paul, Les 337
pay per view 326; *see also* TVOD
payola 90
PBS (Public Broadcasting Service) 108–9; children's programming on 108; founding of 68; funding for 68–9; production for 116
PD (program director) 90, 334, 336–7, 351, 359, 363
PDAs (personal digital assistant) 184–5, 188
Peacock streaming service 114, 313, 327
peer-to-peer networks 314
PEG (public, educational and government) channels 357
Pelosi, Nancy 202
performance rights 45
Permanent Internet Tax Freedom Act 2014 317
per-set tax model 308
personal information 143–5, 175, 177, 266, 268
person-to-person communication 4, 183
petroglyphs *4*
Peyton Place 92
phatic communication 412
physiological testing 291
pilots 109, 213, 290
Pinterest 164–5, 182
Pixar Studios 374
platforms, use of term 15
Plato 216, 234
PlayStation 221–2, 225, 231, 411
Podcast One 46
podcasting 1, 12, 44, 48, 139–40; formats of 14; measuring audience for 302–3; top sites *140*

Podtrac 303
point-to-point communication 25, 36
Pokémon Go 223, 232
police, monitoring 198
political activism, and social media 169–70
political programming, on radio 33, 38
pop-unders 255
pop-ups 255
POTS (plain old telephone service) 190
potted palm music 84
Power of Search 19
powerful-effects models 397–8
PPC (pay per click) 296
PPM (Portable People Meter) 281, *283*, 284
praxinoscope 213
premium channels 69–70, 114; scrambling 72, 360
presence, and XR 217, 234
Presley, Elvis 11, 97–8, 359
primetime 106, 113, 252; advertising costs for 292; network programs 112 during; viewership numbers 287–8
primetime access rule 112
print media: and distance 13–14; and electronic media 16
printing press 3, *4*, 14, 23, 183, 238, 391
privacy, on the Internet 143
private information *see* personal information
product placement 18, 250–1, 258, 288–9
production companies 44, 60, 81, 109–10; independent and network-owned 112–13; and Netflix 361; proposals submitted to 109; for satellite radio 364; and satellite television 360; shifting to UHD 79; and television networks 313; for web sites 142
production department: at cable companies 356–7; in radio stations 333–4; in satellite radio 358; in satellite television 360; in television stations 342, 344–51
profanity 398, 406, 412
profit and loss statement *320*
program clocks 90, *91*
program selection 14–15, 78, 193
programming departments 240, 336–7, 344, 351, 358; *see also* radio programming; television programming
program-syndication companies *see* syndicators
Project Nourished 232
promotions department: at cable companies 357–8; in radio stations 333, 340–1; in television stations 342, 355
propaganda and persuasion theories 392
proprioception 217
PSA (public service announcements) 251
psychographics 290
Public Broadcasting Act 1967 68, 81
public domain 44
public-access television 113; production facilities 357

PUR (People Using Radio) 285
push and pull strategies 254
push technologies 296

Q scores 291
quality score 299
Quinn, Zoe 410
quiz shows 84–5, 98–101, 241, 272, 350

radio 46; becoming mass medium 30–1; characteristics of 12–13; development of 8, 23, 26–9, 47–8; digital *see* digital radio; future of 326–7; indecent language on 397; localism in 16; signal stream of 336; and Telecommunications Act 1996 316; World War I takeover of 32; *see also* broadcast radio; HD radio; Internet radio; over-the-air radio; satellite radio
Radio Act 1912 30–1
Radio Act 1927 36, 47
radio channels, bandwidth of 42–3
radio commercials 240, 248
Radio Conferences 29, 35
radio listenership, measuring 279–85
radio networks 9; early 35, 38, 309; ownership of 310
radio programming: contemporary 87; department responsible for 333–4, 336–9; moving to television 40, 60, 85, 140; regulation of 38; satellite 359; scheduling 90; types of 83–5
radio receivers 8; advertisements for *37*; portable *11*, 41, 85; on ships 29
radio shares 285
radio stations: formats of 85–6, 89, 118; future of 362–3; license renewal 318; local programming 88; low power 41, 46; music licensing fees 317; number of *34*; operations of 333–65; ownership 316; websites of 138–9, 195
Randolph, Joyce *96*
Rather, Dan 40
rating systems: MPAA 372; television *407*, 408; for video games *412*
ratings: early ways of determining 279–80; reporting 287; use of term 285–6
ratings projections 250
RBOCs (regional Bell operating companies) 315–16
RCA (Radio Corporation of America) 29–30, 32; bought by GE 310; color system 59–60; and early television 54; and FM radio 41; and NBC 35–6
Ready Player One 234
RealAudio 137
reality shows 100–3
RealVideo 140
recording, analog and digital 75
Red Channels 63
Red Lion v. FCC 10
The Red Skelton Show 85, 97
Reddit 166, 410
reenactments 321

rehearsals 337, 344, 351, 365
Renfroe, Jay 115
Reno v. ACLU 316
repetition 117
residuals 45
retargeting 263–5, *264*
retransmission 138, 323–5, 327
Reuters 106, 418
Reynaud, Charles-Émile 213
Reynolds, Glenn *132*
RFID (Radio Frequency Identification) 196–7
Rhapsody 138
RIAA (Recording Industry Association of America) 18, 314; and copyright 44–5, 47, 137, 312
rich-media ads 254–5
right to be forgotten 145–7
RKO 95, 381–2, 389
robocalls 326
robots 196, 202–3, 229, 234, 296
rock 'n' roll 86, *88*
The Rocky and Bullwinkle Show 109
Rokid 232
Roku 1, 78, 250, 263
Rolie Polie Olie 109
Romano, Aja 385
Roosevelt, F. D. 36, 40, 54, 240
Roots 96, 107
ROS (run of schedule) 293
routers 125, 351
royalties 35, 45, 47, 194, 314, 385
RSS (Real Simple Syndication) 139
RTDNA (Radio Television Digital News Association) 320
rule of sevens 311
run of network 261
runaway production 385

Saga Communications 315
Salem Media Group 315, 340
sales department: at cable companies 357; cable service's 357; in radio stations 333, 339; in satellite radio 359; in satellite television 360–1; in television stations 342, 355
sales managers 295, 300, 339, 344, 355
sales representatives 318, 339
sampling, in audience measurement 284
Samsung 12, 76, 81, 220–1
Samsung Gear *220*, 221
Sanders, Sarah 202
Sarnoff, David 29, 31, 41, *56*, 308
Satellite Home Viewer Improvement Act 1999 360
satellite radio 14, 17, 42, 45–6, 325, 366; future of 329, 364–5; and Internet radio 191; operations of 358, *359*; royalty payments 47
satellite radio receivers *358*
satellite television 69–71; audience measurement for 290; delivery of 359–60; direct broadcast see DBS; future of 365; home dishes for 72, *73*;
market share of 325; operations of 360; *see also* DBS
satellite uplink 349–50, 358, 365
satellites, types of 70
Saturday Night Live 98, 327
scanning, mechanical and electronic 51–4, *55*, *57*
scatter market 250
scene awareness 224
scheduling frequency 309
Schramm mass communication model 5
Schramm-Osgood communication model 5
Schwarzenegger, Arnold 103, 372
scrambling 72, 359–60
screen sizes 12, 79, 134
scribes 3
SDTV (standard definition television) 74
seamless programming 117
search engines 145, 261, 263
search retargeting 264, *265*
searchability 14
seasons 116
seating, assigned 378
second screening 142, 147
SEGs (special effects generators) 348, 350
Seinfeld 112–14, 288
selective processes 394–5
SEM (search engine marketing) 257
Sensorama *214*, 232
sentient code 233
SEO (search engine optimization) 19
Serial (podcast) 140
serials, on television 92–4
series finales 288
servers 13–14, 124–5
Sesame Street 68, 108–9, *110*, 413
sexploitation and sextortion 414–16
sexually explicit programming 316, 401, 416
The Shadow 84
shares 285–6; reporting 287
Shields, Brooke *244*
show runners 111
Showtime cable channels 15, 18, 72, 74, 110, 320
SHVA (Satellite Home Viewers Act) 73
silent movies 368, 370, 389
The Simpsons 74, 113, 302, 328
simulation, in XR 217
Siri 13, 175, 186, 195–6, 232
Sirius Satellite Radio 14, 42
Sirius XM Radio 14, 42–3, 46, 191, 329, 358–9, 365–6
situation comedies 95; blocking 15, 117; primetime 113
Six Degrees 1, 158, 176
size-based pricing 297
Skype 128, 190; and coronavirus ix, 173
skyscraper/tower advertising 255–6
Slack 174
sleep: impact of digital technology on 409; impact of television on 399, 404–5
Sleeping Beauty (Disney) 374, *375*
SlingTV 193, 263, 325
smart cement 195–6
smart clothing 196
smart glasses 196, 230, 232; *see also* Google Glass
smart jewelry 196
smart watches 196, 217, 272
smartphones x; addiction to 409; and audiences 12; and car audio 147; games on 412; generations of technology 136; listening to radio via 43–6, 48; managing home features 15; ownership of 185, *187*; social impact of 198; and social media 157; top uses of 189; viewing 142; viewing television on 76–9; *see also* mobile apps
SMATV (Satellite Master Antenna Television) 71, *72*
Smith, Buffalo Bob 107, *108*
The Smothers Brothers Comedy Hour 98
The Smurfs 109
Snapchat 15, 145, 167, 223
soap operas 8, 92–3, 113, 350; on radio 84; time slots for 113; women in 402
social contagion effect 169
social media x, xi, 157; advertising on 171, 245, *259*; benefits and consequences of 168–70, 175–6; best practices 175; characteristics as medium 12–13; and crisis communication 173; first platform 1; history of *159*; impact on relationships 170–3; market share *162*; media companies on 17; privacy on 174–5; social impact of 411; subcategorization of 160–1; talking about television on 302; and traditional media 170; use of 158–60; and work 174; *see also* social networking sites (SNS)
social networking: pre-World Wide Web 157; Web-based 158
social networking sites (SNS) 133, 161; and cookies 266; friend connections on 164; media companies on 17; other 167–8; second wave of 158; top 161–7; users of 161–2; and work 174
Social TV 302
Soderbergh, Stephen 18, 379, 381, 384
soft news 39
Somareddy, Veena 225, 226–7
Sony Pictures Television 114
The Sopranos 118
Soul Train 87
Soundcloud 138
SoundExchange 45
South Dakota v. Wayfair Inc. 317
SpaceX mission *21*
spam 267–8, *269*
Spanish American War 24
Spanish-language networks 113
SPAs (station promotional announcements) 344
spatial audio 213, 218, 229, 231
spiders 296
Spielberg, Steven 372, 374

sponsored programs: radio 239–40; for television 241, 244
sponsorship model 241, 308–9, 312, 339
spoofing 326
sports programming 34, 103; and broadcast television 80; on radio 89–90
sports/talk format 89
SPOT (smart personal object technology) 196
spot advertising model 242–3, 309, 339
Spotify 17, 43–4, 138; audience for 302–3; revenues of 47
Sprint 198, 254, 329
Sproull, Bob 214
star model 312, *313*, 315
Star Trek: Discovery 313
Star Wars 224, 328, 372, 374, 383–4, 389
station compensation 60, 324
station manager 287, 295, 334, 340–2
Stempel, Herb 99
stereo sound 213–14, 216, 229; broadcasting 41, 48, 89
stereoscopy 213–14, 216, 376
stereotypes, in advertising 270
storage, and mass media 14
storytelling techniques 14
Storz, Todd 86
streaming services x, 114, 191, 203, 361–2; artist's share of revenue *138*; award-winning programming for 118; business models for 325–6; in coronavirus lockdown ix; cost of 141–2; measuring audiences 302; and music industry 44, 138; number of users *149*; offered by telcos 329; and other forms of television 78–80, 82, 117; profitability of 16; and program selection 15; and radio 139; and satellite delivery 365
streaming studios 382
stress, technology induced 410
stripping 117
Stubblefield, Nathan 26
studio cameras 344–6, 348, 350, 363
studio system xi, 368, 381–5, 389
studio television production 347
stunting 117
subscription model 309, 361
subscription services: public opinion of 328; regulation of 10; Top 10 317
Super Bowl 1; advertising during 251–3; audiences for 104; Janet Jackson's wardrobe malfunction at 407; live stream of 80, 114
supergroups 323
superheterodyne receiver 33
superstations 70–1
superstitial advertising 255, 272
survey research 399
Sutherland, Ivan E. 214, 234
SVOD (Subscription Video-on-Demand) 326
sweeps periods 116, 282
switcher 58, *348*, 349–50
Sword of Damocles 214, *215*

synchronous media 12
syndicated programs 66, 88, 112–14, 117, 287
syndicators 60, 74, 88–9; and broadcast networks 112; and cable companies 113–14; vertical integration of 79

tablets x, 1–2, 12, 188–9; books via 15; Internet via 134; media produced on 18; music via 13; radio via 43, 45–6, 138; screen sizes 12; and second screening 142; and social media 157, 167; television via 76–9, 150
talk radio 48, 310, 340
target audiences, moving 301
task-switching 403–4
taxation, on Internet commerce 317
TCI (Tele-Communications, Inc.) 311
TCP/IP (transmission control protocols/Internet) 125
TD (technical director) 350
Technicolor 371
technographic measures 290
technological change 19, 21
Teenage Mutant Ninja Turtles 109
teenagers: media exposure of 414; and rock 'n' roll 86–7; and social media 158, 160, 176; text messaging 186
telcos 190, 198, 325, 329
Telecommunications Act 1996 11, 14, 38, 315–16; impact on station ownership 16, 48, 81, 311, 318, 333; and obscenity, indecency and profanity 406–7; and telephone companies 325
telecommunications equipment manufacturing 81, 316
telegrams 4, 183
telegraph 183; business model of 307–8; invention of 4, 23–5, *24*
telehealth 226, 326
Telemundo 94, 113, 325
telenovelas 94
telephone: business model of 308; impact of Telecommunications Act 1996 on 315; invention of 4, 25; landline 183–4; *see also* cell phones; Internet telephony; smartphones
telephone surveys 284
telephone wires 5, 56, 183, 312, 336, 351
television: addiction to *415*; audience fragmentation 244–5; debates about programming 67–8; early innovation 51–4; future of 327; instrumental and ritualistic watching 402–3; Internet delivery of *see* Internet television; localism in 16; and movie industry 370–1; and now media 78, 82; origins of 8; production equipment for *346*; and radio 40; signal path for *349*; social effects of 404–8; technological advancements in 11; *see also* broadcast television; cable television; satellite television
television channels: allocation of 57–9; secondary 75

Television City Research Center 291
television commercials 8, 237, 241–3, 249–51, 263; on children's television 68; costs of 113, 249; length of 243–4, 309; managing production of 344; residuals for 45; on streaming services 79; viewers avoiding 193
television networks *see* broadcast networks
television newsrooms, wired *351*
television program testing 291
television programming: in 1960s 67; and advertising 8; cancellation of 113; and digital technology 12–13, 15, 18–19, 80; on DVD 76; educational *see* educational television; HD and UHD 79; network control over 60; operations of 342–51; production and distribution 109–16, 313, 366; scheduling 116; sources of 312–15; strategies 116–18; types of 91–109; use of term 302; *see also* network programming
television quotient data 291
television sets: advertisements for *61–2*, 65; development of 73–4, 76; new forms of 79, 82; smart 78, 141, 192–3, 250, 262, 301–2
television shares 285
television stations: commercial 55–6; experimental 54; frequency allocation 57–9; future of 363; independent *see* independent television stations; job titles in 343; license renewal 318; localism in 16; network-affiliate *see* affiliated stations, television; non-commercial 68; operations of 333, 342–55, 365; ownership rules 81; websites of 76, 191
television studios *52*, *58*, 91, 311, 347–50, 356; floor plan *349*
television viewership: measuring 282–5; reporting 287
television violence *see* violence, media depictions of
tentpoling 116–17
Terminator 2 374
terrestrial radio *see* over-the-air radio
text messaging 12–13, 15, 128, 134; addiction to 409; and driving 409; indicating emotions through 172–3; and instant messaging 128; popularity of 186; in second screening 142; security of 175
theories, use of term 392
third-party monitors 143, 294–8
Thompson, Clive 170
TikTok ix, 13, 157, 167–9
Timberlake, Justin 158
time, and mass media 12
time blocks 40, 281
time spent online 297
time-shifting 74, 80, 194, 289
Titanic 29–31, *30*; 1996 movie about 382
T-mobile 198, 329